Compression for Great Video and Audio

Master Tips and Common Sense

Second Edition

Compression for Great Video and Audio

Master Tips and Common Sense

Second Edition

Ben Waggoner

NEW YORK AND LONDON

First published 2002
This edition published 2013 by Focal Press
70 Blanchard Road, Suite 402, Burlington, MA 01803

Simultaneously published in the UK
by Focal Press
2 Park Square, Milton Park, Abingdon, Oxon OX14 4RN

Focal Press is an imprint of the Taylor & Francis Group, an informa business

Library of Congress Cataloging-in-Publication Data
Waggoner, Ben.
Compression for great video and audio : master tips and common sense / Ben Waggoner.
p. cm.
Rev. ed. of: Compression for great digital video / Ben Waggoner. 1992.
Includes index.
ISBN 978-0-240-81213-7
1. Digital video. 2. Video compression. 3. Multimedia systems. I. Waggoner, Ben. Compression for great digital video. II. Title.

QA76.575.W32 2010
006.7–dc22

2009032833

ISBN-13: 978-0-240-81213-7 (pbk)
ISBN-13: 978-0-080-95905-4 (ebk)

Contents

Color versions of some figures are included in an insert at the back of the book. The black and white versions appear in their respective chapters, and identify which color figure to refer to.

Introduction

It was the fall of 1989, and I was taking a computer animation class at UMass Amherst (I went to Hampshire College, another school in the Five Colleges consortium). We were using long-gone tools like Paracomp Swivel 3D to create a video about robots. This was well before After Effects, so we were using Macromind Director 1.0 as a very basic compositing application. We needed faster playback on our 20 MHz Macintosh IIcx to preview animation timing. I started experimenting with a utility called Director Accelerator, which composited together all layers of the Director project and compressed the final frames with Run-Length Encoding (RLE), a very early form of compression. I was amazed how the resulting files were smaller than the originals. And so I started playing around with optimizing for RLE, and forgot about animation and robots.

I got a D in the class, while the TAs and some other students went on to create Infini-D, a pioneering 3D app for desktop computers. Infini-D lives on as Carrara from DAZ. As for me, I didn't do any animation after that, but darn if that compression stuff didn't stay interesting.

So interesting I've made a career of it, although it took me a while to recognize my destiny. Right after college, I spent a year writing and producing a comedy-horror mini-series with my friends. I wrote an overly long scene that hinged on the compression ratio of Apple Compact Video (later known as Cinepak). I was sure it was interesting! My collaborators never quite believed an audience would enjoy three pages of innuendo-laced banter about kilobits per seconds.

Next, I cofounded a video postproduction company Journeyman Digital in 1994. We started trying to be a standard NLE shop, using a Radius VideoVision Studio card in a PowerMac 8100/80, with a whopping 4 GB RAID drive ($5,000 for the drive, another $1,000 for the controller card). However, due to bugs in that generation Mac motherboard, we couldn't capture more than 90 seconds in sync, making the whole system a better boat anchor than video editor.

Then, one day, someone called us up and asked if we could do video for CD-ROM. They were willing to pay for it, so we said "Yes!" and dove in, trying to figure out what the heck they were talking about. It turned out the VideoVision was perfect for short CD-ROM clips, which never had good sync anyway.

For compression, we used Adobe Premiere 3.0 and Apple's MovieShop. The codecs of choice were Cinepak and Indeo 3.2, with 8-bit uncompressed audio. Data rates for a typical 240 × 180 15 fps movie were around 120 KBytes/sec, more than enough for full-screen web video today. We targeted 80 minutes of encoding time per minute of output, so there was a lot of "bartending."

Still, primitive as the technology was, it enabled some amazing things. Today, that computers can play video is a given, so it's hard to convey the thrill of seeing those first postage-stamp videos play in a small rectangle on the screen. The video might have been small and blocky, but it was interactive! We could actually add video to programs. We worked on all kinds of sales training discs, kiosks, encyclopedias, and multimedia training projects for fun and profit.

By the late 1990s, video on the web was all the rage. Better, faster, cheaper CPUs and Internet connections and new web technologies finally delivered web video that looked like more than pixel soup. The future looked promising. New businesses were springing up like dandelions—everybody wanted to cash in on the opportunities presented by the coming of ubiquitous broadband connectivity. Companies like Akamai, Digital Island, and many others set up content delivery networks and raised billions in their initial public stock offerings.

It was in those heady days I met my future wife, Sacha, at the Portland Creative Conference. I didn't know it when we first met, but she'd been doing digital video about six months longer than I had, and remains the only known person to have made money with the infamously unfinished SuperMac VideoSpigot capture card.

In 1999, I joined Terran Interactive, the original developers of Media Cleaner Pro (the leading professional video compression application at the time) to launch and run Terran's Consulting division. My job was to reap services revenue and to help with marketing and product development. Media 100 went on to acquire Terran, and I was laid off with most of the ex-Terrans when the big Internet bubble "popped" in 2001.

Around this time, we had our first child, and really wanted to stay in Portland to be near family. The local media companies were downsizing, so my wife and I put out our own shingle as a consulting and media services company called Interframe Media. It was a fruitful period, as we expanded our focus beyond compression services to compression tools, and I consulted on the design of many different products, including Windows Media Encoder, Adobe Media Encoder, Telestream's Episode Pro, Canopus ProCoder, Rhozet Carbon Coder, and Sorenson Squeeze.

While things seemed dire in 2001, and a lot of companies went under, there was a reason there had been so much money flowing into the industry. Delivering compressed video over the Internet was going to be big someday. And even if many investors spent too much, too early, that effort laid the foundation for today's industry, and—in an important change—some profitable businesses as well!

Interframe Media had five years of steady growth, and we were trying to figure out if we should actually hire some employees to take on the big projects we were already turning down. As we entered the summer of 2005, the emergence of the HD DVD and Blu-ray war and our third child made us realize that between household and business, we simply didn't have enough time to take on this next generation.

That fall, Microsoft, Amazon, and Google all were recruiting me for their various digital video efforts. All three opportunities were exciting and flattering, but in the end I chose Microsoft, to the surprise of those who'd known me as a Mac user of two decades. But Microsoft had long been at the center of digital media, and in Windows Media 9 Series delivered the first really complete media delivery technology that scaled from phones to HD. After stints on the HD DVD and codec teams, I wound up as the media strategist for Silverlight, working on the whole end-to-end ecosystem of publishing video content. In its own way, that's what made it possible to do a second edition of the book—the format wars were over. Silverlight has joined Flash, QuickTime, and Windows in broad format support, including H.264. The issue of how and what codec you compress to isn't nearly as tied to where it gets played back as it used to be.

The world of compression has exploded in the 20 years since those first lurching frames on a CRT. With the 2009 analog television switch-off in the United States, we're seeing the end of video that *isn't* digitally compressed. Compression's a fact of life for almost everything we see or hear, whether it's delivered via an MP3, DVD, HD broadcast, kiosk in a store, video clip on a phone, voices on a phone video on demand over cable…the list goes on. And while we once fantasized about achieving "VHS quality" on the Web, delivering HD video that looks as good as broadcast and cable is now commonplace.

This book has been a long time in the making. I wrote the outline for the first edition back in 1998, when CD-ROM video was all the rage and web video something we knew would happen "someday." When it was finally published (after one wedding and two baby-induced delays) in 2002, the overall goal hadn't changed a bit, although that someday had become a now. It was all about how to make good compressed video on the many delivery platforms available.

After it had gone to press, I immediately had grand plans for a second edition, but between business, family, and jobs, it always seemed like something I'd get to "in a few months." But, as in most big things in life, when you realize there's never going to be a perfect time, it's probably the right time to do it anyway. In the end, I was inspired by those who kept reading, discussing, and even buying this half-decade-old book. I owed them a new edition for the compression universe of today.

And it really is a new world now. The first edition covered dozens of different codecs and nearly as many media formats and platforms. We've seen a great convergence recently, with the standardized codecs of MPEG-2, VC-1, and H.264 dominating most new content and players. Being a compressionist today allows for more depth, as we don't have to use and master the myriad and varied tools and technologies once required. Conversely, we're seeing much more variety in the places we can play that content back; the first edition was mainly about content that would play back in a particular media player or a browser window on a Mac or Windows PC. But the devices and services available today are much, much more complex, while at the same time the demand for "any content, anywhere, any time" is only increasing.

Who Should Read this Book

This book is for compressionists, people who want to be compressionists, and people who on occasion need to pretend they're compressionists. Most of you will have come from the worlds of video or the web. Video people will be happy to hear that this stuff isn't as alien as it might initially seem. The core skills of producing good video are just as important when dealing with compression, although compression adds plenty of new twists and acronyms. Web folks can find the video world intimidating—it's populated by folks with 20-plus years of experience, and comes festooned with enough jargon to fill a separate volume. And there are plenty of people who just need to get some video compressed, fast, coming from entirely different fields.

That's fine; this book is designed to help you get to where you need to be, regardless of where you're coming from. To that end, I've tried to include definitions and explanations of this jargon as it comes up, and gently introduce concepts from first principles. If you read this book in nonlinear fashion and jump straight to the problem you need to solve *today*, there's a glossary that should help you decode some of the more common terms that crop up throughout.

How This Book is Organized

The chapters are organized in three main sections. The first section (Chapters 1–3) covers general principles of vision, compression, and how compressed video operates. While a fair amount of it will be old hat for practicing compressionists, there should be something of interest for everyone. Even if you skim the first time around, it'd be a fine idea to go back and give it a more careful reading sometime while you're bartending. I've tried to keep it readable—there's no math beyond basic algebra, and I've kept it to details that matter for compression. Since this is "evergreen" stuff for the most part, readers of the first edition will find a lot familiar (and a lot new as well).

Chapters 4–8 cover the fundamentals of video technology and the non-codec parts of the compression workflow. This includes compression-friendly production techniques, video capture, and preprocessing.

Chapters 9-29 cover specific video tools and technologies. Each chapter is self-contained, so it's fine to skip straight to the ones you care about. I include many scenario specific tutorials illustrating how to solve common problems and also how to approach new projects with forethought and experimentation.

The world of compression changes fast enough to give you chronic whiplash. Given the breadth of the topic, I was constantly updating chapters to keep up with new tools and technologies. But we had to stop at some point. Even if the book is a version behind the current release of something, I try to explain the why as much of the how in order to make the skills and mental approach applicable to future versions as well as entirely new tools.

Check http://www.benwaggoner.com/bookupdates.html and http://www.cmpbooks.com/compression for updates, corrections, and other resources.

Acknowledgments from the First Edition

A book like this requires a lot of work. Dominic Milano went far above and beyond the call of duty as technical editor, making sure it was readable, relevant, and right. Jim Feeley edited many articles for DV that were later incorporated here and continues to teach me an enormous amount about how to express complex technical ideas clearly. The long course of this book provided many opportunities to demonstrate the patience, talent, and good humor of Dorothy Cox, Paul Temme, and Michelle O'Neal from CMP Books.

Lots of people taught me the things I said in this book. The founders of Terran Interactive, Darren Giles, Dana LoPiccolo-Giles, and John Geyer answered many dumb questions from me, until I knew enough to ask some smart ones. I learned a lot about compression from them via late night emails.

During the gestation of this one book, my wife Sacha gestated Alexander and Aurora—a much harder job than mine! I'm looking forward to spending more time with them instead of typing in the basement.

My family helped me start a business back when I didn't know enough about what I was doing to even know what I didn't know. I'm sure they bit their lips more than they let on over the past few years. Hopefully, they're able to relax now.

And finally, none of this would have happened without Halstead York, who over the years has talked me into writing scripts, producing video, starting two companies, and learning compression. Without his infectious enthusiasm, vision, and drive, I'd probably still be writing banking software.

Acknowledgments from the Second Edition

Sacha, Alexander, Aurora, Charles

SteveSK, TimHa, AlexZam

Wei-ge Chen for DCT visualizer

Phil Garrett for Microsoft Viewer

Preface: Quick-Start Guide to Common Problems

This book is a hands-on guide to real-world compression. A lot of you are perhaps browsing it right now trying to find a quick solution to a pressing problem. Here are some answers.

My Boss Says I Need to Put Some Video on Our Web Site. Where do I Start?

For simple video on a web page, embedding Flash (Chapter 15) or Silverlight (Chapter 27) is the easiest mechanism today. And generally a web server is fine for delivering short content (see next question).

Do I Need a Streaming Server to Put Video on the Web?

Generally, no, for shorter clips of a single bitrate. It's only long-form content (more than 15 minutes) or when high-quality real-time playback is needed that you'll need a streaming server. For longer content, or to deliver the best quality for uses with a broad range of connection speeds, technologies like IIS Smooth Streaming or Flash's Dynamic Streaming can work well.

My Video has Horizontal Lines Wherever there is Motion

Your video source is interlaced, where even and odd lines contain images from slightly different times, and it wasn't converted to a normal progressive format.

- See page 26 in Chapter 2 for a description of interlacing
- See pages 112–115 in Chapter 6 for how to deinterlace.

If you're shooting your own content, you should be shooting progressive, not interlaced.

My DVD is All Flashing and Stroby Whenever There's Motion

Your source is interlaced like described in the previous answer. You encoded your DVD using the wrong field order, so what should have been the first field displayed is now the second field.

This topic is covered in Chapter 9 on page 179.

My Video Looks Terrible and Blocky

You're probably suffering from some combination of one or more of the following:

- Too big a frame size (see pages 120–121 in Chapter 6).
- Too low a data rate (see page 141 in Chapter 7).
- Sub-optimal encoding settings (see chapter for your format or codec).

My Video is All Stretched or Squished

Your video was probably produced in 4:3 or 16:9 aspect ratio, and then encoded at a different aspect ratio. This is typically addressed by telling your compression tool what shape your source is, and specifying the right frame size for output. See page 120 in Chapter 6 for how to correct.

My Video has Annoying Black Bars in it

Your video has letterboxing (top and bottom) or pillarboxing (left and right). Either the input and output aspect ratios don't match, or there are black bars in the source.

If your source doesn't have the black bars, see the previous question for where to learn about correcting for aspect ratio.

If your source has black bars, you'll need to crop them out before encoding the video.

There's this Annoying Flashing Line at the Top of My Video

That's probably "Line 21," which in analog video specifies a video space set aside for closed captioning.

My Audio is Way Too Quiet

Your audio is probably way too quiet. Most compression tools have a "Normalize" filter that raises the volume of quiet audio while leaving already loud enough audio alone. See "Normalization" on pages 136–137 in Chapter 6.

My Audio Sounds Terrible

You are probably using too low a data rate. Good audio doesn't take up that many bits with modern codecs, so even if your video quality is limited by bandwidth, your audio should never be distractingly bad. See page 66 in Chapter 4 on how to balance video and audio bitrates.

Where Can I Host My Video?

There's no shortage of ways to host a video:

- If you have access to a web server (which you do if you have a web site), you can use that for progressive download media.
- If you don't care about doing your own compression or having another brand on your video, you can use a service like YouTube or Soapbox that will take an uploaded file, compress it, and host it.
- If your organization has an account with a content distribution network (CDN) like Akamai, Limelight, Level 3, or CDNetworks, use that.
- If you don't have needs big enough for a CDN but want to control your compression and brand, there are new lower-cost services like Sorenson 360 that start at $99/month.

How do I make my video come out with the right size file after compression?

If you have a particular file size you want to target, you just need to multiply the total data rate of video and audio by the duration of the clip. However, data rates are typically measured in kilobits per second, while file size is measured in megabytes. To calculate between them:

$$\frac{\textit{Kilobits per second} \times \textit{duration in seconds}}{8000 \textit{ Kilobits per megabyte}}$$

So if you have 30 minutes of 500 Kbps video, you'd get:

$$\frac{500 \textit{ Kilobits/sec} \times 30 \textit{ min} \times 60 \textit{ sec/min}}{8000 \textit{ Kilobits per megabyte}} = \frac{500 \times 30 \times 60}{8000} = \frac{900000}{8000} = 112.5 \textit{ Megabytes}$$

Users Don't Have the Right Player to See My Content

First, find out what players your users have access to. If the video will be played back on a web site, most servers will allow you to log what plug-ins and versions are installed on the browsers that come to your site. It's also helpful to track what operating systems visitors are running.

What players are available depends on the market you're addressing and how you're delivering your video. Good choices include the following:

- Flash: Chapter 15
- Silverlight: Chapter 27

- Windows Media Player (default for Windows users): Chapter 16
- QuickTime (default for Mac users): Chapter 29

How do I Pick the Right Video Format(s)?

The right format depends on the player and version of that player you're targeting. You may need to target more than one player, which means you may need to compress in more than one format.

My Users Keep Getting "Buffering" Errors or Pauses in Playback

This happens when your users aren't able to get your video at the data rate it's encoded at. For generic web video, you probably want to stick to a data rate of 800 Kbps or lower if all users will share a single data rate. Remember, that data rate is based on video and audio—make sure that you're not accidentally leaving the audio bitrate too high.

Adaptive streaming techniques like Smooth Streaming and Dyamic Streaming can offer a much better experience here, because they customize the delivery bitrate on the fly based upon each user's connection speed.

CHAPTER 1

Seeing and Hearing

The rest of this book is about practical compression issues, but it's important to first understand how the human brain perceives images and sounds. Compression is the art of converting media into a more compact form while sacrificing the least amount of quality possible. Compression algorithms accomplish that feat by describing various aspects of images or sounds in ways that preserve details important to seeing and hearing while discarding aspects that are less important.

Understanding how your brain receives and interprets signals from your auditory and visual receptors (your ears and eyes) is very useful for understanding how compression works and which compression techniques will work best in a given situation.

It's not necessary to read this chapter in order to be able to run a compression tool well, but understanding these basics is very helpful when it comes to creating content that will survive compression and understanding how best to tune the tradeoffs in compression. Even if you skip it the first time you're flipping through the book, it's probably worth it to loop back the second time around.

Seeing

What Light Is

Without getting overly technical, light is composed of particles (loosely speaking) called photons that vibrate (loosely speaking) at various frequencies. As the speed of light is a constant, the higher the frequency, the shorter the wavelength. Visible light is formed by photons with wavelengths between 380 to 750 nanometers, known as the visible light spectrum (x-rays are photons vibrating at much higher frequency than the visible spectrum, while radio waves are much slower). This visible light spectrum can be seen, quite literally, in a rainbow. The higher the frequency of vibration, the further up the rainbow the light's color appears, from red at the bottom (lowest frequency) to violet at the top (highest frequency).

Some colors of the visible light spectrum are known as black-body colors (Color Figure C.1 in the color section). That is, when a theoretically ideal black object is heated, it takes on the colors of the visible spectrum, turning red, then yellow, warm white, on through bluish-white, as it gets hotter. These colors are measured in degrees Kelvin, for example 6500 K. What about colors that don't appear in the rainbow, such as purple? They result from combining the pure colors of the rainbow. Everything you can see is made up of photons in the visible light spectrum.

The frequency of the light is measured in terms of wavelength. Visible light's wavelength is expressed in nanometers, or nm. Unless you're looking at something that radiates light, you're actually seeing a reflection of light generated elsewhere—by the sun, lights, computer monitors, television sets, and so on.

What the Eye Does

There's quite a lot of light bouncing around our corner of the universe. When some of that light hits the retina at the back of our eyes, we "see" an image of whatever that light has bounced off of. The classic metaphor for the eye is a camera. The light enters our eye through the pupil, and passes through the lens, which focuses the light (Figure 1.1). Like a camera, the lens yields a focal plane within which things are in focus, with objects increasingly blurry as they get closer or farther than that focal plane. The amount of light that enters is controlled by the iris, which can expand or contract like a camera's aperture. The light is focused by the lens on the back of the retina, which turns the light into nerve impulses. The retina is metaphorically described as film, but it's closer to the CCD in a modern video or digital still camera, because it continually produces a signal. The real action in the eye takes place in the retina.

In high school biology, your teacher probably yammered on about two kinds of light-receptive cells that reside in the retina: rods and cones. We have about 120 million of these photoreceptors in each eye, 95 percent of which are rods. Rods are sensitive to low light and fast motion, but they detect only luminance (brightness), not chrominance (color). Cones detect detail and chrominance and come in three different varieties: those sensitive to blue, red, or green. Cones don't work well in very low light, however. This is why people don't see color in dim light. In 2007, scientists discovered a third photoreceptor type, unpoetically

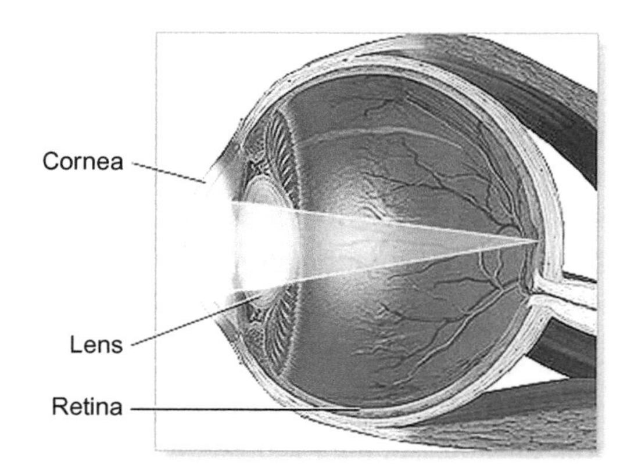

Figure 1.1 A technical drawing of the human eye.

named the "photosensitive ganglion cell" (PGC), that responds to the amount of light available over a longer period.

The good news is you can safely forget about rods and PGCs for the rest of this book. Rods are principally used for vision only when there isn't enough light for the cones to work (below indoor lighting levels)—much lower levels than you'll get out of a computer monitor or television. And the PGCs are involved in wholesale changes of light levels; again, not something that happens during conventional video watching. At normal light levels, we see with just the cones.

Still, the rods make a remarkable contribution to vision. The ratio between the dimmest light we can see and the brightest light we can see without damaging our eyes is roughly one trillion to one! And the amazing thing is that within the optimal range of our color vision, we aren't really very aware of how bright light in our environment is. We see relative brightness very well, but our visual system rapidly adjusts to huge swings in illumination. Try working on a laptop's LCD screen on a sunny summer day, and the screen will be almost unreadably dim even at maximum brightness. Using that same screen in a pitch-black room, even the lowest brightness setting can seem blinding.

There are also some interesting effects when light is in the "mesopic" range, where both rods and cones are active. Rods are most sensitive to blues, which is why colors shift at dawn and dusk.

Folks with normal vision have three different types of cones, each specific to a particular wavelength of light, matching red, green, and blue (as in Figure 1.2, also Color Figure C.2).

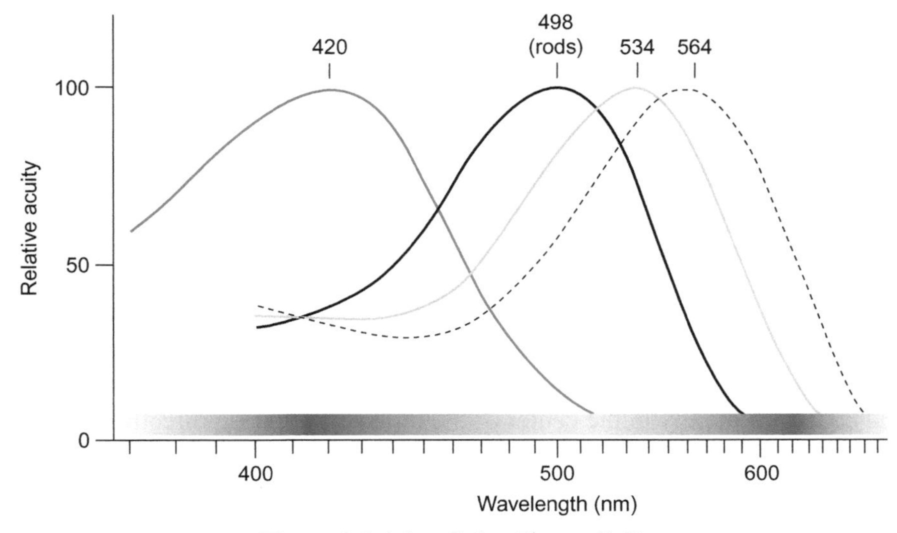

Figure 1.2 (also Color Figure C.2)
Relative sensitivity of the eye's receptors to the different colors. The solid black line is the rods; the other lines are the three cones.

We're not equally sensitive to all colors, however. We're most sensitive to green, less to red, and least of all to blue. Really. Trust me on this. This is going to come up again and again, and is the fundamental reason why we use green screens for "blue screen" video these days, and why red subtitles can get so blocky when going out to DV tape.

Late-Night College Debate Settled Once and for All!

So, if you were prone to senseless philosophizing at late hours as I once was (having children and writing books seem to have effectively displaced that niche in my schedule), you've no doubt heard speculation ad nauseam as to whether or not what one of us calls "red" could be what another perceives as "blue."

Kudos if you've never thought about this, because Color Figure C.2 shows that what you see as red is what every other person with normal version perceives as red. In every culture in the world, in any language, if you ask what the most "red" color is and offer a bunch of samples, almost everyone picks the sample at 564 nm. Thus what you see as red, I see as red, and so does everyone else. The same is true for green and blue. I find it rather pleasing when something that once seemed abstract and unknowable turns out to have an extremely concrete answer like this. I'll be calling out other examples of life's mysteries answered by compression obsession throughout the book.

Also, not everyone has all three types of cones. About seven percent of men (and a much smaller percentage of women) have only two of the three cone types, causing red-green color blindness, the most common type (see Color Figure C.3). Dogs and most other mammals also have only two cone types. Not that it's that much of a problem—many people with red-green color blindness don't even know until they get their vision tested for a driver's license. Red-green color blindness is also why my dad wears rather more pink than he realizes. It's not all bad news, though. People with red-green color blindness are much better at seeing through color camouflage that confuses the three-coned among us.

Another aspect of the retina is that most of the detail we perceive is picked up within a very small area of what the eye can see, in the very center of the visual field, called the fovea. The outer portion of the retina consists mainly of rods for low-light and high-speed vision. The critical cones are mainly in the center of the retina. Try staring directly at a specific point on a wall without letting your eyes move around, and notice how small a region is actually in focus at any given time. That small region of clear vision is what the fovea sees – it covers only 1% of the retina, but about 50% of the visual parts of the brain are devoted to processing information from the fovea. Fortunately, our eyes are able to flick around many times a second, pointing the fovea at anything interesting. And we've got a big chunk of brain that's there for filling in the gaps of what we aren't seeing. At any given time, much of what we think we're seeing we're actually imagining, filling in the blanks from what our brain remembers our eyes were looking at a fraction of a second ago, and guessing at what normally would happen in the environment.

How the Brain Sees

It's hard to draw a real mind/body distinction for anything in neuroscience, and this is no less true of vision. What we call "seeing" takes place partially in the retina and partially in the brain, with the various systems interacting in complex ways. Of course, our sense of perception happens in the brain, but some basic processing such as finding edges actually starts in the retina. The neuroscience of vision has been actively researched for decades, but it's impossible to describe how vision works in the brain with anywhere near the specificity we can in the eye. It's clear that the visual system takes up about a quarter of the total human brain (that's more than the total brain size of most species). Humans have the most complex visual processing system of any animal by a large margin.

Under ideal circumstances, humans can discriminate between about one million colors. Describing how might sound like advanced science, but the CIE (Commission Internationale de l'Eclairage—French for International Commission on Illumination) had it mostly figured out back in 1931. The CIE determined that the whole visible light spectrum can be represented as mixtures of just red, green, and blue. Their map of this color space (Color Figures C.4 and C.5) shows a visible area that goes pretty far in green, some distance in red, and not far at all in blue. However, there are equal amounts of color in each direction; the threshold between a given color and black shows how much color must be present for us to notice. So we see green pretty well, red so-so, and blue not very well. Ever notice how hard it is to tell navy blue from black unless it's viewed in really bright light? The reason is right there in the CIE chart.

Another important concept that the CIE incorporated is "luminance"—how bright objects appear. As noted previously, in normal light humans don't actually see anything in black and white; even monochrome images start with the three color receptors all being stimulated. But our brain processes brightness differently than it does color, so it's very important to distinguish between them. And optimizing for that difference is central to how compression works, and even how color television was implemented a half-century ago. We'll be talking about how that difference impacts our media technologies throughout the book.

Software sometimes turns RGB values into gray by just averaging the three values. However, this does not match the human eye's sensitivity to color. Remember, we're most sensitive to green, less to red, and least to blue. Therefore, our perception of brightness is mainly determined by how much green we see in something, with a good contribution from red, and a little from blue.

For math fans, here's the equation. We call luminance Y because…I really don't know why.

$$Y' = 0.587 \text{ Green} + 0.299 \text{ Red} + 0.114 \text{ Blue}$$

Note that this isn't revealing some fundamental truth of physics; it's an average based on testing of many human eyes. Species with different photoreceptors would wind up with different equations, and there can be some variation between individuals, and quite a lot between species.

Encoding images requires representing analog images digitally. Given our eyes' lack of sensitivity to some colors and our general lack of color resolution, an obvious way to optimize the data required to represent images digitally is to tailor color resolution to the limitations of the human eye.

Color vision is enormously cool stuff. I've had to resist the temptation to prattle on and on about the lateral geniculate nucleus and the parvocellular and magnocellular systems and all the neat stuff I learned getting my BA in neuropsychology. (At my fifteenth reunion, I bumped into the then and still Dean of Hampshire's School of Cognitive Science. He asked why I'd left academic neuropsychology for something so different as putting video on computers. My defense?: "Video compression is just highly applied neuropsychology.")

How We Perceive Luminance

Even though our eyes work with color only at typical light levels, our perception is much more tuned for luminance than color. Specifically, we can see sharp edges and fine detail in luminance much better than we can in color. We can also pick out objects moving in a manner different from other objects around them a lot more quickly, which means we can also notice things that are moving better than things standing still. In optimal conditions with optimal test material, someone with good vision is able to see about 1/60th of a degree with the center of their fovea—the point where vision is most acute. That's like a dot 0.07 inches across at a distance of 20 feet.

A lot of the way we see is, fundamentally, faking it. Or, more gracefully put, our visual system is very well attuned to the sort of things bipedal primates do to survive on planet Earth, such as finding good things to eat without being eaten ourselves. But given images that are radically different from what exists in nature, strange things can happen. Optical illusions take advantage of how our brains process information to give illusory results—a neuropsychological hack, if you will. Check out Figure 1.3. Notice how you see a box that isn't there? And how the lines don't look quite straight, even when they are?

How We Perceive Color

Compared to luminance, our color perception is blurry and slow. But that's really okay. If a tiger jumped out of the jungle, our ancestors didn't care what color stripes it had, but they did need to know where to throw that spear *now*. Evolutionarily, color was likely important for discriminating between poisonous and yummy foods, an activity that doesn't need to happen in a split second or require hugely fine detail in the pattern of the colors themselves.

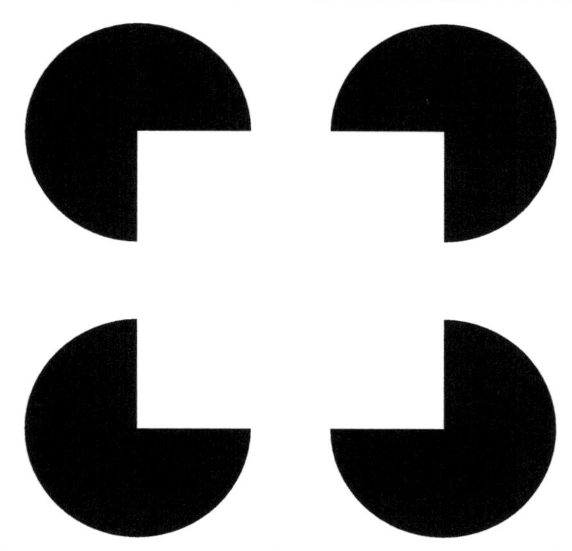

Figure 1.3 Kanizsa Square. Note how the inner square appears brighter than the surrounding paper.

For example, look at Color Figure C.6. It shows the same image, once in full range, once without color, and once without luminance. Which can you make the best sense of? Bear in mind that there is mathematically twice the amount of information in the color-without-luminance image, because it has two channels instead of one. But our brains just aren't up to the task of dealing with color-only images.

This difference in how we process brightness and color is profoundly important when compressing video.

How We Perceive White

So, given that we really see everything as color, what's white? There are no white photons in the visible spectrum; we now know that the absence of colored light is the same as no light at all. But clearly white isn't black—if it was, there wouldn't be zebras. White is the color of a surface that reflects all visible wavelengths, which our vision detects even as illumination changes. In different circumstances, the color we perceive as white varies quite a bit, depending on what's surrounding it. The "color" white can be measured by black-body temperature. As mentioned previously, for outdoors on a bright but overcast day, white is about 6500 K, the temperature of the sun. Videographers know that 6500 K is the standard temperature of white professional lights. But indoor light with incandescent light bulbs will be around 3000 K. CRT computer monitors default to the very high, bluish 9300 K, and many LCDs ship out of the box with that bluish tinge. I prefer to run my displays at 6500 K, as that's what most content is color corrected for.

As you can see in Color Figure C.7, our perception of what white is varies a lot with context. It also varies a lot with the color of the ambient light that you're looking at the image in.

We aren't normally aware of the fact that white can vary so much. Our brains automatically calibrate our white perception, so that the same things appear to have the same color under different lighting, despite the relative frequencies of light bouncing off them and into our eyes varying wildly. This is one reason why white balancing a camera manually can be so difficult—our eyes are already doing it for us.

How We Perceive Space

Now you understand how we see stuff on a flat plane, but how do we know how far things are away from us? There are several aspects to depth perception. For close-up objects, we have binocular vision, because we have two eyes. The closer an object is to us, the more different the image we see with each eye, because they're looking at the object at different angles. We're able to use the degree of difference to determine the distance of the object. But this works only up to 20 feet or so. And you can still have a good idea of where things are even with one eye covered. Binocular vision is important, but it isn't the whole story.

Alas, we don't know all of the rest of the story. There are clearly a lot of elements in what we see that can trigger our perception of space. These include relative size, overlapping, perspective, light and shading, blue shading in faraway objects, and seeing different angles of the object as you move your head relative to the object. See Figure 1.4 for how a very simple image can yield a sense of perspective and Figure 1.5 (also Color Figure C.8) for how a combination of cues can. You don't need to worry about elements that contribute to our depth perception too much for compression, except when compression artifacts degrade an element so much that the depth cues are lost. This is mainly a problem with highlights and shadows in the image.

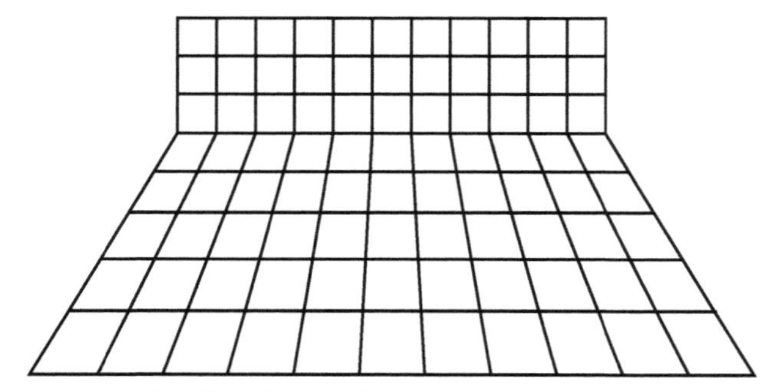

Figure 1.4 This simple grid provides a powerful sense of perspective.

How We Perceive Motion

Our ability to know how things are moving in our environment is critical, and it's crucial to staying alive. We have a huge number of neurons in the brain that look for different aspects of motion around us. These ensure we can notice and respond very quickly. These motion sensor cells are specifically attuned to noticing objects moving in one direction in front of a complex background (like, say, a saber-toothed tiger stalking us along the tree line).

We're able to keep track of a single object among a mass of other moving objects—it isn't hard to pick out a woman with a red hat at a crowded train station if that's who you're looking for. As long as we can focus on some unique attributes, like color or size or brightness, we can differentiate a single item from a huge number of things.

One important facet of motion for compression is persistence of vision. This relates to how many times a second something needs to move for us to perceive smooth motion, and to be something other than a series of separate unrelated images. Depending on context and how you measure it, the sensation of motion is typically achieved by playing back images that change at least 16 times per second or, in film jargon, 16 fps (for frames per second); anything below that starts looking like a slide show.

However, motion has to be quite a bit faster than that before it looks naturally smooth. The original "talkie" movie projectors running at 24 fps had far too much flicker. They were

Figure 1.5 (also Color Figure C.8)
Paris Street: A Rainy Day **by Gustave Caillebotte. This painting uses both shading and converging lines to convey perspective.**

quickly modified so the lamp in the film projector flashes twice for each film frame at 48 Hertz (Hz) to present a smoother appearance. And most people could see the difference between a CRT computer monitor refreshing at 60 Hz and 85 Hz.A higher setting doesn't improve legibility, but a higher refresh rate eventually makes the screen look much more "stable" (it takes at least 100 Hz for me). Modern LCD displays display a continuous image without the once-a-frame blanking of a CRT, and so don't have any flicker at 60 fps.

Hearing

We often forget about audio when talking about compression; we don't see "web video and audio file." But it's important not to forget the importance of sound. For most projects, the sound really is half the total experience, even if it's only a small fraction of the bits.

What Sound Is

Physically, sound is a small but rapid variation in pressure. That's it. Sound is typically heard through the air, but can also be heard through water, Jell-O, mud, or anything elastic enough to vibrate and transmit those vibrations to your ear. Thus there's no sound in a vacuum, hence the *Alien* movie poster tag line: "In space, no one can hear you scream." Of course, we can't hear all vibrations. We perceive vibrations as sound only when air pressure raises and falls many times a second. How rapidly the air pressure changes determines loudness. How fast the pressure changes determines frequency. And thus we have amplitude and frequency again, as we did with video. As with light, where you can have many different colors at the same time, you can have many different sounds happening simultaneously.

Different audio sources exhibit different audible characteristics, which we call timbre (pronounced "tam-ber"). For example, an A note played on a piano sounds very different than the same note played on a saxophone. This is because their sound-generating mechanisms vibrate in different ways that produce different sets of harmonics and other overtones.

Say what? When discussing pitch as it relates to Western musical scales—for example, when someone asks to hear an A above middle C—they're asking to hear the fundamental pitch of that A, which vibrates at a frequency of 440 cycles per second, or 440 Hz. However, almost no musical instrument is capable of producing a tone that vibrates at just 440 Hz. This is because the materials used to generate pitch in musical instruments (strings of various types, reeds, air columns, whatever) vibrate in very complex ways. These complex vibrations produce frequencies beyond that of the fundamental.

When these so-called overtones are whole number multiples of the fundamental frequency, they're called harmonics. The harmonics of A-440 appear at 880 Hz (2 × 440), 1320 Hz (3 × 440), 1760 Hz (4 × 440), 2200 Hz (5 × 440), and so on. The relative way the volumes of each harmonic change over time determines the timbre of an instrument. Overtones that are not simple whole number multiples of the fundamental are said to be enharmonic. Percussion instruments and percussive sounds such as explosions, door slams, and such contain enharmonic overtones. As you'll learn later, enharmonic overtones are a lot more difficult to compress than harmonic overtones, which is why a harpsichord (fewer overtones) compresses better than jazz piano (very rich overtones).

To achieve a "pure" fundamental, folks use electronic devices—oscillators—that produce distortion-free sine waves (a distorted sine wave has harmonics). Clean sine waves are shown in Figure 1.6 (check the DVD-ROM or the web site to hear them as well).

Of course, air doesn't know about overtones or notes or fundamentals or any of that. Sound at any given point in space and time is a change in air pressure. Figure 1.7 shows what the same frequency looks like with piano-like overtones.

When an electronic oscillator is used to produce a square wave (a sine wave with a distortion pattern that makes it look squared off hence its name), it produces a timbre containing the maximum possible number of overtones. It looks like Figure 1.8—and sounds pretty cool, too.

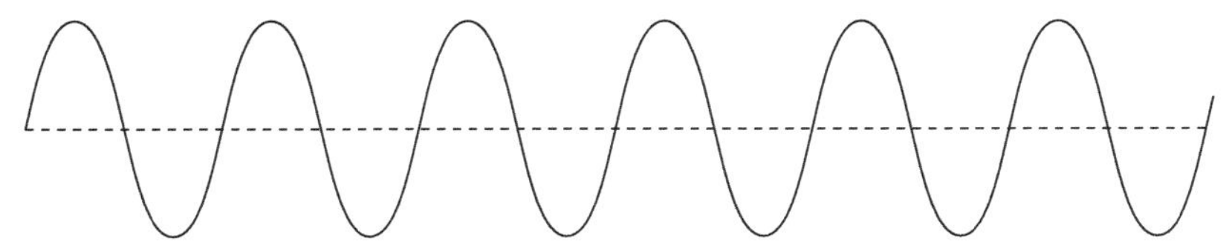

Figure 1.6 The sine wave of a perfectly pure 440 Hz tone.

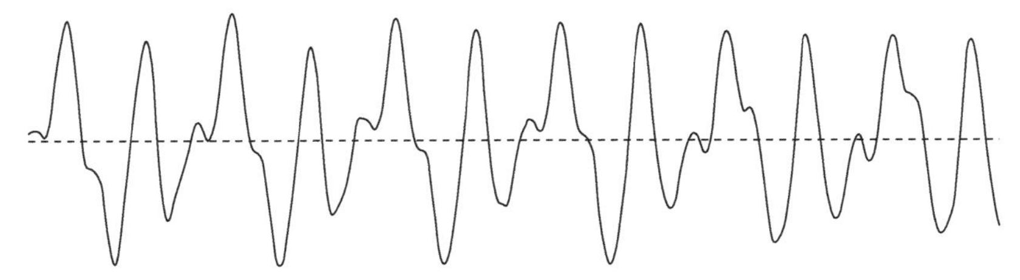

Figure 1.7 This is the waveform of a single piano note at 440 Hz.

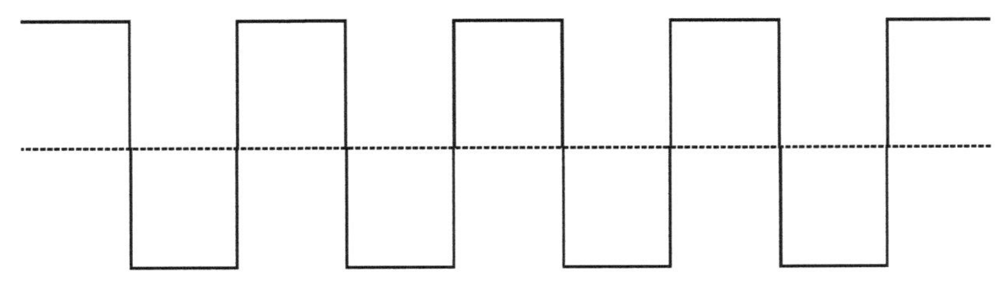

Figure 1.8 A 440 Hz square wave. Even though it looks very simple, acoustically it's the most loud and complex single note that can be produced.

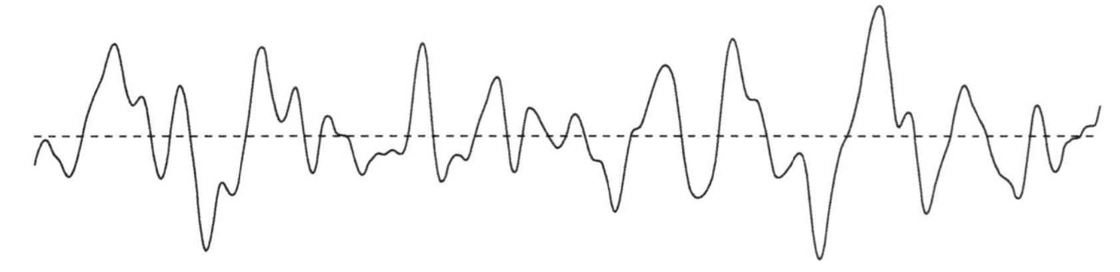

Figure 1.9 A full piano A—major chord at 440 Hz. Note how much more complex it is, but the 440 Hz cycle is still very clear.

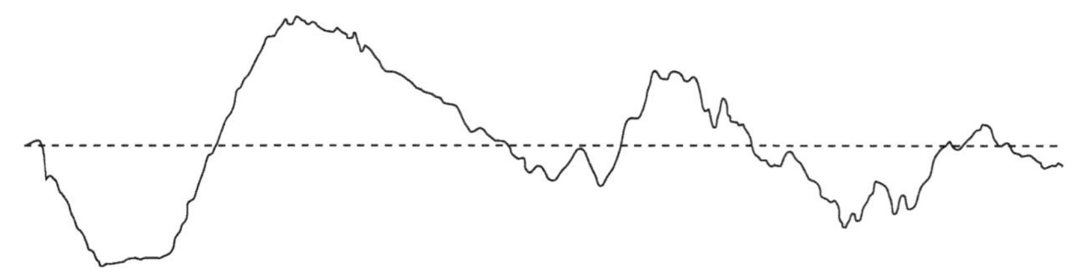

Figure 1.10 The waveform of a rim shot. Note how random the pattern seems.

When there are notes played simultaneously, even if they're just simple sine waves, their waveform produces something like Figure 1.9. It may look like a mess, especially with overtones, but it sounds fine.

Though tonal music is a wonderful thing—and a big part of what we compress—it's far from the only thing. Percussive sounds (drums, thunder, explosions) are made up of random noise with enharmonic spectra. Because there is so little pattern to them, percussive sounds prove a lot more difficult to compress, especially high-pitched percussion instruments like cymbals (Figure 1.10).

How the Ear Works

In the same way the eye is something like a camera, the ear is something like a microphone (Figure 1.11). First, air pressure changes cause the eardrum to vibrate. This vibration is carried by the three bones of the ear to the basilar membrane, where the real action is. In the membrane are many hair cells, each with a tuft of cilia. Bending of the cilia turn vibrations into electrical impulses, which then pass into the brain. We start life with only 16,000 of these hair cells, and they are killed, without replacement by excessively loud sounds. This damage is the primary cause of hearing loss.

Like the rods and cones in our retina, cilia along the basilar membrane respond to different frequencies depending on where along the cochlea they reside. Thus, cilia at specific locations respond to specific frequencies. The cilia are incredibly sensitive—moving the membrane by as little as the width of one atom can produce a sensation (in proportion, that's like moving the top of the Eiffel tower a half inch). They can also respond to vibrations up to 20,000 Hz. And if that sounds impressive, some whales can hear frequencies up to 200,000 Hz!

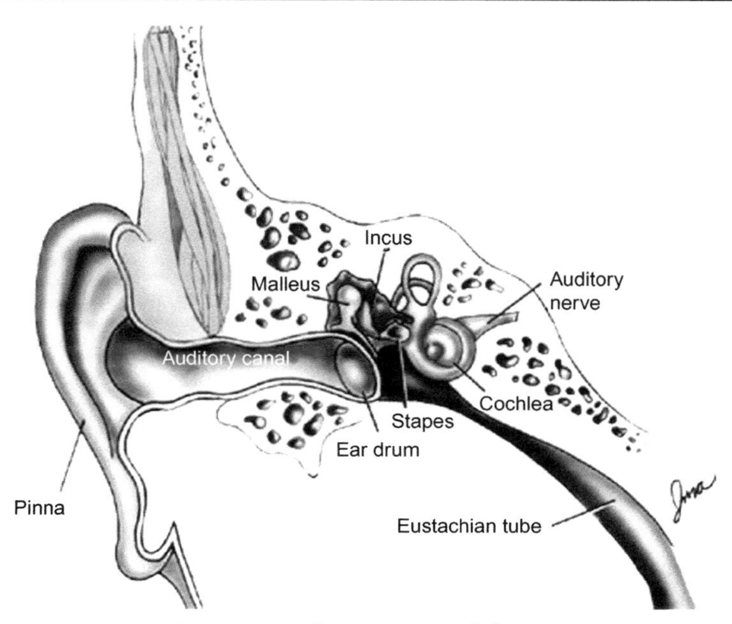

Figure 1.11 The anatomy of the ear.

What We Hear

As Discovery Channel aficionados know, there are plenty of sounds that humans can't hear but other species can. As with visible light, there is a range of vibrations that are audible. Someone with excellent hearing can hear the range of about 20 to 20,000 Hz, or 20 to 20,000 cycles per second. The maximum frequency that can be heard gets lower with age and hearing damage from loud noises. Attending a few thundering rock concerts can dramatically lower your high-frequency hearing (wear your earplugs!). Below 30 to 20 Hz, we start hearing sound as individual beats (20 Hz is the same as 1,200 beats per minute, or bpm—hardcore techno, for example, goes up to around 220 bpm). Below 80 Hz, we feel sound more than we hear it. We hear best in the range of 200 to 4000 Hz, pretty much the range of the human voice. Evolutionarily, it isn't clear whether our speech range adapted to match hearing or vice versa; either way, our ears and brain are highly attuned to hearing human voices well. Hearing loss typically most impacts the 3000–4000 Hz range, overlapping with the higher end of the human speech range. This is a big reason why people with hearing loss have trouble understanding small children—their smaller throats make for higher-pitched voices, and so more of their speech falls into the impaired range.

Sound levels (volume) are measured in decibels (dB). Decibels are expressed in a logarithmic scale, with each increase of 10 dB indicating a tenfold increase in volume. The range between the quietest sound we can hear (the threshold of hearing for those with great ears goes down to about 0 dB) and the loudest sound that doesn't cause immediate hearing damage (120 dB) is roughly 1 trillion to one (interestingly enough, the same ratio as that between the dimmest light we can see and the brightest that doesn't cause damage to the eye).

Less is known about how the brain processes audio than how it processes vision. Still, our auditory system is clearly capable of some pretty amazing things—music, for one huge example. Music is arguably the most abstract art form out there—there's nothing fundamental about why a minor chord sounds sad and a major chord doesn't. And the seemingly simple act of picking out instruments and notes in a piece of music is massively hard for computers to accomplish. A few pieces of music software are capable of transcribing a perfectly clear, pure-tone monophonic (one note at a time) part played on a single instrument fairly well these days. But throw chords or more than one instrument in the mix, and computers can't figure out much of what they're "hearing."

We're also quite capable of filtering out sounds we don't care about, like listening to a single conversation while ignoring other sounds in a crowded room. We're also able to filter out a lot of background noise. It's always amazing to me to work in an office full of computers for a day, and not be aware of the sound—keyboards clacking, fans whirring. But at the end of the day when things are shut off, the silence can seem almost deafening. And with all those things off, you can then hear sounds that had been completely masked by the others—the hum of fluorescent lights, birds chirping outside, and such.

We're able to sort things in time as well. For example, if someone is talking about you at a party, you'll often "hear" their whole sentence, even the part preceding your name, even though you wouldn't have noticed the previous words if they hadn't been followed by your name. This suggests to me that the brain is actually listening to all the conversations at some level, but only bringing things to our conscious attention under certain circumstances. This particular example is what got me to study neuroscience in the first place (although scientists still don't know how we're able to do it).

Psychoacoustics

Although our hearing system is enormously well-suited to some tasks, we still aren't able to use all our sensitivity in all cases. This is critically important to doing audio compression, which is all about not devoting bits to the parts of the music and sound we can't hear, saving bits for what we can. For example, two sounds of nearly the same pitch can sound just like a single note, but louder than either sound would be on its own. In the center of our hearing range, around 200–2000 kHz, we can detect very fine changes in pitch—for example, a 10 Hz shift up or down in a 2 kHz tone. Sounds can also be masked by louder tones of around the same frequency.

Summary

And that was a quick survey of how light becomes seeing and sound becomes hearing. In the next chapter, we dive into the mathematical representation of light and sound in cameras and computers.

CHAPTER 2

Uncompressed Video and Audio: Sampling and Quantization

This chapter is about how we turn light and sound into numbers. Put that way, it sounds simple, but there are a fair amount of technical details and even math-with-exponents to follow. Although it may seem that all the action happens in the codecs, a whole lot of the knowledge and assumptions about human seeing and hearing we talked about in the last chapter get baked into uncompressed video and audio well before it hits a codec.

Sampling and Quantization

In the real world, light and sound exist as continuous analog values. Those values in visual terms make up an effectively infinite number of colors and details; in audio terms, they represent an effectively infinite range of amplitude (loudness) and frequency. But digital doesn't do infinite. When analog signals are digitized, the infinite continuous scales of analog signals must be reduced to a finite range of discrete bits and bytes. This is accomplished via processes known as sampling and quantization. Sampling defines the discreet points or regions that are going to be measured. Quantization defines the actual value recorded.

Sampling Space

Nature doesn't have a resolution. Take a standard analog photograph (one printed from a negative on photographic paper, not one that was mass-produced on newsprint or magazine paper stock via a printing press). When you scan that photo with a scanner, you need to specify a resolution in dots per inch (dpi). But given a good enough scanning device, there is no real maximum to how much detail you could go into. If you had an electron microscope, you could scan at 1,000,000 dpi or more, increasing resolution far past where individual particles of ink are visible. Beyond a certain point, you won't get any more visual detail from the image, just ink particle patterns.

Sampling is the process of breaking up an image into discrete pieces, or samples, each of which represents a single point in 2D space. Imagine spreading a sheet of transparent graph paper over a picture. A sample is one square on the graph paper. The smaller the squares, the more samples. The number of squares is the resolution of the image. Each square is a picture element or pixel (or in MPEG-speak, a pel).

Most web codecs use square pixels, in which the height and width of each sample are equal. However, capture formats such as DV and ITU-R BT.601 often use nonsquare pixels, which are rectangular. Lots more on that later.

Color Figure C.9 demonstrates the effect of progressively coarser sampling on final image quality.

Sampling Time

The previous section describes how sampling works for a single frame of video. But video doesn't have just a single frame, so you need to sample temporally as well. Temporal sampling is normally described in terms of fps or Hertz (cycles per second, abbreviated Hz). It's relatively simple, but explains why video is so much harder to compress than stills; we're doing a bunch of stills many times a second. We sample time more accurately as we capture more frames per second. We generally need at least 15 fps for motion to seem like motion and not a slide show, at least 24 fps for video and audio to appear in sync, and at least 50 fps for fast motion like sports to be clear.

Sampling Sound

Digitized audio, like video, is sampled. Fortunately, audio data rates are lower, and so it's a lot easier to store audio in a high-quality digital format. Thus, essentially all professional audio is uncompressed.

Video has two spatial dimensions (height and width), but audio has only one: loudness, which is essentially the air pressure at any given moment. So audio is stored as a series of measurements of loudness. In the analog world, like when sound is transmitted via speaker cables, those changes are changes in the voltage on the wire. The frequency at which loudness changes are sampled is called the sampling rate. For CD-quality audio, that rate is 44.1 kHz (44,100 times per second). Other common sampling rates include 22.05 kHz, 32 kHz, 48 kHz, 96 kHz, and 192 kHz (96 kHz and 192 kHz are almost exclusively used for authoring, not for delivery). Consumer audio is almost never delivered at more than 48 kHz.

Nyquist Frequency

A concept crucial to all sampling systems is the Nyquist theorem. Harry Nyquist was an engineer at Bell Labs (which will come up more than once in the history of digital media). Back in the 1920s, he proved that you need a sampling rate at least twice as high as the frequency of any signal to reproduce that signal accurately. The frequency and signal can be spatial, temporal, or audible.

The classic example of the Nyquist theorem is train wheels. I'm sure you've seen old movies where wheels on a train look like they're going backwards. As the train speeds up, the wheels go faster and faster, but after a certain point, they appear to be going backwards. If you step through footage of a moving train frame by frame, you'll discover the critical frequency at which the wheel appears to start going backwards is at 12 Hz (12 full rotations)—half the speed of the film's 24 fps.

Figure 2.1 shows what's happening. Assume a 24 fps film recording of that train wheel. I've put a big white dot on the wheel so you can more easily see where in the rotation it is. In the first figure, the wheel is rotating at 6 Hz.At 24 fps, this means we have four frames of each rotation. Let's increase the wheel speed to 11 Hz.We have a little more than two frames per rotation, but motion is still clear. Let's go up to 13 Hz.With less than two frames between rotations, the wheel actually looks like it's going backwards! Once you get past 24 Hz, the wheels look like they're going forward again, but very slowly.

Alas, there is no way to correct for a sampling rate that's too low to accommodate the Nyquist frequency after the fact. If you must have train wheels that appear to be going forward, you need slow wheels or a camera with a fast frame rate!

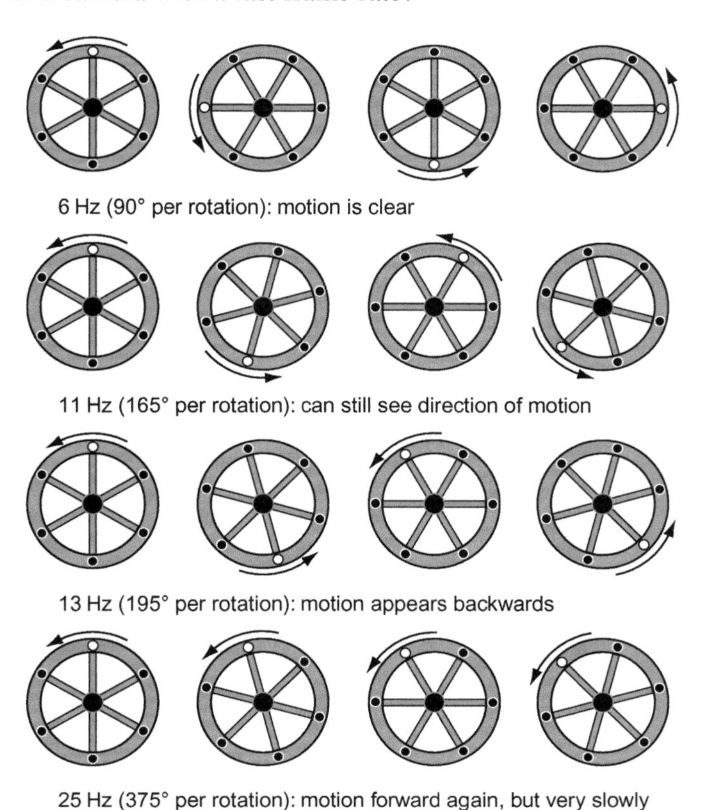

Figure 2.1 The infamous backwards — train — wheel illusion illustrated.

Another way to think about this is with a sine wave. If you sample a sine wave, as long as the source frequency is half or less of the sampling frequency, you can easily draw a new sine wave between the samples. However, if the sample rate is less than twice the frequency of the sine wave, what looks correct turns out to be completely erroneous.

You'll encounter the same issue when working with images—you can't show spatial information that changes faster than half the special resolution. So, if you have a 640-pixel-wide screen, you can have at most 320 variations in the image without getting errors. Figure 2.2 shows what happens to increasing numbers of vertical lines. Going from 320 × 240 to 256 × 192 with 3-pixel-wide detail is okay; 256/320*3 = 2.4; still above the Nyquist limit of 2. But a 2-pixel-wide detail is 256/320*2 = 1.6—below the Nyquist frequency, so we get an image that's quite different from the source.

This problem comes up in scaling operations in which you decrease the resolution of the image, because reducing the resolution means frequencies that worked in the source won't work in the output. Good scaling algorithms automatically filter out frequencies that are too high for the destination output. Algorithms that don't filter out frequencies that are too high

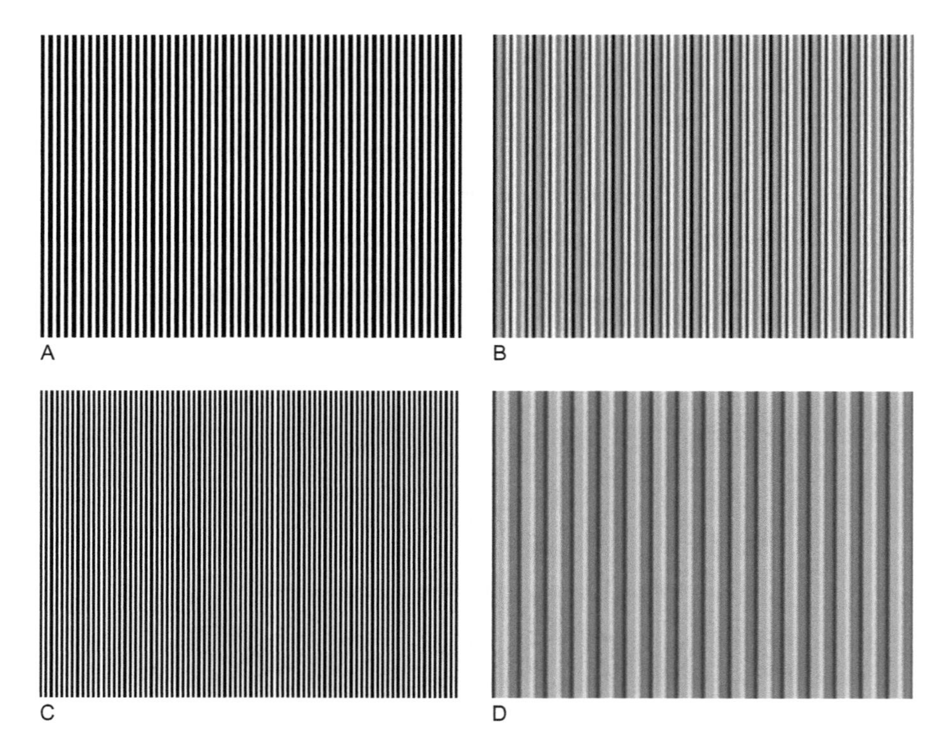

Figure 2.2 Nyquist illustrated. The far left – hand detail is a little over the Nyquist limit after scaling; the far right – hand detail is below it after scaling, resulting in a pattern unrelated to the source.

can cause serious quality problems. Such quality problems can sometimes be seen when thumbnail-sized pictures seem unnaturally sharp and blocky. There's a lot more on this topic in the preprocessing chapter.

Quantization

You've learned that sampling converts analog sounds and images into discrete points of measurement. The process of assigning discrete numeric values to the theoretically infinite possible values of each sample is called quantization. We speak of quantization that has a broader range of values available as being finer, and of that with a smaller range of values available as being coarser. This isn't a simple matter of finer being better than coarser; the finer the quantization the more bits required to process and store it. So a lot of thought and testing has gone into the ideal quantization for different kinds of samples to accurately capture the range we care about without wasting bits on differences too small to make an actual visible or audible difference.

In quantizing, we're making everything into numbers. And because a computer of some flavor is doing all the math (even if it's just a tiny chip in a camera) that math is fundamentally binary math, and made of bits. A bit can have one of two values—one or zero. Each additional bit doubles the sampling resolution. For example, a two-bit sample can represent four possible values; three bits gives eight possible values; eight bits provides 256 possible values. The more bits, the better the sampling resolution.

Bits themselves are combined into bytes, which are 8 bits each. Processing on computers is done in groups of bytes, themselves going up by powers of two. So, typically processing is done in groups of 1 byte (8-bit), 2 bytes (16-bit), 4 bytes (32-bit), 8 bytes (64-bit), et cetera. Something that requires 9-bit is going to really be processed as 16-bit, and so can require twice as much CPU power and memory to process. So, you'll see lots of 8-bit, 16-bit, and 32-bit showing up in our discussion of quantization. We'll also see intermediate numbers show up, like 10-bit, but those are more about saving storage space over the full byte size; it's still as slow as 16-bit in terms of doing the calculations.

Quantizing video

Basic 8-bit quantization, and why it works

Continuing our graph paper metaphor, consider a single square (representing a single sample). If it's a solid color before sampling, it's easy to represent that solid color as a single number. But what if that square isn't a single solid color? What if it's partly red and partly blue? Because a sample is a single point in space, by definition, it can have only one color. The general solution is to average all the colors in the square and use that average as the sample's color. In this case, the sample will be purple, even if no purple appeared in the original image. See Color Figure C.10 for an example of this in practice.

The amount of quantization detail is determined by bit depth. Most video codecs use 8 bits per channel. Just as with 8-bit audio systems, 8-bit video systems provide a total of 256 levels of brightness from black to white, expressed in values between 0 and 255. Unlike 8-bit audio (a lousy, buzzy-sounding mess with no dynamic range, and which younger readers may have been fortunate enough never to hear), 8-bit video can be plenty for an extremely high-quality visual presentation. Note, however, that Y′CbCr (also called, frequently but inaccurately, YUV) color space video systems (more on what that means in a moment) typically use only use the range from 16 to 235 for luminance of typical content, while RGB color space computer systems operate across the full 0 to 255 range. So, RGB and Y′CbCr are both quantized to an 8-bit number, but in practice the range of RGB has 256 values and the range of Y′CbCr is 219.

In the previous chapter, we discussed the enormous range of brightness the human eye is capable of perceiving—the ratio between the dimmest and brightest light we can see is about one trillion to one. Fortunately, we don't need to code for that entire range. To begin with, computer monitors and televisions aren't capable of producing brightness levels high enough to damage vision or deliver accurate detail for dim images. In practice, even a good monitor may only have a 4000:1 ratio, and even that may be in limited circumstances. Also, when light is bright enough for the cones in our eyes to produce color vision, we can't see any detail in areas dark enough to require rods for viewing. In a dark room, such as a movie theater, we can get up to about a 100:1 ratio between the brightest white and the darkest gray that is distinguishable from black. In a more typical environment containing more ambient light, this ratio will be less. We're able to discriminate intensity differences down to about 1 percent for most content (with a few exceptions). Thus, at least 100 discrete steps are required between black and white; our 256 or 219 don't sound so bad.

Note when I say we can distinguish differences of about 1 percent, I'm referring to differences between two light intensity levels. So, if we say if the background has 1,000 lux of light, we can tell the difference between that 1,000 and an object at 990 (1 percent less) or 1,010 (1 percent more). But if the background is 100 we can distinguish between 99 and 101. It's the ratio between the light levels that's important, not the actual amount of light itself. So, the brighter the light, the bigger a jump in absolute light levels we need to perceive a difference.

We take advantage of this by making a quantization scale that uses fixed ratios instead of fixed light levels, which we call "perceptually uniform." In a perceptually uniform scale, we code the differences in terms of how it would be perceived, not how it's produced. With a perceptually uniform luma scale, the percentage (and thus the perceived) jump in luminance looks the same from 20 to 21 as from 200 to 201, and the difference from 20 to 30 the same as from 200 to 210. This means that there is an exponential increase in the actual light emitted by each jump. But it makes the math a lot simpler and enables us to get good results in our 219 values between 16 and 235, which is good; this means we can do our video in 8 bits, or one byte, which is the optimal size to be able to process and store the bits.

The exponential increase used to make the scale perceptually uniform is called gamma, after the Greek letter γ. The nominal value of video codecs these days is a uniform 2.2. This was a hard-won battle; in the first edition of this book I had to talk about codecs that used gammas from 1.8 to 2.5, and video often had to be encoded differently for playback on Mac and Windows computers! It was a pleasure to delete most of that material for this edition.

One cool (and sadly historical) detail is that good old CRT displays had essentially the same 2.2 relationship between input power and displayed brightness as does the human eye. This meant that the actual electrical signal was itself perceptually uniform! And thus it was really easy to convert from light to signal and back to light without having to worry.

In the LCD/Plasma/DLP era, the underlying displays have very different behavior, but they all emulate the old CRT levels for backwards compatibility. Unfortunately, they can't all do it as well, particularly in low black levels, particularly with LCD. We'll talk about the implications of that when we get to preprocessing.

Gradients and Beyond 8-bit

With sufficiently fine quantization, a gradient that starts as total black and moves to total white should appear completely smooth. However, as we've learned, digital systems do not enjoy infinite resolution. It's possible to perceive banding in 8-bit per channel systems, especially when looking at very gradual gradients. For example, going from Y' = 20 to Y' = 25 across an entire screen, you can typically see the seams where the values jump from one value to another. But you aren't actually seeing the difference between the two colors; cover the seam with a little strip of paper, and the two colors look the same. What you're seeing is the seam—revealed by our visual system's excellent ability to detect edges. We solve that problem in the 8-bit space by using dithering (more on that in the discussion on preprocessing). But dithering is a lossy process, and only should be applied after all content creation is done. It's also why I didn't include a visual image of the previous example; the halftone printing process is an intense variety of dithering, and so would obscure the difference.

This is why most high-end professional video formats and systems process video luma in 10 bits. Although those two extra bits increase the luma data rate by only 25 percent, their addition quadruples the number of steps from 256 to 1,024 (and makes processing more computationally intensive).

We also see some high-end 12-bit systems used professionally. And in Hollywood-grade film production, perceptually uniform gamma can be left behind for logarithmic scales that better mimic the film process. And we have reasonably priced tools now that use 32-bit floating-point math to avoid any dithering. We'll cover those subjects in the chapters on video formats and preprocessing.

But that's all upstream of compression; there are no delivery codecs using more than 8 bits in common use today. There are specs for 10- and 12-bit video in the H.264 spec, which we'll discuss more in the H.264 chapter. However, that's not likely to matter until we see consumer displays and connections that support more than 8-bit become more common. The decades of development around digital video delivery we have today are all based around 8-bit, so it would be a slow and gradual process for more than 8-bit delivery to come to the fore.

Color Spaces

So far, we've been describing sampling and quantization of luminance, but haven't mentioned color—which, as you're about to discover, is a whole new jumbo-sized can of worms.

It's not a given that chroma and luma get quantized and sampled the same way. In fact, in most video formats they're handled quite differently. These different methods are called subsampling systems or color spaces.

As discussed in the previous chapter, all visible colors are made up of varying amounts of red, green, and blue. Thus, red, green, and blue can be sampled separately and combined to produce any visible color. Three different values are required for each sample to store color mathematically. And as we also discussed, human visual processing is very different for the black-and-white (luma) part of what we see than for the color (chroma) part of what we see. Specifically, our luma vision is much more detailed and able to pick up fast motion, while our chroma vision isn't great at motion or detail, but can detect very slight gradations of hue and saturation.

Back in the 1930s, the CIE determined that beyond red, green, and blue, other combinations of three colors can be used to store images well. We call these triads of values "channels," which can be considered independent layers, or planes, of the image. But these channels don't have to be "colors" at all; a channel of luma and two of "color difference" work as well. Because luma is mainly green, the color difference channels measure how much red and blue there is relative to gray. This allows us to sample and quantize differently based on the different needs of the luma and chroma vision pathways, and allows many awesome and wonderful things to happen.

So, let's examine all the major color spaces in normal use one by one and see how they apply to real-world digital image processing.

RGB

RGB (Red, Green, Blue) is the native color space of the eye, each being specific to one of the three types of cones. RGB is an additive color space, meaning that white is obtained by mixing the maximum value of each color together, gray is obtained by having equal values of each color, and black by having none of any color.

All video display devices are fundamentally RGB devices. The CCDs in cameras are RGB, and displays use red, green, and blue phosphors in CRTs, planes in LCDs, or light beams in projection displays. Computer graphics are almost always generated in RGB. All digital video starts and ends as RGB, even if it's almost never stored or transmitted as RGB.

This is because RGB is not ideal for image processing. Luminance is the critical aspect of video, and RGB doesn't encode luminance directly. Rather, luminance is an emergent property of RGB values, meaning that luminance arises from the interplay of the three channels. This complicates calculations enormously because color and brightness can't be adjusted independently. Increasing contrast requires changing the value of each channel. A simple hue shift requires decoding and re-encoding all three channels via substantial mathematics. Nor is RGB conducive to efficient compression because luma and chroma are mixed in together. Theoretically, you might allocate fewer samples to blue or maybe red, but in practice this doesn't produce enough efficiency to be worth the quality hit.

So, RGB is mainly seen either in very high-bitrate production formats, particularly for doing motion graphics with tools like Adobe After Effects, or in still images for similar uses.

RGBA

RGBA is simply RGB with an extra channel for alpha (transparency) data, which is used in authoring, but content isn't shot or compressed with an alpha channel. Codecs with "alpha channel support" really just encode the video data and the alpha channel as different sets of data internally.

Y′CbCr

Digital video is nearly always stored in what's called Y′CbCr or a variant thereof. Y′ stands for luminance, Cb for blue minus luminance, and Cr for red minus luminance. The ′ after the Y is a prime, and indicates that this is nonlinear luma; remember, this is gamma corrected to be perceptually uniform, and we don't want it to be linear to actual light levels.

Confusing? Yes, it is. Most common first question: "where did green go?" Green is implicit in the luminance channel!

Let's expand the luminance equation from the previous chapter into the full formula for Y′CbCr.

$$
\begin{aligned}
Y' &= 0.299\ \text{Red} + 0.587\ \text{Green} + 0.114\ \text{Blue} \\
Cb &= -0.147\ \text{Red} - 0.289\ \text{Green} + 0.436\ \text{Blue} \\
Cr &= 0.615\ \text{Red} - 0.515\ \text{Green} - 0.100\ \text{Blue}
\end{aligned}
$$

So, Y′ mainly consists of green, some red, and a little blue. Because Cb and Cr both subtract luminance, both include luminance in their equations; they're not pure B or R either. But this matches the sensitivities of the human eye, and lets us apply different compression techniques based on that. Because we see so much more luma than chroma detail, separating out chroma from luma allows us to compress two of our three channels much more. This wouldn't work in RGB, which is why we bother with Y′CbCr in the first place.

The simplest way to reduce the data rate of a Y′CbCr signal is to reduce the resolution of the Cb and Cr channels. Going back to the graph paper analogy, you might sample, say, Cb and Cr once every two squares, while sampling Y′ for every square. Other approaches are often used—one Cb and Cr sample for every four Y′s, one Cb and Cr sample for every two Y′ and so on.

Formulas for Y′CbCr sampling are notated in the format x:y:z. The first number establishes the relative number of luma samples. The second number indicates how many chroma samples there are relative to the number of luma samples on every other line starting from the first, and the third represents the number of chroma samples relative to luma samples on the alternating lines. So, 4:2:0 means that for every four luma samples, there are two chroma samples on every other line from the top, and none on the remaining lines. This means that chroma is stored in 2 × 2 blocks, so for a 640 × 480 image, there will be a 640 × 480 channel for Y′, and 320 × 240 channels for each of Cb and Cr.

The quantization for Y′ runs as normal—0 to 255, perceptually uniform. Color is different. It's still 8-bit, but uses the range of –127 to 128. If both Cb and Cr are 0, the image is monochrome. The higher or lower the values are from 0, the more saturated the color. The actual hue of the color itself is determined by the relative amounts of Cb versus Cr. In the same way that the range of Y in video is 16 to 235, the range of Cb and Cr in video is limited to –112 to 112.

Terminology

In deference to Charles M. Poynton, who has written much more and much better about color than I have, this book uses the term Y′CbCr for what other sources call YUV. YUV really only refers to the native mode of NTSC television, with its specific gamma and other attributes, and for which broadcast has now ended. Y′U′V′ is also used sometimes, but isn't much more relevant to digital video. When used properly, U and V refer to the subcarrier modulation axes in NTSC color coding, but YUV has become shorthand for any luma-plus-subsampled chroma color spaces. Most of the documentation you'll encounter uses YUV in this way. It may be easier to type, but I took Dr. Poynton's excellent color science class at the Hollywood Postproduction Alliance Tech Retreat, and would rather avoid his wrath and disappointment and instead honor his plea for accurate terminology.

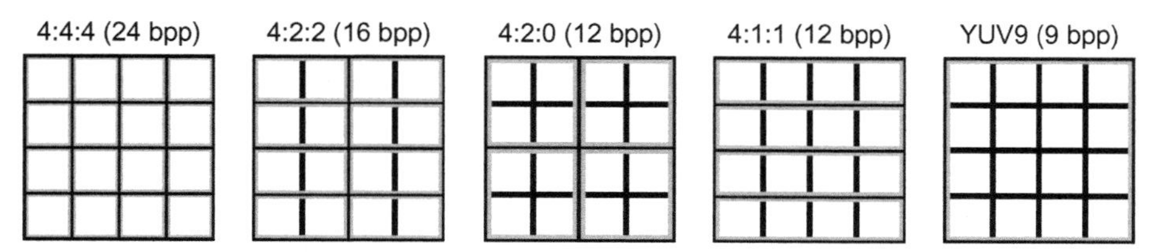

Figure 2.3 Where the chroma subsamples live in four common sampling systems, plus the highly misguided YUV − 9. The gray boxes indicate chroma samples, and the black lines the luma samples.

4:4:4 Sampling

This is the most basic form of Y′CbCr, and the least seen. It has one Cb and one Cr for each Y value. Though not generally used as a storage format, it's sometimes used for internal processing. In comparison, RGB is always 4:4:4. (See Figure 2.3 for illustrations of 4:4:4, 4:2:2, 4:2:0. 4:1:1, and YUV-9 sampling.)

4:2:2 Sampling

This is the subsampling scheme most commonly used in professional video authoring: one Cb and one Cr for every two Y′ horizontally. 4:2:2 is used in tape formats such as D1, Digital Betacam, D5, and HDCAM-SR. It's also the native format of the standard codecs of most editing systems. It looks good, particularly for interlaced video, but is only a third smaller than 4:4:4.

So, if 4:2:2 results in a compression ratio of 1.5:1, why is 4:2:2 called "uncompressed"? It's because we don't take sampling and quantization into account when talking about compressed versus uncompressed, just as audio sampled at 44.1 kHz isn't said to be "compressed" compared to audio sampled at 48 kHz.So we have uncompressed 4:2:2 codecs and compressed 4:2:2 codecs.

4:2:0 Sampling

Here, we have one pixel per Y′ sample and four pixels (a 2 × 2 block) per Cb and Cr sample. This yields a 2:1 reduction in data before further compression. 4:2:0 is the ideal color space for compressing progressive scan video, and is used in all modern delivery codecs, whether for broadcast, DVD, Blu-ray, or web. It's often called "YV12" as well, referring to the specific mode that the samples are arranged.

4:2:0 is also used in some consumer/prosumer acquisition formats, including PAL DV, HDV, and AVCHD.

4:1:1 Sampling

This is the color space of NTSC DV25 formats, including DV, DVC, DVCPRO, and DVCAM, as well as the (thankfully rare) "DVCPRO 4:1:1" in PAL. 4:1:1 uses Cb and Cr samples four pixels wide and one pixel tall. So it has the same number of chroma samples as 4:2:0, but in a different (nonsquare) shape.

So why use 4:1:1 instead of 4:2:0? Because of the way interlaced video is put together. Each video frame is organized into two fields that can contain different temporal information—more about this later. By having chroma sampling be one line tall, it's easier to keep colors straight, between fields. However, having only one chroma sample every 4 pixels horizontally can cause some blocking or loss of detail with the edges of saturated colors. 4:1:1 is okay to shoot with, but you should never actually encode to 4:1:1; it's for acquisition only.

YUV-9 Sampling

YUV-9 was a common colorspace in the Paleolithic days of compression, with one chroma sample per 4 × 4 block of pixels. A break from the 4:x:x nomenclature, YUV-9's name comes from it having an average of 9 bits per pixel. (Thus 4:2:0 would be YUV-12, hence YV12.) 4 × 4 chroma subsampling wasn't enough for much beyond talking head content, and sharp edges in highly saturated imagery in YUV-9 looks extremely blocky.

This was the Achilles heel of too many otherwise great products, including the Indeo series of codecs and Sorenson Video 1 and 2. Fortunately, all modern codecs have adopted 4:2:0.

Lab

Photoshop's delightful Lab color mode is a 4:4:4 Y′CbCr variant. Its name refers to Luminance, chroma channel A, and chroma channel B. And luma/chroma separation is as good for still processing as it is for video, and I've been able to wow more than one old-school Photoshop whiz by jumping into Lab mode and fixing problems originating in the chroma channels of a JPEG file.

CMYK Color Space

CMYK isn't used in digital video, but it's the other color space you're likely to have heard of. CMYK stands for cyan, magenta, yellow, and black (yep, that final letter is for blacK).

CMYK is used in printing and specifies the placement and layering of inks on a page, not the firing of phosphors. Consequently, CMYK has some unique characteristics. The biggest one is adding a fourth channel, black. This is needed because with real-world pigments, mixing C + M+Y yields only a muddy brown, so a separate black channel is necessary.

In contrast to additive color spaces like RGB, CMYK is subtractive; white is obtained by removing all colors (which leaves white paper).

There are some expanded versions of CMYK that add additional inks to further improve the quality of reproduction, like CcMmYK, CMYKOG, and the awesomely ominous sounding (if rarely seen) CcMmYyKkBORG.

Color spaces for printing is a deep and fascinating topic that merits its own book. Fortunately, this is not that book. CMYK is normally just converted to RGB and then on to Y′CbCr in the compression workflow; all the tricky stuff happens in the conversion to RGB. And as CMYK is only ever used with single images, this is really part of editing and preprocessing, well before compression. In general it's a lot easier to convert from CMYK than to it; getting good printable images out of a video stream can be tricky.

Quantization Levels and Bit Depth

Now that you've got a good handle on the many varieties of sampling, let's get back to quantization. There are a number of sampling options out in the world, although only 8-bit per channel applies to the codecs we're delivering in today.

8-bit Per Channel

This is the default depth for use in compression, for both RGB and YUV color space, with our canonical 8-bits per channel. This works out to be 24 bits per pixel with 4:4:4, 16-bits per pixel with 4:2:2, and 12 bits per pixel with 4:2:0. 8-bit per channel quantization is capable of excellent reproduction of video without having to carry around excess bits. And because our entire digital video ecosystem has been built around 8-bit compression, transmission, and reproduction, more bits would be hard to integrate.

Almost all of the examples in this book wind up in 8-bit 4:2:0. I document the other quantization methods, since they'll come up in source formats or legacy encodes.

1-bit (Black and White)

In a black-and-white file, every pixel is either black or white. Any shading must be done through patterns of pixels. 1-bit quantization isn't used in video at all. For the right kind of content, such as line art, 1-bit can compress quite small, of course. I mainly see 1-bit quantization in older downloadable GIF coloring pages from kids' web sites.

Indexed Color

Indexed color is an odd beast. Instead of recording the values for each channel individually, an index of values specify R, G, and B values for each "color", and the index code is stored for each pixel. The number of items in the index is normally 8-bit, so up to 256 unique colors are possible. The selection of discrete colors is called a palette, Color Look Up Table (or CLUT, which eventually stops sounding naughty after frequent use, or eighth grade), or index.

8-bit is never used in video production, and it is becoming increasingly rare for it to be used on computers either. Back in the Heroic Age of Multimedia (the mid-1990s), much of your media mojo came from the ability to do clever and even cruel things with palettes. At industry parties, swaggering pixel monkeys would make extravagant claims for their dithering talents. Now we live in less exciting times, with less need for heroes, and can do most authoring with discreet channels. But 8-bit color still comes up from time to time. Indexed color's most common use remains in animated and still GIF files. And I even use it for PNG screenshots of grayscale user interfaces (like the Expression products) on my blog, as it's more efficient than using full 24- or 32-bit PNG. And it somehow crept into Blu-ray! BD discs authored with the simpler HDMV mode (but not the more advanced BD-J) are limited to 8-bit indexed PNG for all graphics. Go figure.

Depending on the content, 8-bit can look better or worse than 16-bit. With 8-bit, you can be very precise in your color selection, assuming there isn't a lot of variation in the source. 16-bit can give you a lot more banding (only 32 levels between black and white on Mac), but can use all the colors in different combinations.

Making a good 8-bit image is tricky. First, the ideal 8-bit palette must be generated (typically with a tool such as Equilibrium's DeBabelizer). Then the colors in the image must be reduced to only those appearing in that 8-bit palette. This is traditionally done with dithering, where the closest colors are mixed together to make up the missing color. This can yield surprisingly good results, but makes files that are very hard to compress. Alternatively, the color can be flattened—each color is just set to the nearest color in the 8-bit palette without any dithering. This looks worse for photographs, but better for synthetic images such as screen shots. Not dithering also improves compression.

Many formats can support indexed colors at lower bit depths, like 4-color (2-bit) or 16-color (4-bit). These operate in the same way except with fewer colors (see Color Figure C.11).

8-bit Grayscale

A cool variant of 8-bit is 8-bit grayscale. This doesn't use an index; it's just an 8-bit Y′ channel without chroma channels. I used 8-bit grayscale compression for black-and-white video often—Cinepak had a great-looking 8-bit grayscale mode that could run 640 × 480 on an Intel 486 processor. You can also force an 8-bit index codec to do 8-bit grayscale by giving it an index that's just 0 to 255 with R = G=B. The same effect can be done in Y′CbCr codecs by just putting zeros in the U and V planes. H.264 has an explicit Y′-only mode as well, although it isn't broadly supported yet.

16-bit Color (High Color/Thousands of Colors/555/565)

These are quantization modes that use fewer than 8 bits per channel. I can't think of any good reason to make them anymore, but you may come across legacy content using them. There are actually two different flavors of 16-bit color. Windows uses 6 bits in G and 5 bits each in R and B. "Thousands mode" on the Mac is 5 bits each for R, G, and B, with 1 bit reserved for a rarely used alpha channel.

16-bit obviously had trouble reproducing subtle gradients. Even with a 6-bit green (which is most of luminance) channel, only 64 gradations exist between black and white, which can cause visible banding.

No modern video codecs are natively 16-bit, although they can display to the screen in 16-bit. The ancient video codecs of Apple Video (code named Road Pizza–its 4CC is "rpza") and Microsoft Video 1 were 16-bit. But hey, they didn't have data rate control either, and ran on CPUs with a tenth the power of the cheapest cell phone today. You could probably decode them with an abacus in a pinch.

The Delights of Y′CbCr Processing

Warning: High Nerditry ahead.

One of the behind-the-scenes revolutions in editing tools in the last decade is moving from using RGB native processing to Y′CbCr-native processing for video filters.

But though the mainstream editing tools like Adobe Premiere and Apple Final Cut Pro are now Y′CbCr native, compositing tools (I'm looking at you, After Effects!) retain an RGB-only color model. One can avoid rounding errors from excess RGB < > Y′CbCr conversion by using 16-bit or even 32-bit float per channel, but that's paying a big cost in speed when all you want to do is apply a simple Levels filter to adjust luma.

There are three reasons why Y′CbCr processing rules over RGB.

Higher Quality

The gamut of RGB is different than Y′CbCr, which means that there are colors in Y′CbCr that you don't get back when converting into RGB and back to Y′CbCr (and vice versa; Y′CbCr can't do a 100-percent saturated blue like RGB can). Every time color space conversion is done, some rounding errors are added, which can accumulate and cause banding. This can be avoided by using 32-bit floating-point per channel in RGB space, but processing so many more bits has an impact on performance.

Higher Performance

With common filters, Y′CbCr's advantages get even stronger. Filters that directly adjust luminance levels—like contrast, gamma, and brightness—only need to touch the Y′ channel in Y′CbCr, compared to all three in RGB, so in 8-bit per channel that's 8 bits versus 24 bits in RGB. And if we're comparing 10-bit versus 32-bit float, that's 10 bits versus 96 bits. And chroma-only adjustments like hue or saturation only need to touch the subsampled Cb and Cr channels. So in 8-bit, that'd be 4 bits per channel for

(Continued)

Y′CbCr compared to 24 bits for RGB, and in 10 bit Y′CbCr versus 32-float, that's 8 bits against 96 bits.

What's worse, for a number of operations, RGB must be converted into a luma/chroma-separated format like Y′CbCr anyway, then converted back to RGB, causing that much more processing.

Channel-specific Filters

One of my favorite ways to clean up bad video only works in Y′CbCr mode. Most ugly video, especially coming from old analog sources like VHS and Umatic, is much uglier in the chroma channels than in luma. You can often get improved results by doing an aggressive noise reduction on just the chroma channels. Lots of noise reduction adds a lot of blur to the image, which looks terrible in RGB mode, because it's equally applied to all colors. But when the chroma channels get blurry, the image can remain looking nice and sharp, while composite artifacts seem to just vanish. Being able to clean up JPEG sources like this is one big reason I love Photoshop's Lab mode, and I wish I had Lab in After Effects and other compositing/motion graphics tools.

■

10-bit 4:2:2

Professional capture codecs and transports like SDI and HD SDI use Y′CbCr 4:2:2 with 10 bits in the luma channel (the chroma channels remain 8-bit). This yields 1,024 levels of gradation compared to 8-bit's 256. While there's nothing about 10-bit which requires 4:2:2, all 10-bit luma implementations are 4:2:2.

I'd love to see 10 bits per channel become a minimum for processing within video software. It breaks my heart to know those bits are in the file but are often discarded. Even if you're delivering 24-bit color, the extra two bits per channel can make a real difference in reducing banding after multiple passes of effects processing. While the professional nonlinear editing systems (NLEs) now can do a good job with 10-bit, few if any of our mainstream compression tools do anything with that extra image data.

12-bit 4:4:4

Sony's HDCAM-SR goes up to RGB 4:4:4 with 12 bits per channel at its maximum bitrate mode. While certainly a lot of bits to move around and process, that offers a whole lot of detail and dynamic range for film sources, allowing quite a lot of color correction without introducing banding. 12-bit 4:4:4 is mainly used in the high-end color grading/film timing workflow. We video guys normally get just a 10-bit 4:2:2 output after all the fine color work is done.

16-bit Depth

16-bit per channel was introduced in After Effects 5, and was the first step towards reducing the RGB/Y′CbCr rounding issues. However, it didn't do anything to deal with the gamut differences.

32-bit Floating-point

So far, all the quantization methods we've talked about have been integer; the value is always a whole number. Because of that limitation we use perceptually uniform values so we don't wind up with much bigger steps between values near black, and hence more banding. But floating-point numbers are actually made of three components. The math-phobic may tune out for this next part, if they've made it this far, but I promise this is actually relevant.

A Brief Introduction to Floating-Point

The goal of floating-point numbers is to provide a constant level of precision over a very broad range of possible values.

First off, let's define a few terms of art.

- Sign: whether the number is positive or negative (1-bit)
- Exponent, which determines how big the number is (8 bits in 32-bit float)
- Mantissa, which determines the precision of the number (23 bits in 32-bit float)

By example, we can take Avogadro's number, which is how many atoms of carbon there are in 12 grams. As an integer, it is awkward:

602,200,000,000,000,000,000,000

That's a whole lot of zeros at the end that don't really add any information, but which would take a 79-bit binary number! Which why it's normally written in scientific notation as

$$6.022 \times 10^{23}$$

Here, 6.022 is the mantissa and 23 is the exponent. This much more compact, and provides a clear indication of its precision (how many digits in the mantissa) and its magnitude (the value of the exponent) as different values.

With floating-point numbers, we get 23 bits worth of precision, but we can be that precise over a much broader number range. And we can even go negative which isn't useful for delivery, of course, but can be very useful when stacking multiple filters together—a high contrast filter can send some pixels to below 0 luma, and then a high brightness filter can bring them back up again without losing detail.

Floating-point numbers are generally not perceptually uniform either. To make the calculations easier, they actually use good old Y, not Y′, and calculate based on actual light intensity values.

Modern CPUs are very fast at doing 32-bit float operations, and by not having to deal with being perceptually uniform, 32-bit float may not be any slower than 16-bit per channel. And since 32-bit float can always return the same Y′CbCr values (where those values weren't changed by any filters). For that reason, I normally just use 32-bit float over 16-bit per channel.

Quantizing Audio

So, given all that, how do we quantize audio? As with sampling, things in the audio world are simpler.

Each bit of resolution equals six decibels (dB) of dynamic range (the difference between the loudest and softest sounds that can be reproduced). The greater the dynamic range, the better the reproduction of very quiet sounds, and the better the signal-to-noise ratio.

8-bit Audio

Early digital audio sampling devices operated at 8-bit resolution, and it sounded lousy. And to make it even passably lousy instead of utterly incomprehensible required specialized software or hardware.

16-bit Audio

Audio CD players operate at 16-bit resolution, providing 65,536 possible values. And using our 6 dB-per-bit rule, we get 96 dB of dynamic range. This is substantially higher than even the best consumer audio equipment can deliver, or that would be comfortable to listen to; if it was loud enough to hear the quietest parts, it'd be painfully loud at the loudest parts.

There are many near-religious debates on the topic of how much quantization is too much for audio. My take is that 16-bit audio is like 8-bit video; I'd like to start and work with more than that, but properly processed, it can produce very high-quality final encodes.

And as with making good 8-bit video, dithering is the key to 16-bit audio to soften the edges between fast transitions. I'm of the opinion that the CD's 44.1 KHz 16-bit stereo is quite sufficient for high-quality reproduction on good systems for people with good hearing, assuming the CD is well mastered. The main advantage to "HD audio" is in adding additional channels, not any increase in sample rate or bit depth.

20-bit Audio

20-bit offers only 4 more bits than 16-bit, but that's another 24 dB, which is more than a hundred times difference in dynamic range. At 20-bit, dithering becomes a nonissue, and even the finest audio reproduction system with an industry recognized "golden ear" can't tell the difference between 20-bit and higher in an A/B comparison. If you're looking for overkill, 20-bit is sufficiently murderous for any final encode.

24-bit Audio

Codecs like DD+ and WMA 10 Pro offer 24-bit modes. These make for trouble-free conversion from high-bit sources without having to sweat dithering. They're in practice the same quality as 20-bit. It's 24-bit, since that's three bytes (3 × 8), the next computer-friendly jump from 16-bit (2 × 8).

24-bit is also often used for recording audio, where all that headroom can be handy. Still, even 16-bit offers far more dynamic range than even the best microphone in the world can deliver. 24-bit and higher really pays off when mixing dozens of audio channels together in multiple generations.

32-bit and 64-bit Float

Floating-point is also available for audio production, although nothing is ever delivered in float. It offers the same advantages of high precision across a huge range like 32-bit float in video does.

Quantization Errors

Quantization (quant) errors arise from having to choose the nearest equivalent to a given number from among a limited set of choices. Digitally encoding colors inevitably introduces this kind of error, since there's never an exact match between an analog color and its nearest digital equivalent. The more gradations available in each channel—that is, the higher the color depth—the closer the best choice is likely to be to the original color.

For 8-bit channels that don't need much processing, quantization error doesn't create problems. But quant errors add up during image processing. Every time you change individual channel values, they get rounded to the nearest value, adding a little bit of error. With each layer of processing, the image tends to drift farther away from the best digital equivalent of the original color.

Imagine a Y value of 101. If you divide brightness in half, you wind up with 51 (50.5 rounded up—no decimals allowed with integers). Later, if you double brightness, the new value is 102. Look at Figure 2.4, which shows the before (left) and after (right) doing some typical gamma and contrast filtering. Note how some colors simply don't exist in the output, due to rounding errors; instead doubling the amount of an adjacent value.

Quant errors also arise from color space conversion. For example, most capture formats represent color as Y′CbCr 4:1:1 or 4:2:2, but After Effects filters operate in RGB.

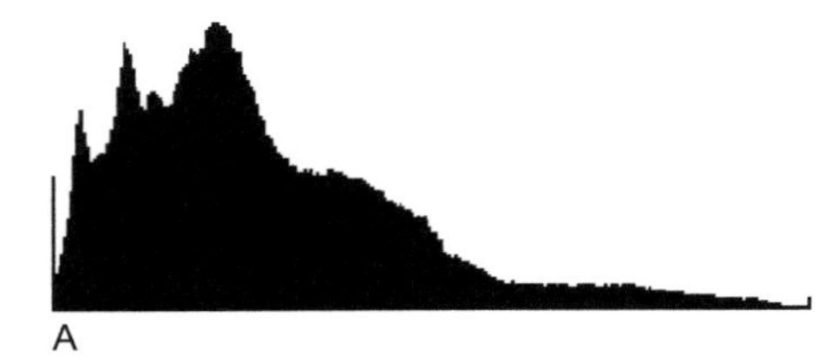

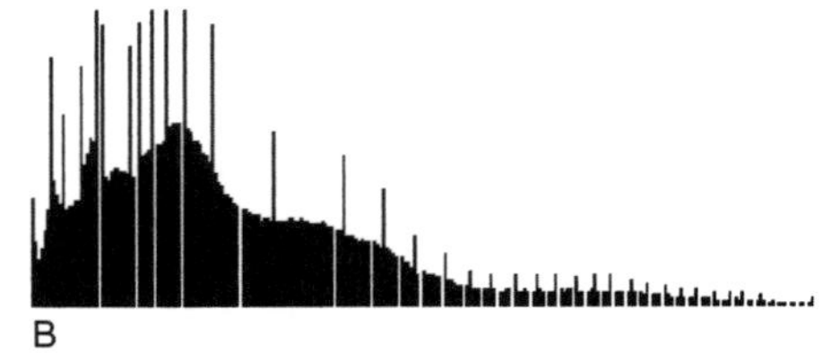

Figure 2.4 Even a slight filter can cause spikes and valleys in the histogram, with some values missing from the output and others overrepresented.

Transferring from tape to tape via After Effects generally involves converting from Y′CbCr to RGB and back to Y′CbCr—at the very least. What's more, filters and effects often convert color spaces internally. For instance, using Hue, Saturation, and Lightness (HSL) filters involves a trip through the HSL color space. Going between Y′CbCr and RGB also normally involves remapping from the 16–235 to the 0–255 range and back again each time, introducing further rounding errors.

Another trouble with color space conversions is that similar values don't always exist in all color spaces. For example, in Y′CbCr, it's perfectly possible to have a highly saturated blue-tinted black (where Y′ = 16 and Cb = 112). However, upon conversion to RGB, the information in the Cb and Cr channels disappears—black in RGB means R, G, or B are all 0; no color information at all. If the RGB version is converted back to Y′CbCr, the values in the Cb and Cr channels are no longer fixed. As Y gets near black or white, the Cb and Cr values approach zero.

Conversely, RGB has colors that can't be reproduced in Y′CbCr: for example, pure R, G, or B (like R = 255, G = 0, B = 0). Since each of Y′, Cb, and Cr is made of a mix of R, G, and B, it simply can't have that degree of separation at the extremes. It can be "quite blue" but there's a range of "super blues" that simply can't be quantized in video.

And that's how we get video and audio into numbers we can process on computers. And while we may think of uncompressed as being "exactly the same as the source", a whole lot of tuning for the human visual and auditory systems is implicit in the quantization and sampling modes chosen.

Now, on to compression itself.

CHAPTER 3

Fundamentals of Compression

This chapter is about the fundamental techniques used to get from uncompressed digital video and audio to compressed video and audio. There's a lot of math implied in this chapter, but I'll keep the equations to a minimum and describe what they mean for the more narratively inclined. While it's not essential to master every detail of how codecs work, a working knowledge of what's happening under the hood helps you understand the results you're getting, spot problem areas, and plan strategies for overcoming challenges.

Even if you skip this chapter the first time around, you might want to try again when while you're waiting for that big encode to finish.

Compression Basics: An Introduction to Information Theory

The fundamentals of information theory were thought up almost entirely by a single guy—Bell Labs Researcher Claude E. Shannon. His "A Mathematical Theory of Communication" remains the cornerstone of the entire field. Published in the *Bell System Journal* in 1948, it addressed an important question: "How much information can we actually send over these telephone and telegraph lines?" and then defined a mathematical definition of information and its limits. Shannon might not be a household name like Edison or Einstein , but the field he founded is as central to our modern world as indoor lighting and nuclear physics.

Tempted as I am to deliver a book-length discourse on information theory, I'll just summarize a few critical insights underlying video and audio compression.

Any Number Can Be Turned Into Bits

This is an obvious point, but bears repeating. The bit, a single 0 or 1, is the basis of all math on computers, and of information theory itself. Any number can be turned into a series of bits, be it an integer like "3," a fraction like 7/9ths, or an irrational constant like pi (3.1415926535…). This means we can use bits as the basis for measuring everything that can be sampled and quantized.

The More Redundancy in the Content, the More It Can Be Compressed

English in plain text is rather predictable. For example, I'm much more likely to use the word "codec" at the end of this sentence than XYZZY. The more you understand the distribution of elements within the information you need to compress, the more efficient the compression can be. For example, the distribution of letters in English isn't random: the letter "e" appears a lot more often than "q," so it is more efficient to design a system where "e" can be coded in fewer bits than "q." A table that translates from the original characters to shorter symbols is called a *codebook*. Shannon's grad student David A. Huffman devised a method for generating an optimal codebook for a particular set of content (for a term paper, no less). Huffman coding uses shorter codes for more common symbols and longer codes for less common ones. Different codebooks work better or worse for different kinds of content. A codebook for English won't work well for Polish, which has a very different distribution of letters. For an English codebook, the Polish distribution of letters would require more bits to encode, possibly even more bits than the uncompressed original file. The more you know about what's going to be compressed, the better a codebook can be generated.

Taken to the next level, the order of letters is also not random. For example, "ee" isn't as common a sequence in English as "th," even though "t" and "h" individually appear less often than "e." So a coding scheme can be expanded to take account of the probability of a particular letter given previous letters, assign symbols to a series of letters instead of a single one, or having a different codebook based on past letters.

Codebooks can be agreed upon in advance, so the decoder knows how to best decode the content in question. Multiple codebooks can be used, with a code at the start of the file or even changing throughout the file to indicate which codebook is to be used. Or a codebook can be dynamically created while the file is being compressed.

The converse of "more redundant content compresses better" is that less redundant content doesn't compress as well. Truly random data won't compress at all; there isn't any greater probability for any given number, so there isn't a useful Huffman code. Thus the codebook plus the compressed data can wind up larger than the original data. Modern compressors leave that kind of uncompressed, so the output file won't be larger than the input file.

In fact, we use randomness as a measure of compressibility. Compression is sometimes called "entropy coding," since what you're really saving is the entropy (randomness) in the data, while the stuff that could be predicted from that entropy is what gets compressed away to be reconstructed on decode.

The More Efficient the Coding, the More Random the Output

Using a codebook makes the file smaller by reducing redundancy. Because there is less redundancy, there is by definition less of a pattern to the data itself, and hence the data itself looks random. You can look at the first few dozen characters of a text file, and immediately see what language it's in. Look at the first few dozen characters of a compressed file, and you'll have no idea what it is.

Data Compression

Data compression is compression that works on arbitrary content, like computer files, without having to know much in advance about their contents.

There have been many different compression algorithms used over the past few decades. Ones that are currently available use different techniques, but they share similar properties.

The most-used data compression technique is Deflate, which originated in PKWare's .zip format and is also used in .gz files, .msi installers, http header compression, and many, many other places. Deflate was even used in writing this book—Microsoft Word's .docx format (along with all Microsoft Office ".???x" formats) is really a directory of files that are then Deflated into a single file.

For example, the longest chapter in my current draft ("Production, Post, and Acquisition") is 78,811 bytes. Using Deflate, it goes down to 28,869 bytes. And if I use an advanced text-tuned compressor like PPMd, (included in the popular 7-Zip tool), it can get down to 22,883 bytes. But that's getting pretty close to the theoretical lower limit for how much this kind of content can be compressed. That's called the Shannon limit, and data compression is all about getting as close to that as possible.

Well-Compressed Data Doesn't Compress Well

As mentioned earlier, an efficiently encoded file looks pretty much random. This means there isn't any room for further compression, assuming the initial compression was done well. All modern codecs already apply data compression as part of the overall compression process, so compressed formats like MP3, or H.264 typically won't compress further. The big exception is with authoring formats, especially uncompressed formats. You might get up to a 2:1 compression with an uncompressed video file, or a intraframe codec where several frames in a row are identical.

A time-honored Internet scam/hoax is the compression algorithm that can reduce the size of random data, and thus even its own output. Information theory demonstrates that such systems are all bunk.

General-Purpose Compression Isn't Ideal

Generating a codebook for each compressed file is time-consuming, expands the size of the file, and increases time to compress. Ideally, a compression technology will be able to be tuned to the structure of the data it gets. This is why lossless still image compression will typically make the file somewhat smaller than doing data compression on the same uncompressed source file, and will do it faster as well. We see the same thing as with the text compression example.

Small Increases in Compression Require Large Increases in Compression Time

There is a fundamental limit to how small a given file can be compressed, called the Shannon limit. For random data, the limit is the same as the size of the source file. For highly redundant data, the limit can be tiny. A file that consists of the pattern "01010101" repeated a few million times can be compressed down to a tiny percentage of the original data. However, real-world applications don't get all the way to the Shannon limit, since it requires an enormous amount of computer horsepower, especially as the files get larger. Most compression applications have a controlling tradeoff between encoding speed and compression efficiency. In essence, these controls expand the amount of the file that is being examined at any given moment, and the size of the codebook that is searched for matches. However, doubling compression time doesn't cut file size in half! Doubling compression time might only get you a few percentages closer to the Shannon limit for the file. Getting a file 10 percent smaller might take more than 10 times the processing time, or be flat-out impossible.

Lossy and Lossless Compression

Lossless compression codecs preserve all of the information contained within the original file. Lossy codecs, on the other hand, discard some data contained in the original file during compression. Some codecs, like PNG, are always lossless. Others like VC-1 are always lossy. Others still may or may not be lossy depending on how you set their quality and data rate options. Lossless algorithms, by definition, might not be able to compress the file any smaller than it started. Lossy codecs generally let you specify a target data rate, and discard enough information to hit that data rate target. This really only makes sense with media—we wouldn't want poems coming out with different words after compression!

Spatial Compression Basics

I'm going to spend a lot of time on spatial compression, because it forms the basis of all video compression.

Spatial compression starts with uncompressed source in the native color space, typically 4:2:0 at 8-bits per channel. Each channel is typically compressed by itself, so you can think of it as three independent 8-bit bitmaps being compressed in parallel. For simplicity, I'll mainly be talking about single-channel compression and hence grayscale (Y′-only) images.

Spatial Compression Methods

There are a ton of spatial compression algorithms and refinements to those algorithms. In the following, I briefly touch on the most common and best-known. Most (but not all) modern video codecs use Discrete Cosine Transformation (DCT) for images, so we'll focus on its description, and talk about wavelets some as well. First, though, a quick look at the lossless compression techniques used in codecs.

Run-Length Encoding

The simplest technique used to compress image data is Run-Length Encoding (RLE). In essence, the data is stored as a series of pairs of numbers, the first giving the color of the sequence and the second indicating how many pixels long that line should be. This works very well for content like screenshots of Windows 2000 era operating systems, with long lines of the same value (although with modern antialiasing, textures, and backgrounds). It can also work well for text, since a sharp edge isn't any harder to encode than a soft one, although a soft, antialiased edge can require several samples. However, RLE is terrible at handling natural images such as photographs (or a modern OS like Vista/Windows 7's Aero Glass or Mac OS X). Just a slight amount of random noise can cause the image to be as large as an uncompressed file, because if each pixel is even slightly different from its neighbor, there aren't any horizontal lines of the same value. With a seemingly simple "01010101" repeating pattern, RLE wouldn't be able to compress the file at all, since there is never more than one identical value in a row.

Advanced Lossless Compression with LZ77 and LZW

As mentioned earlier, the more the structure of data is predictable, the more efficient the compression can be.

The first highly compressed file formats were based on either the LZ77 (Lempel Ziv 1977) or the LZW (Lempel-Ziv-Welch) algorithms (Abraham Lempel and Jacob Ziv have had a huge impact on digital media, and Terry Welch is no slouch either). I'll spare the details, but they're both essentially combinations of advanced RLE with Huffman coding. Both build a codebook as they go, and then look for repeated strings of the same symbols and assign those codes as well.

LZW was adopted by Adobe for TIFF and CompuServe for GIF (predating JPEG). Both were pretty straightforward implementations; they just started in the upper-left corner and encoded pixel-by-pixel. But after some years Unisys, which had wound up with the LZW patents, started to demand patent licensing fees for anyone using GIF files, which was pretty much the entire web at the time.

The PNG image format was born out of this kerfuffle, as an alternate image format based on the older LZ77, for which all patents had since expired. PNG offers dramatically better compression than either LZW-based codecs, not because the fundamental compression algorithm is that much more efficient, but because the format supports lossless filtering to better arrange image data for more efficient compression. Specifically, for each line, each pixel can be described as the difference from the pixel above, the pixel to the left, or the average of those two pixels. Unlike RLE, which relies on identical pixels, PNG is then able to use a few bits to encode slight differences. And the resulting pattern itself can have a lot of redundancy (a black line under a white line will all have the same difference, and so the final lossless encode will be very efficient).

This core concept of optimizing the structure of the data to make the final lossless encode more efficient is a critical one used in every delivery codec.

Arithmetic Coding

We have one last lossless technique to introduce: arithmetic coding. With a traditional Huffman codebook, the best you can hope for is for one symbol to be encoded in one bit. But each symbol requires a whole number of bits. Arithmetic coding allows for the average number of bits per symbol to be a fraction of a bit, reflecting their statistical probability.

The math is awesome but obtuse; the core fact here is arithmetic encoding allows for more efficient lossless encoding, getting much closer to the Shannon limit.

The most common use of arithmetic coding for lossless spatial image encoding is the Lagarith codec. Compared to the older Huffyuv lossless codec, Lagarith can do 30 percent better with typical video content and better yet with less noisy content by:

- Encoding each channel independently (the Y′ values have a lot more in common with their neighbors than the Cb/Cr values of the same pixel)
- Applying arithmetic encoding to the result

With arithmetic coding of video, we can sometimes get down to 4:1 lossless compression! That's getting pretty close to the Shannon limit for typical film/video content with some noise. And that noise is the biggest part of that limit. With 8-bit video, if the last bit is essentially random, which means the best possible compression would be 12.5 percent (1 out of 8) even if the image is otherwise a completely flat color.

Still, 4:1 compression isn't much. At DVD data rates (up to 9 Mbps peak for video), that would only give us enough bits for a 416 × 240 image. But DVD regularly uses half that as an average bitrate, and with 350% as many pixels. And modern codecs can do 3–4x better than MPEG-2.

Lossless image compression doesn't get us near where we need to be for video.

Discrete Cosine Transformation (DCT)

And so we leave the world of mathematically lossless compression of pixels and enter the world of frequency transformation and lossy compression, where we get the huge compression ratios required for modern media delivery.

The most common basic technique used for lossy image compression is DCT. While there have been plenty of other techniques proposed, nothing has come close to matching DCT and its derivatives for video encoding, and it's extremely competitive for still images as well.

Before we dive into the details, the core concept behind DCT comes straight from Chapter 1—it's the edges we see, not the flat area in between. Since it's the changes between pixels that matter, DCT rearranges the image so that bits get spent on those changes instead of the pixels themselves, and so that those changes are what's best preserved as lossy compression is applied.

But how?

Let's start with a sketch of how classic JPEG compression works.

Fourier

The basic mathematical idea behind DCT was developed by Joseph Fourier a couple of centuries ago (he also discovered how greenhouse gases influence global temperature). Fourier proved that any series of numbers can be produced by a sufficiently complex equation. So for any sequence of pixel values, there's an equation that can describe that sequence. This was a revolutionary concept leading to many further discoveries, including all digital video and audio processing.

The DCT

DCT is an implementation of Fourier's discovery, and is a simple way to take a series of numbers and calculate coefficients for a series of cosines that produce the input series. A cosine itself is a simple oscillating wave: a frequency like the classic illustration of a radio wave. The coefficients of the DCT measure how tall that wave is (amplitude) and how quickly it goes from one peak to the next (frequency).

In classic JPEG and other codecs, the basic unit of compression is the 8×8 block. Figure 3.1 shows how a few different 8×8 blocks images turn into frequencies.

Mathematically, you get one coefficient on output for each sample pixel, so putting 64 pixels of source video into the DCT will output 64 coefficients. The coefficients that result won't fall into the 0–255 8-bit range the source was coded in; larger numbers can exist, requiring higher precision. So, just converting from pixel values to cosine coefficients didn't yield any compression at all; in fact, we've lost ground so far.

Name	Source Image	Q 90 image	Matrix	Total of coefficients	Comments
Left to Right gradient			−7 −73 2 −2 0 0 0 0 0 0 0 0 0 0 0 0 0 0 0 0 0 0 0 0 0 0 0 0 0 0 0 0 0 0 0 0 0 0 0 0 0 0 0 0 0 0 0 0 0 0 0 0 0 0 0 0 0 0 0 0 0 0 0 0	84	With a strong horizontal gradient, we see all coefficients on the horizontal axis. The first non-DC is a very high-73, and is responsible for most of the image.
Top to Bottom gradient			−7 0 0 0 0 0 0 0 −72 0 0 0 0 0 0 0 1 0 0 0 0 0 0 0 −3 0 0 0 0 0 0 0 0 0 0 0 0 0 0 0 0 0 0 0 0 0 0 0 0 0 0 0 0 0 0 0 0 0 0 0 0 0 0 0	83	We have essentially a 90 degree rotation in the quant matrix when the image is also rotated 90 degrees.
Diagonal Gradient			−6 −30 0 −2 0 0 0 0 −28 1 1 0 0 0 0 0 0 1 0 0 0 0 0 0 −2 0 0 0 0 0 0 0 0 0 0 0 0 0 0 0 0 0 0 0 0 0 0 0 0 0 0 0 0 0 0 0 0 0 0 0 0 0 0 0	71	With a diagonal, we actually get the high amplitude values in horizontal and vertical.
Same, with noise			−10 −54 6 −2 0 0 0 0 −53 3 2 −2 0 0 1 0 0 3 −1 −1 0 0 0 0 −3 0 −3 0 0 0 0 0 −2 2 −1 0 0 0 0 0 −2 0 0 0 0 0 0 −1 0 1 1 0 1 0 0 −1 −1 1 0 0 0 −1 0 0	159	Once we add noise, we get a whole lot more coefficients, and still lose a lot of the noise texture.
Diagonal Sharp			8 62 0 5 0 1 0 0 62 −14 −23 0 −2 0 0 0 0 −25 10 8 0 1 0 0 5 0 9 −5 −2 0 0 0 0 −3 0 −2 2 1 0 0 1 0 0 0 1 −2 0 0 0 0 0 0 0 0 1 0 0 0 0 0 0 0 0 −2	258	Far more coefficients yet, even though this is the simplest image visually. But even with the most data and low compression, that very sharp line yields a lot of ringing. Note all the coefficients along the diagonal.
Just Noise			−15 −27 18 −8 −3 0 8 −2 8 −9 7 −1 5 −1 −2 0 1 −7 −8 −4 −2 1 −1 3 −2 8 0 −2 0 0 1 5 10 2 −1 −2 −1 −2 1 −1 0 1 1 3 −1 −1 −1 0 0 1 0 −1 1 −2 2 1 −4 −2 −1 1 −2 −2 1 0	208	Random noise is actually easier to encode in visible quality loss and coefficients than a sharp edge. The matrix is less sparse, though. While pixel-by-pixel it's not incredibly accurate, when it's really noise, it's the sense of texture that matters.

Figure 3.1 A variety of source blocks and their resulting quant matricies and images using light compression.

But we have set the stage for the magic to happen.

As we look at the resulting DCT matrices, there are a few things to note:

- The coefficient in the upper-left corner square is the DC coefficient: that's the base value for the whole matrix. A flat 8 × 8 block would have the luma of the block as the coefficient there, and all other values 0. The other blocks are AC* coefficients, and record frequencies across the block.

- The further we get away from the DC* coefficient, the higher frequency is being coded. The upper-right coefficient is the highest horizontal frequency, and the lower-left coefficient is the highest vertical frequency.
- The coefficients not on the top or right are frequencies that are at a diagonal to the image.
- The coefficient value in each position is the amplitude of that frequency. A 0 means that there's none of that frequency.

Since the human eye looks for changes and notices edges, the most important parts of the image have the highest amplitude (biggest changes). But the bigger those changes, the less the precise value of the change is critical. High frequencies (how quickly the changes happen) are very visible as well, but precise sharpness of the edge isn't that big a factor; the difference between lines 1.1 pixels and 1.2 pixels wide is too subtle for us to notice in a moving image.

For a number of our sample matrices, we see bigger numbers in the upper left, going down as we head to the lower right. This reflects that most images don't have worst case detail. And since that's a predictable pattern, we're now able to apply some data compression on this structure.

First, we need to go from the 2D array to a single series of values. Starting from upper right, samples are pulled out of the matrix in a zigzag pattern (Figure 3.2).

In JPEG, simple RLE compression is then applied (not very effective in most of these cases, given there's almost always a little detail in the real-world images, but it'll become more useful in the next step). And then Huffman coding is applied. The distribution of frequency coefficients is more predictable than for the original pixels, which should yield an improvement in compression.

But we also went from our original 8-bit values to higher precision, so we might not have gained anything in net. There are two obvious ways to further improve compression in this scheme:

- Find a way to get back to 8-bit or even lower precision
- Get more of the coefficients to have the same value, in order for the RLE to be more effective

But we must also make sure that we're preserving the visually important detail as much as we can.

And this is where we move into lossy compression, and the point of this exercise. We can now start reducing the bitrate required.

---

* DC comes from "Direct Current" since the concept was originally developed in talking about electrical circuits. DC power, like from a battery, is continuous, like the flat color of the block. Similarly, AC comes from "Alternating Current." AC power alternates its voltage in a cosine pattern, like the cosines that the AC coefficients define.

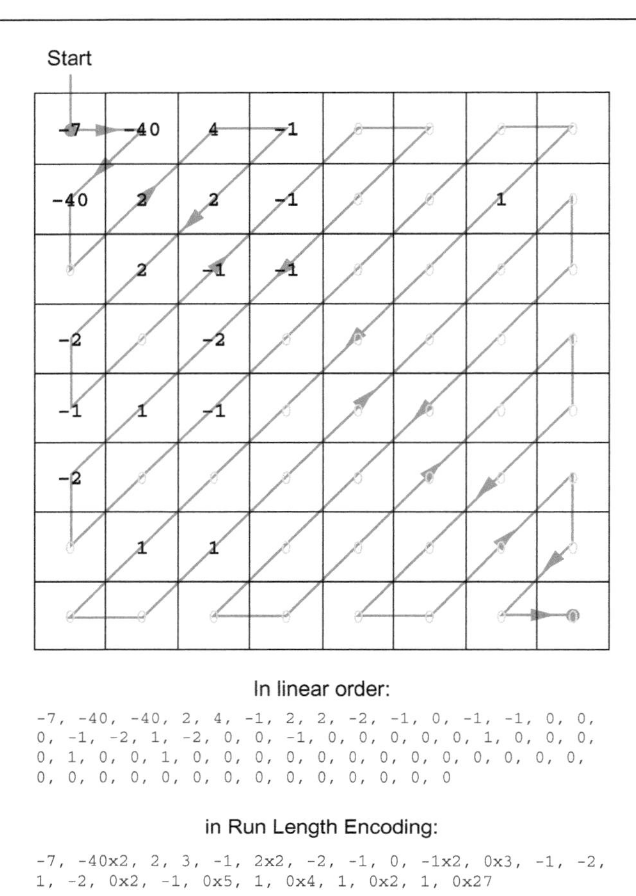

Figure 3.2 The zigzag scan pattern with a DCT on top. Note how the scan pattern results in efficient RLE, with most of the last half of samples 0.

DCT quantization

Now we want to start eliminating the data we don't need visually, so that we're only spending bits on "differences that make a difference."

The first thing we're going to do is reduce the precision of our matrix so we need fewer bits to store each value. Also, there are frequencies with such low amplitude that they aren't visible; we can eliminate them as well. But the relative value of different frequencies can vary quite a bit:

- The DC coefficient is really important.
- The higher the frequency, the less important small differences in amplitude are.
- We're visually drawn to right angles, so frequencies that are fully horizontal or vertical matter more than diagonals.

$$\begin{bmatrix} 16 & 11 & 10 & 16 & 24 & 40 & 51 & 61 \\ 12 & 12 & 14 & 19 & 26 & 58 & 60 & 55 \\ 14 & 13 & 16 & 24 & 40 & 57 & 69 & 56 \\ 14 & 17 & 22 & 29 & 51 & 87 & 80 & 62 \\ 18 & 22 & 37 & 56 & 68 & 109 & 103 & 77 \\ 24 & 35 & 55 & 64 & 81 & 104 & 113 & 92 \\ 49 & 64 & 78 & 87 & 103 & 121 & 120 & 101 \\ 72 & 92 & 95 & 98 & 112 & 100 & 103 & 99 \end{bmatrix}$$

Figure 3.3 The standard JPEG quant matrix.

Our tool here is the quantization matrix.

The classic JPEG matrix is shown in Figure 3.3.

The least-lossy compression in JPEG is simply to divide the coefficients in the source matrix by the value in the same position in the quant matrix, rounding to the nearest integer. A few things to note here:

- The DC coefficient is 16. This compensates for the increase in precision past 8-bit yielded by the DCT. Using a coefficient of 16 will take 12-bit precision back down to 8-bit, as $2^4 = 16$, and there are 4 fewer bits in 8 than 12. So this wouldn't have a net impact on image fidelity.
- The other quant matrix values increase with increasing frequency, and on the diagonal.
- The 99 in the bottom right is lower than the surrounding quant values. This allows a sharp diagonal line to retain a little more accuracy.
- Within those general rules, the numbers can seem pretty random. These tables are the result of a whole lot of simulation, and wind up being what works best on the content they were tested with more than any abstract theory. Don't sweat the why of any particular value too much.

Figure 3.1 shows what happens to the image with lightweight JPEG compression, and what the resulting coefficients turn out as.

Note how many of the low amplitude samples wind up as 0, which are obviously very easy to compress. The range of values that turn into a 0 in the output is called the "dead zone," which is half the quant matrix coefficient for that point.

And we get more samples in a row, so RLE encoding is more efficient. And with the smaller range of more consistent numbers, the Huffman coding is more efficient as well.

And the images still look pretty much the same. Even though we've removed a whole lot of the mathematical detail, most of that wasn't visual detail.

Quality	Clean Total	Clean Matrix	Clean Gradient	Noise Gradient	Noise Matrix	Noise total	Comments
Source							
100	128	–11 –51 1 –3 0 0 0 0 –51 1 2 0 0 0 0 0 1 2 0 0 0 0 0 0 –4 0 0 0 0 0 0 0 0 0 0 0 0 0 0 0 –1 0 0 0 0 0 0 0 0 0 0 0 0 0 0 0 0 0 0 0 0 0 0 0			–10 –54 6 –2 0 0 0 0 –53 3 2 –2 0 0 1 0 0 3 –1 –1 0 0 0 0 –3 0 –3 0 0 0 0 0 –2 2 –1 0 0 0 0 0 –2 0 0 0 0 0 0 –1 0 1 1 0 1 0 0 –1 –1 1 0 0 0 –1 0 0	159	Quality with minimum quantization is quite good. Note how many more coefficients there are in the noisy image.
90	95	–8 –38 1 –2 0 0 0 0 –38 1 2 0 0 0 0 0 0 2 0 0 0 0 0 0 –3 0 0 0 0 0 0 0 0 0 0 0 0 0 0 0 0 0 0 0 0 0 0 0 0 0 0 0 0 0 0 0 0 0 0 0 0 0 0 0			–7 –40 4 –1 0 0 0 0 –40 2 2 –1 0 0 1 0 0 2 –1 –1 0 0 0 0 –2 0 –2 0 0 0 0 0 –1 1 –1 0 0 0 0 0 –2 0 0 0 0 0 0 0 0 1 1 0 0 0 0 0 0 0 0 0 0 0 0 0	113	Quite a visible change with our first step, with much of the noise detail gone, although there is certainly a sense of texture remaining.
80	71	–6 –30 0 –2 0 0 0 0 –28 1 1 0 0 0 0 0 0 1 0 0 0 0 0 0 –2 0 0 0 0 0 0 0 0 0 0 0 0 0 0 0 0 0 0 0 0 0 0 0 0 0 0 0 0 0 0 0 0 0 0 0 0 0 0 0			–6 –32 3 –1 0 0 0 0 –29 2 1 –1 0 0 1 0 0 2 –1 –1 0 0 0 0 –1 0 –1 0 0 0 0 0 –1 1 0 0 0 0 0 0 –1 0 0 0 0 0 0 0 0 1 1 0 0 0 0 0 0 0 0 0 0 0 0 0	87	Not much change here; the matrix isn't more sparse, although the values are more quantized.
70	60	–5 –25 0 –2 0 0 0 0 –23 1 1 0 0 0 0 0 0 1 0 0 0 0 0 0 –2 0 0 0 0 0 0 0 0 0 0 0 0 0 0 0 0 0 0 0 0 0 0 0 0 0 0 0 0 0 0 0 0 0 0 0 0 0 0 0			–4 –27 3 –1 0 0 0 0 –24 1 1 –1 0 0 0 0 0 2 –1 –1 0 0 0 0 –1 0 –1 0 0 0 0 0 –1 1 0 0 0 0 0 0 –1 0 0 0 0 0 0 0 0 1 0 0 0 0 0 0 0 0 0 0 0 0 0 0	72	The most compression that still includes noise texture.
60	51	–4 –22 0 –1 0 0 0 0 –19 1 1 0 0 0 0 0 0 1 0 0 0 0 0 0 –2 0 0 0 0 0 0 0 0 0 0 0 0 0 0 0 0 0 0 0 0 0 0 0 0 0 0 0 0 0 0 0 0 0 0 0 0 0 0 0			–4 –23 2 –1 0 0 0 0 –20 1 1 –1 0 0 0 0 0 1 –1 0 0 0 0 0 –1 0 –1 0 0 0 0 0 –1 1 0 0 0 0 0 0 –1 0 0 0 0 0 0 0 0 0 0 0 0 0 0 0 0 0 0 0 0 0 0 0	60	The texture is gone, even though the matrix is still less sparse.
50	43	–4 –18 0 –1 0 0 0 0 –17 0 1 0 0 0 0 0 0 1 0 0 0 0 0 0 –1 0 0 0 0 0 0 0 0 0 0 0 0 0 0 0 0 0 0 0 0 0 0 0 0 0 0 0 0 0 0 0 0 0 0 0 0 0 0 0			–3 –19 2 –1 0 0 0 0 –18 1 1 –1 0 0 0 0 0 1 0 0 0 0 0 0 –1 0 –1 0 0 0 0 0 –1 1 0 0 0 0 0 0 –1 0 0 0 0 0 0 0 0 0 0 0 0 0 0 0 0 0 0 0 0 0 0 0	52	Other than that "dip" in the middle, looking increasingly like a gradient, although matrix itself continues to be less sparse.
40	37	–3 –16 0 –1 0 0 0 0 –14 0 1 0 0 0 0 0 0 1 0 0 0 0 0 0 –1 0 0 0 0 0 0 0 0 0 0 0 0 0 0 0 0 0 0 0 0 0 0 0 0 0 0 0 0 0 0 0 0 0 0 0 0 0 0 0			–3 –17 2 –1 0 0 0 0 –15 1 1 –1 0 0 0 0 0 1 0 0 0 0 0 0 –1 0 –1 0 0 0 0 0 0 0 0 0 0 0 0 0 –1 0 0 0 0 0 0 0 0 0 0 0 0 0 0 0 0 0 0 0 0 0 0 0	45	
30	31	–2 –13 0 –1 0 0 0 0 –12 0 1 0 0 0 0 0 0 1 0 0 0 0 0 0 –1 0 0 0 0 0 0 0 0 0 0 0 0 0 0 0 0 0 0 0 0 0 0 0 0 0 0 0 0 0 0 0 0 0 0 0 0 0 0 0			–2 –13 1 0 0 0 0 0 –12 1 1 0 0 0 0 0 0 1 0 0 0 0 0 0 –1 0 –1 0 0 0 0 0 0 0 0 0 0 0 0 0 –1 0 0 0 0 0 0 0 0 0 0 0 0 0 0 0 0 0 0 0 0 0 0 0	34	

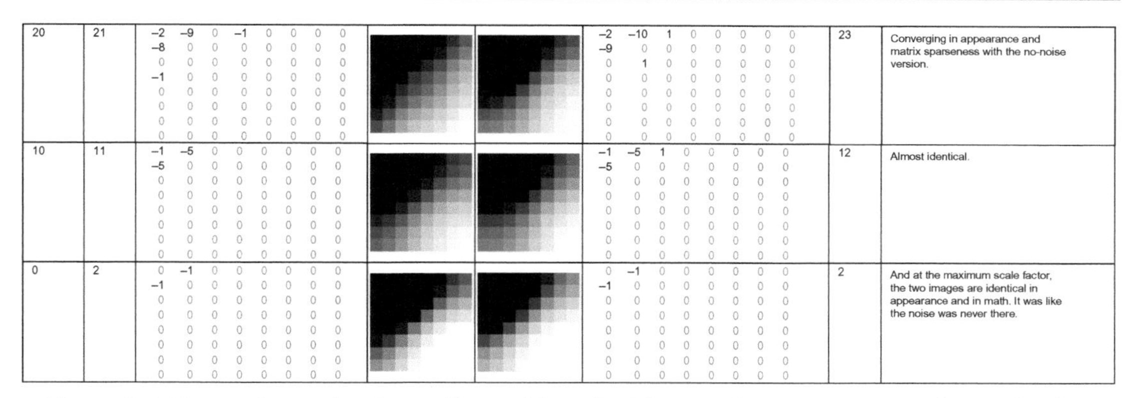

20	21	−2 −9 0 −1 0 0 0 0 −8 0 0 0 0 0 0 0 0 0 0 0 0 0 0 0 −1 0 0 0 0 0 0 0 0 0 0 0 0 0 0 0 0 0 0 0 0 0 0 0 0 0 0 0 0 0 0 0 0 0 0 0 0 0 0 0			−2 −10 1 0 0 0 0 0 −9 0 0 0 0 0 0 0 0 1 0 0 0 0 0 0 0 0 0 0 0 0 0 0 0 0 0 0 0 0 0 0 0 0 0 0 0 0 0 0 0 0 0 0 0 0 0 0 0 0 0 0 0 0 0 0	23	Converging in appearance and matrix sparseness with the no-noise version.
10	11	−1 −5 0 0 0 0 0 0 −5 0 0 0 0 0 0 0 0 0 0 0 0 0 0 0 0 0 0 0 0 0 0 0 0 0 0 0 0 0 0 0 0 0 0 0 0 0 0 0 0 0 0 0 0 0 0 0 0 0 0 0 0 0 0 0			−1 −5 1 0 0 0 0 0 −5 0 0 0 0 0 0 0 0 0 0 0 0 0 0 0 0 0 0 0 0 0 0 0 0 0 0 0 0 0 0 0 0 0 0 0 0 0 0 0 0 0 0 0 0 0 0 0 0 0 0 0 0 0 0 0	12	Almost identical.
0	2	0 −1 0 0 0 0 0 0 −1 0 0 0 0 0 0 0 0 0 0 0 0 0 0 0 0 0 0 0 0 0 0 0 0 0 0 0 0 0 0 0 0 0 0 0 0 0 0 0 0 0 0 0 0 0 0 0 0 0 0 0 0 0 0 0			0 −1 0 0 0 0 0 0 −1 0 0 0 0 0 0 0 0 0 0 0 0 0 0 0 0 0 0 0 0 0 0 0 0 0 0 0 0 0 0 0 0 0 0 0 0 0 0 0 0 0 0 0 0 0 0 0 0 0 0 0 0 0 0 0	2	And at the maximum scale factor, the two images are identical in appearance and in math. It was like the noise was never there.

Figure 3.4 Comparing a simple gradient with and without noise over a range of quantization levels.

Scale factor and the quantization parameter

So we had some real gains with our initial quantization step. But what if the file still wasn't small enough (and it probably isn't)? We can simply scale up the values in the quant matrix. This is controlled by a "scale factor" which multiplies all the values in the quant matrix by a particular value.

The control for the scale factor is called the Quantization Parameter, or QP, which I'll be talking about a few more hundred times in this book. Any time you've got a "quality" control in a codec, be it JPEG or VC-1, what it really controls is QP, and hence the scale factor used in the quant matrix. The classic "0–100" quality slider used in JPEG encoders the world over is setting the quantization parameter, with "100" the least quantized and "0" the most.

As the scale factor matrix get bigger, several things happen:

- The dead zone gets bigger, so there are more zeros.
- The remaining values wind up more "coarsely" quantized, with a wider range of different input amplitudes winding up with the same value.
- Smaller numbers mean more will be identical, for better RLE, and the number of different possible values goes down, making for more efficient Huffman encoding.
- Quality will start visibly degrading past a certain point, as the differences between the input and output frequencies, and the pixels that come out, grow bigger.

So, how does different content respond to different scale factors in compression efficiency and output quality?

Quality	Black and white diagonal	Matrix	Total of coefficients	Comments
100		11 83 0 7 0 1 0 0 83 –18 –33 0 –3 0 0 0 0 –33 14 11 0 1 0 0 7 0 12 –7 –3 0 0 0 0 –4 0 –3 3 1 0 0 1 0 1 0 1 –2 –1 0 0 0 0 0 0 –1 2 0 0 0 0 0 0 0 0 –2	349	Even at the minimum scale factor, the default quantization matrix still yields some ringing. Coefficients are mainly on the diagonal, as expected.
90		8 62 0 5 0 1 0 0 62 –14 –23 0 –2 0 0 0 0 –25 10 8 0 1 0 0 5 0 9 –5 –2 0 0 0 0 –3 0 –2 2 1 0 0 1 0 1 0 1 –2 0 0 0 0 0 0 0 0 1 0 0 0 0 0 0 0 0 –2	258	Increasing ringing.
80		6 50 0 4 0 0 0 0 45 –10 –18 0 –2 0 0 0 0 –19 8 6 0 1 0 0 4 0 7 –4 –2 0 0 0 0 –2 0 –2 2 1 0 0 1 0 1 0 1 –1 0 0 0 0 0 0 0 0 1 0 0 0 0 0 0 0 0 –1	199	Now we start to see a pattern to the ringing, as the coefficients masking the strong diagonal cosine become coarser or even drop to 0.
70		5 42 0 3 0 0 0 0 38 –8 –15 0 –1 0 0 0 0 –16 6 5 0 0 0 0 3 0 5 –3 –2 0 0 0 0 –2 0 –1 1 0 0 0 1 0 0 0 1 –1 0 0 0 0 0 0 0 0 1 0 0 0 0 0 0 0 0 –1	161	Pattern is actually a bit weaker at this level. Strange things can happen with the default quant matrix.
60		4 36 0 2 0 0 0 0 31 –7 –13 0 –1 0 0 0 0 –13 5 4 0 0 0 0 3 0 4 –3 –1 0 0 0 0 –1 0 –1 1 0 0 0 1 0 0 0 1 –1 0 0 0 0 0 0 0 0 1 0 0 0 0 0 0 0 0 –1	135	
50	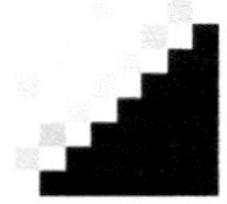	4 29 0 2 0 0 0 0 28 –6 –11 0 –1 0 0 0 0 –11 5 4 0 0 0 0 2 0 4 –2 –1 0 0 0 0 –1 0 –1 1 0 0 0 0 0 0 0 0 –1 0 0 0 0 0 0 0 0 1 0 0 0 0 0 0 0 0 –1	116	Pattern is back and stronger.
40	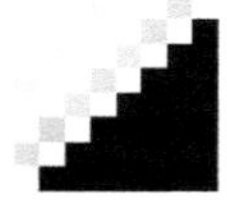	3 26 0 2 0 0 0 0 24 –5 –9 0 –1 0 0 0 0 –10 4 3 0 0 0 0 2 0 3 –2 –1 0 0 0 0 –1 0 –1 1 0 0 0 0 0 0 0 0 –1 0 0 0 0 0 0 0 0 1 0 0 0 0 0 0 0 0 –1	101	Stronger pattern yet.
30	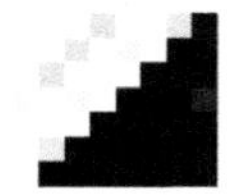	3 21 0 1 0 0 0 0 19 –4 –8 0 –1 0 0 0 0 –8 3 3 0 0 0 0 2 0 3 –2 –1 0 0 0 0 –1 0 –1 1 0 0 0 0 0 0 0 0 0 0 0 0 0 0 0 0 0 0 0 0 0 0 0 0 0 0 –1	83	Cosine pattern becomes really obvious in black as well.
20	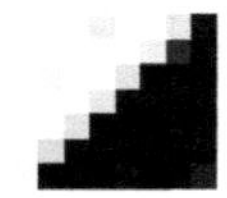	2 15 0 1 0 0 0 0 14 –3 –5 0 –1 0 0 0 0 –6 2 2 0 0 0 0 1 0 2 –1 –1 0 0 0 0 –1 0 –1 1 0 0 0 0 0 0 0 0 0 0 0 0 0 0 0 0 0 0 0 0 0 0 0 0 0 0 0	59	Getting very odd looking.

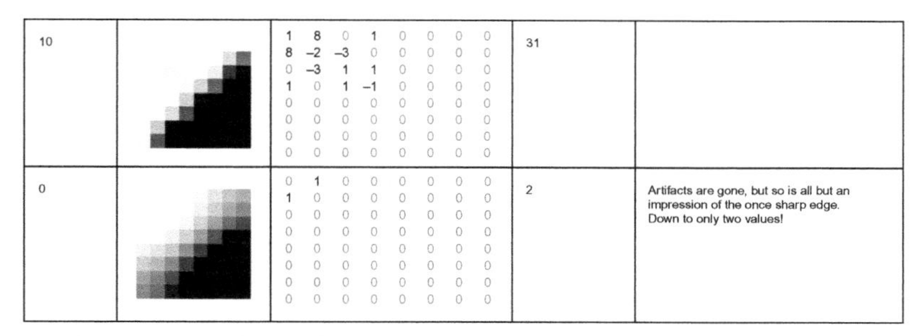

Figure 3.5 A sharp diagonal line is extremely hard to compress.

Ringing and blocking

A gradient that varied in luma exactly like a cosine would be very easy to encode, since a single coefficient would do the job. But real-world complex images are more difficult to compress, because they require more coefficients and have a broader distribution of detail.

When quantization is pushed too high, there are two main types of artifacts—visible distortions—that can affect the image.

Blocking is just that–the 8 × 8 block structure (sometimes called a "basis pattern") becomes visible, because the decoded values on the edges of adjoining blocks don't match. This is because the quantization in the blocks changed or removed the frequencies that would have left the pixels at the edge of the block the same as the adjacent block. At high quantizations, you might wind up with adjoining blocks just the shade of the DC coefficient. Unfortunately, that kind of sharp straight edge is exactly the kind of high-frequency, high-amplitude image that we'd expect to be very visible with human vision. And it is. Minimizing blocking is one of the core goals of the compessionist.

Ringing (Figure 3.5) is an artifact at edges. A sharp edge is actually the hardest thing to encode with a DCT, since it's a high-frequency, high-amplitude change. And a high-amplitude coefficient at a high frequency to capture that edge will oscillate over the whole block—remember, these are cosines. At low quantization, the codec uses coefficients at other frequencies to balance out those high frequencies. But as quantization goes up, the balancing frequencies become less accurate, and can even fall into the dead zone, Thus you'll see weird distortions around sharp edges, like text. This is also called "mosquito noise."

Chroma Coding and Macroblocks

So far we've really just been talking about the luma channel, but most content has color as well. Our visual acuity for chroma is less than for luma, and even subsampled chroma normally has much lower frequencies in general. So a different, flatter quant matrix is normally used for chroma; the most widely used is shown in Figure 3.6.

$$\begin{bmatrix} 17 & 18 & 24 & 47 & 99 & 99 & 99 & 99 \\ 18 & 21 & 26 & 66 & 99 & 99 & 99 & 99 \\ 24 & 26 & 56 & 99 & 99 & 99 & 99 & 99 \\ 47 & 66 & 99 & 99 & 99 & 99 & 99 & 99 \\ 99 & 99 & 99 & 99 & 99 & 99 & 99 & 99 \\ 99 & 99 & 99 & 99 & 99 & 99 & 99 & 99 \\ 99 & 99 & 99 & 99 & 99 & 99 & 99 & 99 \\ 99 & 99 & 99 & 99 & 99 & 99 & 99 & 99 \end{bmatrix}$$

Figure 3.6 The standard JPEG chroma quant matrix.

Note that the chroma matrix is heavily weighted towards lower frequencies. Chroma and luma will share the same scale factor, and with this matrix, relatively more of chroma will fall into the dead zone than luma. Both chroma channels together add up to only half as many samples as luma with 4:2:0 subsampling, but chroma will wind up with an even smaller portion of bits in the encoded file due to more aggressive quantization and lower complexity.

Because color channels are subsampled, in 4:2:0 their 8 × 8 blocks actually cover a 16 × 16 area of the original image. This is called a "macroblock" (Figure 3.7), which is a combination of four 8 × 8 luma blocks and the corresponding 8 × 8 chroma blocks (one each for Cb and Cr).

This macroblock structure is why codecs compress in blocks of 16 × 16 pixels. Formats that allow arbitrary resolutions encode full macroblocks, and just pad the pixels at the edge with blank data. So, a 240 × 180 file is actually coded as 240 × 192, with the bottom 12 pixels filled with bogus data and not drawn on playback. Ideally, keep resolution divisible by 16 to squeeze out the maximum bang for the bit.

Finishing the Frame

And that's how spatial compression is applied to a single block and then assembled into a macroblock. To make a full image, we just make an array of macroblocks.

And that, dear reader, is your introduction to JPEG and still image coding.

There are a few takeaways you might mull over:

- DCT is all about putting the bits where they make a difference: frequencies.
- Different frequencies matter more than others, hence the quant matrix.
- More compression yields fewer and less accurate frequencies.
- The classic visual distortions of DCT compression are blocking and ringing.
- Easier-to-encode content can use a lower QP to hit the same file size compared to more difficult content, and hence will have higher quality.

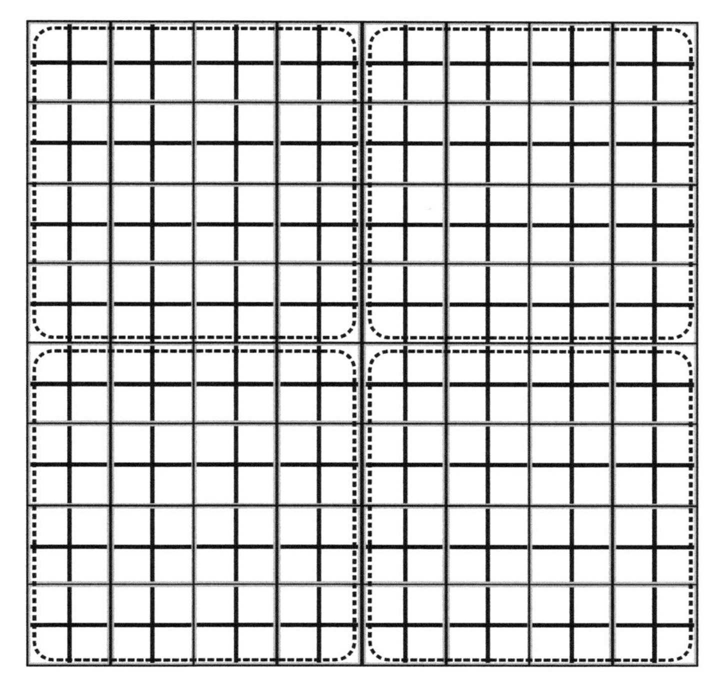

Figure 3.7 The layout of blocks and samples within a macroblock. The dotted lines indicate the position of the four luma blocks in the macroblock. The subsampled chroma blocks cover the entire 16 × 16 area.

For another taste of this in action, see Color Figure C.12, which compares the two images of different encoded to with a low QP (high quality), a high QP (low quality), and QPs that yields the same file size.

Temporal Compression

So, we've got a frame. But we want a lot of frames. How?

The simplest approach is just to make a file where each frame is an independently coded still image. And that's pretty much Motion JPEG right there.

But we still need more compression, and there's an obvious next area of redundancy to target: most of the time, most of a frame is like the frame before it.

Prediction

Imagine a little two-frame video. The first frame used classic JPEG-style compression. We call that an intra-coded frame, or I-frame. An I-frame is self-contained and isn't based on any other frames.

We need to start with an I-frame, but they're not very efficient, so next we want to make a new frame based on (predicted from) the I-frame. And that predicted frame is called a P-frame.

To take the absurdly simplest case of prediction, a macroblock in the P-frame that's exactly the same as the same macroblock on the I-frame can simply be coded as a "skip"—just copy the same pixels over.

In a more common case, the previous macroblock can be very similar, but have a slightly different noise pattern or something (Figure 3.8). In that case, the P-frame can start with the difference between the I-frame's macroblock and its version. If they're close, it'll be a pretty good match, and the difference would have lower frequencies and hence will encode better.

It's also possible that there's just no good match at all. In that case, the P-frame can use an intra-coded macroblock, just like an I-frame.

Motion Estimation

Simple prediction is a start, but it's too basic to be used in mainstream codecs. In real-world video, there's often a ton of motion. If the camera panned left 10 pixels between the I- and P-frame, none of the macroblocks would be good matches as is, even though most of the P-frame's image is seen in the I-frame.

Enter the motion vector. A motion vector specifies a block of pixels in the reference frame that's copied into the P-frame; see Color Figure C.13. So in the case where the whole frame moved 10 pixels, each macroblock can grab the pixels 10 to the left in the reference frame, and have a very good match.

And since each P-frame macroblock can have its own motion vector, there's no need for the whole frame to move at once. Multiple objects can be moving in multiple directions, with each part of the frame having an appropriate motion vector.

Of course, this all presumes that the codec knows where the matches are. And that's harder than it sounds; human vision is really adept at picking out motion, but a computer has to do it one little bit of math at a time. Trying to find good motion vectors is where most of the CPU cycles of compression go, and the accuracy of that effort is the big difference between fast and slow-but-good encoding.

There's one big downside of P-frames: to jump to a particular frame you'll need to decode all the P-frames between it and the previous I-frame. Imagine an encode with an I-frame every 100 frames; Frame 1 and 101 are I-frames, while frames 2-99 are P-frames. If you want to play just frame 78, it's predicted from frame 77, so you'll need to decode that. And frame 77 is predicted from 76, and 76 is predicted from 75, all the way back to 1. So doing random access to a particular frame can be a lot slower, particularly when the I-frames are far apart.

Figure 3.8 In low–motion sequences, most parts of the image are pretty similar. Even with high motion, much of the frame matches the previous one.

Bidirectional Prediction

So, I- and P-frames are all that are needed for interframe compression, but it was discovered in the early days that having a frame based on the previous and next frame added further efficiency.

And thus the B-frame. A B-frame can be based on the previous *and* the next frame. And it can use motion vectors from either. That yields better efficiency, which is a big reason B-frames are used.

Note that since a B-frame is based on the next reference frame, the display order is different than the encoded order. So if the video is displayed as:

1:I 2:B 3:P 4:B 5:P 6:B 7:P

it would be actually encoded, transmitted, and decoded as:

1:I 3:P 2:B 5:P 4:B 7:P 6:B

So the P-frame following the B-frame is encoded and then decoded before the B-frame, even though that B-frame is temporally first in order.

Another property of B-frames is that they're not actually needed for playback. That means they can be dropped from transmission under bandwidth stress, dropped from decode under CPU stress, or skipped when doing random access. Also, since B-frames aren't reference frames, they can be given just enough bits to not have distracting artifacts without worrying about being a good basis for future frames. This saves bits that can be used to improve the reference frames, thus providing B-frames more accurate references and thus higher quality (Figure 3.9).

And that's it for our high-level overview of interframe compression. Again, a few take-homes:

- The more motion in the video is on a 2D plane, the more efficient motion vectors are.
- Noise will make motion estimation less efficient, since the matches will be less accurate.
- Random access is an interplay between frequency of I-frames and B-frames; the more B-frames you have, the longer you can go without I-frames.
- B-frames are also very helpful for scalability.

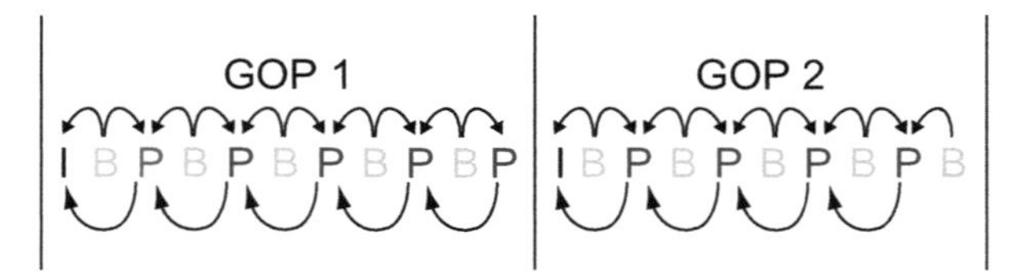

Figure 3.9 An example of GOP structure. I–frames are self contained, P–frames reference the previous P– or I–frame, and a B–frame can reference the previous and next I– or P–frame, but nothing references them.

Rate Control

There's one last part to our fundamental video coding exercise: rate control. So far, we've talked about how quantization controls frame size, but only indirectly. Very complex frames will use a lot more bits than very simple frames, even with the same QP.

The simplest style of rate control would be to have every frame be the same size, varying the QP. But this could yield really horrible quality: the most interesting frames would look the worst, while we might use more bits than needed on the easy ones. What we want is to distribute the bits so that quality is consistent, and as consistently high as possible.

So, with rate control, you'll see the I-frames being biggest, P-frames smaller, and B-frames smaller yet. And within the allowed buffer of the encoding mode (more about those in Chapter 7), harder frames will get more bits than easier frames.

There's a whole academic discipline around how to do this, called Rate Distortion Theory. Yeah, Claude Shannon thought of this too. Fortunately, you don't really need to tune this yourself; PhDs bake it into the encoders. But good rate control is one of the things that makes a good codec; two implementations can seem generally the same, but the one that spends the bits where they can best be spent will make much better-looking video.

An example of more advanced rate control is when a codec discovers that spending more bits on a particular I-frame lets it be a better match for following frames, reducing the bitrate required for those. Simple constant quality would have wound up being less efficient, with the I-frame lower quality and later frames spending more bits to get more accurate matches.

Beyond MPEG-1

The previous discussion was a very high-level description of a very basic encoder like that used in MPEG-1. The ones you'll use in the real world will have many more features and unique twists. Each description of the major codecs will include an overview of its critical differences and enhancements from the description in this chapter.

Perceptual Optimizations

But there's one last critical concept to introduce, even though it's not really applicable to the simple model we've described so far: perceptual optimizations.

A perceptual optimization is any adjustment to improve the perceived quality of the image instead of its mathematical accuracy. A simple example used in some advanced codecs is Differential Quantization (DQuant). Some codecs allow different blocks to use different quantization parameters. From a pure math perspective, this is the most efficient, as it minimizes the difference between source and output. But visually, some differences matter a lot more than others! So, as shown in Color Figure C.14, DQuant would pick out blocks that include flatter gradients or sharper edges, and give them more bits and less quantization. The flip side of that is more textured blocks will get fewer bits and more quantization. But as we saw in our DCT examples, quantization impacts the appearance of those sharp edges a lot more than it does noisy texture detail.

Alternate Transforms

I'm pretty well sold on DCT-like transforms as fundamentally a lot better than any of the alternative technologies once interframe coding comes into play.

The other transforms, particular wavelets, are being put to good use in still image formats.

Wavelet Compression

Wavelet compression uses a very different transform from DCT that's popular for still image compression (notably JPEG2000), and is used in some experimental video codecs.

The fundamental idea behind wavelets is similar to DCT—turn a sequence of samples into an equation—but that's where the similarities stop. Wavelet images are built out of a series of images, starting small and doubling in height and width until the final image is built at the file's resolution. Each successively higher resolution is predicted from the previous, storing the additional detail of that band. The space savings from this technique can be tremendous—each band is predicted from the previous sub-band, so bits are only spent where there's detail not in the sub-band. Thus for flat areas or gradients, like clouds or ripples in water, the later bands need to add very little data. Conversely, more detailed images, such as faces in a crowd or film grain, might require a lot of detail in each new band.

One useful aspect of these bands is only as many bands as desired need to be decoded; the process can stop at any band. So a low-resolution proxy view is pretty much a free feature.

Good wavelet implementations don't have nearly the problems encoding text as do classic JPEG-like DCT codecs (although more modern DCT-derived codecs like H.264 do much better in that regard). They also don't block in the same way, as there's no 8 × 8 basis pattern. Thus wavelets tend to have less objectionable artifacts at lower rates, though the images can still get quite soft.

Alas, wavelets aren't nearly so promising for use in interframe codecs, because there's not really an efficient way to use motion vectors to predict a future wavelet. There have been many attempts at making wavelet interframe video codecs over the years, and several are ongoing like the BBC's Dirac, but none have ever been competitive.

So far, the primary uses of wavelets have been in JPEG2000 and Cineform, with interest in the "VC-2" I-frame-only subset of Dirac. None use interframe encoding.

Fractal Compression

Fractals were the rotary engine of the 1990s—the technology that promised to make everything wonderful, but never actually turned into mainstream delivery technologies. The idea behind fractals is well-known to anyone who breathlessly read James Gleick's *Chaos: Making a New Science* (Penguin, 1988). In its classic definition, a fractal image is able to reuse parts of itself, transformed in many ways, in other parts of the image. This concept is based on self-similar objects, so in fractal compression, things like ferns or trees can compress very well. Fractal compression isn't so useful with people—our fingers don't have smaller fingers hanging off their ends. Fractal compression can be thought of as extending wavelet style sub-band prediction to rotation, scaling, and other kinds of transformations. This mathematical approach is also called an Iterated Function System, after which Iterated Systems, the company that tried to bring the technology to market, is named.

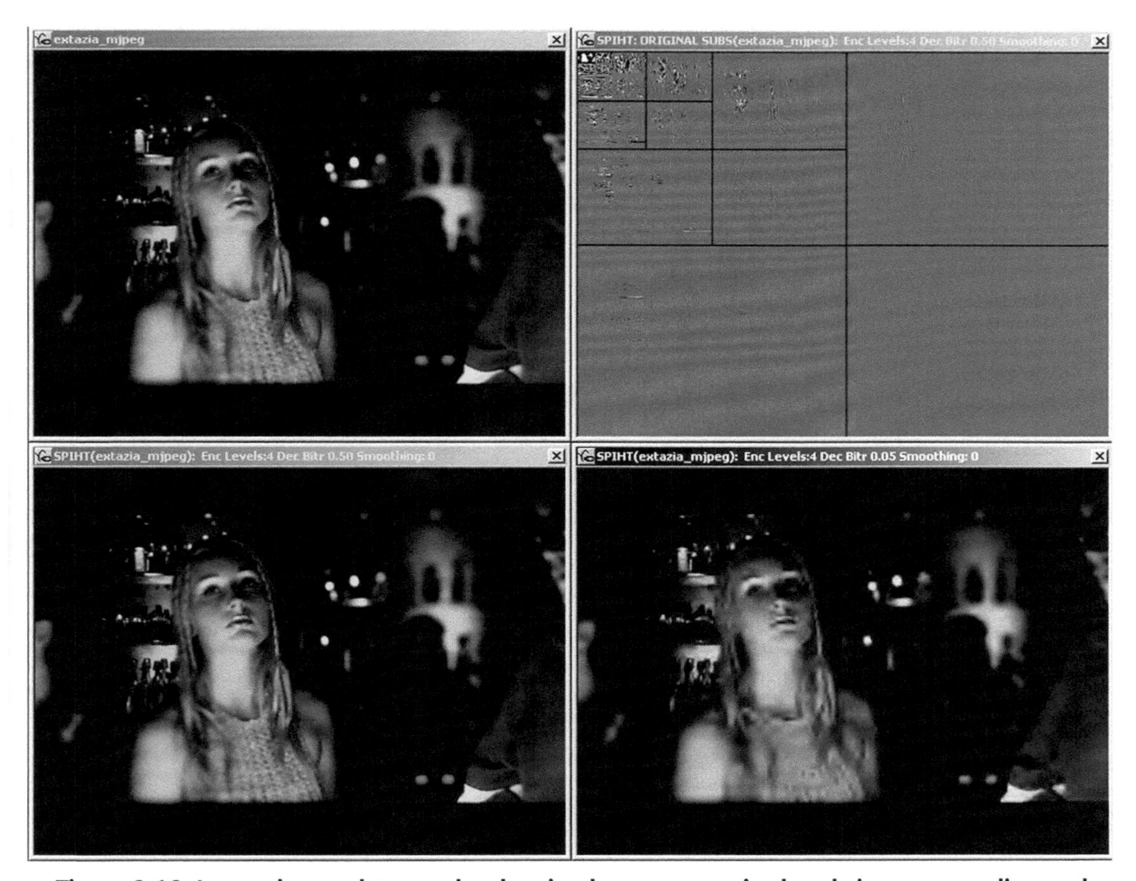

Figure 3.10 A sample wavelet encode, showing how progressive bands improve quality, and (in the upper–right corner) what detail gets added in each band.

The problem is that actually compressing a fractal image is hugely difficult. A fractal codec needs to see whether one part of an image matches another after being rotated, scaled, flipped, and otherwise looked at from every which way. All of which requires an insane amount of CPU power, and led to the concept of the Graduate Student Algorithm, wherein a grad student is locked in a room with a workstation for a few days to find the best compression for the image. Needless to say, the Graduate Student Algorithm wasn't very practical in the real world.

Iterated Systems did ship a fractal video codec called ClearVideo; it was never particularly successful, and was eclipsed by Sorenson Video. Iterated's still image fractal technology, FIF, has gone through a couple of corporate owners, and is now mainly pitched as an algorithm that offers better scaling.

The real advantage of these fractal compressors was that they offered resolution independence (the image could be decompressed at a higher resolution with fewer artifacts than other algorithms), and that a lower-quality image could be generated from the beginning of the file, *à la* wavelets.

Is a Picture Worth a Thousand Words?

We've all heard the old saying that it is. And now, with the power of compression, we check to see how good a picture we can make with the information in a thousand words. First, I'm going to take the first thousand words of this chapter, and save it as a text file. To make it a fair comparison, I'm going to take the redundancy out of the file via Deflate at default compression. This gives a file size of 2793 bytes. Not a lot to play with, but we can definitely make some kind of image with it.

Then I'll use the same number of bytes to create a grayscale JPEG instead. (I'm using grayscale because "a picture is worth a thousand words" was generally associated with newspapers back in the day, when newspapers were black-and-white.)

So, is Figure 3.11 worth a thousand words?

Audio Compression

Audio compression uses a lot of the same principles as video compression, but with twists, mainly due to the differences in how we hear instead of see.

Sub-Band Compression

In the same way that DCT video compression turns samples into frequencies and removes frequencies that are redundant or hard to see, sub-band codecs convert samples to frequencies and remove frequencies that are redundant or hard to hear.

Figure 3.11 The picture worth a thousand words...worth of bytes.

First the audio is broken up into discrete "frames" between 20 to 500 fps. (These don't have any relationship to video frames the audio might be attached to.) Then, each frame is turned into a series of from anywhere between two to more than 500 frequency bands. Different bands have varying complexity, and lower complexity bands take up less space; thus, a lossless compression stage is able to take good advantage of redundancy and predictability.

Psychoacoustic models can then remove any masked sounds. As discussed in the previous chapter, you can't hear a quiet sound next to a loud sound of the same frequency. Nor will you hear a quiet sound of the same frequency as a loud sound that precedes it closely in time. Using these psychoacoustic techniques, the codec is able to remove further unnecessary information.

Another important technique is used with stereo audio. Left and right channels almost always contain redundant information, so it would be inefficient to encode them separately. For example, the left channel and right channels are mixed together and encoded, then the differences between left and right channels are encoded separately. This substantially reduces the data rate requirements for stereo audio.

There are a lot of variations, tweaks, and flavors of these techniques. Most general-purpose audio codecs are based on them.

Of course, beyond a certain point, audio compression degradation will become audible. At that point, the codec tries to provide the best possible quality given the available bits.

Percussive sounds, which have relatively random waveforms, and hence sub-bands with rapidly changing values, are the hardest for most codecs to compress. Cymbals and hi-hats are typically the most difficult to encode, being random and high-pitched. Sometimes very clean, well-produced audio is difficult to compress as well, since even slight distortions are audible.

Audio Rate Control

Unlike video codecs, most audio codecs are used as constant bitrate (CBR)—each frame takes up the same amount of bandwidth. Some codecs like MP3 allow a bit-buffer, where extra bits from previous frames can be used in later ones. There are also many codecs with quality-based variable bitrate (VBR) modes, where a fixed quantization is used, and bitrate will vary as needed.

There are plenty of VBR 1-pass audio codecs. LAME for MP3 and Ogg Vorbis are probably the best-known of these. The VBR of MP3 audio means the data rate of each frame can vary within the allowed range of 8 Kbps and 320 Kbps. So, when the audio is silent or extremely simple in nature, the frame will be 8 Kbps, and with the data rate rising as the content becomes more complex. Apple's AAC-LC includes a pretty good 1-pass VBR mode as well. So far, Microsoft's WMA codecs have the only broadly used full rate-controlled 2-pass VBR modes that deliver predictable file sizes.

CHAPTER 4

The Digital Video Workflow

We've talked about the fundamentals of compression. Now we're going to dive into how to use this stuff in the real world for real projects. While all the issues covered in this chapter will be gone over in loving detail over the course of the rest of this book, it's important to establish a general understanding of the digital video workflow to see how the different processes fit together.

Planning

The first phase of any production is planning. Don't underestimate its importance. The best way to achieve good results is to know what "good" is for your project before you start. Even if you're in charge of just a small portion of the overall project, knowing how it fits into the overall structure and having a solid understanding of the goals of the project will help you deliver optimal results on time and on budget.

You should define three basic, interdependent, elements as early as possible: content, communication goals, and audience. If you have a good grip on these three, the process will go much more smoothly.

Content

What is the nature of your content? Music video? Movie trailer? Feature film? Training video? Annual report? Produced on film or video? Letterboxed or full frame? HD or SD? PAL or NTSC? How long is it?

In an ideal world, you'd be able to design content that'll compress with high quality on your target platforms. But very often we have to start with existing content produced for other purposes. Either way, understanding the source as well as possible is critical, as you'll need to finesse the process to play to your content's strengths and mitigate its weaknesses. The right compression settings can vary quite a bit with different types of content, even when targeting the same format and data rate. And there's often plenty of processing required before the content hits the codec that can have a big impact on quality.

In general, the lower the bitrate and simpler the device you're targeting, the more impact different types of content have on compression.

Communication Goals

Most of us don't compress video just for the fun of it. We're making compressed media for someone to watch, and thus we have a communication goal that compression is meant to achieve. Having a clear understanding of that goal is essential. However, it's all too common for folks to not define explicitly what they're trying to achieve with a project.

But how we encode a particular clip can vary radically with different communication goals. Take a music video, for example. If you were a member of the band and are embedding the video into your web site, with the goal of selling albums and concert tickets, you'd make sure to use enough bits for the audio to sound great, and would use technologies that most users wouldn't need an extra install in order to watch. Conversely, if you were the cinematographer of the video, and you were putting the video on your online demo reel, you'd emphasize video quality over audio. And since you're aiming for a more technical audience, you might use a technology with higher system requirements to make sure the experience is very high quality. If you're trying to sell your ability to make beautiful images, it's better for someone to not see your reel at all than to see it presented badly; there's never a second chance at a first impression.

Audience

We also need to know who's going to try to watch the content, what experience they're hoping for, and what their playback device is capable of.

The first order of business is to predict what formats the targeted device can play back. If it's a mobile phone, there's a limited number of formats it can play, so you need to use one of those. If it's a PC (be it Windows, Mac, or Linux) format support can be a lot broader but with more variation in what a particular PC may support in formats, performance, and bandwidth. And, of course, there's the possibility of requiring the user to update or install software for an improved experience.

If your target audience is a particular corporate intranet with the same identical system image and a known minimum hardware spec, count your blessings. Your choice of media player and format will be an easy one (almost always Windows Media), and you'll know how much bandwidth you can have.

But if your audience is anyone who accesses a general interest web site via the Internet, things can get very complex. Making a single file that 90 percent of the audience can watch is pretty easy, but getting to 99 percent is a lot more complicated, as you have to start including Linux systems and a variety of mobile devices, plus ancient slow computers. You can get quickly into cases where you have a single or a few encodes (perhaps WMV and H.264 .MP4), but need to do a lot of web-side programming to dynamically embed the file in the available media player, whether it be Windows Media Player, Silverlight, Flash, Moonlight, VLC, Safari, or another player.

The other critical concern is bandwidth—how many bits a second can the user get? And knowing bandwidth isn't just knowing what grade of DSL or cable they're connected to. Even if they get the full rated bandwidth, a household could have multiple computers and game systems in use, with bandwidth split among them. And a single user can be running multiple applications that are all consuming bandwidth at once; that system update in the background is using bits that could otherwise be used for video. So you want to know the real-world range of bandwidth is that users are likely to have, and from that decide what you want to support. Fortunately, we now have good adaptive streaming technologies that can detect and select streams based on real-world bandwidth, so things are easier if you're using one of those.

Be warned that a client may say, "It's critical that everyone sees this video!" But embedding an http://foo.mpg link to a MPEG-1 file is as universal as you can get. Almost every computer shipped since the mid-1990s can play that out of the box. But no one uses that, because it launches an external player, and good quality requires a much higher bitrate than modern codecs; you'd be trading a lower bar of codec support for a higher bar of bandwidth required. For every project, you need to draw the line on how ubiquitous it really needs to be to meet the communication goals, and that's a tradeoff between cost and time in authoring, complexity of management, and the minimum experience you're not willing to deliver below. That gets down to devices (no phones?), operating systems and player combos (no Linux without Moonlight, Flash, or VLC installed?), and bandwidth (nothing below 200 Kbps?).

After all, there are still millions of people using dial-up, and the industry has abandoned delivering compressed media to them; no one with dial-up has any expectations of getting a decent video experience. And it's not worth sweating how to stream full-length movies to phones; people just don't consume content that way. Better to understand how they would want to experience that content, and figure out how to deliver it in a way that suits the user and device.

Balanced Mediocrity

So how do you balance content, communication goals, and audience? You're trying to achieve what I call, tongue-only-partially-in-cheek, balanced mediocrity.

For video on the web, you'll often not have as much bandwidth to deliver or as much CPU power to play back as much resolution, frame rate, or detail as your luscious source may contain. Compromises must be made.

Finding the right balance of elements can feel like making everything mediocre, so you're not spending bits making one aspect perfect at the expense of everything else getting worse.

Balanced means "in the right proportion." So, given our music video example, you want to make the ratio between video and audio data rates be such that each seems about as good as the other, *relative to your goals*. The relative part there is critical—you'd rate "quality" differently between the version for the cinematographer and the version for the band. So, even though you'd go through the same decision-making process, bitrate and format choices would be quite different in the end.

A good test for balanced mediocrity is to determine whether changing any particular parameter diminishes the overall experience—for example, when raising the data rate hurts more by slowing download times than it helps by improving video quality, and reducing data rate hurts video quality more than it helps by reducing download times. Or increasing resolution hurts more by increasing artifacts than it helps by increasing detail, and reducing resolution would hurt detail more than it would help by reducing artifacts. The bitrate requirement, video quality, resolution, and artifacts may not all be quite good as you'd like them to be, but if you've got the right balance of mediocrity for your project, you'll have done what you can.

Production

Production is the actual shooting of the film or video, creating the animation, or otherwise creating the raw assets of the project. This is often the most expensive and most difficult part of any project, so getting it right is critical. And remember that the phrase "We'll fix it in post" should be black humor, not an actual plan.

Professional video and film production is much, much harder than video compression, and it requires an experienced crew. One of the biggest mistakes many companies and individuals make in creating web video is thinking that because the content is going to be delivered on the web, production standards don't matter. This is completely untrue. Amateur video on the web looks at least as bad as amateur video on television.

There are many important, irrevocable decisions made during production. For video production, the content needs to be shot as progressive or interlaced, at 4:3 or 16:9 aspect ratios, in black-and-white or in color; at a certain frame rate; with a cinematographic style that can run the gamut from *Oprah* to *Battlestar Galactica*. Of course, there is no one correct answer to such questions. It all depends on the nature of each project. (Well, shooting interlaced is never the correct answer anymore.)

Production is covered in Chapter 5.

Postproduction

Postproduction is when the snippets of images and sounds created in production are turned into a final product. Postproduction includes editing, compositing, audio sweetening, and so on.

Much of the style of a production comes during post. Different editing styles can have dramatic effects on the final results, both in terms of the uncompressed project and in terms of the compressed version—MTV-style fast cutting yields a totally different viewing experience than long, slow cuts. Likewise, MTV-style cutting is enormously more difficult to compress than more conventional, sedate editing.

Postproduction issues are discussed in Chapter 5.

Acquisition

If the source is a live video signal or on a tape, it'll need to be converted to a form that can be compressed. The good news is that more and more, an all-digital workflow is possible, where the output of postproduction is a file that can be used directly rather than being captured.

But, capture *is* required when all you have access to is a tape camera, flash card, and so on. Tape is the most challenging and expensive method, particularly with professional formats, as the decks can cost into six figures. More and more cameras shoot in a compressed format like DV, DVCPROHD, HDV, or AVCHD that can be losslessly captured as a bitstream using USB, FireWire, or a card reader. Things get harder with analog sources and formats that offer only uncompressed digital output. These require capture cards and sufficient storage speed and capacity. Digitization isn't particularly difficult, but it does require the right equipment. Luckily, getting DV format video off a MiniDV or DVCAM tape is trivially easy with modern computers. Capture and Digitization is covered in Chapter 6.

Preprocessing

Preprocessing is everything done between the source video and audio and what is finally handed off to the codec. This includes cropping, scaling, deinterlacing, image adjustment, noise reduction, and audio normalization. While this may happen behind the scenes in consumer products, it's always being applied. The output of preprocessing is the final series of images in the native sampling and quantization (typically 8-bit 4:2:0) that the codec will take.

The goal of preprocessing is to take content optimized for whatever it was created for and transform it into the best version to be compressed. In essence, it's about making the content as much like itself as possible, given its playback environment. We want blacks to be black and the image shape to match the display. Preprocessing requires the most craft and subjectivity of any part of the compression workflow. On quality-critical projects, I find myself spending far more keyboard-and-mouse time doing preprocessing than on codec settings. In the end, the codec can never output anything better than the source it's given—problems in preprocessing can ruin quality that the best compression can't fix.

Preprocessing is covered in Chapter 7.

Compression

If you cared enough to buy this lovely book, you already know compression is the Big Kahuna. In the compression process, the preprocessed content is turned into the final video and audio bitstreams and the file that contains them.

Before compression, many decisions need to be made, including choosing formats, codecs, data rates, and frame rates. While picking the right compression settings seems daunting, it becomes much easier when you you've got good answers to our three questions:

- What's the content?
- What's the communication goal?
- Who's the audience?

The bulk of this book is about the compression phase, although we'll continue to talk about the earlier stages throughout. Make sure to read the chapters specific to the formats you're using.

Delivery

Delivery is how the files get to the user. This can be on a DVD, Blu-ray, or CD-ROM, downloaded from a service, streamed to a media player, embedded in a web page, sideloaded onto a phone, and so on. The files can be standalone or integrated into a larger project such as a web site or computer game.

More complex delivery schemes often have their own, more stringent issues. Delivery issues with particular technologies are covered within the chapters about those technologies.

CHAPTER 5

Production, Acquisition, and Post Production

Introduction

From 1995–1999 I was the Chief Technologist and partner in a small (maxed out at eight full-timers) company that offered what we called "digital media production, post and compression services." This was a good time, and Portland, Oregon was a good market for that stuff. We had a lot of technology companies using digital media really early, like Intel, Tektronix, and Mentor Graphics, plus the double-footed shoe titan of Nike and Adidas Americas. During that period, most of our content was delivered via kiosks or CD-ROM, with the web getting big near the end.

And compression was hard back then, and very content-specific. There was simply no way to encode a frenetic music video that would look or sound good played back on the technology of the era. So, our business model integrated the whole workflow, so that we'd script and storyboard knowing what the final compression would require, and make the creative choices up front that would yield great-looking video on the final display. And it delivered. We used and even pioneered a lot of techniques I talk about here, particularly fixed cameras, avoiding cross-dissolves, and keeping motion graphics in 2D.

The good news is that we've had big gains in decoding horsepower, codecs, and bandwidth. So it's less necessary to sweat these things than it used to be; it's rare to have content that becomes actually incomprehensible after compression. But it can still matter, particularly when targeting lower bitrates. Sure, H.264 degrades into being soft more than blocky, but the fine detail can still go missing. Knowing that will happen in advance lets you pick techniques to retain fine detail, or avoid having important fine detail to lose.

And some of this is just general advice on best practices for production and post that applies to any content, no matter how it will be delivered.

I also want to warn about, and arm you against, the pernicious assumption that "it's just for the web, so why worry about quality." In fact, the opposite is true. In any video destined for viewing in some compressed form, be it online or on low-bitrate wireless devices or elsewhere, all the tenets of traditional production must be adhered to while addressing the additional constraints imposed by compression. Well-lit, well-shot, well-edited content will compress better and show fewer artifacts than a poorly done source. And there's more than one axis of bad in this world. There's nothing about a small screen that makes bad white balance look better or bad audio sound better.

Broadcast Standards

Globally, there are two major broadcast standards that encompass every aspect of video production, distribution, and delivery: NTSC (National Television Systems Committee) used primarily in the Americas and Japan and PAL (Phase Alternating Line), used throughout Eurasia and Africa, plus Brazil and Argentina. Signals produced from one will be incompatible for delivery on the other, because NTSC and PAL have different native frame rates and numbers of lines. Luckily, a lot of video tools—especially postproduction software—can handle both formats. Alas, it's not as common for tape decks and monitors to be format-agnostic. Most video production facilities are either PAL- or NTSC-centric. When called upon to deliver content in a non-native format, they do a format conversion. This conversion is never perfect, as going from PAL to NTSC results in a loss of resolution and going from NTSC to PAL loses frames. In both cases, motion is made less smooth.

NTSC

The first National Television Systems Committee consisted of 168 committee and panel members, and was responsible for setting standards that made black-and-white analog television transmission a practical reality in the United States. The standards the NTSC approved in March 1941 were used for broadcast in the United States until 2009 and remain in use in other countries. Those standards serve as the framework on which other broadcast television formats are based. Even though analog broadcast has ended, lots of our digital video technology echoes and emulates analog standards.

Some aspects of the NTSC standard—the 4:3 aspect ratio and frame rate—had been recommended by the Radio Manufacturers of America Television Standards Committee back in 1936. One of the most important things the NTSC did in 1941 was ignore the arguments of manufacturers who wanted 441 lines of horizontal resolution and mandate 525-line resolution—then considered "High Definition."

The original frame rate of NTSC—30 interlaced frames per second—yielded 60 fields per second, and was chosen because U.S. electrical power runs at 60 Hz. Electrical systems could interfere quite severely with the cathode ray tubes of the day. By synchronizing the display with the most likely source of interference, a simple, nonmoving band of interference would appear on the screen. This was preferable to the rapidly moving interference lines that would be produced by interference out of sync with the picture tube's scan rate.

When color was later added to NTSC, it was critical that the new color broadcasts be backwards compatible with existing black-and-white sets. The ensuing political battle took more than a decade to resolve in the marketplace (from 1948, when incompatible color TV was developed, through 1953 when a backward-compatible technique was devised, until 1964 when color TV finally caught on with the public); but that's a tale for another book. Politics aside, backward compatibility was a very challenging design goal. The NTSC worked around it by leaving the luminance portion of the signal the same, but adding color information in the gaps between adjoining channels. Since there was limited room for that information, much less bandwidth was allocated to chroma than luma signals. This is the source of the poor color resolution of NTSC, often derided as Never Twice the Same Color. The frame rate was also lowered by 0.1 percent, commonly but imprecisely truncated to 29.97 (59.94 fields per second).

The canonical number of lines in NTSC is 525. However, many of those lines are part of the vertical blanking interval (VBI), which is information between frames of video that is never seen. Most NTSC capture systems grab either 480 or 486 lines of video. Systems capturing 486 lines are actually grabbing six more lines out of the source than 480 line systems.

PAL

PAL is the color-coding system used in Europe. By reversing the phase of the reference color burst on alternate scan lines, PAL corrects for shifts in color caused by phase errors in transmission signals. For folks who care about such things, PAL is often referred to as Perfection At Last. Unfortunately, that perfection comes at a price in circuit complexity, so critics called it Pay A Lot.

PAL was developed by the United Kingdom and Germany and came into use in 1967. European electrical power runs at 50 Hz, so PAL video runs at 50 fields per second (25 frames per second) for the same interference-related reasons that NTSC started at 60 Hz. Nearly the same bandwidth is allocated to each channel in PAL as in NTSC, and each line takes almost the same amount of time to draw. This means more lines are drawn per frame in PAL than in NTSC; PAL boasts 625 lines to NTSC's 525. Of those 625 lines of horizontal resolution, 576 lines are captured. Thus, both NTSC and PAL have the same number of lines per second, and hence equal amounts of information (576×25 and 480×30 both equal exactly 14,440).

The largest practical difference between NTSC and PAL is probably how they deal with film content. NTSC uses convoluted telecine and inverse telecine techniques. To broadcast film source in PAL, they simply speed up the 24 fps film an additional 4 percent so it runs at 25 fps, and capture it as progressive scan single frames. This makes it *much* easier to deal with film source in PAL. This is also why movies on PALDVD are 4 percent shorter than the NTSC version.

SECAM

SECAM (Système Électronique pour Couleur Avec Mémoire) is the French broadcast standard. It is very similar to PAL, offering better color fidelity, yet poorer color resolution. SECAM today is only a broadcast standard—prior to broadcast, all SECAM is made with PAL equipment. There's no such thing as a SECAM tape. SECAM is sometimes jokingly referred to as Something Essentially Contrary to the American Method.

ATSC

The Advanced Television Standards Committee created the U.S. standard for digital broadcasting. That Digital Television (DTV) standard comes in a somewhat overwhelming number of resolutions and frame rates. As a classic committee-driven format, it can provide incredible results, but its complexity can make it extremely expensive to implement. And even though Moore's Law continues to bring chip prices down, the expense necessary to build display devices (be they projection, LCD, or CRT) that can take full advantage of high-definition quality remains stubbornly high.

Resolutions at 720 × 480 and below are called Standard Definition (SDTV). Resolutions above 720 × 480 are called High Definition (HDTV).

With the transition to HD digital broadcasting underway around the world, production companies have increasingly been shooting HD to future-proof their content.

DVB

Digital Video Broadcasting is basically the PAL countries' version of ATSC.

Preproduction

As with any production, proper planning is critical to shooting compression-friendly video. Beyond the usual issues of location, equipment, cast, crew, script, and so on, with compressed video you need to be aware of where your content will be viewed. Is it going to DVD? CD-ROM? Broadband web video? 56 kbps dial-up modem connection? Wireless handheld devices? All of the above? Is it going to be evergreen content that needs to work on future devices not yet conceived of? Knowing your delivery platform helps you understand what you can and can't get away with in terms of aspect ratios, backgrounds, framing, and the like.

It's always better to plan ahead, to know in advance what you're trying to do and how you're going to do it. Even if you're making a documentary in which you'll find the story as you edit your footage, knowing your target delivery platform will dictate how you'll light and shoot your interview subjects, capture your audio, select archival footage, and so on.

Things that should be decided and communicated to your crew in advance of your shoot include what aspect ratio you'll be shooting, what frame rate the camera is running at, your shooting style (locked-down cameras or handhelds), and whether you'll only use tight close-ups for talking heads or opt for a mix of close-ups and wide angle shots.

Production

Production is generally the most expensive phase of any project, and hence the most costly to get wrong. The standard rules of producing good video apply to producing video that compresses well: good lighting, quality cameras, and professional microphones are always required.

Professional video production is difficult, and involves a lot of people doing their own difficult jobs in coordination. And a lot of web content is shot with Handycams by novice users who don't know enough to know they don't know what they're doing. I don't mean to sound elitist by that; the democratization of production has been one of the best things to happen during my career. But even though a $2,000 camera today can outperform a $100,000 camera of a decade ago, video shot by hobbyists still looks like video shot by hobbyists. In general, it takes several years of experience working in a particular production discipline before you can call yourself a pro—it's not something you pick up over a weekend after reading a book, even one as fabulous as this. The explosion of great, affordable equipment gives many more people the chance to learn professional skills, but it's not a substitute for them.

Last, whenever anyone on the set says "We'll just fix it in post," be afraid. As someone who's been on both ends of that conversation, I can tell you that phrase contains one of the least accurate uses of the word "just." As said previously, "just fix it in post" is more often black humor than a good plan. Repairing production problems in post is invariably more expensive than getting it right the first time, and the resulting quality is never as good as it would be if the shot had been done right the first time. Post is the place to be doing the things that couldn't be done on the set, not fixing the things that should have been.

Production Tips

There are a number of simple and not-so-simple things you can do to improve the quality of your video. Optimizing it based on a single target delivery medium is relatively simple. However, we live in a world in which "write once, deliver everywhere" has become a business necessity. Shooting for multiple delivery platforms forces you to make choices, but there are things that shouldn't be optional, no matter what your target delivery platform.

Interlaced is dead. Shoot progressive

Interlaced video was a great hack that had a great half-century run, but it has no place in the flat-panel era. No one makes interlaced displays anymore, and tons of devices have no ability to play back interlaced content at all. Anything you shoot interlaced is going to get converted to progressive at some point, either before compression or in the display. And the earlier in the chain you do it, the better the final results will be. I'll talk in great detail in Chapter 6 about how to do deinterlacing, but we'll both be a lot happier if you don't give yourself a reason to read it.

Even if someone says you need to shoot interlaced because the content is going to go out as 480i/576i broadcast or otherwise has to work with legacy displays, you've got two fine options:

- Shoot 50/60p. If you shoot using a frame rate of your target field rate (so, double the interlaced frame rate), you can convert to interlaced *after* post for interlaced deliverables. Going from 60p to 30i and 50p to 25i delivers all the motion doing 30i/25i originally could have, and you've got a higher quality progressive version as well. Interlacing is a lot easier than deinterlacing!
- Shoot 24p. For film/entertainment content, 24p serves as a great universal master. You can leave it as native 24p for most deliverables, and then convert to 3:2 pulldown for NTSC and speed up 4% to 25p for PAL.

If there's a lot of motion, 720p60 is better than 1080p24 or 1080p30. So: only progressive. No interlaced, ever.

Use pro gear

If you want professional quality, use gear that's of professional caliber, operated by someone who knows how to use it. The good news is that pro-caliber gear is now dirt-cheap compared to only a few years ago. And when we talk about pro gear, don't focus mainly on format support and imaging spec. When a camera's marketing materials start talking about the lens more than the recording format, that's a good sign of a good camera. Good glass beats lots of megapixels. You want to be able to control focus, zoom, depth of field, and other parameters with fine precision.

And don't forget the importance of quality microphones. Avoid using the built-in microphone on video cameras—use a high-quality external mic. A shotgun or boom is great in the hands of a skilled operator. A lavaliere (clip-on) microphone is more trouble-free, and will record voice well. Remember, even if a playback device has a tiny screen, it's also going to have a headphone jack, and that sound can be as big as the world. Since audio has the best chance of making it through the compression process with uncompromised quality, getting the sound right will really pay off.

Produce in the right aspect ratio

16:9 versus 4:3? These days, I'd generally default to 16:9 (and of course all HD cameras are always 16:9). I only consider 4:3 if I know 4:3 devices are going to be targeted for playback; mainly phones (although the iPhone uses a 3:2 aspect ratio, which lies between 4:3 and 16:9). While letterboxing works well for larger displays, being able to use the full 320 × 240 on that size screen gives a lot more picture than 320 × 176. Shooting hybrid is a great way to hedge, with the critical action constrained to the 4:3 center, but including the full-width shot and checked for boom mics, and so on. Then the video can be delivered as-is for 16:9,

Figure 5.1 Guides for shooting hybrid for 16:9 and 4:3. The innermost vertical lines show where action should be kept to for easy cropping to 4:3.

and cropped ("center-cut") for 4:3. This requires careful production to make sure that both framing modes offer good visuals. See Figure 5.1.

Bright, soft lighting

Lighting plays an enormous role in determining how easily your video will compress. Soft, diffuse, even lighting is easier to compress than lighting that produces deep shadows and hard edges. A three-light kit and some diffusers will usually be enough for a simple on-location shoot. For outdoors, bounce cards can be used to reduce shadows.

More important than softness, there should be enough light that there's not extra noise coming from using a higher gain in the CCD, or from using a faster film stock. Grain and gain noise creates codec-killing small details that change every frame. While it can be removed in post, it's hard and slow to do without hurting detail.

Bear in mind that modern flat-panel displays don't have perceptually uniform gamma like most codecs assume, which often makes compression issues in shadow detail very visible and can lead to distracting blocky artifacts in the not-quite-blacks that wouldn't have been visible on a calibrated CRT display. This can be particularly hard with mobile device screens, far too many of which use 16-bit processing without any dithering. So beware of dark scenes if you're targeting handheld device playback. They can look great on set, but can wind up like a mosaic on the final display.

Record DV audio in 48 kHz 16-bit

DV supports two audio recording modes: 4-channel 32 kHz 12-bit and 2-channel 48 kHz 16-bit. Always use the 48 kHz mode. 32 kHz isn't great, but 12-bit audio can be a quality killer.

Control depth of field

Compared to film, most video cameras use a much deeper depth of field. So a scene shot by a film camera (or a digital camera that optically emulates film) could show foreground and background objects as blurry when the action in the middle is in focus, while the video camera leaves everything in focus over the same range of distances. That may sound more like a bug than a feature, but depth of field control has long been a big part of the film vocabulary, and is required for a convincing "film look" if that's what desired. Also, using a narrow depth of field forces the codec to spend bits on just the elements you've focused on. If producing in an environment where the background can't be controlled, such as outdoors or on a convention floor, making sure your talent is in focus and irrelevant details aren't means that the subject will get most of the bits. Plus, background details that won't matter won't distract the viewer.

Controlling what's in focus and what's not may require some preplanning in terms of camera location, as you'll typically need to be farther away than you might expect.

Watch out for foliage!

One of the most difficult things to compress is foliage blowing in the wind, which makes sense mathematically: it's green, and thus mainly the nonsubsampled Y′ channel, and it's highly detailed. And it has lots of subtle random motion. You can be amazed by how many more artifacts the talent's face will get when in front of a hedge in a gentle breeze. I've had cases where I had to quadruple bitrate for talking-head video based just on background foliage.

Use a slow shutter speed

If you're shooting slower motion content, stick to a shutter speed of 1/60 (NTSC), 1/50 (PAL), or 1/48 (24p) of a second. These are the defaults for most cameras. Slow shutter speed produces very natural-looking motion blur, meaning that when the image is detailed, it's not moving much, and when it's moving a lot, it's not very detailed (see Figure 5.2). Thus, the

A

B

Figure 5.2 With motion blur, moving parts of the image blur along the axis of motion (5.2A), while static parts of the image remain sharp (5.2B).

codec has to handle only one hard thing at a time. The very slow shutter of film is one reason film content encodes so nicely. Slow shutter speeds also let in more light per frame, which reduces the amount of light needed to avoid grain and gain.

If you want viewers to see it, make it big

A long panoramic shot may look great in the viewfinder, but the talent in the middle of the screen may vanish into insignificance on a phone. An initial establishing shot may be effective, but the bulk of your shots should be close up on what you want the viewer to focus their attention on.

Limit camera motion

Handheld camera work can be extremely difficult to compress, since the whole frame is moving at once, and likely moving in a different direction than the action. Most codecs work on a 2D plane, so if the camera is moving parallel to the action, it compresses great. But if the camera is rotating (a "dutch"), zooming, or moving towards the action, it's a lot harder for the codec to make good matches from frame to frame.

The worst case here is when the camera is mounted on something that is moving. Perhaps the hardest single thing to compress is the helmet-mounted camera giving a point of view (POV) of someone mountain biking. The motion is fast and furious, the camera is whipping around with constant angular changes, plus it's rapidly changing position. And remember that thing about foliage? You'll need to have radically more bits per pixel for that kind of content.

The tripod and dolly are classics that still work great. If handheld camera work is absolutely required, try to use a Steadicam or similar stabilization system. A skilled camera operator will be better at keeping the camera still.

Mr. Rogers Good, MTV Bad

As you learned in Chapter 4, the two things that use up the most bits in compressed video are detail and motion. The more complex your image and the more it moves, the more difficult it is for a codec to reproduce that image well. At lower data rates, this difficulty manifests as increasingly poor image quality.

As a rule of thumb, classic, sedate shooting and editing compress well. Jump cuts, wild motion graphics, and handheld (shaky-cam) camera work are difficult or impossible to compress well at lower data rates. "Mr. Rogers good, MTV bad" says it all.

When designing for low-data-rate delivery, avoid using elements that aren't essential to what you're trying to communicate. If a static logo works, don't use a moving one. The lower the target data rate, the simpler the video needs to be. If you want to do live streaming to handsets, Spartan simplicity is your goal. But if you're going for Blu-ray, you can get away with almost anything.

Picking a Production Format

Even though we've seen a great winnowing down of delivery codecs and formats, it seems you can't sneeze without some new production codec or format popping up; there's far more now than ever.

Types of Production Formats

Interframe vs. intraframe

The biggest difference between production formats is whether they encode each frame independently (intraframe) or use motion vectors and encode a number of frames together in a GOP (interframe).

Both can work, but interframe codecs trade substantially increased pain in postproduction for bitrate savings when recording. Bear in mind that the postproduction pain can be high enough that the interframe codecs may just get re-encoded to an intraframe codec on import, causing a generation loss and throwing away any efficiency savings.

Interframe is mainly seen in the consumer HDV and AVCHD formats. They're great from a cost-of-hardware and cost-of-capture media perspective, but can be pretty painful when trying to ingest a lot of content. Interframe encoding makes lossless editing much harder, since all edits need to be at I-frames. Any edits requiring more than a half-second or so precision would require re-encoding frames around the edit point.

DCT vs. wavelet

Classic video delivery codecs all used DCT as discussed in Chapter 3. But we're seeing a whole lot of wavelet codecs used in various production formats. Wavelet has some key advantages:

- It's more efficient than simple DCT for intraframe encoding (although worse for interframe).
- It degrades into softness, not blockiness.
- It's easily extended to different bit depths, number of channels, etc.
- It supports decoding only to a sub-band. So just the 960 × 540 band of a 1920 × 1080 frame could be decoded to display a 960 × 540 preview. They typically have bands at 2x ratios for width/height, so even thumbnails are easy to get.

Subsampled vs. full raster

The higher-end formats will encode a pixel for every pixel of the video format, so a full 1920 × 1080 or 1280 × 720. Others use subsampling, like encoding just 1440 × 1080 or 960 × 720. You'll want to capture as many pixels wide as you'll want to display.

4:2:2 vs. 4:4:4 vs. 4:2:0 vs. 4:1:1 vs. "RAW"

Most pro video formats use 4:2:2 chroma subsampling. This is a fine choice; it's a superset of 4:2:0 and 4:1:1, and means that any interlaced production (not that you would produce one!) doesn't have to share chroma samples between fields.

One drawback to some tape formats is that it's hard to get access to the native bitstream. So even though you have a nice compact version on the tape, the deck converts it back to uncompressed, making capture more expensive and difficult, and potentially requiring recompression. I don't like this approach.

One recent emerging choice is "Video RAW." Like a still camera RAW mode, this actually captures the sensor image data directly. With a CCD, that's a single color per sensor element, with more precision and in a nonlinear gamma. This winds up replicating some of the classic film workflow. The RAW video, like a film negative, needs to be processed and color-timed in order to make something that's editable. But the RAW approach preserves all the detail the camera is capable of; there's no baked-in contrast, color correction, and so on. So it preserves the maximum flexibility for creative color correction in post.

Tape vs. solid state vs. cable

This last subsection is about what medium the content is recorded to.

Back in the day, it was always tape. Sure, in the end it was just rust glued to Mylar, but it worked and was generally cheap and reliable. But it's also linear; capturing an hour of tape takes an hour of transfer time. And if you want to find just that one scene? You've got to spin the spindles to the right part of the tape. With device control this is automatable, but it's a slowdown in the workflow.

Now we have an increasing number of solid-state capture mechanisms using flash memory. These write a file to a flash card, and so can be popped straight into a computer when done. No fast-forward/rewind; it's like accessing any other file. Initially there was a price premium for the convenience, but flash prices are dropping at a Moore's Law–like pace, while tape price remains pretty static, so it won't be long until solid state is cheaper than tape. Flash memory also is much more reusable than tape; thousands of writes unlike the dozen or so at most for typical video tape before reliability starts going down.

The last option is to just capture the bits live to a hard drive, either connected to the camera or to a PC. The workflow is great here as well, although it requires a very fast computer and a good backup mechanism. Most cameras that support PC capture can do it in parallel with tape or card, for a real belt-and-suspenders experience.

If you're not familiar with the professional video industry, you might find this list of options perplexing. It's all just glass, CCD, and encoded bitstream, so why so many different products with such massive variability in price points? And why are the "consumer" formats missing critical features that would be trivial to add? Well, the companies that historically have dominated this industry have worked hard to make sure that equipment from the consumer divisions doesn't get so good that they eat into sales of the professional products. This segmentation has led to some very annoying frustrations in the industry.

Fortunately, the competition between the consumer divisions of different companies causes some upward feature creep ever year. And the good news is that as new companies without legacy business, notably RED, enter the market, they don't have to worry about segmentation and are shaking things up.

Production codecs

DV

DV was the first affordable digital video format and remains in wide use. It is a simple 8-bit DCT intraframe codec running at 25 Mbps (thus the nickname DV25). SD only, it's always 720 × 480 in NTSC and 720 × 576 in PAL. Depending on the camera, it can support interlaced or progressive, 4:3 or 16:9, and even 24p. 24p on DV is really 24PsF (Progressive segmented Frame)—it encodes 24 unique frames per second and 6 duplicate frames in order to maintain the 30i frame rate. This wastes some bits, but has no impact on quality and is automatically corrected for inside NLEs, so all you'll normally see is the 24p.

The biggest drawback is that the NTSC version uses 4:1:1 chroma subsampling, which makes for a lot of blocking with saturated colors. The PAL version is a more sensible 4:2:0 (except for the largely abandoned PAL 4:1:1 DVCPRO variant).

DVCPRO and DVCAM are JVC and Sony's brands for their Pro DV versions, respectively. They're both the standard DV25 bitstream but offer larger higher capacity cassettes that won't fit consumer DV cameras.

DVPRO50

DVCPRO50 is JVC's Pro variant of DV. Essentially two DV codecs in parallel, it uses 4:2:2 instead of 4:1:1 and doubles the bitrate to 50 Mbps. It looks great and offers the easy postproduction of DV. DVCPRO50 uses normal DV tapes running at double speed, so you only get half the run time that DV25 has.

DVCPRO-HD

DVCPRO-HD is sometimes called DV100, as it's essentially four DV codecs ganged together. Thus, it's intraframe 4:2:2 with up to 100 Mbps (although it can go down to 40 Mbps depending on the frame size and frame rate). It's available in both tape and flash (P2) versions. The tape runs at 4x the speed and so has 1/4 the duration as standard DV.

The biggest drawback to DVCPRO-HD is that it uses aggressive subsampling. But the high and reliable per-pixel quality can be well worth it.

DVCPRO-HD has several modes:

- 960 × 720p50/60
- 1280 × 1080i30
- 1440 × 1080i25

DVCPRO-HD is used in VariCam, which allows highly variable frame rate recordings for special effects.

HDV

HDV is an interframe MPEG-2 HD format. Designed to be a straightforward upgrade from DV, it uses DV tapes and the same 25 Mbps bitrate.

When HDV works, it's fine. While earlier NLEs had a hard time playing it back, modern machines are easily fast enough. But since it's an interframe MPEG-2 codec, once motion or detail starts getting too high, it can get blocky. So production technique matters; HDV is not a good choice for our helmet-cam mountain biking example. The quality of the MPEG-2 encoder chip varies between cameras (though across the board they're better today than when the format launched), so trial and error is required to figure out what's possible to get away with. Even if it is recorded with the blockies, there are ways to reduce that on decode (as we'll discuss later), but it slows down the process. Compression artifacts are particularly problematic in the 1080i mode; see Color Figure C.15 for an example. Shooting progressive—particularly 24p—improves compression efficiency and thus reduces compression artifacts.

There are two main modes of HDV:

- HDV 720p is a full 1280 × 720 at 24, 25, 50, or 60 fps at 20 Mbps. This has only been seen in JVC cameras to date. 720p24 is easy to do at that bitrate, and can yield high-quality results.
- HDV 1080i is 1440 × 1080 at 25 Mbps. Although originally meant to be a progressive format, Sony added 1080i to the format, and has that as the default mode in their products.

This has led to a frustratingly large variety of attempts to get 24p back out again.

- Cineframe was just a not very good deinterlacer applied after the native-interlaced signal processing. It's horrible; don't use it, or a camera that has it.
- "24" uses 3:2 pulldown, so two out of five frames are coded interlaced. It can be a pain in post, but result can be good.
- 24A uses PsF, where there are six repeated frames every second. This works fine in practice; bits aren't spent on the repeated ones.

True progressive models with a native 24/25/30 bitstream are now available. However, some still use interlaced sensors; make sure that any camera you get supports progressive lens to bitstream. Since HDV is a 4:2:0 format, shooting interlaced can result in some color challenges.

AVCHD

AVCHD is a H.264 bitstream in a MPEG-2 program stream on solid-state memory. It's conceptually similar to HDV—use a commodity consumer codec at the highest bitrate. The big difference is that it has followed the photography market in using flash memory instead of tape.

H.264 is a much more efficient codec than MPEG-2, and also degrades more gracefully, so quality is generally better. The bigger challenge with AVCHD is it takes a lot of horsepower to decode. These bitrates can be challenging to decode in software on their own, and it can take GPU acceleration to be able to scrub or edit crisply. Some editing products, including Final Cut Pro 6, automatically recompress from AVCHD to another simpler format right away.

AVCHD is definitely the best consumer video format at this point, but have a postproduction plan before you use it.

IMX

IMX was an early Sony attempt at a DVCPRO50 competitor, using intraframe MPEG-2 from 30–50 Mbps. It's still supported in many cameras, and is a fine choice for SD-only productions.

XDCAM

XDCAM is Sony's pro post-tape format. It originally supported the Professional Disc optical format, with support for SxS and SDHC cards added later. There are a number of variants and modes, but all are interframe MPEG-2. Since they target solid-state, they can use VBR, raising bitrate for challenging scenes; this makes the risk of HDV-like blockiness much lower. There are 19 and 25 Mbps modes, but you should just use XDCAM at the HQ 35 Mbps top bitrate; storage is cheap these days, so there's no reason to run the risks of artifacts.

HDCAM

HDCAM was Sony's first pro HD offering. It's a nice intraframe DCT codec, but with very odd subsampling. For luma, it's a pretty typical 1440 × 1080. But for chroma, it's a unique 3:1:1 chroma subsampling. Yep. It has 1440/3 = 480 chroma samples horizontally. This is only half the chroma detail of a full 1920 × 1080 4:2:2 encode (1920/2 = 960). Not bad for acquisition, but pretty lossy for post, particularly for animation and motion graphics. Being Sony, it started out as 1080i only, but since has gained 24/25p modes (used in Sony's highend CineAlta cameras).

RED

The RED cameras are one of those things like Moby Dick and the Arctic Monkeys: everyone talks about how religiously awesome it is, to the point where it seems reality has to disappoint, But winds up being pretty awesome anyway. It's been the focus of incredible buzz at NAB show for many years, and is being used with great real-world results. Beyond some great and promising products, it's also exciting to have a new entrant into the camera business without a vested interest in the way things have always been done, and willing to explore new price points and configurations.

Their main product now is the RED ONE, which uses a 4096 × 2304 imager (4 K). That's actual sensor count for individual red, green, and blue sensors, however, not full pixels. But it's fully capable of making great 2 K and higher images, with frame rates up to 120 Hz. A feature film grade camera can be had for under $30 K, without lab fees—far less than a good Pro SD digital video camera was in the 1990s.

"REDCODE RAW" is the native RED bitstream, a mathematically lossy wavelet codec recording the actual values from the 4 K receptors. This offers the most flexibility in post, but requires a whole lot of processing to turn into editable video as well.

At this point, RED is really best thought of as a 35 mm film replacement technology. It can radically outperform existing video cameras in capture detail, but it also requires a much more intensive ingestion process.

Post-only formats

The post-only formats won't ever be used to create content, but you may receive content on those formats. As much as I love that the HD future has become the HD present, there's a huge legacy of content out there that you may need to access.

VHS

VHS (Video Home System) was a horrible video standard. Back in the day we talked about "VHS" quality, but that was mainly an excuse for our grainy, blocky 320 VHS × 240 Cinepak clips. VHS has poor luma detail, much worse chroma detail, and is extremely noisy. People just watching a VHS movie for fun may not have noticed how nasty VHS looked, but when the content is captured and displayed on a computer screen, the results are wretched.

Sometimes content is only available on VHS, and if that's all you've got, that's all you've got. But any kind of professionally produced content was originally mastered on another format. It's almost always worth trying to track down a copy of the original master tape, be it on Beta SP or another format. There's some aggressive preprocessing that can be done to make VHS less terrible, but that's lurching towards mediocrity at best.

VHS supports several speed modes, the standard SP, the rare LP, and the horrible (and horribly common) EP. Because EP runs at one-third the speed of SP, you can get three times as much video on a single tape, with triple the amount of noise. Tapes recorded in EP mode are more likely to be unplayable on VCRs other than the one they were recorded with, and most professional decks can read only SP tapes. If you have a tape that won't work in your good deck, try putting it into a consumer VHS deck and see if it's in EP mode.

VHS can have linear and HiFi audio tracks. The HiFi ones are generally good, even approaching CD quality. Linear tracks may be encoded with good old analog Dolby—if so recorded, playing back in Dolby mode can help quality and fidelity.

S-VHS

S-VHS (the "S" stands for "super") handles saturated colors better than VHS through improved signal processing. S-VHS was never a popular consumer standard, but was used for corporate- and wedding-grade production. You'll need a real S-VHS deck to capture at full quality, though; some VHS decks with a "pseudo S" mode can play S-VHS, but only at VHS quality.

D-VHS

D-VHS was an early attempt at doing digital HD for the home, led by VHS creator JVC. The technology was a straightforward extension of ATSC and VHS—a MPEG-2 transport stream on a VHS tape. The quality was fine when encoded well (it handled up to 25 Mbps). And it was DRM-free with IEEE 1394/FireWire support, so the MPEG-2 bitstream could be directly captured! D-VHS also supports recording encrypted cable content, since it's just a bit pump. A CableCARD or similar decryption device is required upstream.

Prerecorded D-VHS was called D-Theater, and it saw quite a few releases from 2002–2004, ending as HD DVD and Blu-ray were emerging. No one is releasing new content for it anymore, but it was a popular way to easily access HD source content.

8 mm

8 mm was a popular camcorder format before DV. It's better than VHS, although not by a whole lot.

Hi8

Hi8 is an enhanced 8 mm using similar signal processing as S-VHS to improve quality. However, the physical tape format itself wasn't originally designed for carrying information as dense as the Hi8 calls for. Thus, the tape itself is very sensitive to damage, often resulting in *hits*—long horizontal lines of noise lasting for a single frame. Every time the tape is played, the wear can cause additional damage. If you get a Hi8, first dub the source to another tape format, and then capture from the fresh copy. *Do not* spend a lot of time shuttling back and forth to set in- and out-points on a Hi8 tape. Lots of Hi8 tapes are going to be too damaged to be useful this many years later.

3/4 Umatic

3/4 (named after the width of its tape) was the default production standard pre-Betacam. Quality was decent, it didn't suffer much from generation loss, and the tapes themselves were very durable. There's lots of archival content in 3/4. The original UMatic was followed by Umatic SP with improved quality.

Betacam

Betacam, based on the Betamax tape format, provided much higher quality by running several times faster, increasing the amount of tape assigned to each frame of video. The small tape size of Betacam made it a hit with ENG (Electronic News Gathering) crews.

There are two sizes of Betacam tapes, one a little smaller (though thicker) than a standard VHS tape meant for cameras, the other about twice as large meant for editing.

Betacam SP was a later addition that quickly took over, and was the last great analog tape format. SP offered low noise, tape durability, and native component color space. It actually had different Y′, Cb, and Cr tracks on the tape. The first wide use of component analog was with Betacam SP.

If the tape is in good shape, you can get pretty close to DVD quality out of SP. It's the analog format you hope for.

Digital Betacam

Digital Betacam (often referred to as D-Beta or Digi-Beta) was a lightweight SD digital codec, and the first wide use of 10-bit luma. It's captured via SDI (described below).

Betacam SX

This weird hybrid format uses interframe MPEG-2. It was used by some diehard Sony shops, but was eclipsed by DV in the broader market.

D1

D1 is an uncompressed 4:2:2 8-bit per channel SD format. D1 was extremely expensive in its day, and offers great SD quality.

D2

D2 is an anomaly—it's a digital *composite* format. D2 just digitally sampled the waveform of the composite signal itself. This meant that the signal had to be converted to composite first, throwing a lot of quality out. There are still some of these tapes floating around in the world, although it can be a challenge to find a working deck.

D5

D5 was Panasonic's attempt at reaching the D-Beta market with a 10-bit, 4:2:2 format. The same tape format was used in the much more popular D5 HD (see following section).

D5 HD

D5 HD was the leading Hollywood post format for several years, as the first broadly used tape format with native 24p, 8 or 10-bit 4:2:2, and no horizontal subsampling. It's great, although is being displaced by HDCAM-SR. D5 uses 6:1 compression in 8-bit and 8:1 in 10-bit. Given the intraframe DCT codec used, I've seen it introduce some blocking in highly detained chroma, like a frame full of green text.

HDCAM-SR

HDCAM-SR almost makes up for all of Sony's sins in pushing interlaced and offering way too many segmented formats over the years. HDCAM-SR supports a full 10-bit 4:2:2 1920 × 1080 without subsampling and without artifacts. It's the only major use of the MPEG-4 Studio Profile. HDCAM-SR also has a high-end 12-bit 4:4:4 mode suitable for mastering film or animation prior to color correction.

D9 (Digital-S)

Originally named Digital-S, D9 from JVC was originally touted as an upgrade path for S-VHS users and was the first use of the DV50 bitstream.

Acquisition

Acquisition, also called "capture," is how the source content gets from its current form to something you can use in your compression tool. For live video, acquisition is grabbing the video frames as they come down the wire. Of course, the goal is for your source to look and sound exactly the same after capture and digitization as it did in its original form. Capture is probably the most conceptually simple aspect of the entire compression process, but it can be the most costly to get right, especially if you're starting with analog sources.

Digitization, as you've probably guessed, is the process of converting analog source to digital form—our old friends sampling and quantization. That conversion may happen during capture or it may occur earlier in the production process, say during a video shoot with DV format cameras or during a film-to-video transfer. When you capture video off a DV or HDV tape with a 1394 connector, or import an AVCHD or RED file off a memory card, you aren't digitizing, because you aren't changing the underlying data. The same is true if you're capturing uncompressed video off SDI or HD-SDI.

Let's look at some of the technologies and concepts that come into play with capture and digitization.

Video Connections

While analog video is intrinsically YUV and hence Y′CbCr, the manner in which the video signal is encoded can have dramatic effects on the quality and range of colors being reproduced. Both the type of connection and the quality of the cabling are important. Things are simpler in the digital world, where connection types are more about workflow and cost. The Holy Grail is the high-quality digital interconnect that allows direct transfer of the high-quality bitstream initially captured.

Coaxial

Anyone with cable television has dealt with coaxial (coax) connections. Coax carries the full bandwidth and signal of broadcast television on its pair of copper wires surrounded by shielding. Coax not only carries composite video signals; audio signals are mixed with the video signal for broadcast television. This can cause a lot of interference—bright video can leak into the audio channel, and loud audio can cause changes in the video.

There are plenty of consumer capture boards that include tuners and can capture analog over coax. The good news is that analog coax is going away. Coax itself lives on, but it mainly carries digital cable or ATSC signals.

Composite

The composite format was the first better-than-coax way to connect a VCR and TV. At least it doesn't mix audio with video or multiple channels into a single signal. But in composite video, luma and chroma are still combined into a single signal, just like for broadcast. And the way they are combined makes it impossible to separate luma and chroma back perfectly, so information can erroneously leak from one domain to the other. This is the source of the colored shimmer on Paul's coat in *Perry Mason* reruns. You may also see strange flickering, or a checkerboard pattern, in areas of highly saturated color.

How well equipment separates the signals plays a big part in how good video acquired via composite can look, with higher-end gear including 3D comb filters as your best bet. But it's all gradations of bad; no video encoding or compression system should use composite signals. Throw out all your yellow-tipped cables.

Professional systems use a BNC connector without color-coding that has a wire protruding from the center of the male end and a twist top that locks on. But that doesn't help quality at all; it doesn't matter how good the cable is if it's still composite.

S-Video

S-Video (also called Y/C) is a leap in quality over composite. With S-Video, luma and chroma information are carried on separate pairs of wires. This eliminates the *Perry Mason* effect when dealing with a clean source, as chroma doesn't interfere with luma. The chroma signals can interfere with each other, but that's much less of a problem.

Consumer and prosumer analog video tapes (VHS, S-VHS, Hi8) recorded luma and chroma at separate frequencies. When going to a composite output, luma and chroma are mixed back together. So, even with a low-quality standard like VHS, you can see substantial improvements by capturing via S-Video, avoiding another generation of mixing and unmixing luma and chroma.

Component Y′CbCr

Component video is the only true professional analog video format. In component, Y′, Cb, and Cr are stored and transmitted on different cables, preserving separation. Betacam SP was the main tape format that was natively component analog.

Component analog was also the first HD signal, and supports all the classic HD resolutions and frame rates, up to 1080i30. More recent component analog gear supports 1080p60.

When color-coding cables for Y′CbCr, Y′ = Green, Cb = Blue, and Cr = Red. This is based on the major color contributor to each signal, as described in Chapter 3.

Component RGB

As its name implies, Component RGB is the color space of component video. It isn't used as a storage format. It is a transmission format for CRT computer monitors, game consoles, and some high-end consumer video gear. Unless you're trying to capture from one of those sources, there is generally no need to go through Component RGB.

VGA

VGA carries Component RGB. VGA cables can be available in very high quality shielded models, and can support up to 2048 × 1536p60 with some high-end CRT displays. VGA is sensitive to interference, so unshielded cables, noisy environments, low quality switchers, or long runs can cause ghosting or other visual problems. Higher resolutions and frame rates increase the risk of interference, so if you capture from VGA, use the lowest resolution and frame rate that works for your content. Better yet—capture DVI.

DVI

DVI was the first consumer digital video format, and has largely (and thankfully) replaced VGA for modern computers and displays. It is 8-bit per channel RGB and can go up to 1920 × 1200p60. DVI cables can also carry a VGA signal over different pins.

There's also a dual-link version (still using a single cable) that goes up to 2560 × 1600p60, which is a glorious thing to have in a monitor, particularly when authoring HD content.

There are a variety of capture cards that can grab single-channel DVI, but I don't know of any that can do dual-link yet.

There are four types of DVI connectors seen in the wild:

- DVI-I: Supports both DVI-A and DVI-D, so can be used with analog and digital displays. Make sure that digital-to-digital connections are using the digital output.
- DVI-D: Carries only the digital signal.
- DVI-A: Carries only the analog signal. Most often seen on one end of a DVI to VGA cable.
- DVI-DL: Dual Link connector for resolutions beyond 1920 × 1200. Works fine as as a DVI-D cable.

HDMI

HDMI is a superset of single-channel DVI with embedded audio, worse connector (no screws; it can fall out a lot), and a guarantee of HDCP (see sidebar). As it's electrically identical with the video portion of DVI, simple passive adaptors can be used to convert from one to the other. However, HDMI supports Y′CbCr color as well as RGB, including rarely used "high-color" modes like 4:4:4 16-bit per channel. HDMI equipment normally doesn't support all the "PC" resolutions like 1024 × 768 and 1280 × 960; HD is typically limited to 1280 × 720 and 1920 × 1080.

HDMI has become the standard for home theater, as it's cheap and of high quality. Its only real drawback is that darn connector.

HDMI without HDCP is easy to capture as video using products like the BlackMagic Intensity.

DisplayPort

Oh, so very many ports! DisplayPort is aimed at replacing both DVI and HDMI, combining support for greater-than-1080p resolutions with HDCP. It's designed for easy expandability to new display types like 3D. It's used in just a few products, and there isn't a way to capture it yet.

HDCP

DVI devices can but are not required to use HDCP content protection, while HDMI devices always support HDCP. HDCP is a content protection technology that prevents playback if later devices in the single chain could allow an unauthorized capture to take place.

This is a big limitation with DVI to HDMI converters, since often trial and error is required to find out if a particular DVI device supports HDCP.

In some markets, boxes that strip HDCP data from DVI and HDMI are available, but they're illegal to sell in the U.S.

In general, Hollywood content available in digital HD, be it Blu-ray or other future format, will require HDCP if it is going to be displayed digitally. Most other content won't use it; for example, game output from the PlayStation 3 and Xbox 360 never uses HDCP, but playback of premium HD video on those devices does.

FireWire

FireWire (also called IEEE 1394, just 1394, or i.Link by Sony) is a high-speed serial bus protocol developed by Apple. The first version supported up to 400 Mbps. The newer 1394b supports 800 Mbps, with the spec going up to 3200, but uses a different connector.

FireWire was a big part of what made DV so revolutionary, as it supports bidirectional digital transmission of audio, video, and device control information over a single cable. The plug-and-play simplicity was a huge improvement over older formats. A single Beta SP deck uses more than 10 cables to provide the same functions, each of which costs more than that single FireWire cable.

The original 1394 connectors come in two flavors: 4-pin and 6-pin. Most DV cameras require 4-pin connectors, while computers and other 1394-equipped devices use the 6-pin variety. Camera-to-computer connections require a 4-pin-to-8-pin cable. Note the extra two pins on 6-pin cables are used to carry power to devices chained together via 1394. The 1394a 800 MHz version came with a third connector type called "beta" that supports power; however, it hasn't been adopted widely apart from the Mac, and even there mainly for storage.

Lots of video formats use FireWire as the interconnect, including DV25, DV50, DV100, HDV, and D-VHS.

SDI

Serial Digital Interface (SDI) connections transmit uncompressed 720 × 486 (NTSC) or 720 × 576 (PAL) video at either 8-bit or 10-bit per channel resolution at up to 270 Mbps. SDI can handle both component and composite video, as well as four groups of four channels of embedded digital audio. Thus, SDI transfers give you perfect reproduction of the content on the tape. Due to the wonders of Moore's Law, you can now buy an SDI capture board for less than a good component analog card.

HD-SDI

HD-SDI is the primary high-end interconnect these days. It uses the same cabling as SDI, but provides uncompressed video up to 10-bit 4:2:2 up to 1080i30. Thankfully, all flavors support 24p natively.

There's an even higher-end variant called dual-link HD-SDI, which uses two connectors to deliver up to 1080p60, and up to 12-bit 4:4:4 Y′CbCr and RGB. 3G SDI is intended to replace that with a single connection capable of the same.

Unfortunately, consumer electronics are not allowed to include SDI or HD-SDI interfaces as a piracy prevention measure.

Audio Connections

Like most things in the video world, audio often gets the short shrift in the capture department. However, it is critically important to pay attention to audio when targeting compressed video. Audio is sometimes the only element you have a hope of reproducing accurately on small screens or in limited bandwidth. Plus, extraneous noise can dramatically hurt compression, so it's even more important to keep audio captures noise-free.

Unbalanced

Unbalanced analog audio is the format of home audio equipment, where RCA plugs and jacks are the connectors of choice. In audio gear designed for home and project studios, unbalanced signals are often transmitted over cables fitted with 1/4-inch red (right) and white (left) phone plugs. Those audio inputs on computer sound cards are unbalanced mini jacks, like those used for portable headphones. Note mini connectors come in two sizes. It's nearly impossible to see the slight difference in size, but they aren't compatible.

Unbalanced audio cables come in both mono and stereo flavors. Mono mini and 1/4-inch plugs use tip (hot) and ring (ground) wiring. Stereo mini and 1/4-inch plugs use the same configuration, but sport two rings.

You've no doubt noticed that stereo RCA cables use separate wires for left and right channels. Their connectors are generally color-coded red for right and black or white for left.

So what's an unbalanced audio signal? Unbalanced cables run a hot (+) signal and a ground down a single, shielded wire. Such cables have the unfortunate ability to act as antennas, picking up all sorts of noise including 60 Hz hum (50 if you're in a country running 50 Hz power) and RF (radio frequency interference). Shielding is supposed to cut down on the noise, but the longer the unbalanced cable, the more it's likely to behave like an antenna. Keep unbalanced cables as short as possible. And when confronted with the choice, buy shielded cables for audio connections between devices.

Jacks and Plugs

It's easy to confuse these two terms, and they're often incorrectly used interchangeably. So for the record: *plugs* are male connectors. *Jacks* are female connectors that plugs are "plugged" into.

Balanced

Balanced, like unbalanced, sends a single signal over a single cable; however, that cable houses three wires, not two. The signal flows positive on one wire, negative on another, and to ground on the third. I won't burden you with the details, but this design enables balanced wiring to reject noise and interference much, much better than unbalanced.

Balanced connections are highly desirable in electromagnetically noisy environments, such as live production environments or video editing studios. They're also helpful where long cable runs are required.

Balanced audio connections are most often thought of as using XLR connectors, but quarter-inch phone connectors can also be wired to carry balanced signals. XLRs are also called

Cannon connectors after the company that invented them—ITT-Cannon. The company made a small X series connector to which they added a latch. Then a resilient rubber compound was added to the female end, and the XLR connector was born.

All professional video cameras feature balanced audio connectors.

Levels

It's worth noting that matching levels plays an important role in getting the best audio quality. Professional audio equipment typically runs at +4 dBm (sometimes referred to as +4 dBu) and consumer/prosumer gear runs at −10 dBV. Without getting immersed in the voodoo of decibels, suffice it to say that these two levels (though measured against different reference points) represent a difference in voltage that can lead to signals that are either too quiet (and thus noisy) or too loud (and very distorted) if equipment with different levels is used.

S/PDIF

S/PDIF (Sony/Philips Digital InterFace) is a consumer digital audio format. It's used for transmitting compressed or uncompressed audio. Because it's digital, there's no quality reason to avoid it in favor of the professional standard AES/EBU, although that format offers other useful features for audio professionals that don't relate to capture.

S/PDIF cables are coax-terminated in either phone or BNC connectors. For most purposes, S/PDIF and AES/EBU cables are interchangeable. If, however, you're connecting 20- or 24-bit devices, you'll likely want to differentiate the cables by matching their impedance rating to that of the format: 75 ohm for S/PDIF; 110 ohm for AES/EBU.

AES/EBU

AES/EBU, from the Audio Engineering Society and the European Broadcast Union, is the professional digital audio format. Of course, its actual content is no different than that of S/PDIF. The advantage of AES/EBU comes from being able to carry metadata information, such as the timecode, along with the signal. If you're using SDI for video, you're probably using AES/EBU for audio.

AES/EBU 110-ohm impedance is usually terminated with XLR connectors.

Optical

Optical audio connections, under the name TOSLink, are proof of P. T. Barnum's theory that a sucker is born every minute. The marketing behind optical proclaimed because it transmitted the actual optical information off a CD, it would avoid the distortion caused by the D/A conversions. Whatever slight advantage this may have once offered, audio formats like S/PDIF offer perfect quality from the digital source at a much lower cost.

That said, lots of consumer equipment offers only TOSLink for digital audio output.

Frame Sizes and Rates

A compressed bitstream comes in at a particular frame size and rate, of course. But when digitizing from analog or uncompressed to digital, the frame size can be selected.

Capturing Analog SD

For analog, the resolution of the source is explicit vertically, since each line is a discrete unit in the original, and only so many contain visual information. But resolution is undefined horizontally, since each line itself is an analog waveform. So, 640 × 480 and 720 × 480 capture the same number of lines. But 720 × 486 captures six more lines than 720 × 480—each line of the 480 is included in the 486, and six total must be added to the top or bottom to convert from 480 to 486. Capturing 720 pixels per line doesn't mean you're capturing more of each line than in a 640 pixels per line capture—you are just capturing more samples per analog line, and thus have anamorphic pixels. In most analog sources, the 8 pixels left and right are often blank; it's almost always fine to capture as 704 wide instead (which is a smaller area than 720). It really gets down to how big your biggest output frame is going to be.

The most important thing in capturing analog SD is to use component analog or at least S-Video. As mentioned previously, composite cables are death to quality, and coax is worse yet.

Analog SD other than component is always an interlaced signal, even if some of those frames happen to have aligned fields so that it looks progressive.

Plenty of capture cards offer analog preprocessing, like noise reduction, deinterlacing, and cropping. These are great for live encoding. When capturing for later preprocessing, you have to weigh the ease of capturing optimized source with the loss of flexibility in being able to apply more advanced software preprocessing. If you know you're never going to do any per-file tweaking and don't ever want to go back out to 480i/576i, you can preprocess while capturing.

If you don't know what you're going to do, capturing the full 720-width interlaced gives you the maximum flexibility to tweak things later.

Capturing Component Analog

Capturing component analog is generally a lot better, since you don't have the composite quality hit, and the content is often natively progressive. And, of course, it can have more pixels, which are always welcome. While we normally think of component analog in terms of being HD, it also supports SD, including progressive SD resolutions.

Table 5.1 Component Analog Modes.

Common Name	Frame Size	Aspect Ratio	Frames Per Second	Frame Mode	Images Per Second	Bitrate for Uncompressed 10-bit 4:2:2
480i	Typically 640, 704, or 720 wide; height 480 or 486	4:3 or 16:9	29.97 (30/1.001)	Interlaced	59.94	187 Mbps
576i	720, 704, 640 wide, height 576	4:3 or 16:9	25	Interlaced	50	187 Mbps
480p	640, 704, 720 wide; height 480	4:3 or 16:9	59.94 or 60	Progressive	Same as fps	373 Mbps
576p	720, 704, 640 wide; height 576	4:3 or 16:9	50	Progressive	Same as fps	373 Mbps
720p	1280 × 720	16:9	50, 59.94, 50	Progressive	Same as fps?	50p: 829 Mbps 60p: 995 Mpbs
1080i	1920 × 1080	16:9	25/29.97	Interlaced	Same as fps	25i: 933 Mbps 30i: 1120 Mbps
1080p	1920 × 1080	16:9	24, 25, 50, 60	Progressive	Same as fps	24p: 896 Mbps 25p: 933 Mbps 50p: 1866 Mbps 60p: 2239 Mbps

Component analog can be either RGB or Y′CbCr depending on the source device. If the content is originally video, Y′CbCr output may skip a color correction in the player and be more accurate. If it's a game console, RGB is going to be the native color space. Note game consoles often have multiple options for output luma level.

Capturing Digital

A digital capture is more straightforward, since the source frame size, frame rate, and other qualities are known. So an uncompressed capture simply needs to copy the pixel values straight from SDI to disc.

Capturing from Screen

Capturing screen activity from a computer is important in a lot of markets. Screen captures get used often for game demos, PowerPoint presentations, application tutorials, and so on.

Capturing via VGA

There are a variety of ways to capture via VGA. The simplest is to use a VGA to component fan out (an adaptor that goes from the VGA connector's pins to separate component cables), and capture with a component analog device capable of handling component RGB. The biggest limitation to this approach is it only supports the component progressive frame sizes, typically 1920 × 1080 or 1280 × 720. Since few laptops have those native resolutions, this normally requires the demo machine to be reconfigured, which can be impractical in some demo environments.

There are dedicated VGA capture systems like Winnov's CBox 3 and Digital Foundry's TrueHD that are capable of handling arbitrary VGA frame sizes. For historical reasons, VGA may allow full-resolution HD output for copy-protected content in cases where that is now blocked over analog or non-HDCP DVI.

Capturing via DVI or HDMI

As DVI can be easily adapted to HDMI, HDMI capture boards like Black Magic's Intensity can be used to capture computer output, albeit again restricted to standard video sizes. And systems with native HDMI are easier yet. Capturing digitally obviously produces higher-quality results than analog, and should be done when possible. TrueHD is the only available system I'm aware of that can handle DVI capture using all the PC resolution options.

Capturing via screen scraping

Capturing what's on the display from within the computer has long been a popular technique, and is certainly simpler than having to mess with a video device. This has been done with applications like TechSmith's Camtasia on Windows and Ambrosia Software's SnapzProX. This technique is often called "screen scraping," as the apps are basically taking a screenshot periodically, and then saving them with an interframe lossless codec tuned to screen captures.

This continues to work well for lower-motion, lower-resolution content, particularly if there's a simple background and (in Windows) the Classic theme. But once you need full-motion or high frame sizes, or to include natural images or (worse) video on the screen, the RLE-like codecs just can't keep up. And Aero Glass isn't supported at all (Figure 5.3 and Color Figure C.15).

If you're using a classic screen scraper, make sure to use the smallest capture region that works for your image. Ideally, you can capture just an application instead of the whole display. The fewer pixels per second you need, the less of a CPU hit you'll get from the capture process, and the better performance you'll get from what you're capturing.

For that reason, a multicore system is very helpful—the capture process doesn't have to share as much CPU resources with the rest of the system.

Capturing via windows' desktop window manager

Windows Vista introduced a new display architecture—the Desktop Window Manager (DWM). The DWM is used with the Aero Glass theme, enabling the transparencies and other effects. DWM works by using the GPU's 3D engine to handle compositing the image; it essentially turns the user interface into a 3D game. Contrary to rumors, this "eye candy" actually improves performance, by offloading the math required to draw the screen from CPU to GPU.

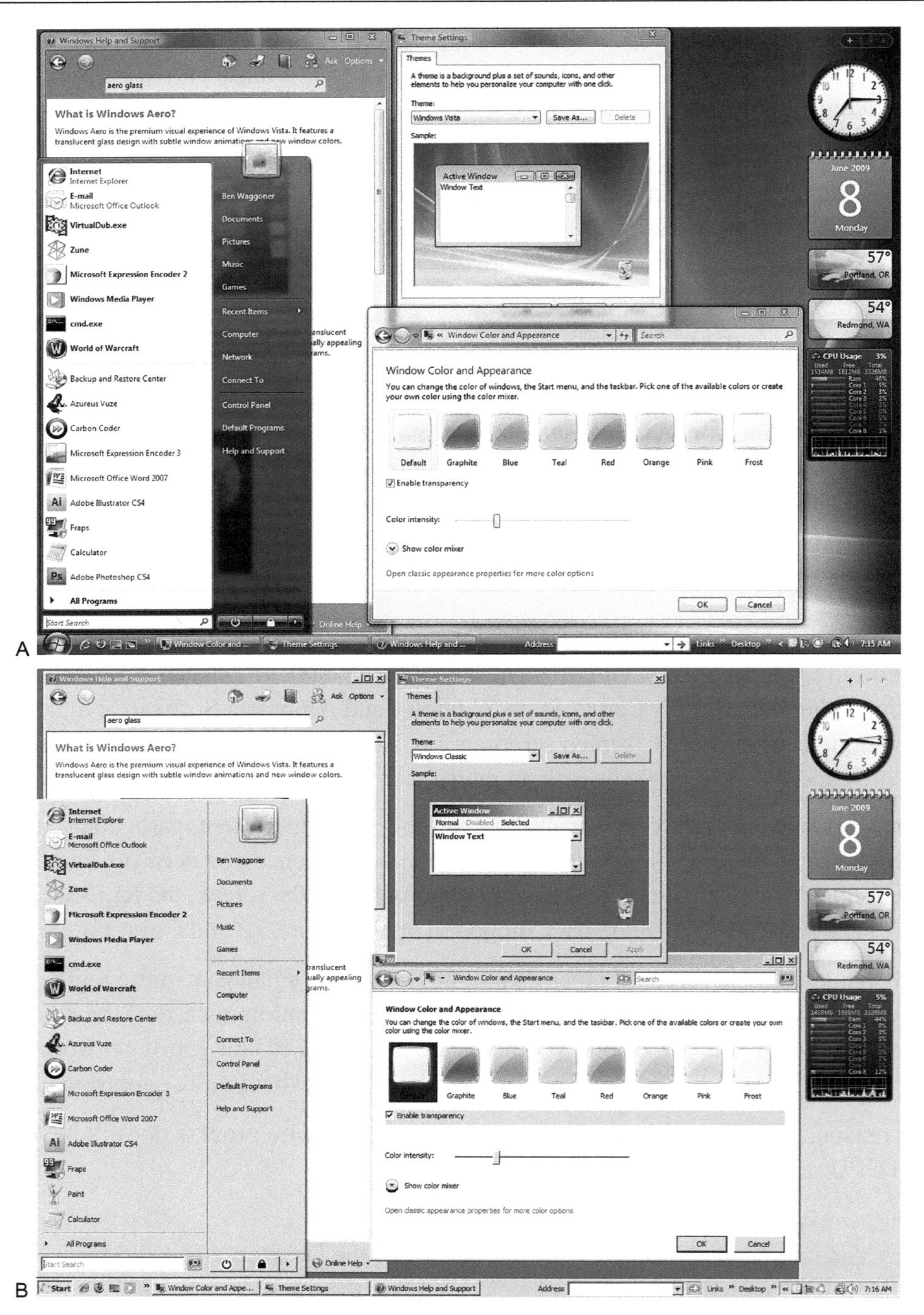

Figure 5.3 (also Color Figure C.15) The same screen rendered in the Aero Glass (5.3A) and Classic (5.3B) themes. Aero Glass looks much nicer, while Classic has lots of flat areas for easy RLE-like compression.

Because it's using the 3D card, the final screen is resident as a "3D surface" on the GPU. We now have modern capture technologies that are capable of grabbing that directly without interfering with the normal operation of the display, allowing for much higher frame rates and frame sizes to work.

FRAPS was the first product to take advantage of this. It was originally created for game captures under Windows XP, and so has a lightweight (but space-hungry) codec that handles very high motion with aplomb. I've captured up to 2560 × 1600p60 with FRAPS (albeit on an 8-core system).

Microsoft's Expression Encoder 3 includes a new Expression Encoder Screen Capture (EESC) application, including DWM capture. This uses a new codec that is extremely efficient for screen activity, with a controllable space/quality level. EESC can simultaneously capture screen activity and a live video input, to enable synchronized recording of a presenter and presentation.

Capturing via MacOS's quartz

Another approach to capturing screen activity is to actually capture the commands being sent to the display manager, and then be able to play those back. This is the technique used by Telestream's ScreenFlow for Mac OS X, which captures the input of the Mac's Quartz rendering engine. While it's not exactly a codec, it basically makes a movie out of all the commands sent to the screen, and can convert that to an editable or compressible file after the fact. It's cool stuff.

Capture Codecs

Uncompressed

Uncompressed is the simplest to capture, although the most expensive in terms of space. But even uncompressed video might not always be exactly the same as the source. For example, 8-bit 4:2:2 can be captured from 10-bit 4:2:2. Some hardware, like AJA's capture boards, have a nice 10-bit to 8-bit dither, and so it can make sense to do that conversion on capture if no image processing is planned and there isn't a good dither mode later in the pipeline.

DV

I'm not a fan of digitizing to DV25. 4:1:1 color subsampling is just too painful. There are lots of cheap composite/S-Video-to-DV converter boxes, but I don't consider them capable of production-quality video.

Motion JPEG

Motion JPEG and its many variants (like the Media 100 and Avid AVR codecs) were popular for intraframe encoding in NLEs for many years, and are still supported in many cards. They give decent lightweight compression; the main drawback is that they're limited to 8-bit.

DN x HD

Avid's DNxHD is another DCT intraframe 4:2:2 codec offering good quality and speed, which is well-suited to real-time capture. I prefer to use the higher 220 Mbps mode, which can provide 10-bit and doesn't use horizontal subsampling. DNxHD is being standardized as VC-3. Free decoders are available for Mac and Windows.

ProRes

Similar to DNxHD, Apple's ProRes codec is also a 4:2:2 intraframe DCT codec. Its video modes have a maximum bitrate of 220 Mbps with 10-bit support, with a new 12-bit 4:4:4 mode added as of Final Cut 7. Encoding for ProRes is available only for the Mac, although there's a (slow) decoder available for Windows.

Cineform

Cineform is a vendor of digital video workflow products, centered around their wavelet-based Cineform codec. The codec itself is available in a variety of flavors, and can handle RGB/Y′CbCr, up to 4 K, 3D, up to 16-bit, mathematically or visually lossless, and a variety of other modes. There is a free cross-platform decoder, and the spec for the bitstream has been published.

Their Prospect plug-ins for Adobe Premiere heavily leverages wavelet sub-banding for fast effects and previews. It's long been popular as an intermediate codec for conversion from HDV and AVCHD sources.

Huffyuv

Huffyuv is a lightweight, mathematically lossless codec for Windows, using Huffman encoding. It's easily fast enough to use in real-time capture. It offers 8-bit in RGB and 4:2:2.

Lagarith

Lagarith is an updated version of Huffyuv including multithreading and arithmetic coding. It typically offers about 30 percent lower bitrates than Huffyuv, but requires somewhat more CPU power to encode and decode.

FFV1

FFV1 is a new lossless open-source codec somewhere between Huffyuv and Lagarith in compression efficiency and decode performance. It's included in ffmpeg, and so available for many more platforms than the Windows-only Lagarith.

MPEG-2

There's a wide variety of capture technologies based around MPEG-2. These can be High Profile (4:2:2) or Main (4:2:0), interframe- or intraframe-coded, and can use a variety of different wrappers. Most MXF files are compressed with MPEG-2.

An intraframe coded MPEG-2 is quite similar to all the other DCT codecs.

AVC-Intra

AVC-Intra is an attempt to let H.264 provide similar capture and post functionality to I-Frame MPEG-2, with more efficient encoding. It supports 10-bit luma, and has two standard modes:

- 50 Mbps with 75 percent anamorphic scaling and 4:2:0 coding
- 100 Mbps without anamorphic scaling and 4:2:2 coding

AVC-Intra can produce high quality, certainly, although it's somewhat slower to decode than older codecs. The 100 Mbps variant uses the lighter-weight CAVLC entropy encoding.

Data Rates for Capture

One critical issue with all capture is the data rate. With a fixed codec like DV25, it's easy: DV is locked in at 25 Mbps. And capturing uncompressed is easy: bits per pixel × height × width × frame rate. But other codecs can have a variety of configurable options, controlling bitrate directly or by a quality level (typically mapping to QP). Fixed bitrates made sense when storage speed was a big limitation of performance, but that's much less of an issue these days of fast, cheap RAID. Even with a fixed bitrate, the actual output data rate may be less if some frames are very easy to compress, like black frames. This is a good thing, as it's just free extra capacity.

The simplest thing to do if you don't care about file size is to just use the highest available data rate/quality combination and call it a day. If you're trying to find a balance between good enough quality and efficient storage, capture a few of your favorite, challenging clips at a variety of bitrates, and compare them on a good monitor. Pick the lowest bitrate/quality combination where you can't see any difference between that and the source.

The three limiting factors in data rate are drive space, drive speed, and capture codec. Drive space is self-explanatory, and since we discuss capture codecs at length elsewhere, we don't need to look at them here. But let's look at drive speed.

Drive Speed

If access to the hard drive isn't fast enough to capture all the video, frames may be dropped. This can be a serious, jarring quality problem. Always make sure your capture program is set to alert you when frames have been dropped.

If you have decent drives in a RAID configuration, even uncompressed 1080i is pretty easy these days. A few rules of thumb:

- Video goes down in long sequences, so rpm doesn't matter. Don't pay extra for low-capacity 15000 rpm drives; 7200 is cheaper and offers better capacity.

- Solid state drives are great for netbooks, but they're not set up for the sustained writes of video capture today. They might work for DV25, but won't give anything close to the write throughput of a good HD RAID.
- Don't let your drives get too full (under 80 percent full is a good rule), and defragment them occasionally.
- If RAID seems expensive or complex, know that my first RAID cost me $6,000, was 4 GB, and couldn't do more than about 50 Mbps sustained.

Postproduction

Postproduction (or just "post") is where the raw footage gets turned into finished video, through editing, the addition of effects and motion graphics, voice-overs, audio sweetening, and so on. Compared to production, postproduction is marvelously flexible—with enough time and big enough budget, a project can be tweaked endlessly. This can include the luxury of test encodes to locate and adjust trouble spots.

Again, classic editorial styles and values prove to be good for compression as well. And as a general matter of craft, it's good to match the production techniques and parameters in post where possible.

Postproduction Tips

Use antialiasing

Antialiasing is the process of softening the edges of rendered elements, like text, so they don't appear jagged. By eliminating those jagged edges, antialiasing looks better when compressed. Most apps these days either have antialiasing turned on by default or make it an option, often as part of a high-quality mode. If importing graphics made in another app, such as Photoshop, make sure that those files are antialiased as well. See Figure 5.4.

Use motion blur

Earlier, you learned how codecs spend bandwidth on complex images and motion, and hence the most difficult content to compress is rapidly moving complex images. Shooting on film or video with a slower shutter speed gives us motion blur, a great natural way to reduce this problem. With motion blur, static elements are sharp and moving elements become soft or blurred. Thus, there are never sharp *and* moving images at the same time. If you're importing graphics or animation from a tool such as After Effects, make sure those images use motion blur. If you're compositing graphics over video, the graphics' motion blur value should match the shutter speed used in the camera. Beyond compressing well, it also looks much more natural; lots of otherwise good animation or compositing has been ruined by a mismatch in motion blur between foreground and background.

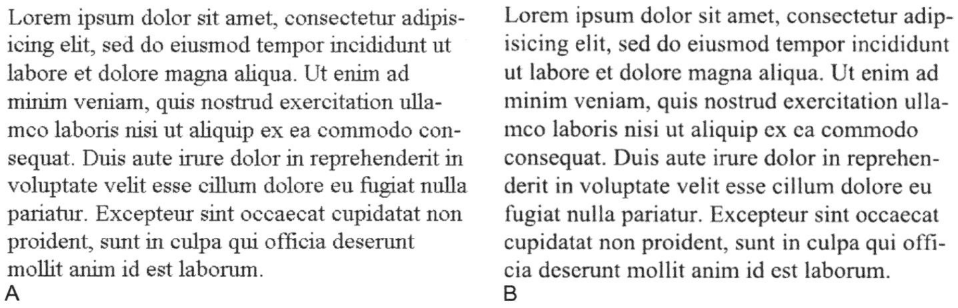

Lorem ipsum dolor sit amet, consectetur adipisicing elit, sed do eiusmod tempor incididunt ut labore et dolore magna aliqua. Ut enim ad minim veniam, quis nostrud exercitation ullamco laboris nisi ut aliquip ex ea commodo consequat. Duis aute irure dolor in reprehenderit in voluptate velit esse cillum dolore eu fugiat nulla pariatur. Excepteur sint occaecat cupidatat non proident, sunt in culpa qui officia deserunt mollit anim id est laborum.
A

Lorem ipsum dolor sit amet, consectetur adipisicing elit, sed do eiusmod tempor incididunt ut labore et dolore magna aliqua. Ut enim ad minim veniam, quis nostrud exercitation ullamco laboris nisi ut aliquip ex ea commodo consequat. Duis aute irure dolor in reprehenderit in voluptate velit esse cillum dolore eu fugiat nulla pariatur. Excepteur sint occaecat cupidatat non proident, sunt in culpa qui officia deserunt mollit anim id est laborum.
B

Figure 5.4 The same text encoded to a 30 K JPEG, one without anti – aliasing (5.4A, left), the second with it (5.4B, right).

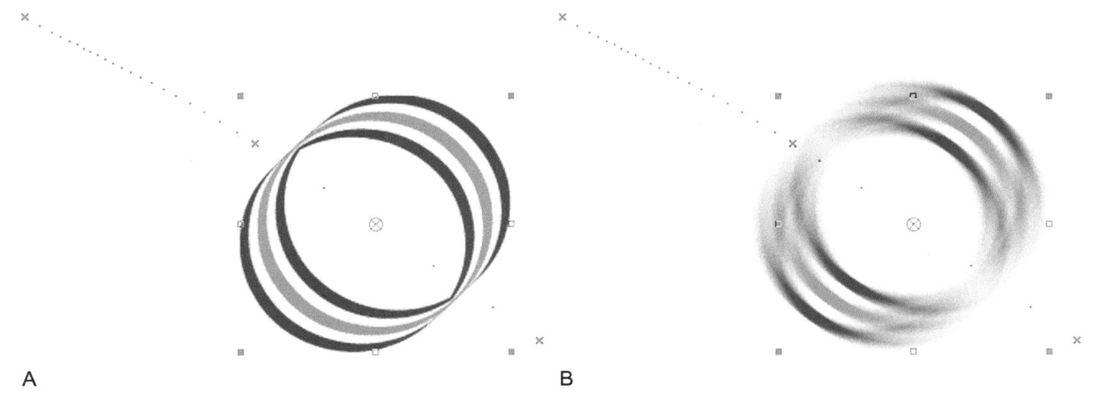

Figure 5.5 Starting with the image in 5.5A, rendered motion blur (5.5B, right) should look just like that from a camera, with blurring along the axis of motion (seen here as the dotted line).

Some motion blur effects spread one frame's motion out over several frames or simply blend several frames together. Avoid these—you want natural, subtle motion blur without obvious banding. The NLE may have a configurable pattern for how many samples to run during motion blur. Higher values are slower to render, but reducing banding/mirroring. See Figure 5.5 for a low-banding example.

Render in the highest-quality mode

By default, you should always render at highest quality, but it's doubly important to render any effects and motion graphics in the highest-quality mode when you're going to deliver compressed footage. Depending on the application, the highest-quality rendering mode activates a number of important features, including motion blur and antialiasing. This removes erroneous high-frequency data; softening the image where it should be soft, leaving more bits to be spent on the real image.

If you're creating comps in a faster mode, don't make any decisions based on test compressions made with those comps—the quality can be substantially worse than you'll get from the final, full-quality render.

Avoid rapid cutting

At each edit, the change in the video content requires a new I-frame to account for the big change in the content, starting a new GOP. A cut every few seconds is generally okay at moderate or higher bitrates, but a Michael Bay–style, hyperkinetic montage with an edit every few frames can be a codec killer, and turn into a blurry or blocky mess.

Avoid cross-dissolves

A cross-fade is one of the most difficult kinds of content to compress, as it's a blend of two different frames, each with their own motion, and with the intensity of each changing frame-to-frame. This keeps motion estimation from being as helpful as normal; some codecs can wind up using many I-frames during a cross-dissolve. It's much more efficient to use hard cuts. Fades to/from black are only slightly easier to compress than a cross-fade, and should be avoided if possible. Wipes and pushes work quite well technically, but aren't appropriate stylistically for many projects. A straight cut is often your best bet.

Use codec-appropriate fades

A fade to/from black or white may seem like the simplest thing in the world to encode, but the changing intensity is very hard on traditional DCT motion estimation. Older codecs like MPEG-2 or MPEG-4 part 2 can require a bitrate spike or yield lower quality during these fades. VC-1 and H.264 have specific fade compensation modes that make this much easier, although implementations vary quite a bit in how well they handle fades.

No complex motion graphics

Complex motion graphics, where multiple onscreen elements change simultaneously, are extremely difficult to encode. Try to make each element as visually simple as possible, and move only things that you have to. Flat colors and smooth gradients compress more easily than textures. Two-dimensional motion is very easy for motion estimation; if you want to have complex moving patterns, take complex shapes and move them around on the x and y planes without touching the z axis. And always remember motion blur.

Don't rotate

Codecs don't rotate as part of motion estimation, so that animated radial blur can be rather problematic, as would be that tumbling 3D logo.

Use simple backgrounds

Keeping the background elements simple helps with compression. The two things to avoid in backgrounds are detail and motion. A static, smooth background won't be a problem. A cool particle effects background will be. Motion blur can be your friend if you want to have background motion without eating a lot of bits. See Figure 5.6.

A

B

Figure 5.6 A complex background (5.6A) sucks bits away from the foreground, reducing quality of the important parts of the image compared to a simple background (5.6B).

Lorem ipsum dolor sit amet, consectetur adipisicing elit, sed do eiusmod tempor incididunt ut labore et dolore magna aliqua. Ut enim ad minim veniam, quis nostrud exercitation ullamco laboris nisi ut aliquip ex ea commodo consequat. Duis aute irure dolor in reprehenderit in voluptate velit esse cillum dolore eu fugiat nulla pariatur. Excepteur sint occaecat cupidatat non proident, sunt in culpa qui officia deserunt mollit anim id est laborum.

A

Lorem ipsum dolor sit amet, consectetur adipisicing elit, sed do eiusmod tempor incididunt ut labore et dolore magna aliqua. Ut enim ad minim veniam, quis nostrud exercitation ullamco laboris nisi ut aliquip ex ea commodo consequat. Duis aute irure dolor in reprehenderit in voluptate velit esse cillum dolore eu fugiat nulla pariatur. Excepteur sint occaecat cupidatat non proident, sunt in culpa qui officia deserunt mollit anim id est laborum.

B

Figure 5.7 A thick, simple font (5.7A) survives a lot better than a complex, thin one (5.7B) when encoded to the same data rate.

No fine print

The sharp edges and details of small text are difficult to compress, especially if the video will be scaled down to smaller resolutions. Antialiasing can help preserve the legibility of text, but it can't work miracles if the text is scaled down too much. For best results, use a nice thick sans serif font. See Figure 5.7.

Watch that frame rate

Animators often work at whole number frame rates, like 24 and 30. But video formats run at the NTSC versions which are really 24/1.001 (23.976) and 30/1.01 (29.97). Make sure that everyone's using the exact same frame rate for assets, so no unfortunate mismatches happen.

PAL rates are the nice round numbers of 25 and 50, and so should not have a 1.001 correction applied.

Know and stick to your color space

This is last not because it's least important, but because I want to make sure that anyone skimming finds it.

As I mentioned back when discussing sampling and quantization, there are a few variants in formulas used for the Cb and Cr for converting between RGB and Y′CbCr, codified in Recommendations 601 and 709 from the International Telecommunications Union (ITU). Specifically, there are three:

- SMPTE C (NTSC [less Japan] Rec. 601)
- EBU (Japan and PAL Rec. 601)
- Rec. 709 (HD)

Mathematically, 709 is right between SMPTE C and EBU, making a conversion between them straightforward. And the luma calculations are the same either way. But you can get a slight but perceptual color shift if 601 is decoded as 709 or vice versa. One hit of that shift isn't that noticeable that often, but it's enough to cause a noticeable seam in compositing.

More problematic is multigenerational processing that winds up applying multiple generations of wrong color correction. For example, using a tool that exports to 709 correctly, but assumes source files are 601, and so each round of processing causes another 601 to 709 color change. See Color Figure 16 for the effects of incorrect color space conversion.

Getting this right is normally the simple matter of making sure that each file is tagged correctly on import and that the export color mode is specified.

If a workflow makes it hard to track what particular assets are, appending a "_709" or a "_601" to the filename is a handy shorthand.

CHAPTER 6

Preprocessing

Preprocessing encompasses everything done to the original source between the time it's handed to you and the time you hand it off to a codec in its final color space. Typical preprocessing operations include cropping, scaling, deinterlacing, aspect ratio correction, noise reduction, image level adjustment, and audio normalization. It's during preprocessing that whatever the source is becomes optimized for its delivery target.

Some projects don't need preprocessing at all; if the source is already 640 × 480 4:2:0, and that's what it'll be encoded to, there's nothing that needs to be done. Other times the preprocessing is implicit in the process; converting from DV to DVD requires a conversion from 4:1:1 to 4:2:2, but that's done silently in most workflows. And in still other instances (generally for challenging sources needed for critical projects), preprocessing is a big operation, involving dedicated tools and maybe even some plug-ins.

The more experience I get as a compressionist, the more and more preprocessing is where my keyboard-and-mouse time goes. In the end, codecs are a lot more straightforward and scientific; once you know the technical constraints for a particular project and the optimum settings to get there, there's only so much room for further tuning.

But preprocessing is as much art as science; one can always spend longer to make things ever so slightly better. I might be able to get 80 percent of the way to perfect in 15 seconds with a few checkboxes, and then spend another week getting another 18 percent, with that last gap from 98 percent to 100 percent just out of reach.

And for those hard, important projects, preprocessing can actually take a whole lot more CPU time than the final encode will. Give me some noisy 1080i HDV source footage that'll be seen by a million people, and my basement office will be kept nice and toasty all weekend with many CPU cores going full blast.

For a quick sample of how preprocessing matters, check out Color Figure C.18. You'll see good preprocessing at 800 Kbps can soundly beat 1000 Kbps without preprocessing.

General Principles of Preprocessing

Sweat Interlaced Sources

If you're not compressing for optical disc or broadcast, you'll probably need to compress as progressive. If you have interlaced source, you've signed up for a lot of potential pain in massaging that into a high-quality progressive output. So when you see fields in your source, make sure to budget for that in your time and quality expectations.

Use Every Pixel

You want every pixel that'll be seen to include real video content you want to deliver. You want to make sure anything in your source that's not image (letterboxing, edge noise, tearing, Line 21) doesn't make it into the final output frame.

Only Scale Down

Whenever you increase height or width more than the source, you're making up visual data that doesn't actually exist, so there's mathematically no point. Avoid if possible.

Mind Your Aspect Ratio

Squares shouldn't be rectangles and circles shouldn't be ovals after encoding. Make sure you're getting the same shapes on the output screen as in the source.

Divisible by 16

For the most efficient encoding with most codecs, the final encoded frame's height and width should be divisible by 16 (also called "Mod16"). The cropping, scaling, and aspect ratio may need to be nudged to get there. Some codecs like MPEG-1 and often MPEG-2 require Mod16.

Err on the Side of Softness

Lots of preprocessing filters can add extra erroneous details, like scaling artifacts and edges from deinterlacing. That's basically adding incorrect high frequency details that'll suck bits away from your image. At every step, you want to avoid adding wrong detail, and if the choice is between too sharp and too soft, be soft and make things easier for the codec.

Make It Look Good Before It Hits the Codec

Hopefully you're using a tool that gives a way to preview preprocessing so you can verify that the output looks good before it hits the codec. Preview before you hit encode.
It's frustrating to get problems in the output and not know whether they're coming from source, preprocessing, or compression.

Think About Those First and Last Frames

Lots of players show the first frame for a bit while they queue up the rest of the video, and then end with the last frame paused on the screen. Having something interesting in the first and last frames is a valuable chance to communicate, or at least not bore.

Decoding

Some older codecs have more than one decoder mode. Perhaps the most common of these is Apple's DV codec in QuickTime. Each QuickTime video track has a "High Quality" flag that can be on or off. That parameter is off by default in most files, and ignored by most codecs other than DV. But if the DV codec is left with HQ off, it makes for a lousy-looking single field half-size decode (so 360 × 240 for NTSC and 360 × 288 for PAL). Most compression tools will automatically force the HQ flag on for all QuickTime sources, but if you are seeing poor quality from DV source, you need to go in and set it manually. This can be done via QuickTime Pro.

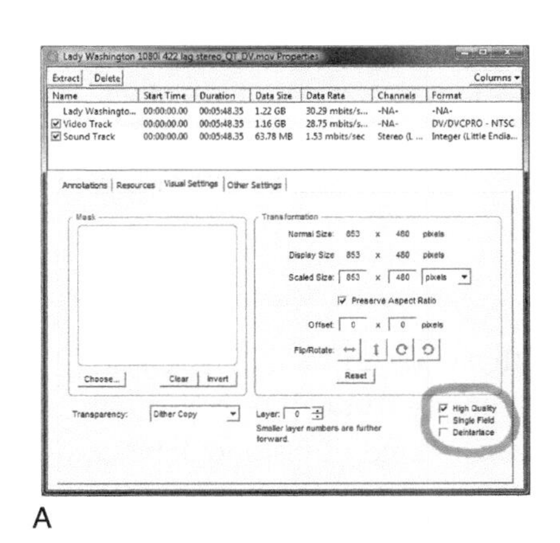

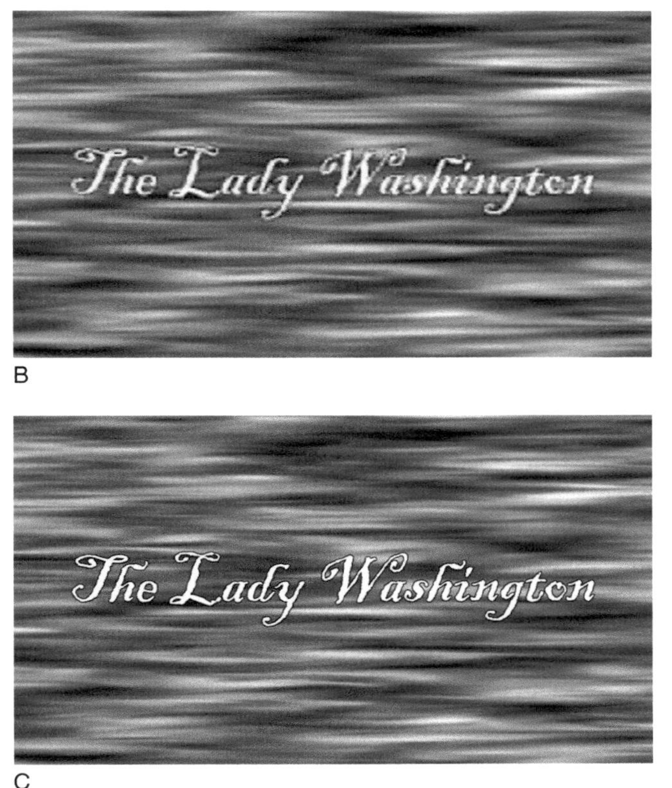

Figure 6.1 Here's the magic checkbox (6.1A) that you need to hit to make QuickTime's DV codec look good (only available in QuickTime Pro). The quality difference is dramatic; 6.2B shows the before, and 6.1C shows the after.

In most cases, we want sources to be uncompressed or in a production grade codec. However, those files can be far too big for easy transport via FTP, and so more highly compressed files are often provided. These are called mezzanine files, and should be encoded at high enough bitrates that they don't have visible artifacts. But that's not always the case.

If your source uses lossy compression, there's the potential of artifacts. Blocking and ringing in the source are false details that'll suck away bits from the real detail. When source and encode are macroblock-aligned (e.g., if no cropping or scaling was done, so the 8 × 8 block edges are on the exact same pixels in source and output) the artifacts are not as bad, since the detail is between blocks where it won't mess up any particular DCT matrix. But when resizing loses macroblock alignment, those ugly sharp (high-frequency) edges land right in the middle of blocks in the new codec, really hurting compression. Worse, those artifacts have a fixed position, ignoring the actual motion in the source. So they also confuse motion estimation, due to a mismatch of static and moving pixels inside each block. Nightmare. Generally, any source that has visible compression artifacts is going to be significantly harder to encode than a cleaner version of the same source, and will wind up with even more compression artifacts after encoding.

The first thing to do is try to fix whatever is causing artifacts in the source. Way too many video editors don't know how to make a good mezzanine file, and just use some inappropriate web bitrate preset from their editing tool.

But if the only source you have has artifacts, you've got to make the best of it.

While there are some decent (if slow) filters that can be applied to video after decoding to clean up DCT blocks, it's much more effective and much faster to apply deblocking in the decoder. Since the decoder knows how compressed the video is and exactly where block boundaries are, it can make a much quicker and more accurate determination of what's an artifact and what's not.

MPEG-2

MPEG-2 is by far the most common source format I get with source artifacts. There's so much MPEG-2 content out there that's blocky, whether it's coming from HDV, lower bitrate XDCAM, DVD, ATSC, or DVB. And MPEG-2 doesn't have any kind of in-loop deblocking filter to soften edges.

There's a bunch of MPEG-2 decoders out there, of course. Unfortunately, the consumer-focused decoders tend to be the ones with integrated postprocessing. Those bundled into compression tools never seem to, darn it. Hopefully we'll see compression products take advantage of the built-in decoders.

When I get problematic MPEG-2, I tend to go to AVISynth using DGMPGDec, which includes a decoder .dll (DGDecode) and an indexing app (DGIndex) that makes a index file required for DGDecode from the source. It's complex, but it works. If I have source

Figure 6.2 High – motion HDV source with deblocking decoding turned on (6.2A) and off (6.2B). The deblocked version will be much easier to encode.

with visible blocking, I go ahead and turn on cpu = 4 mode, which applies horizontal and vertical deblocking, and if it's bad enough to have obvious ringing, I'll use cpu = 6, which adds deranging to luma and chroma. It can be a little slow if trying to play full HD, but is well worth it for decoding. Using idct = 5 specifies IEEE 64-bit floating-point mode for the inverse DCT, minimizing the chance of decode drift at the sacrifice of some speed. Here's an example script that will deblock a HDV file:

```
# Created by AVSEdit
# Ben Waggoner 6/8/2009
LoadPlugin("C:\Program Files\AviSynth 2.5\plugins\DGDecode.dll")
MPEG2Source("Hawi_1920 × 1080i30_HDV.d2v",cpu = 4,idct = 5)
```

DGDecode also includes a filter called BlindPP filter that will deblock and dering already decoded frames. It's not as accurate, but can be quite useful in a pinch.

VC-1

Windows Media's VC-1 is a more efficient codec than MPEG-2 and has an in-loop deblocking filter available that helps somewhat (but used to be automatically turned off for encodes beyond 512 × 384 or so in size). When VC-1 is used as a mezzanine file, it's typically at a 1-pass VBR quality level high enough to be clean.

But if not, Microsoft's built-in WMV/VC-1 decoders will automatically apply deblocking and deranging during playback as needed for lower bitrate content, when sufficient extra CPU power is available (Figure 6.3). Whether this gets turned on or not when doing file-to-file transcoding can be unpredictable. Fortunately, there is a registry key that turns on full

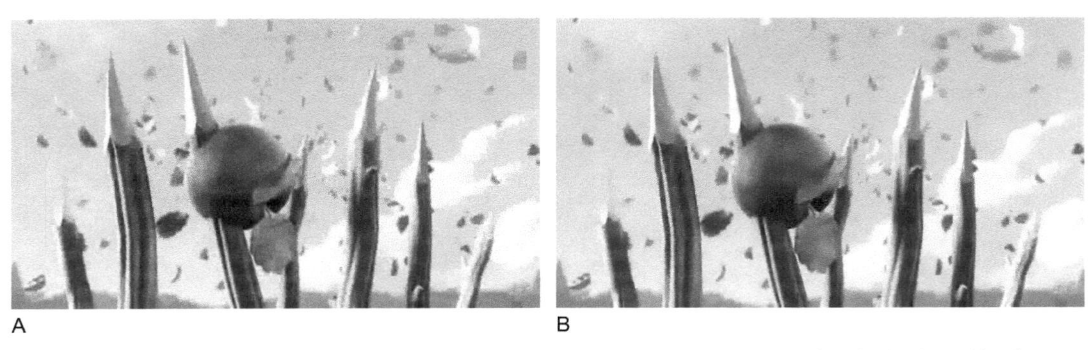

Figure 6.3 WMV source files encoded in recent years shouldn't have this kind of quality issue, but if you get some, postprocessing with strong deblocking and deringing can be a big help.

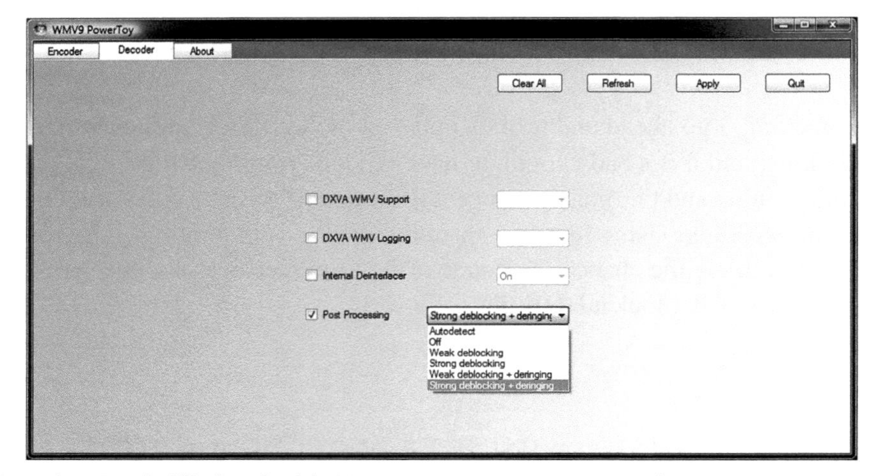

Figure 6.4 Alex Zambelli's invaluable WMV PowerToy, set to maximum VC-1 postprocessing.

deblocking and deringing for all WMV/VC-1 decoding. Since it's an adaptive filter, it won't soften the image when frames are high quality, so it's safe to leave on for all transcoding. The only drawback is that it will make playback somewhat more CPU-intensive.

The simplest way to set the key is to use WMV 9 PowerToy (Figure 6.3). I recommend at least strong deblocking it be left on anytime you're trying to transcode from WMV or VC-1 files, with deringing used as needed.

H.264

The good/bad news is that H.264's strong in-loop deblocking filter generally means it's had about as much filtering as it can handle. Generally, H.264 gets soft as it gets more compressed. Thus there's not much that can be done on playback to improve H.264. It'll take general noise reduction and other image processing steps described later in this chapter.

Color Space Conversion

601/709

Ideally, making sure that your file is flagged correctly with the same color space as the source will take care of things. However, many devices ignore the color space flag in the file and always decode as one or the other, or else base it on frame size.

If there's a difference in 601/709 from your source and what you need in the output, you should make sure that conversion is applied if color is critical to the project. Some compression tools like ProCoder/Carbon include a specific conversion filter for this, as do all the video editing tools. And if the source is RGB, you want to make sure it's converted to the correct color space as part of RBG > Y′CbCr Conversion.

Refer to Color Figure C.16 for the risks of poor 601/709 conversion.

Chroma Subsampling

When converting from higher subsampling like 4:2:2 to 4:2:0, a simple averaging is applied almost universally and almost universally works well. It's when you need to interpolate on one axis that potential quality issues can intrude. The classic case of this is coming from NTSC DV 4:1:1 to 4:2:0, particularly for DVD.

There's no way around the fact that 4:1:1 doesn't provide enough detail for the sharp edge of a saturated color. But when the 4-pixel wide chroma sample's value is simply copied twice into two 2-pixel wide sample, you get aliasing as well, with a sharp edge. This is essentially using a nearest neighbor scale on the chroma channel. The simplest thing to do is use a DV decoder that integrates a decent interpolation filter, easing the transition between adjacent samples. It doesn't get back the missing color detail, but at least it isn't adding bad high-frequency data on top of that.

Dithering

When converting from 10-bit or higher sources down to 8-bit, you've got a lot more luma resolution to use to determine what's signal and what's noise. But in the end, you're going to take your 64–940 (10-bit Y′CbCr) or 0–255 (RGB) range and get a 16–235 range. And you want a few particular characteristics in that output:

- The histogram should match the pattern of the source; no spikes or valleys due to different amounts of input values mapping to particular output values.
- There shouldn't be any banding in output that's not present in the source.

The key process here is called dithering or random error diffusion. The core concept is that there's not a perfect way to translate the luma values. If one simply truncates the two least-significant bits of a 10-bit value, you'll see $Y' = 64$ turn into $Y' = 16$, but $Y' = 62$, $Y' = 63$, and $Y' = 65$ become $Y' = 16$ too. And thus, what was smooth in the source can wind up banded with a sharp boundary between $Y' = 16$ and $Y' = 17$ in the output.

This truncation is probably the worst approach. $Y' = 65$ should really come out as $Y' = 16.25$; but have to end with a whole number. Instead we could randomly assign 75% of pixels of $Y' = 65$ to become $Y' = 16$ and 25% $Y' = 65$ pixels to $Y' = 17$. And instead of a big region of $Y' = 16$ with an obvious seam between it and a big region of $Y' = 17$, we get a slowly changing mix of $Y' = 16$ and $Y' = 17$ across the boundary. Big improvement! There's lots of more advanced ways to do dither than this process, but that's the basic idea. Though adding a little randomness may seem like it would hurt compression, it's very slight compared to even moderate film grain.

Unfortunately, very few compression tools include dithering at all. This is generally the province of high-end Blu-ray/DVD encoder tools; Inlet's high-end Fathom and Armada products are the only I know of that can use good dithering with web formats. And for once AVISynth is of no help to us; it's built from the ground up as an 8-bit-per-channel technology. So, for my readers who work for compression tool vendors, get on it!

And if you do, may I suggest Floyd-Steinberg dithering with a randomized pattern? That seems to do the best job I've seen of creating as little excess noise as needed while providing a uniform texture across the frame, and is the secret behind many of the best-looking HD DVD and Blu-ray titles.

See Color Figure C.19 for a demonstration of what dithering can accomplish.

Deinterlacing and Inverse Telecine

As you learned in Chapter 5, there's way too much interlaced video in the world. This means that the even and the odd lines of video were captured 1/59.94th of a second (or 1/50th in PAL) apart. If you're going from interlaced source to an interlaced display format without changing the frame size, you should leave the video interlaced if your target playback device is DVD, broadcast, or any other natively interlaced format. However, most other delivery formats either require or support progressive content.

Depending on the nature of your source, you will need to do one of four things to preprocess your footage for compression—deinterlace, bob, inverse telecine, or nothing. Failure to deinterlace an interlaced frame that is then encoded as progressive results in severe artifacts for all moving objects, where a doubled image is made of alternating lines. This looks really bad on its own merits, and it then hurts compression in two ways. First, the high detail in the interlaced area will use a whole bunch of your bits, starving the actual content. Second, moving objects split into two and then merge back into one as they start and stop moving, wreaking havoc with motion estimation. Failure to deinterlace when needed is among the worst things you can for your video quality.

Deinterlacing

You can tell if content was originally shot on video, because every frame with motion in it will show interlacing in the moving parts. Samples of the various deinterlacing modes can be found in Color Figure C.20.

Field elimination

The most basic form of deinterlacing simply throws out one of the two fields. This is very fast, and is often used by default on low-end products. The drawback is that half of the resolution is simply thrown away. For 720 × 480 source, this means you're really starting with just a 720 × 240 frame. Throw in a little cropping, and you'll actually have to scale up vertically to make a 320 × 240 image, even though you're still scaling down horizontally. This produces a stair-stepping effect on sharp edges, especially diagonals, when going to higher resolutions. Still, it's a whole lot better than not deinterlacing at all.

Spatial adaptive

A spatial adaptive deinterlacer compares a given pixel to the pixels around it to see if there's a better match between the neighboring area in the other field (the lines above and below) versus its own field (the lines two above and two below). If the neighboring area is at least a good match, that part of the image probably isn't moving, and so can be deinterlaced. The net effect is that static parts of the image are left alone and moving parts are deinterlaced. So where you can easily see detail, it's there, and when the jagged edges come in, it's where motion would mask it at least somewhat.

Spatial adaptive used to be significantly slower than a basic deinterlace, but it's not a significant hit on modern PCs. It's really the minimum needed for a half-way decent compression tool.

Motion adaptive

A motion adaptive deinterlacer compares the current frame with a past and future frame to get a better sense of where motion is taking place, and is essentially an improved spatial adaptive. It can be slower to preview, as multiple frames need to be decoded to show any particular one. But it's not much worse for encoding when the input and output frame rates are the same, since all frames need to be decoded anyway.

Motion search adaptive

A motion search adaptive deinterlacer tracks the motion between frames. This enables it to reassemble the even and odd lines of moving objects, much improving their edges.

Unfortunately, motion search adaptive deinterlacing is rarely seen in mainstream compression tools. A big part of this is that it's really slow, particularly with HD. It can take longer to deinterlace than to actually encode the video. It's possible with several tools:

- Compressor, using the Advanced Image Processing pane.
- Episode, via Create new Fields By: Motion Compensation.
- AVISynth, with a variety of plug-ins. I'm a fan of the EEDI2 and TIVTC combination.

Blend (a.k.a. "I hate my video, and want to punish it")

One technique I normally avoid is a "blend" deinterlace. These simply blur the two fields together. This blending eliminates the fine horizontal lines of deinterlacing, but leaves different parts of the image still overlapping wherever there's any motion. And it softens the image even where there isn't motion. Though better than no deinterlacing at all, this blending makes motion estimation very difficult for the codec, substantially reducing compression efficiency. Plus, it looks dumb. Please don't do it.

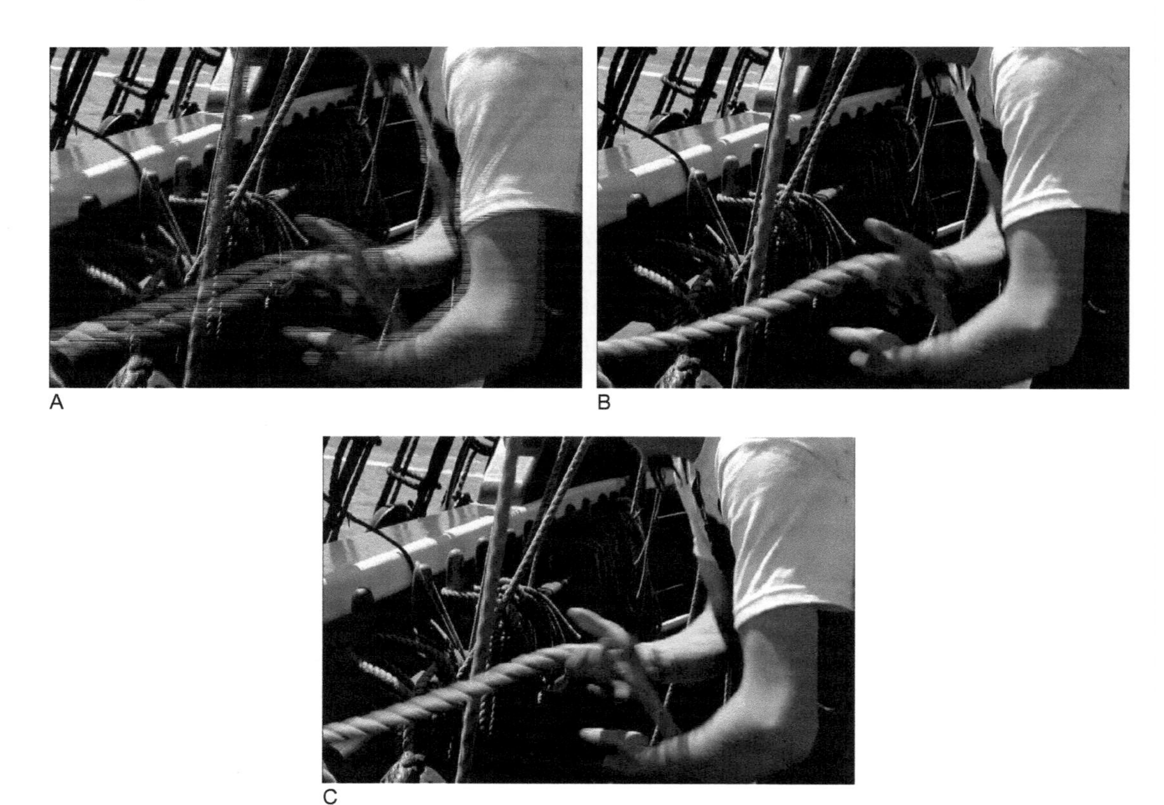

Figure 6.5 Bob deinterlacing turns each field from an interlaced frame into individual progressive frames.

Bob

In normal deinterlacing, we take an interlaced frame in and get a progressive frame out, eliminating any of the "off" field with motion that doesn't match. With a bob deinterlace, we generate one progressive frame out of each field of the source; so 25i to 50p or 30i to 60p. It's called "bob" because the up-and-down adjustment applied to where the active field lands in each output frame looks like something bobbing up and down, (and isn't named after some legendary compression engineer named Robert).

When bitrate isn't of paramount concern, this is the best way to preserve high-action content, particularly sports. And doubling framerate doesn't require doubling bitrate; since the closer the frames are in time the less has changed between them, and thus fewer bits are needed to encode the difference. About 40–50 percent higher bitrate is typically sufficient to keep the same perceptual quality when doubling frame rate.

Bobbing is one way to extract some more value out of lower-quality interlaced sources. Since the temporal detail of interlaced is perfect, more frames gives the encode a more realistic feeling, and at the higher frame rate, any frame's random noise is visible for only half as long, distracting less from the content.

Bob requires excellent deinterlacing quality, particularly with horizontal or shallow diagonal details. Since the serif of a capital "H" may only have been in one of the two fields in the first place, the deinterlacer will need to be able to reconstruct that accurately. Otherwise, those small details will flicker in-and-out every frame, which can be extremely distracting. This requires a motion adaptive deinterlacer, and ideally a motion search-adaptive one. Once again, Compressor, Episode, and AVISynth are good products for this, due to having good motion search adaptive deinterlacing, and all support bob.

Telecined Video—Inverse Telecine

Content originally produced on film and transferred to NTSC video via the telecine process should be run through an inverse telecine process. As you learned in Chapter 4, telecine takes the progressive frames of film and converts them into fields, with the first frame becoming three fields, the next frame two fields, then three, then two, and repeating. See Color Figure 21 for an illustration of how they are assembled.

The great thing about telecined content is that reversing it restores a 24p sourced content to its native resolution and speed, eliminating the need for deinterlacing and making motion accurate. It's basically as good as having 24p source in the first place. You can recognize telecined content by stepping through the source one frame at a time. You'll see a repeating pattern of three progressive frames followed by two interlaced frames. This 3:2 pattern is a good mnemonic to associate with 3:2 pulldown.

Bear in mind that the film and its corresponding audio are slowed down by 0.1 percent, just as 29.97 is slowed down from 30 fps. This means the 3:2 pulldown pattern can be continued indefinitely, without any adjustments for the 29.97. However, it also means the inverse telecine process leaves the frame rate at 23.976, not 24. And note some compression tools require the encoded frame rate to be manually set to 23.976 when doing inverse telecine.

One "gotcha" to watch out for is with video that was originally telecined, but was then edited and composited in a postproduction environment set up for interlaced video. The two big problems are cadence breaks and graphics mismatches. In a cadence break, a video edit doesn't respect the 3:2 pulldown order, and might wind up with only a single field remaining from a particular film frame. This breaks the stride of the 3:2 pattern. You'll need to be using an inverse telecine algorithm that can find and correct for cadence breaks.

In a graphics mismatch, the graphic overlay doesn't match the movement of the film source underneath. This could be due to the video having been field-rendered, so you could have a title that moves every field, while the underlying film source only changes every two or three fields. The result of this is an odd mish-mash of both types. There isn't really any straightforward way to deal with this kind of source beyond having an inverse telecine algorithm smart enough to reassemble the source based on the underlying film motion, and then automatically deinterlacing the motion graphics part of the image. Delightfully, film transferred to PAL doesn't have to run through any of that 3:2 rigmarole. Instead, the 24 fps source is sped up 4 percent, and turned into 25 fps progressive. Yes, it's progressive! And sports no decimals. The only warning is that if someone does compositing or other processing to that source with fields turned on, it'll ruin the progressive scan effect. Generally, adaptive deinterlace filters do an excellent job with this kind of footage, restoring almost all the original 25p goodness. This is only one of the many delights of PAL that leave me fantasizing about moving to Europe.

Mixed Sources

What about content where different sections use different field patterns? This is pretty common with film documentaries that'll mix 24p source of clips with 30i source of interviews.

If the video content is predominant, your most likely choice is to treat all of your source as interlaced video. If the film content is predominant, you could try to use an inverse telecine filter. Also, the judder of converting to a different frame rate is much more apparent with fast motion than a talking head. Ideally, a compression tool would be able to vary the output frame rate between sections when going to a format such as QuickTime that supports arbitrary frame timing. Expression Encoder will do this if the source has variable frame rate, but isn't able to dynamically figure out where to apply 3:2 pulldown by section.

Again, it gets down to some specialized AVISynth filters to reliably be able to switch frame rates by section. Unfortunately, AVISynth itself requires a fixed frame rate per project, so to get real 24/30/60 adaptibility, AVISynth needs to run at 120 fps (5×24, 4×30, 2×60), and then you have to hope that the encoder won't encode duplicate frames.

Progressive Source—Perfection Incarnate

The best option, of course, is to get your video in progressive scan in the first place. Remember what I said about never shooting interlaced?

When working with progressive source, make sure that any compositing and rendering is done in progressive scan as well. All it takes is one field-rendered moving title to throw the whole thing off. Generally, most tools will do the right thing when working with compressed bitstreams that include metadata indicating picture structure. But when you capture uncompressed video to .mov or .avi, that data isn't always indicated, so a given tool might assume interlaced is progressive or vice versa. It's always good to double-check.

Cropping

Cropping specifies a rectangle of the source video that is used in the final compression. While we can think of it as "cutting out" stuff we don't want, it's really just telling the rest of the pipeline to ignore everything outside the cropping box. There are a few basic things we always want to crop out for computer playback:

- Letterboxing/pillarboxing. Drawing black rectangles is easy; we don't need to waste bits and CPU power baking them into our source and then decoding them on playback.
- Any "blanking" or other noise around the edge of the video.

Having unwanted gunk in the frame occurs much more frequently than you might think. Content originally produced for analog television typically doesn't put image all the way to the edge, while what we create on a computer or modern digital camera typical does.

Edge Blanking

Broadcast video signals are overscanned—that is, they're larger than the viewable area on a TV screen. This is done to hide irregularities in the edges of aging or poorly adjusted picture tubes on old analog TVs, and to account for the distortion caused by the curvature at the edges of every CRT. Bear in mind that the flat panel era is less than a decade old and follows 50 years of the CRT; there's a huge library of content shot for CRT, and plenty new stuff that still needs to work well on CRT. The area of the video frame that carries the picture is called the active picture. The areas outside the active picture area are called blanking areas.

When describing video frame sizes, for example 720 × 480 or 720 × 486, vertical height (a.k.a. the 480 and 486 in these example sizes) is expressed as active lines of resolution. The number of pixels sampled in the active length in our two examples is 720. Active lines and line length equate one-for-one to pixels in the digital domain.

So what's all that mean in practical terms? Examine a full broadcast video frame and you'll see the blanking quite clearly (Figure 6.6). It appears as black lines bordering both the left

Figure 6.6 VHS can have a lot of extreme edge noise, including Line 21, horizontal blanking, and tearing at the bottom. For DVD output that should be blanked out, but cropping should be used for formats that don't have a fixed frame size.

and right edges, as well as the top and bottom of the frame. Sometimes you'll also see black and white dashes at the top of the screen: that's Line 21, which contains the closed captions in a sort of Morse Code-like pattern.

On the other hand, computer monitors and similar digital devices show every pixel. CRT monitors are naturally analog, but instead of cutting off the edges of the video, the whole image is shown with a black border around the edge of the screen. So, when displayed on a computer monitor or in a media player's screen, blanking from the analog source is extraneous and distracting, and you should crop it out. The active image width of 720-pixel wide source is typically 704 pixels, so you can almost always crop eight pixels off the left and right sides of the video. Different cameras and switching equipment can do different things around those edges, so you may see the width of the blanking bars vary from shot to shot; you'll need to crop enough to get inside all of them.

If you're unfortunate enough to have VHS source, you'll encounter tearing. This is a dramatic distortion in the lower portion of the frame. Such distortion usually occurs outside the viewable area of a television screen, but is ugly and quite visible when played back on a computer monitor. Crop it out as well.

You don't have to crop the same amount from every side of the video—if there's a thick black bar on the left and none on the right, and there's VHS tearing at the bottom but the top looks good, crop only those problem areas. One important special case for cropping is if you're authoring content that could go back out to an analog display, like DVD. If you have some tearing or whatever in a 720 × 480 source, you're generally better off blanking out the

Figure 6.7 Cropped (6.7A) and blanked (6.7B) versions of Figure 6.6.

edges instead of cropping, since cropping requires the video to be then scaled as DVD has a fixed frame size. What you want is the video to be framed so that it looks good if someone's watching on an overscan display, but there's no distracting junk (especially Line 21!) if they're watching the full raster (Figure 6.7).

Letterboxing

For their first several decades, feature films were mainly produced in the 4:3 (1.33:1) aspect ratio that broadcast television then adopted. But once TV caught on, film studios decided they needed to differentiate themselves, and so adopted a number of widescreen formats (including the storied Cinemascope, Vista-Vision, and Panavision) for film production, with the most popular aspect ratios being 1.85:1 and 2.4:1. Modern HDTVs are now 16:9 (1.78:1), which is a little squarer than 1.85:1. Outside of televisions, the most common computer monitor aspect ratio today is 16:10. Lots of portable devices are 4:3 or 16:9, but some—like the iPhone—are 3:2 (1.5:1).

Thus, we have a bunch of potential mismatches between content and display. We could have source at anything from 4:3 to 2.4:1 (almost double the width), with displays (outside of theaters) between 4:3 and 16:9.

So, how do you display video of a different shape than the screen? There are three basic strategies: pan-and-scan, letterboxing, and stretching. Pan-and-scan entails chopping enough off the edges of the film frame for it to fit the 4:3 window. On the high end, operators spend a lot of time trying to make this process look as good as possible, panning across the frame to capture motion, and scanning back and forth to catch the action. If you receive pan-and-scan source, you can treat it as any other 4:3 content in terms of cropping

But, especially when dealing with source beyond 1.85:1, film purists do not find pan-and-scan acceptable. The initial solution was letterboxing, which displays the full width of the film frame on a 4:3 display and fills the areas left blank above and below the 16:9 image with black bars. This preserves the full image but makes the image smaller by not filling the available screen real estate. Starting with Woody Allen's *Manhattan*, some directors began insisting that their films be made available letterboxed. Letterboxing has since been embraced by much of the public, especially when DVD's higher resolution made the loss of image area less of a problem. DVDs can also store anamorphic content at 16:9, which reduces the size of the black bars needed for letterboxing, and hence increases the number of lines of source that can be used. When compressing letterboxed source for computer delivery, you'll generally want to crop out the black bars and deliver the content at its native aspect ratio; you can't know in advance what the display aspect ratio is going to be and so have to trust the player to do the right thing. The last approach used is stretching: simply distorting the image to the desired shape. Distortions of less than 5 percent or so aren't very noticeable. But going 16:9 to 4:3 is a very visible 33 percent stretch! Yes, some people do watch TV like that at home, but many other viewers (including myself) get *very* annoyed if that is done at the authoring stage. Some users will always mess up their presentations, but you want to make sure that viewers who care to optimize their experience can have a good one.

If a client requests you deliver the video at 4:3 or some other specific frame aspect ratio, you may have no choice but to either pan-and-scan the video yourself or deliver the video with the black bars intact. The pan-and-scan process is inevitably labor-intensive if you want to do it well. I've found that clients who insist all content be 4:3 tend to quickly change their minds when they see a realistic pan-and-scan bid (or, horror of horrors, just insist on a 4:3 center cut, and woe betide any action happening at the edges of the frame).

If you need to deliver in a set window size for browser-embedded video, ideally you'll encode the video at the correct frame width and reduce height with wider aspect ratios. By scripting the web plug-in, you can usually specify that the video always appears vertically centered in the frame. This will keep the frame size constant, while still preserving the full aspect ratio.

Flash and Silverlight make it easy for the player to automatically resize to match the content with the right player tweaking.

When you have to add letterboxing

If you need to leave the bars in for a format like DVD or Blu-ray with a fixed frame size, make sure the bars are a single, mathematically uniform $Y' = 16$ for more efficient encoding.

If you have to letterbox, try to get the boundary between the black and the image to fall along the 8 × 8 grid most DCT codecs use for luma blocks. And for a slight improvement beyond that, align along the 16 × 16 grid of macroblocks. This will allow complete blocks of flat color, which are encoded with a single DC coefficient. While most standard resolutions are already Mod16, there's one glaring exception: 1920 × 1080. 1080 is only Mod8. When encoding as 1920 × 1080, the codec is actually internally coding at 1920 × 1088 with eight blank pixels beyond the edge of the screen. So, when letterboxing for 1080i/p, you'll want the

Table 6.1 Recommended Matte Sizes.

Encoded	Width	Height	Screen	Content	Top Crop or Matte	Bottom Crop or Matte	Left & Right Crop or Matte	Active Image
720 × 480	720	480	4 × 3	4 × 3	0	0	0	720 × 480
720 × 480	720	480	4 × 3	16 × 9	64	64	0	720 × 352
720 × 480	720	480	4 × 3	1.85:1	64	64	0	720 × 352
720 × 480	720	480	4 × 3	2.39:1	112	112	0	720 × 256
720 × 480	720	480	16 × 9	4 × 3	0	0	96	528 × 480
720 × 480	720	480	16 × 9	16 × 9	0	0	0	720 × 480
720 × 480	720	480	16 × 9	1.85:1	16	16	0	720 × 448
720 × 480	720	480	16 × 9	2.39:1	64	64	0	720 × 352
720 × 576	720	576	4 × 3	4 × 3	0	0	0	720 × 576
720 × 576	720	576	4 × 3	16 × 9	80	80	0	720 × 416
720 × 576	720	576	4 × 3	1.85:1	80	80	0	720 × 416
720 × 576	720	576	4 × 3	2.39:1	128	128	0	720 × 320
720 × 576	720	576	16 × 9	4 × 3	0	0	96	528 × 576
720 × 576	720	576	16 × 9	16 × 9	0	0	0	720 × 576
720 × 576	720	576	16 × 9	1.85:1	16	16	0	720 × 544
720 × 576	720	576	16 × 9	2.39:1	80	80	0	720 × 416
1280 × 720	1280	720	16 × 9	4 × 3	0	0	160	960 × 720
1280 × 720	1280	720	16 × 9	16 × 9	0	0	0	1280 × 720
1280 × 720	1280	720	16 × 9	1.85:1	16	16	0	1280 × 688
1280 × 720	1280	720	16 × 9	2.39:1	96	96	0	1280 × 528
1920 × 1080	1920	1080	16 × 9	4 × 3	0	0	240	1440 × 1080
1920 × 1080	1920	1080	16 × 9	16 × 9	0	0	0	1920 × 1080
1920 × 1080	1920	1080	16 × 9	1.85:1	16	24	0	1920 × 1040
1920 × 1080	1920	1080	16 × 9	2.39:1	128	144	0	1920 × 808

height of your top matte to be Mod16, but the height of your bottom matte to be Mod16 ± 8. And yes, it won't be quite symmetric; I prefer to shift it up slightly, to leave more room for subtitles to fit outside of the image at the bottom. But don't worry, no one will notice.

Table 6.1 shows my recommended crop and matte sizes for typical combinations of frame sizes, source aspect ratios, and content aspect ratios.

Safe Areas

While blanking is used to accommodate the foibles of television picture tubes, video safe areas were designed to further protect important information from the varying ravages of analog televisions. CRTs vary in how much of the active picture they show, both between models and over the life of a set. Also, portions of the image that appear near the edge of the screen become distorted due to curvature of the tube, making it difficult to see text or other fine details. Because of this, video engineers defined "action-safe" and "title-safe" areas

Figure 6.8 Aspect ratios in common use.

Figure 6.9 The classic safe areas for 4:3 video. No important motion will be outside of motion safe, and nothing needing to be visible without distortion will be outside of title safe.

where visually important content would be retained across the CRTs in the wild. The action safe area is inset five percent from the left/right/top/bottom edges of the frame. Assume anything falling within this zone will not be visible on all TV sets.

Title-safe comes in 10 percent from the active image area. Important text is not placed outside the title-safe area, as it may not be readable by all viewers. This is where the phrase "lower thirds" comes from—when the bottom of the text or graphics needs to be 10 percent up from the bottom of the screen, the top of the text/graphics naturally starts about 1/3 of the way from the bottom.

Any television should be able to accurately display motion in the action-safe area, and display text clearly in the smaller title-safe area. As you can see, these areas limit the overall video-safe area by a significant amount.

Whether you crop out any of the safe area depends on your targeted output resolution. As discussed in potentially mind-numbing detail in the forthcoming section on scaling, we don't want to scale the frame size up in either height or width. Thus, when you're targeting higher resolution video, you want to grab as much of the source video as possible—if the output is 640 × 480, you'll want to grab all of the 720 × 480 you can. However, if you're targeting a lower resolution than the source, a more aggressive crop will make the elements in the middle of the screen larger. This will make them easier to see at low resolutions, particularly on phones and media players. If your footage was originally shot for video, the camera operators made sure any important information was within the safe areas, so a 5 percent crop around the edge to action-safe will be generally fine. Title-safe is riskier, but possible if you scrub through the video first.

When working with letterboxed source, you'll want to do this before cropping the letterbox, of course. Content up to the edge of the matte is important to preserve. So with a 4:3 frame at 720 × 480 with 16:9 matted in (thus 720 × 352 image area), you could crop 18 pixels left/right and then crop out letterboxing, leaving 684 × 352. That'd make a nice 608 × 352 converted to square pixel, and thus would nicely fill an iPhone 3G display with around a 480 × 288 rectangle. Had we started with the full 720 × 352 area, we'd have gotten 480 × 272.

Scaling

Cropping defines the source rectangle, and scaling defines the size and the shape of the output rectangle and how to get there. It sounds simple enough, but there are a number of goals here:

- Have a frame size that'll offer good quality with the content, codec, and bitrate.
- Get displayed at the correct aspect ratio: circles aren't ovals, and squares aren't rectangles.
- Be the optimal frame size for compression, typically with height and width divisible by 16.
- Use an algorithm that provides a high-quality, clean image without introducing scaling artifacts.

Aspect Ratios

First, a few definitions of art terms to help keep things straight:

- Pixel Aspect Ratio (PAR): The shape of each pixel. Square pixel is 1:1.
- Display Aspect Ratio (DAR): The shape of the frame overall. Like 16:9 or 4:3.
- The Nameless Ratio (no acronym): The actual height/width in pixels of nonsquare pixel video. So 720 × 480 would be 3:2.

That 3:2 is a particularly useless and misleading number, as there's nothing 3:2 about the entire frame. If you plug that into Excel trying to figure out your frame size, you're likely to wind up doing something crazy (and very common) like encoding at 360 × 240. The Zen trick is to tune out the algebra and just make sure the output aspect ratio matches the input aspect ratio. When encoding for a square pixel output (which will be true for most PC and device playback) just make sure the height/width of the output frame size matches that aspect ratio of the source image (after cropping out any letterboxing). Divide the output resolution of each axis by the ratio for that axis, and make sure the numbers are identical or nearly so. So, with 4:3 going to 320 × 240, divide 320 by 4 and 240 by 3. In this case, both values are exactly 80. Perfect. See Figure 6.10 for examples of correct and incorrect scaling, as well as the settings in Expression Encoder and Compressor.

When the output frame size is anamorphic (nonsquare pixel), it's generally because you're using the same aspect ratio as the source.

Downscaling, Not Upscaling

Anyone who has tried to take a low-resolution still photograph (like a DV still) and blow it up knows scaling can hurt image quality. Even a 2x zoom can look painfully bad with detailed content. There's really almost never a good reason to do any upscaling in compression. Doing it well is really a creative process, and thus should be handled upstream in post.

The good news is that scaling down, when done correctly, looks great

Scaling Algorithms

Different scaling algorithms provide different scaling quality. And unfortunately, many tools use the same name for what are different implementations of the algorithms with different implications. And although it sounds crazy, the choice of a scaling algorithm can have a bigger impact on your video quality than all the codec tuning in the world.

There are a number of tradeoffs in scaling algorithms. A big one is speed; the bad ones are used because they're fast. The ones that yield better quality, particularly when coming from HD source, can be painfully slow. Quality is next; you don't want to see aliasing or ringing, even with big scaling ratios (see Figure 6.11). Going from 1920 × 1080 to 320 × 176 is a pretty typical encode these days. Algorithms also vary in how much they err on the side of softness versus sharpness as they scale down. Sharper can "pop" but also have more high-frequency data and hence are harder to encode.

Here are some of the common scaling types you'll see.

Nearest-neighbor

Also called a box filter. Each output pixel is a copy of a particular input pixel. So if there's a gradient over 5 pixels in the source, but the video is scaled down 5x, only one of those pixels gets picked for the output. This leads to crazy aliasing and some horrible-looking video. Shun it.

Bilinear

Bilinear is the first semidecent scaling algorithm, made somewhat confusing by highly varying implementations. A basic bilinear is generally good as long as the output is within

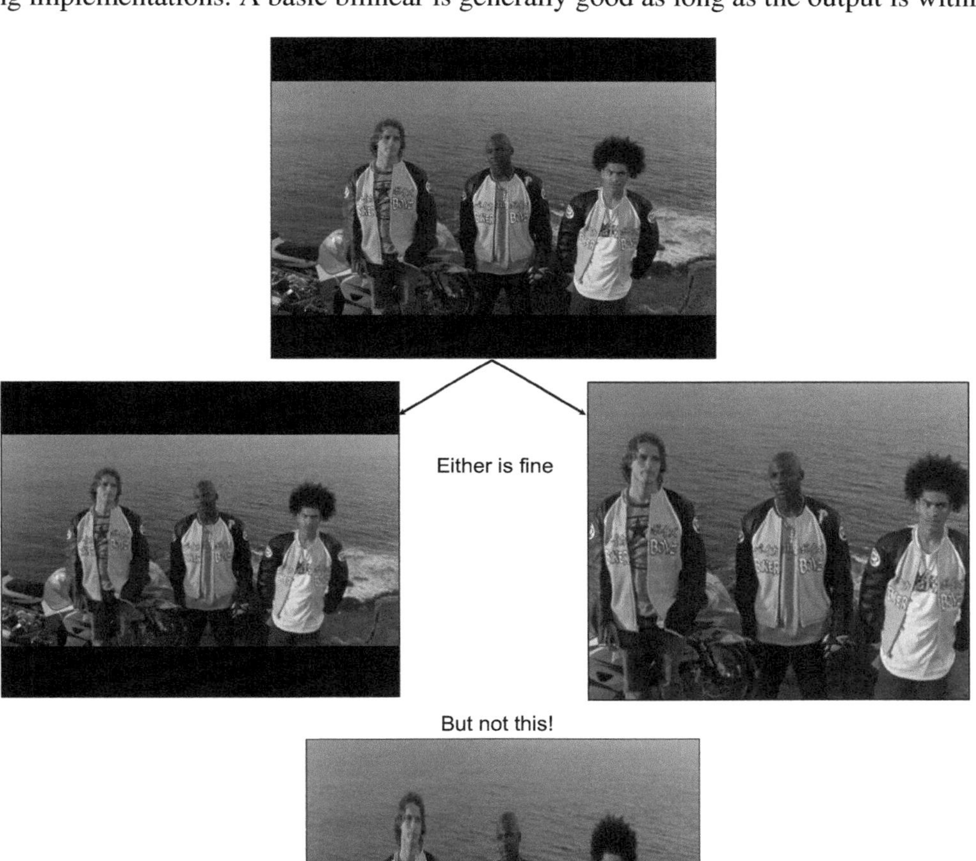

Figure 6.10 Depending on your content and playback environment, either pan – and – scan or letterboxing can make sense to fit 16:9 content into a 4:3 screen (6.10A).

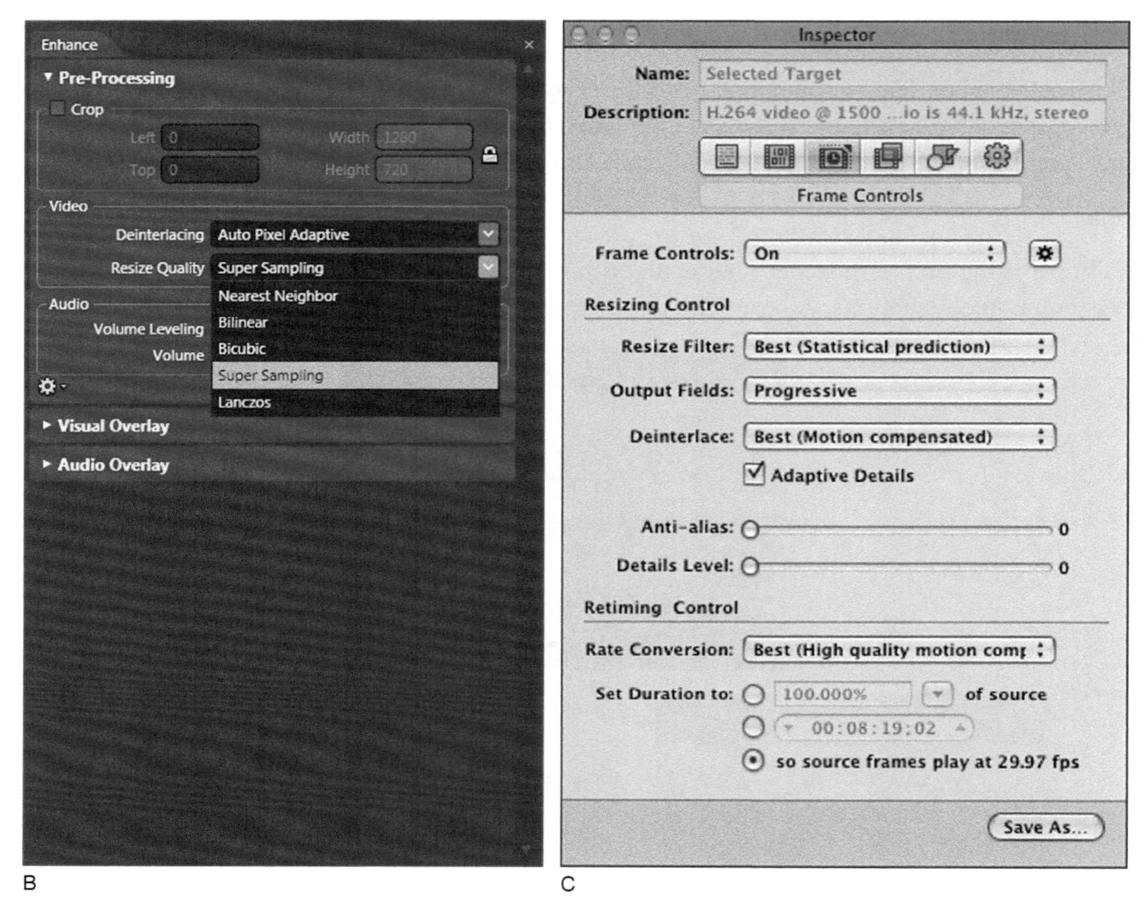

Figure 6.10 (continued) But don't just stretch it! The correct settings are shown for Expression Encoder (6.10B) and Compressor (6.10C).

50–200 percent of the size of the original on each axis. However, it can start showing bad aliasing beyond that.

Some bilinear implementations, like that in AVISynth, include filtering that prevents aliasing and making it arguably the best downsampling filter, as it adds the least sharpness, and hence has less high-frequency data to throw off the codec.

Bicubic

Bicubic is the "good" choice in a lot of tools, including Adobe's. It's slower than bilinear, but offers good quality well below 50 percent of the original size. Different implementations vary a lot in sharpness; the VirtualDub version is a lot sharper than the AVISynth default.

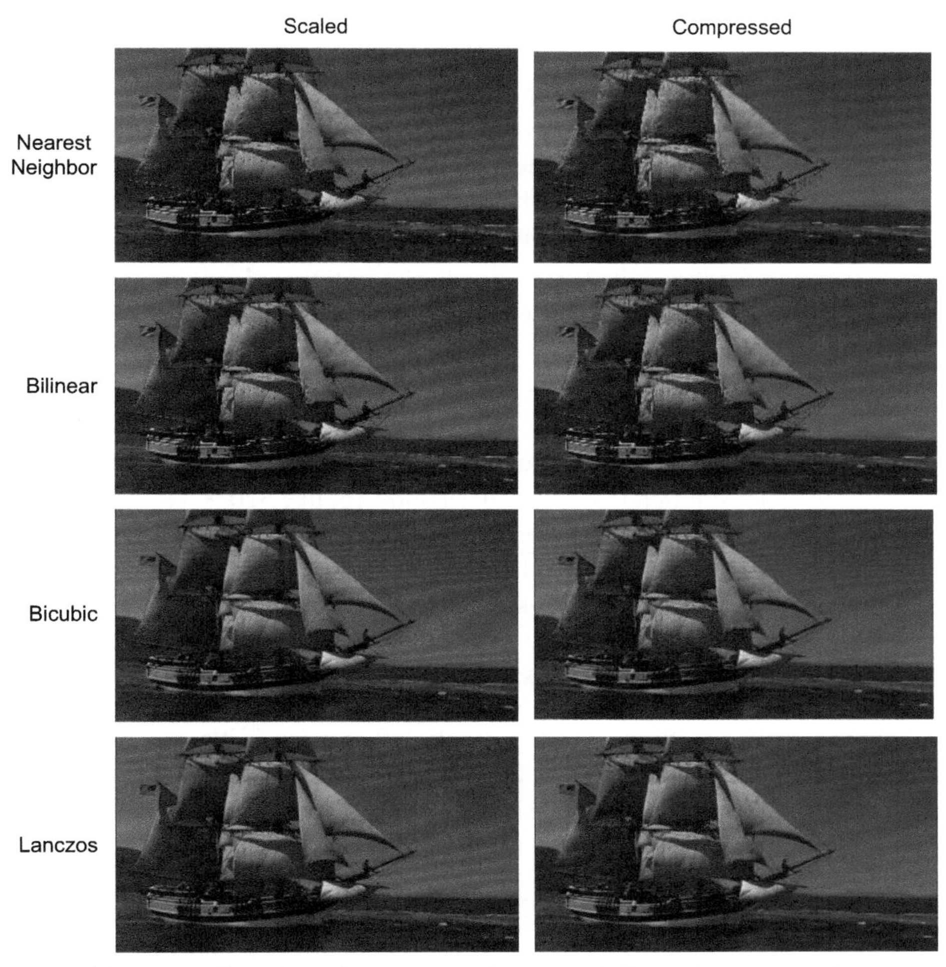

Figure 6.11 Different scaling algorithms can result in very different images. And poor – quality algorithms can yield artifacts that are exaggerated by compression.

Lancsoz

Lancsoz is a high-quality filter, but definitely errs on the side of sharpness; you'll need more bits per pixel to avoid artifacts with it. But if you've got bits to burn and are looking to pop off that phone screen, it does the job very well.

Super sampling

Super sampling is the best scaling algorithm in Expression Encoder, and perhaps my favorite. It's very neutral between sharpening and softening, without significant ringing. It's not the fastest, but the source is better enough that it can be worth using the slower scaling mode and a lower complexity on the codec; better source leaves the codec with less to do.

Scaling Interlaced

The biggest problem we're likely to face with scaling is when taking interlaced source to progressive output. As discussed earlier, when deinterlacing with a nonadaptive system, or on a clip with high motion where adaptive deinterlacing isn't able to help, in NTSC you're really starting with 240 lines of source at best. This is the big reason progressive scan and film sources are so well suited to computer delivery. 16:9 is also good for this reason, as you spend more of your pixels horizontally than vertically. If you've only got 240 real lines, 432 × 240 from 16:9 can be a lot more impressive than the 320 × 240 you get from 4:3.

Mod16

All modern codecs use 16 × 16 macroblocks, and so really work in 16 × 16 increments. When you encode to a resolution that's not divisible by 16, like 320 × 180 (16:9), it rounds up to the next 16 internally, and then doesn't show those pixels. So, 320 × 180 is really processed as 320 × 192, with the bottom 12 lines filled with dummy pixels that aren't shown. It's not a disaster, but it makes those bottom 4 lines less efficient to encode and decode. It would be better to use 320 × 176 in that case. As mentioned previously, this padding technique is used in 1080 video, where it's really encoded as 1088 with 8 dummy lines.

Rounding out to Mod16 can throw your aspect ratio off by a bit. If it's only a bit, don't worry about it. A distortion to aspect ratio really doesn't become really noticeable until around 5 percent, and even the fussy won't notice below 2 percent. This can also be addressed somewhat by tweaking safe area cropping to closer match the final shape.

Noise Reduction

Noise reduction refers to a broad set of filters and techniques that are used to remove "bad stuff" from video frames. These errors encompass all the ways video goes wrong—composite noise artifacts, film grain and video grain from shooting in low light, film scratches, artifacts from prior compression, and so on. While noise reduction can help, sometimes a lot, it's never as good as if it were produced cleanly in the first place. But when all the video editor hands you is lemons, you've got to make lemonade, even if there never is enough sugar.

There's not always a fine line between noise and style. Film grain, for example, can be hard to encode at lower bitrates, but it really is part of the "film look" and the character of the source, and should be preserved when possible. Removing it may make for a cleaner encode, but may also be less true to the original. This is part of what makes preprocessing an art; you need to know what in the image is important to preserve.

The noise reduction features in most compression-oriented tools tend to be quite basic—they're generally simple low-pass filters and blurs. Some codecs, like Microsoft's VC-1 implementations, also include their own noise reduction methods, although they're typically not that advanced.

The drawback in all noise reduction filters is that they make the video softer. Of course, softer video is easier to compress, but you want to keep the right details while getting rid of the bad ones.

Sharpening

I don't recommend applying sharpening to video. Sharpening filters do make the image sharper, but sharper images are more difficult to compress. And sharpening filters aren't selective in what they sharpen—noise gets sharpened as much as everything else. And at low-to-moderate bitrates, all that high-frequency data can make the compression a lot harder, pushing up the QP, and thus yielding blocks or softening of the new detail you tried to add anyway.

For the usual strange historic reasons, Unsharp Mask is another name for sharpening, typically seen in applications whose roots are in the world of graphic arts.

Blurring

The most basic and worst noise reduction filter is blur. There are many kinds, the most common being the Gaussian blur, which produces a very natural out-of-focus look. The only time a straight blur really makes sense if the source is flat-out horrible, like it had previously been scaled up nearest-neighbor, or is interlaced video that was scaled as progressive.

Some blur modes, like that in Adobe Media Encoder, allow separate values for horizontal and vertical. This is nice when dealing with deinterlacing artifacts, as those only need vertical blurring.

Low-Pass Filtering

A low-pass filter lets low frequency information pass through (hence "low-pass") and cuts out high frequencies. For video, frequency is a measure of how fast the video changes from one pixel to the next. Thus, a low-pass filter softens sharp edges (including film grain), but not smoother parts of the image.

Still, low-pass is a pretty blunt instrument.

Spatial Noise Reduction

A spatial noise reduction filter uses a more complex algorithm than blur/low-pass, attempting to find where on a frame noise is and isn't, and only trying to remove the noise. In general, spatial noise reduction will try to preserve sharp edges, and remove or reduce noise from flatter areas with a lot of seemingly random fine detail, like grain/gain. The challenge is they can't always discriminate between static textures, like cloth, and actual noise. This problem is generally more of a false negative; noise not being removed instead of detail lost, but the image can get weirdly soft in patches if values get set high enough.

You also need to watch out for a "falloff" effect when there's texture on an object at an angle to the camera. You can get odd-looking cases where the texture in some parts of the image is within the noise reduction threshold and is outside of it in others. This can look very weird with motion.

Temporal Noise Reduction

For real grain/gain removal, you need a temporal filter. That compares a pixel with its spatial neighbors in the current frame and its temporal neighbors in adjacent frames. Since grain/gain is random on every frame, the filter can pick out flecks of detail that change from frame-to-frame (noise) and those that don't (detail), and thus be much more aggressive about eliminating the former. And since that random noise is the biggest compression killer (fine detail without any motion vectors to take advantage of), this can pay off very nicely.

AVISynth

I keep talking about AVISynth as having high-quality algorithms, and this is truer than anywhere in noise reduction, where they have an embarrassment of riches. They even have filters designed to clean up bad VHS sources. There's a huge and ever-evolving variety of AVS noise reduction filters, but here are a few of my favorites:

FFT3DFilter

FFT3DFilter is, for those with the right kind of decoder ring, a Fast Fourier Transform 3D noise reduction Filter. Specifically, it turns the entire frame into frequency data (like DCT, but without 8 × 8 blocks) and then applies further processing based on those frequencies. Like most AVS filters, it has a huge number of different modes. Myself, I like this for cleaning up typical film grain a little while preserving some of the texture:

```
FFT3DFilter(sigma = 1.5, plane = 4, bt = 5, interlaced = false)
```

The sigma value controls the strength of the denoise. I'll use up to 2.5 for somewhat grainer but still Hollywood-caliber content. Higher is only useful for rawer sources.

Once nice thing about FF3D3Filter is that is can also operate on interlaced frames. Since the gain noise in an interlaced video is localized to the field, this is generally better. For very noisy sources, I've even applied it twice; once before deinterlacing, and then using interlaced = false after deinterlacing.

The big caveat to FFT3DFilter is that it can be incredibly slow. However, nothing really beats it for cleaning up noisy video for cleaner compression.

VagueDenoiser

VagueDenoiser applies a wavelet transform to the video, and then will filter out small details based on that. It's a spatial filter, and offers a more low-pass-like effect than FFT3DFilter. I only use it with highly troubled sources, like HDV.

BlindPP

BlindPP is another filter included in DGDecode (mentioned in Chapter 5) that removes blocking and ringing artifacts from content that's been through a visually lossy 8 × 8 DCT encode (including MPEG-1/2, MPEG-4 part 2, VC-1 and JPEG). It requires that the content hasn't been scaled or cropped in order to preserve the original macroblock structure. Still, it's not as good as using postprocessing in a decoder, since it doesn't know what the source was, but it's a lot better than nothing in a pinch.

Deblock

Deblock is also included with DGDecode. It's an MPEG-2 specific filter, and wants the source not to have been scaled/cropped since decode, in order to accurately line up with the original 8 × 8 block pattern. If your content fits that description, it'll work better than BlindPP. It's great for fixing content sourced from HDV or DVD and then converted to another codec.

Luma Adjustment

As its name implies, luma adjustment is the processing of the luma channel signal. This is often called color correction, though it's often not color (chroma) that we're worried about. Most tools that offer any image level controls offer some luma-only filters like brightness, contrast, and gamma.

One big change in QuickTime since version 6 is that it's adopted industry standard gamma, so you no longer have to worry about encoding with different gamma for Mac and Windows.

Another great change is the predominance of all-digital workflows. Analog video almost always requires at least some luma level tweaking to get blacks normalized. But most all-digital projects don't have any noise in the blacks of titles and credits, and so are likely to need no image processing at all.

Normalizing Black

An important aspect of the remapping is getting parts of the screen that are black down to Y′ = 16 or RGB = 0. Noise that wouldn't have been noticed on an old CRT TV can be painfully obvious on a flat-panel TV or even a CRT computer monitor. And LCDs have a pretty bright black, so we want our blacks to be the lowest value we can get to the screen. The net effect is that a slightly noisy black region composed of a jumble of Y′ = 16, 17, 18, and 19 may look black on a calibrated professional CRT display, but on a consumer set it can look like someone is boiling rice in oil. If we get that all down to Y′ = 16, all that erroneous motion goes away. Also, a large field of the same number is extremely easy to compress compared to a jumble of 16, 17, 18, and 19, which looks "black" but not black. So in addition to looking better, it saves bits, letting other parts of the frame look better.

Lastly, getting blacks to nominal black makes it possible to embed the media player inside of a black frame, without an obvious seam between video and frame when a black frame is up. It also avoids the annoying appearance of PC-drawn letterboxing bars not matching block frames.

Getting black levels to Y′ = 16 can be difficult with noisy source, as many pixels that should be black have values well above 16. The solution is to get those brighter pixels down to Y′16 as well. First, noise reduction helps by reducing the variance between the average brightness and the brightest pixels of noise. Beyond that, a combination of luma mapping, decreased brightness and increase contrast will get you where you need to go. However, care must be taken that parts of the image that should truly be dark gray, not black, don't get sucked down as well.

Brightness

Brightness filters adjust the overall intensity of the light by raising or lowering the value of each pixel by a fixed amount. So, Brightness-5 would typically reduce Y′ = 205 to Y′ = 200 and Y′ = 25 to Y′ = 20.

If brightness is used in compression, it is normally used to make the video darker overall, instead of lighter, since we don't want to do anything to shift blacks higher than Y′ = 16.

Contrast

Contrast increases or decreases the values of each pixel by an amount proportional to how far away it is from the middle of the range. So, for an 8-bit system, 127 (the middle of the range) stays the same, but the farther away from the center value a pixel is, the larger the change from contrast; it makes black blacker and white whiter.

One classic use of contrast is to expand or contrast the luma range to fix a RGB 0-255/Y′CbC4 16–235 range mismatch; for example, video that comes in with blacks at Y′ = 32 and whites at Y′ = 215. Adding contrast so the values land at the right spots helps a lot.

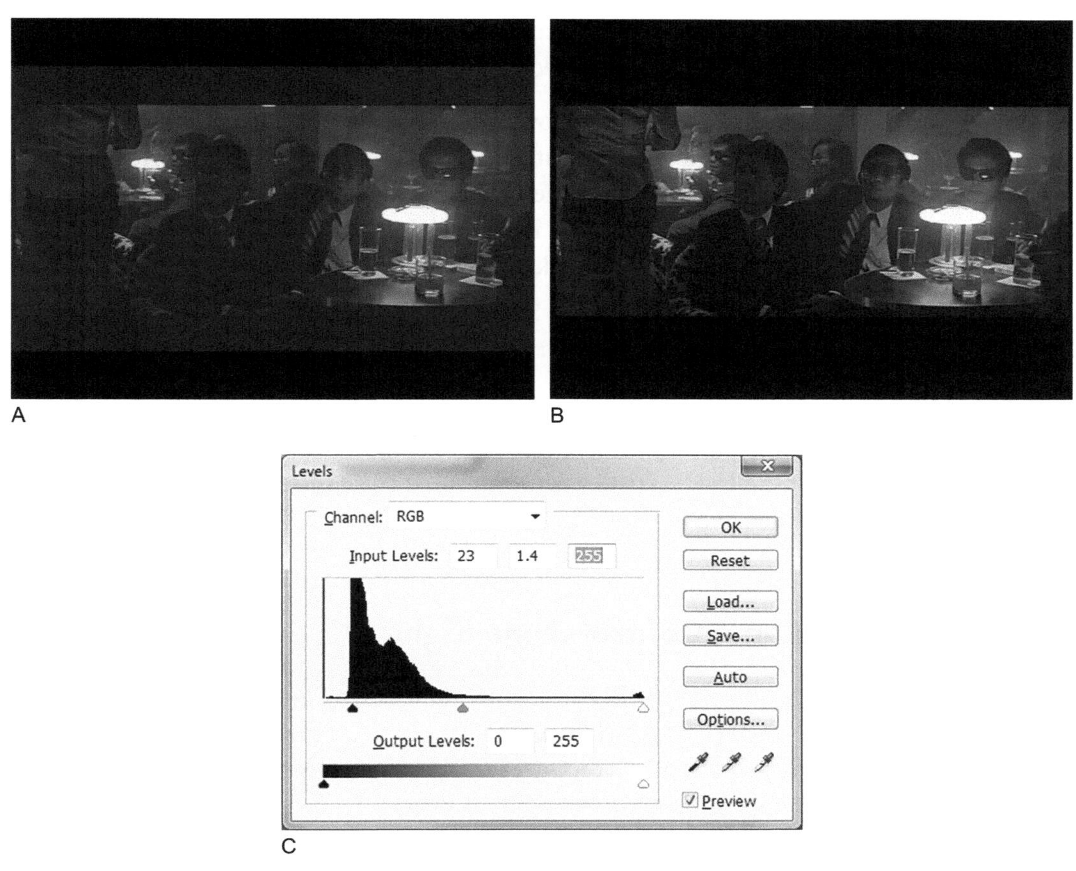

Figure 6.12 Before (6.12A) and after (6.12B) luma correction and black restoration, and the histogram that made it possible (6.12C).

With noisy source, it can often help to use a somewhat higher contrast value to lower black levels. This technique is often mixed with a corresponding decrease in brightness to keep the whites from getting blown out, and to get a combined effect in getting the dark grays down to black without hurting the rest of the image too much.

Gamma Adjustment

Gamma adjustment was the bugaboo of compression from CD-ROM until QuickTime 6. Back then, QuickTime and its codecs didn't do any gamma correction, and Macs and Windows machines used a different gamma value (1.8 on Mac, and video-standard 2.2 on Windows). Thus if you encoded video so it looked good on a Mac, the midtones were too dark on Windows, and Windows-tuned video was too bright on Macs. However, all codecs introduced since QuickTime 6 (so MPEG-4 pt 2 and H.264) do automatic gamma correction. There's still some older Sorenson Video 3 content around the web here and there that has this property, but you won't face it in any content you create today. Mac OS X 10.6 finally brought standard 2.2 gamma to the Mac for everything, not just video.

Gamma adjustment can be a useful tool in its own right. If the blacks and whites look just right, but the midtones seem a little washed out or dark, adjusting gamma can often improve quality significantly. And if you want to change the subjective "brightness" of a clip, gamma is a much more useful tool than the brightness itself, as gamma gives you brightness where it's most visible without messing up your black and white levels.

Processing Order

I want to make a little comment about luma processing order. It should always go in the order of Brightness, Contrast, Gamma. This is so blacks and whites are normalized before they hit a gamma filter, so that you don't need to readjust gamma every time you change blacks/whites.

Adobe's Levels: Luma Control Done Right

Sure, we can have separate sliders for luma control. But why are we still bothering? Photoshop nailed the perfect UI for this nearly 20 years ago (see Figure 6.12c). It has an integrated histogram to show the distribution of luma values in the source, and handy little triangles to set the black and white points. So easy!

Chroma Adjustment

Chroma adjustment, as you've already guessed, refers to tweaks made to color. The good news is that with most projects, you won't need to make chroma adjustments. And when you are required to do so, it likely won't be with the complexity that luma adjustments often require.

Saturation

Saturation controls affect changes in the intensity of color, analogous to luma. Saturation changes affect both color channels equally, but have no impact on luma.

It's pretty unusual to tweak saturation at the compression stage, particularly with a digital workflow. Analog video may wind up a little undersaturated, in which case raising it can help make thing a little more punchy. And video with a lot of chroma noise can sometimes benefit by some desaturation to reduce that noise.

Hue

Hue is determined by the ratio between Cb and Cr, and determines the mix of wavelengths of light emitted by the pixel. Incorrect hue can be due to white balance not being set correctly when the video was shot, from capturing through badly calibrated composite video systems, or messed-up 601 / 709 conversions. These should be fixed in postproduction, but sometimes are only caught at the compression stage.

Typically the problem shows up when flesh tones and pure white don't look correct. The fix is to adjust the hue until whites and skin look correct. Ideally, you'll be able to use a full-featured color-correction tool when such problems abound in your source. The hue of flesh tones falls along a very narrow range, A vectorscope will show you the distribution of colors in a frame and a line indicating ideal skin color tone (see Color Figure C.22).

Life's Mysteries Solved by Compression #2

And here's another college debate topic solved by image science. Question: "Aren't we all really the same color in the end?" In fact, yes, we're all the same hue. All ethnicities have skin the same hue even though there's lots of variation in luma and saturation. This is why a getting the hue of skin tones even a little off can look very weird.

Frame Rate

While frame rate is typically specified as part of the encoding process, conceptually it really belongs with preprocessing. The critical frame rate rule is the output frame rate must be an integer division of the source frame rate. Acceptable output frame rates are shown in Table 6.2.

And that's it! After so many subjective concepts, I'm happy to finally present a hard and fast rule.

Table 6.2 Acceptable Frame Rates.

Frame Rate	Full	Half	Third	Quarter	Fifth
23.976	23.976	11.988	7.992	5.994	4.795
24.000	24.000	12.000	8.000	6.000	4.800
25.000	25.000	12.500	8.333	6.250	5.000
29.970	29.970	14.985	9.990	7.493	5.994
30.000	30.000	15.000	10.000	7.500	6.000
50.000	50.000	25.000	16.667	12.500	10.000
59.940	59.940	29.970	19.980	14.985	11.988
60.000	60.000	30.000	20.000	15.000	12.000
120.000	120.000	60.000	40.000	30.000	24.000

We need to use these frame rates to preserve smooth motion. Imagine a source that has an object moving an equal amount on each frame. At the source frame rate, everything looks fine. If you drop every other frame, the amount of movement in each frame doubles, so it is consistent. Same goes for dropping two frames out of every three, or three out of every four. But if you use any other frame rate, you'll be dropping an uneven number of frames. Take, for example, going from 24 to 18 fps. That will drop one frame out of every three, so you'll have two frames that show equal motion, then one with double motion, yielding a "judder" of jerky motion.

Of course, there are some complexities. For one, how do you treat source that is mixed, like a promotional piece for a movie in which video interviews are intermixed with telecined film scenes? If you're outputting QuickTime, Windows Media, or MPEG-4, which all allow variable frame rates, the frame rate could go up or down as appropriate (if the tool supports that). But MPEG-1/2 and AVI require a fixed frame rate (although frames can be dropped), meaning a given frame can have its duration doubled, tripled, and so on, filling the space where several frames would have normally gone.

I include 120 fps on the chart to show how it can hit all the NTSC frame rates with subdivisions: 60, 30, and 24. So if an authoring tool can process at 120 fps internally, it should be able to handle any kind of variable frame rate you throw at it.

As for conversion between the NTSC 1.001 rates, that difference is imperceptible. If you're using a compression tool like ProCoder/Carbon that supports audio time resampling preprocessing, you can easily convert between 29.97/30 and 23.976/24 by speeding up/slowing down that 0.1%.

PAL/NTSC conversion is even feasible with this mechanism. It involves a 4 percent frame rate change between 24 and 25 fps, but for most content that looks fine (and people in PAL countries are very used to it; they've seen the 4 percent speedup as much as NTSC consumers have seen 3:2 pulldown).

Audio Preprocessing

If all goes well, you won't need to do much audio preprocessing, because problematic audio can be a nightmare. Fortunately, well-produced audio aimed for home stereo or television delivery will already be normalized and clean, and shouldn't need anything further for successful compression except potentially normalization. Raw captured audio might need quite a bit of sweetening, but again that's normally a post-production task.

Normalization

Normalization adjusts the level of an audio clip by finding the single loudest sound in the entire file, and then raising or lowering the volume of the whole clip so the loudest sound matches whatever level has been specified. This doesn't change the relative volume at all, just the absolute overall volume.

Normalization is of vital importance to the overall listening experience. I'm sure everyone has tried to listen to some lousy, overly quiet web audio, cranking up the volume to hear it clearly. And then, in the middle of the clip the next "You've got mail" arrives as a thunderous, ear-splitting BOOM! Not pleasant. Normalizing audio tracks eliminates this problem, and is an important courtesy to your audience. As a rule, the loudest sound in the audio should be around as loud as a system beep.

Dynamic Range Compression

In a source of perennial confusion, there is an audio filter called a compressor. In audio, extreme transients (very loud, very short noises), such as explosions or cymbal crashes, can cause distortion in audio playback and recording devices. Likewise, extremely quiet passages can get lost in the noise floor (the always present hums and hisses). Audio compressors act on dynamic range, smoothing out peaks and valleys of a signal making more consistent and thus "louder"—audio compression, when used properly, can tame wild level changes.

Audio compression was critically important in the 8-bit audio era, when already lousy CD-ROM audio played through tiny internal speakers. Today, it's thankfully much less of an issue. If you're dealing with source mixed for television delivery, it will likely have audio compression on it already. Soundtracks mixed for home theater may also have too broad a dynamic range, with the average volume level quite low to leave headroom for big explosions.

Audio Noise Reduction

Noise is no better for audio than video. And audio can have quite a lot of it: tape hiss, air conditioners in the background, hum from power lines, wind in the mic, and so on. A number of professional audio tools are available to reduce this noise, and the results can be pretty miraculous.

The scope and use of such audio noise reduction systems is broad, well beyond our purview here. However, if you have bad-sounding audio, there is often (but not always) something a good audio engineer can do to help the quality. The results are always better when the source is clean from the get-go.

Few compression tools have integrated audio noise reduction.

CHAPTER 7

Using Video Codecs

You've probably noticed that there are roughly six bazillion video codecs out there. While the rest of this book offers specifics on how to use each of those individually, there are many aspects of video codecs and compression we can investigate in general terms. Be aware that many of these technologies use different terms to describe the same feature, and sometimes call different features by the same name.

For this chapter, I'm speaking of a codec as a particular encoder implementation of a particular bitstream (dubbed an "encodec" in a memorable typo on Doom9). So, Main Concept H.264 and x264 are different codecs that both make H.264 bitstreams. So even when you know the bitstream you need to make, there are still plenty of choices between implementations of the major codec standards.

Bitstream

The most basic codec setting is the setting that determines which codec is being used. Some formats and platforms, such as MPEG-1, offer but one codec. QuickTime, at the other extreme, is a codec container that can hold any of literally dozens of codecs, most of which are wildly inappropriate for any given project.

Picking the right codec is of critical importance. You'll find details on each codec's strengths and weaknesses in the chapters devoted to each format. When evaluating new codecs, compare their features to those described in the following to get a sense of where they fit into the codec ecology.

Profiles and Level

Profile and Level define the technical constraints on the encode and the features needed in the decoder. They're generally described together as *profile@level*, for example, "H.264 High 4:2:0 Profile @ Level 2.1."

Profile

A profile defines the basic set of tools that can be used in the bitstream. This is to constrain decoder requirements; the standardized codecs have a variety of profiles tuned for different complexities of devices and decoder scenarios. For example, H.264 has a Baseline Profile for playback on low-power devices, which leaves out features that would take a lot of battery life, a lot of RAM, or a lot of silicon to implement. Conversely, PC and consumer device with a plug instead of a battery typically use High Profile, which requires more watts/pixel but fewer bits/pixel. And MPEG-2 has a Main Profile used for content delivery with 4:2:0 color and a High Profile used for content authoring that does 4:2:2.

Profiles define the maximum set of tools that can be used on encode, and the minimum that need to be supported on decode. For example, Main Profile H.264 encoder adds CABAC entropy coding and, B-frames, (described in the chapter on H.264). QuickTime's H.264 exporter doesn't use CABAC or multiple reference frames, but can use B-frames. Since it doesn't use any tools not supported in Main Profile, most notably 8 × 8 block size, its output is still Main Profile legal. Conversely, any Main Profile decoder must support all of the tools allowed in Main (and QuickTime can play back Main and High Profile content using features QuickTime doesn't export).

Particular devices often have additional constraints beyond the published Profiles. For example, many MPEG-4 pt. 2 encoders and decoders use "Simple Profile + B-Frames," which would normally be Advanced Simple Profile (ASP), but the other tools in ASP are of questionable real-world utility but add a lot of decoder complexity.

Level

Level defines particular constraints within the profile, like frame size, frame rate, data rate, and buffer size. These are often defined somewhat esoterically, like in bytes of video buffering verifier (VBV) and total macroblocks per second. I provide tables of more applicable definitions for each level in the chapters covering the various codecs.

Like Profile, Level is a maximum constraint on encoder and minimum constraint on decoder. Typically the maximum bitrate allowed by a Level is much higher than you'd use in practice, and many devices have further constraints that must be kept to.

Encoders targeting the PC market, and decoders in it, tend not to focus on Level. With software decode, lower resolution, very high-bitrate clips can be hard to decode; thus specifying levels for software playback is generally insufficient detail.

Data Rates

Data rate is typically the single most critical parameter when compressing, or at least when compression is hard.

Modern formats measure bitrate in kilobits per second (Kbps) or megabits per second (Mbps). I think we've purged all the tools that used to think that K = 1024 (see sidebar), so this should be reasonably universal now.

For lots of content, bitrate is the fundamental constraint, particularly for any kind of network-based delivery. You want to use the fewest bits that can deliver the experience you need; spending bits past the point at which they improve the experience is just increasing costs and connectivity requirements.

Within a codec's sweet spot, quality improves with increased data rate. Beyond a certain point (at the same resolution, frame rate, and so on), increasing the data rate doesn't appreciably improve quality. The codec or its *profile@level* may also have a fixed maximum bitrate. Below a certain data rate (again, with all parameters fixed), the codec won't be able to encode all the frames at the requested bitrate; either frames are dropped (better) or the file just comes out bigger than requested (worse). And some codecs also reach a point where slight drops in data rate lead to catastrophic drops in quality. MPEG-1 is especially prone to this kind of quality floor, as it lacks a deblocking filter.

Compression Efficiency

Compression efficiency is the key competitive feature between codecs and bitstreams. When people talk about a "best codec" they're really talking "most efficient." When a codec is described as "20 percent better than X" this means it can deliver equivalent quality at a 20 percent lower data rate, even if it's used to improve quality at the original data rate.

Compression efficiency determines how few bits are needed to achieve "good enough" quality in different circumstances. Codecs vary wildly in typical compression efficiency, with modern codecs and bitstreams being able to achieve the same quality at a fraction of the data rates required by older codecs and bitstreams (today's best video and audio codecs are easily 10x as efficient as the first). And different codecs have different advantages and disadvantages in compression efficiency with different kinds of sources and at different data rates. The biggest

A

B

Figure 7.1 16 years of compression efficiency on display right here. Both are 800 Kbps. (A) The ancient Apple Video codec. 5-bit per channel RGB and no rate control! Could only get down to 800 Kbps at 160 × 120. (B) H.264 High Profile via a quality-optimized 3-pass x264 encode. Nearly transparent compression at 640 × 480 in 800 Kbps.

differences are in the range of "good enough" quality—the higher the target quality, the smaller the differences in compression efficiency. H.264 may only need one-third the bitrate of MPEG-2 to look good at low bitrates, but it'll have less of a relative advantage at Blu-ray rates.

Some authoring and special-use codecs don't offer a data rate control at all, either because the data rate is fixed (as in DV) or because it is solely determined by image complexity (like PNG, Lagarith, and Cineform).

A Plea for Common Terminology

The computer industry got into a bad habit a few decades ago. Back in 1875, the metric system was fully codified, and it defined the common kilo-, mega-, and giga- prefixes, each 1000x greater than the one before. These are power-of-ten numbers, and thus can be written in scientific notation. So, 2 terabytes (2 TB) would be 2×10^9, the 9 indicating that there are nine zeros.

However, computer technology is based on binary calculation and hence uses power-of-two numbers. Ten binary digits (1×2^{10}) is 1024, very close to three decimal digits (1×10^2) 1000. And so computer folks started calling 1024 "kilo." And then extended that to mega, tera, and on to penta and so on.

But that slight difference may not be so slight, particularly when it's the difference between "just fits on the disc" and "doesn't fit on the disc." So there is now a new nomenclature for the power-of-two values, sticking an "ib" after each number. Thus:

- Computer industry numbers
  - "K" = Kib = 2^{10} = 1024
  - "M" = Mib = 2^{20} = 1,048,576
  - "G" = Gib = 2^{30} = 1,073,741,824
- Correct numbers
  - K = 10^3 = 1000
  - M = 10^6 = 1,000,000
  - G = 10^9 = 1,000,000,000
- Difference between values
  - K v. Ki = 2.4 percent
  - M v. Mi = 4.8 percent
  - G v. Gi = 7.37 percent

Rate Control

All that 1-pass, 2-pass, CBR, VBR terminology describes data rate modes. These represent one of the key areas of terminology confusion, with the same feature appearing under different names, and different features being given identical names. We can classify data rate control modes in a few general categories.

VBR and CBR

The variable bitrate (VBR) versus constant bitrate (CBR) distinction can be confusing. All interframe compressed video codecs are variable in the sense that not every frame uses the same number of bits as every other frame. This is a good thing—if every frame were the same size, keyframes would be terrible compared to delta frames. Even codecs labeled "CBR" can vary data rate quite a bit throughout the file. The only true CBR video codecs are some fixed frame-size authoring codecs like DV. But even there, the trend for solid-state capture is to use variable frame sizes to make sure hard frames get enough bits.

So, what's the difference here? From a high level, a CBR codec will vary quality in order to maintain bitrate, and a VBR codec will vary bitrate in order to maintain quality.

VBV: The fundamental constraint

Another way to think about CBR and VBR is how the decoder does, instead of how the encoder does. What the decoder really cares about is getting new video data fast enough that it's able to have a frame ready by the time that frame needs to be displayed, but doesn't get so much video data that it can't store all the frames it hasn't decoded yet. So the two things that can go wrong for the decoder are a buffer underrun, when it runs out of bits in its buffer and so doesn't have a frame it can decode by the time the next frame is supposed to display, or a buffer overrun, where it has received so many bits it can't store a frame that has yet to be displayed more.

Thus, every decoder has a video buffering verifier (VBV) that defines what a decoder has to be able to handle for a particular *profile@level*, and hence how much variability the encoder could face.

So, really, the core issue is the VBV; an encoder can vary allocation of bits between frames all it wants to as long as it doesn't violate the VBV. The VBV defines how many bytes can be in the pipeline. The VBV is thus the same as the bitrate (or peak bitrate) times the buffer duration. So, a 4-second buffer at 1000 Kbps would be 4000 Kbits, or 500,000 bytes.

And, from that understanding, we can think of a CBR as encoded to maximize use of the VBV, while a VBR encode is one where the average bitrate can be lower. The highest bitrate a VBR encode can go up to is the same as a CBR encode, and looks very much like a CBR encode for that section. With a five-second buffer, the data rate for any five seconds of the video must not be higher than the requested average. This isn't a question of multiple, discrete five-second blocks, but that any arbitrary five seconds plucked at random from the file must be at or under the target data rate.

CBR for streaming

The name "streaming" is a good one—a stream is a constant flow of water. But even though it is nominally constant, it isn't always possible for the encoder to use every last bit. If the video fades to black for a few seconds, there simply isn't enough detail to spend it on. For web streaming applications, fewer bits get sent; no biggie. Fixed-bandwidth broadcast like ATSC actually needs to use the full bandwidth, and will introduce "padding bits" of null data.

That said, adaptive streaming techniques can support much more variability, since it's the size per chunk that matters, not size per stream. A number of products support the new "VBR VC-1" Smooth Streaming SDK from Microsoft, where the right "stream" for the client's bitrate gets assembled out of the available chunks from different encoded bitrates. So even though the encoded files are VBR, the client is receiving data at an essentially constant rate.

VBR for download

We generally use VBR for video that's going to be downloaded or stored on disc. The fundamental definition of VBR is that the average bitrate is less than the peak bitrate, while those are identical with CBR. Thus, for a VBR encode, we specify the average bitrate as well as the buffer (either via VBR or by defining peak bitrate and buffer duration).

The goal of a VBR encode is to provide better overall compression efficiency by reducing bitrate and "surplus" quality in the easier portions of the video as it does in the more complex portions.

Depending on format, it's perfectly possible to use VBR files for streaming. This is sometimes used as a cost-saving measure in order to reduce total bandwidth consumption. Its peak bandwidth requirements would be the same of a CBR file with identical buffer. For example, encoding with a 1500 Kbps peak and a 1000 Kbps average would still require 1500 Kbps for playback, but would save a third on per GB delivery costs.

The average and peak rates are really independent axes. Average bitrate gets down to how big a file you want, and the peak is based on how much CPU you need to play it back. Take a DVD with video at a pretty typical 5 Mbps average 9 Mbps peak. If a bunch of content gets added, meaning more minutes need to be stuffed into the same space, the average bitrate may need to be dropped to 4 Mbps, but the 9 Mbps maximum peak will stay the same. Conversely, if the project shifts from using replication to DVD-R without any change in content, the peak bitrate may be dropped to 6.5 Mbps for better compatibility with old DVD players, but the average wouldn't need to change.

Quality-limited VBR

A quality-limited VBR is a special case of VBR when data rate isn't controlled at all; each frame gets as many bits as it needs in order to hit the quality target. In some cases, this can be a fixed-QP encode. Other models can be a little more sophisticated in targeting a constant perceptual quality; for example allowing B-frames to have higher QP. In those cases, you can think of a bitrate-limited VBR as basically a mechanism to find the QP that'll give the desired file size.

Some codecs, more common in audio than video, offer a sort of hybrid quality/bitrate VBR mode with an overall target bitrate and target quality, and the actual bitrate per file will go up

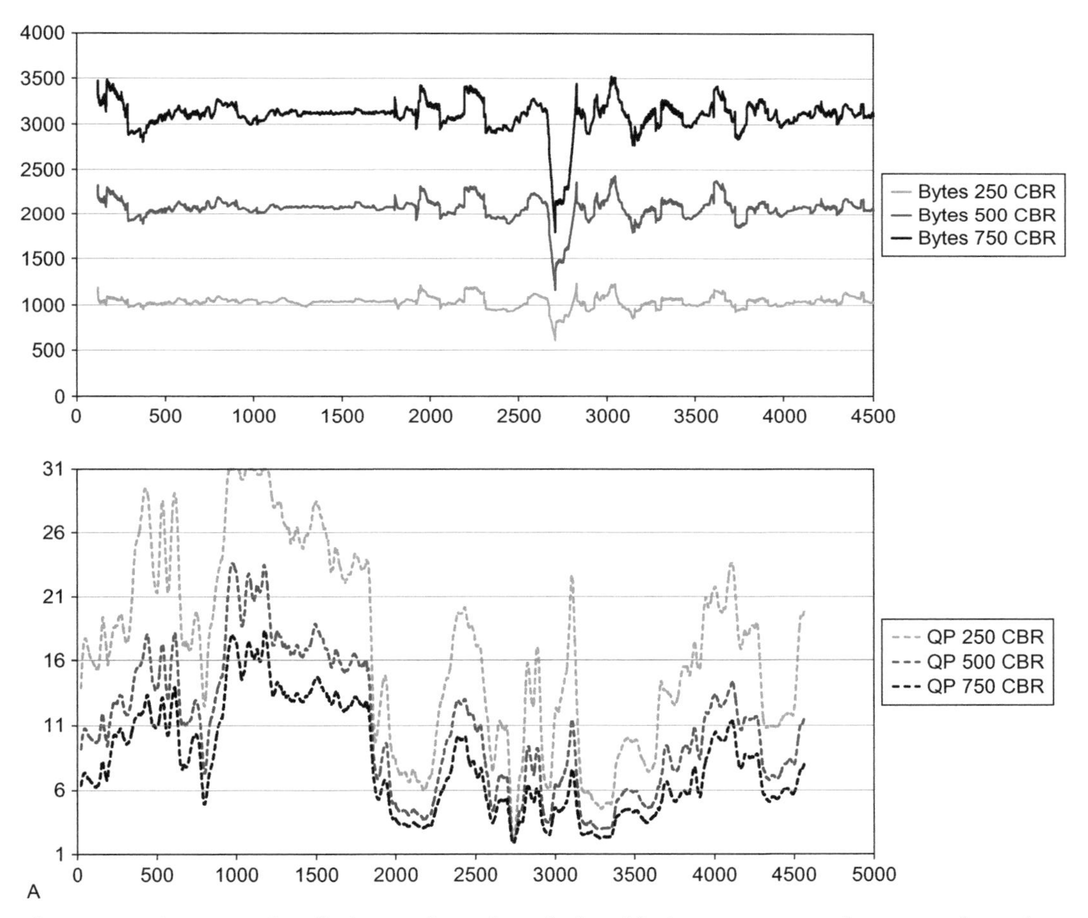

Figure 7.2 These graphs all aim to show the relationship between two values over time: data rate and the quantization parameter (QP). As you may remember from Chapter 3, a higher QP drives a higher scale factor, and hence more compression and lower quality. So a higher QP means lower quality and a lower QP means higher quality. (A) Three CBR encodes with a 4-second buffer. As we'd expect from a constant bitrate encode, the general pattern of data rate QP is similar across all three encodes, with the lower bitrate having a higher QP. There's some variability in data rate, of course, although it follows the VBV constrained accurately. There's lots more variation in QP, since harder scenes can't spend more bits and thus look at lot worse.

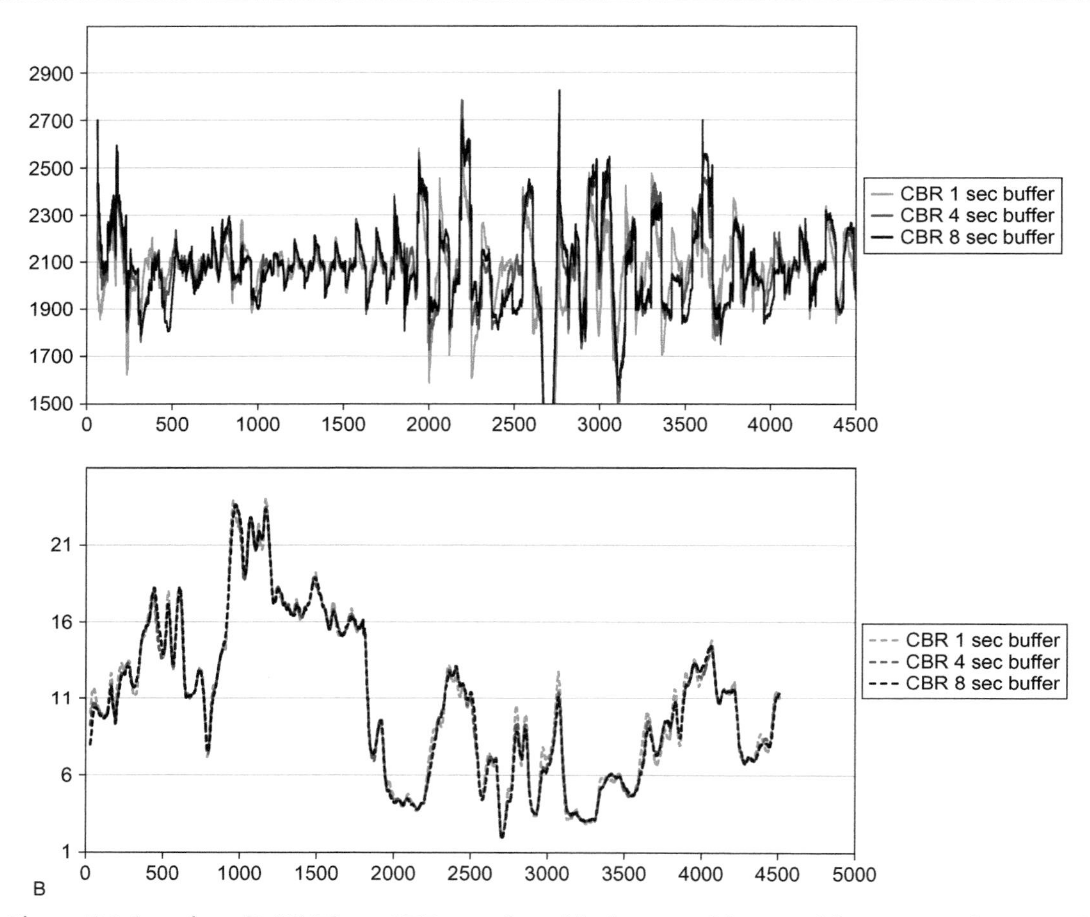

Figure 7.2 (continued) (B) Three CBR encodes with the same bitrate, with a 1, 4, and 8 second buffer. The larger buffer results in a little more variability in bitrate (higher spikes) and a slight reduction in variability of QP (fewer high QP, redistributing bits from increasing QP on easy frames).

and down within a band to hit that quality. In video codecs, this is supported in x264's CRF mode, and in ProCoder/Carbon's quality-with-VBR MPEG-2 mode.

1-Pass versus 2-Pass (and 3-Pass?)

1-Pass

By default, most codecs function in 1-pass mode, which means compression occurs as the file is being read (although generally with a buffer the same size as the VBV). Obviously, any live encoding will require 1-pass encoding.

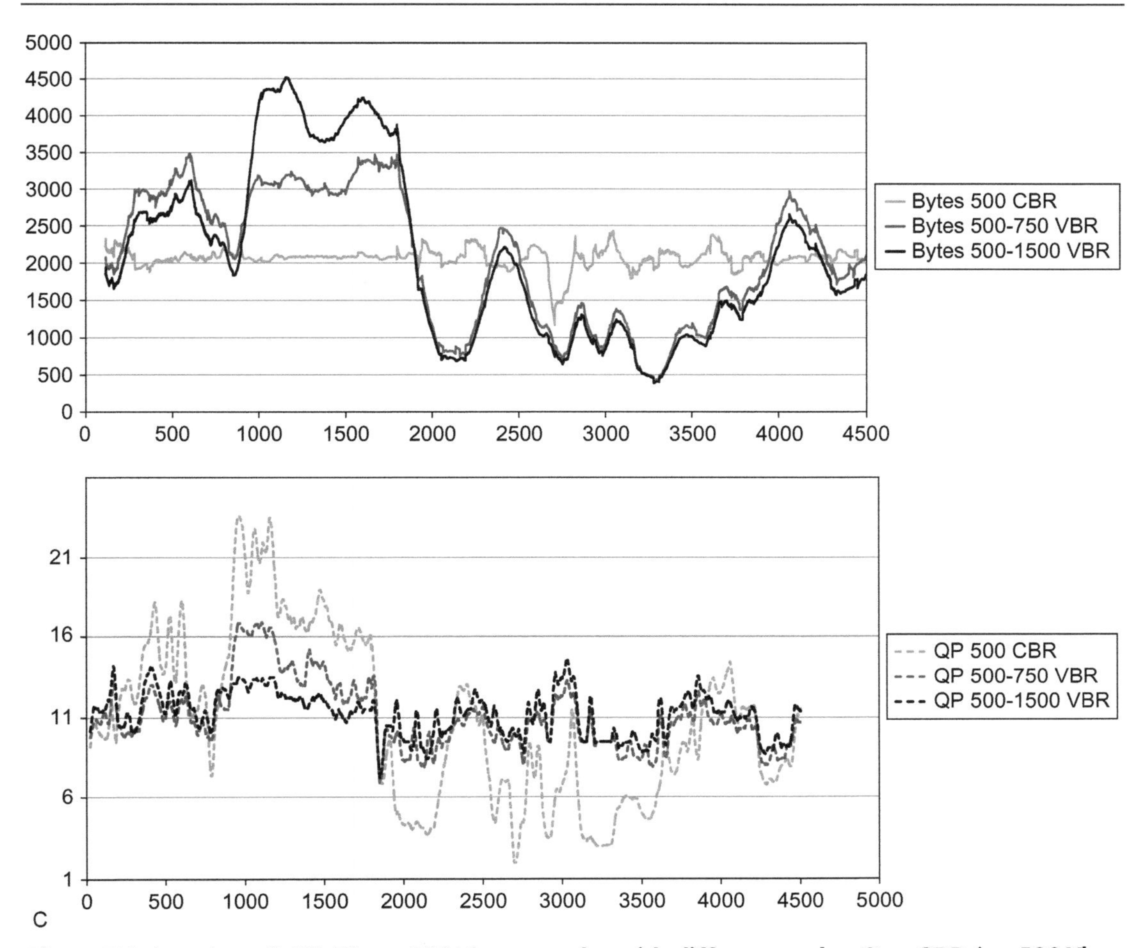

Figure 7.2 (continued) (C) Three 500 Kbps encodes with different peaks: One CBR (so 500 Kbps peak), one VBR at 750 Kbps peak, the last VBR with 1500 Kbps peak. As bitrate variability increases, we see bigger changes in bitrate, but much smaller changes in QP. The highest QPs are a lot lower with even a 750 Kbps peak, and getting almost flat at 1500 Kbps.

The limitation of traditional 1-pass codecs is that they have no foreknowledge of how complex future content is and thus how to optimally set frame types and distribute bits. This is generally less challenging for CBR encodes, since there is limited ability to redistribute bits anyway, but is a bigger deal for VBR. Although a lot of work has gone into finding optimal data rate allocation algorithms for 1-pass codecs, there will always be files on which different implementations will guess wrong, either over-allocating or under-allocating bits. An over-allocation can be especially bad, because when the data rates go up above the average they will have to eventually go below the average by the same amount. If this happens when the video is getting still more complex, quality can get dramatically awful.

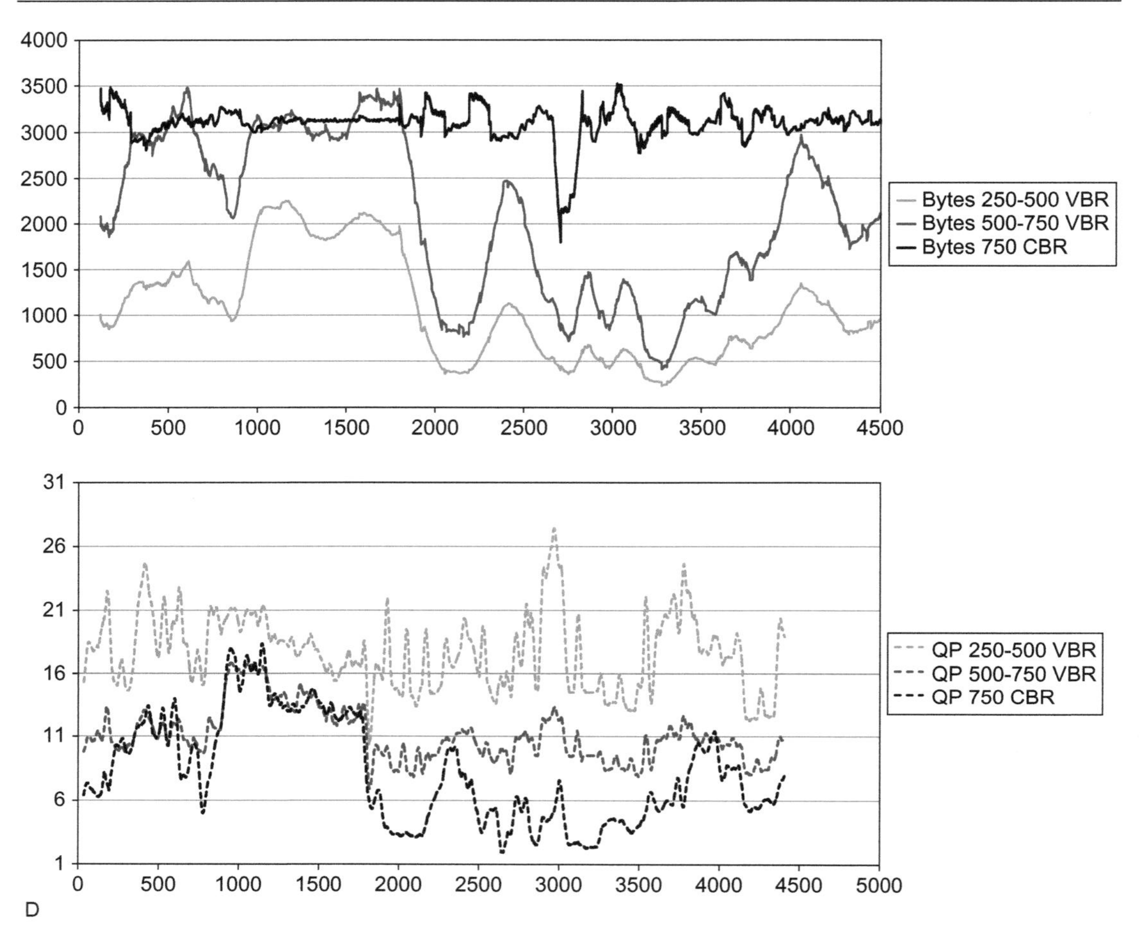

Figure 7.2 (continued) (D) Three encodes with the same peak bitrate of 750 Kbps, but different averages of 250, 500, and 750. As you'd expect, the CBR encode has a flat bitrate and a pretty variable QP, always lower than the streams with a lower average. But the 500 and 750 Kbps streams match closely in terms of the hardest part of the video, where they both use the full 750 Kbps peak and hit the same QP. The CBR is able to make the easier frames better. The 250 Kbps stream has a pretty consistent QP and a more variable data rate; the big ratio between peak and average lets it act almost like a fixed quality encode.

1-Pass with lookahead

Some codecs are able to use "lookahead" techniques to provide some of the quality advantages of 2-pass encoding in a 1-pass encode. For example, most VC-1 implementations are able to buffer up to 16 frames to analyze for scene changes, fades to/from black, and flash frames, and then set each frame for the optimum mode. Some also support lookahead rate control, where bitrate itself is tuned based on future frames in the buffer. Telestream's Episode is actually doing 1-pass with lookahead in its "two-pass" modes, with the final encode pass following up to 500

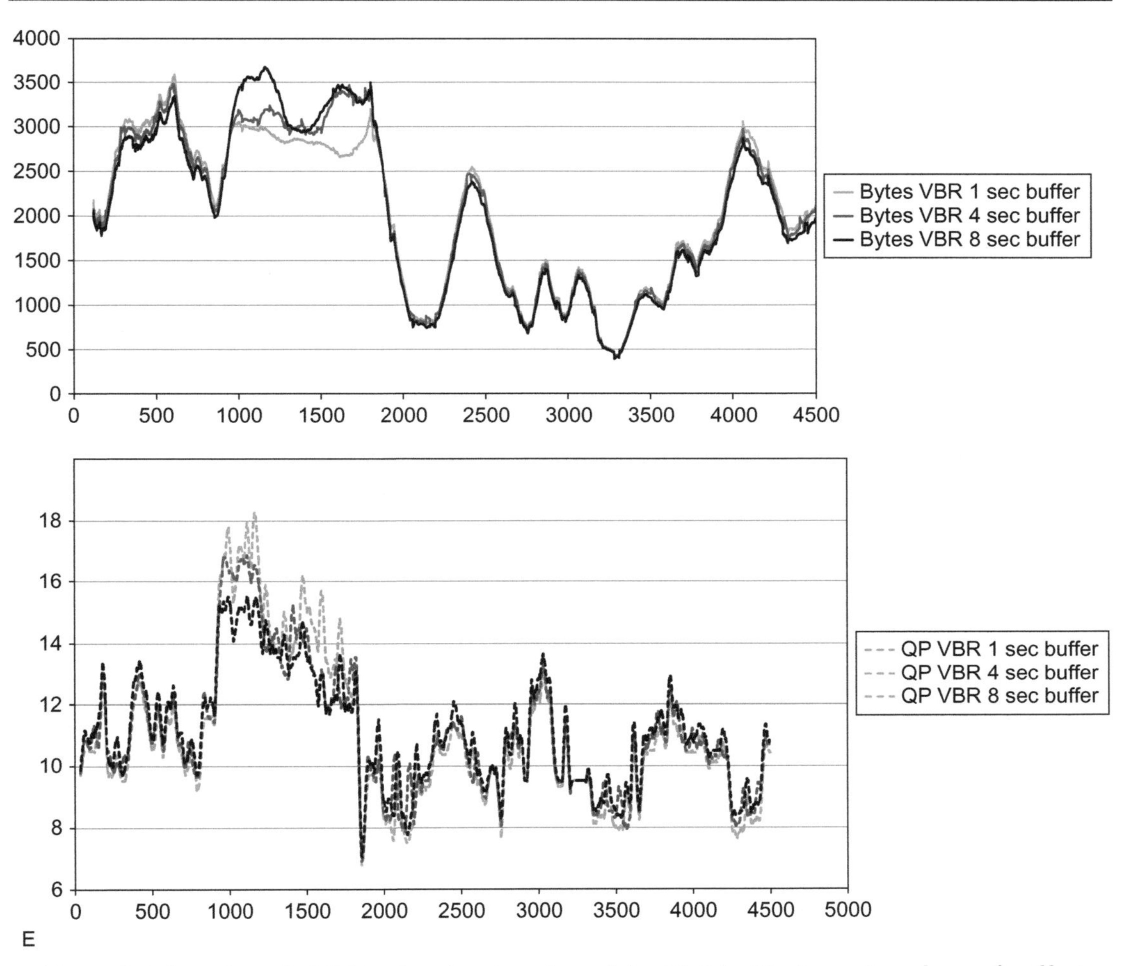

Figure 7.2 (continued) (E) Varying the duration of the VBR buffer has a less dramatic effect than increasing the peak, but it's still very helpful. The larger buffers result in peak QPs getting a lot lower.

frames behind the first. This is excellent use of highly multicore systems, since different cores can working on the lookahead and primary at the same time.

2-Pass

2-pass codecs first do an analysis pass, where they essentially do a scratch encode to figure out how hard each frame is. Once the entire file has been analyzed, the codec calculated an optimal bit budget to achieve the highest possible average quality over the file. In essence, this lets the codec see into the future, so it always knows when and how much to vary the data rate. 2-pass compression can yield substantial improvements in compression efficiency for VBR encodes, easily 50 percent with highly variable content. For example, it knows how many bits it can save from the end credits, and apply them to the opening title sequence.

2-pass is less helpful for CBR in general, but can vary quite a bit by codec type. Quite a few codecs don't offer a 2-pass CBR at all, reserving 2-pass for VBR modes. Going to 2-pass doesn't always mean that compression time will be doubled. Many codecs use a much quicker analysis mode on the first pass.

3-Pass and beyond

A few encoders, including x264 and CinemaCraft, offer a third or even more passes. Much like the first pass is a temp encode that is refined, the third pass is able to take the output from the first pass and further refine it. Most of the value of multipass is achieved in the second pass, but for highly variable content or a big average/peak ratio, a third is sometimes further help.

Segment re-encoding

Segment re-encoding is when the encoder is able to encode just specific sections of the video, leaving others alone. This is sometimes called third pass, but there's no reason why there couldn't be three or one passes first, or that the tweaking will only take one additional pass. A few encoders can do this automatically, like QuickTime's H.264 encoder.

More commonly, segment re-encoding is a manual process, with a compressionist picking particular shots that aren't quite perfect and adjusting encoding settings for them. This is the domain of high-end, high-touch compression products typically targeting Hollywood-grade optical discs, like CineVision PSE and CinemaCraft. The main product targeting streaming and other formats with segment re-encoding is Inlet's Fathom. And it is glorious.

It can also be a real time rathole, since there's no clear definition of "done." The highest-paid compressionists do this kind of work for Hollywood DVD and Blu-ray titles.

Frame Size

While not typically listed as a codec setting, resolution is one of the most important parameters to consider when choosing a codec. With a particular source and frame size, a given codec needs a certain number of bits to provide adequate quality, so there is an important balancing act between resolution and bitrate.

Some codecs can use any resolution. Others may require each axis be divisible by two, by four, or by 16 (and some decoders offer better performance with Mod16 even if it's not strictly required). Others may have maximum resolutions, or support only one or a few resolutions.

Note the relationship between frame size and bitrate isn't linear. < *Math Warning!* > There's an old rule of thumb called the "Power of 0.75" that says data rate needs to be changed by the power of 0.75 of the relative change in frame size. By example, assume a video where 640×360 looks good at 1000 Kbps, and a 1280×720 version is needed. There are four times as many pixels in the new frame: $(1280 \times 720)/(640 \times 360) = 4$. And $4^{0.75} = 2.828$. Times the

1000 Kbps, that suggests 2828 Kbps would be about right for the same quality. This is a rule of thumb, and will vary in practice depending on how much detail is in the source.

That's a handy rule when figuring out the frame sizes for the different bitrates when doing multiple bitrate encoding. It's also why my most used apps in compression are probably Excel and Calc.exe.

Aspect Ratio/Pixel Shape

Traditionally, web video always used square pixels, and hence the aspect ratio is solely determined by the frame size. But that's not a hard-and-fast rule, and there's a place for doing anamorphic encoding. Here are the times I consider using anamorphic:

- When the output formats require it. There's no square-pixel mode allowed for DVD.
- When I'm limited by the source resolution. If I've got enough bitrate for 1080p, but my source is an anamorphic 1440 × 1080p24, I'll encode matching that frame size. There's no point in encoding more pixels than are in the source!
- When motion is very strongly along one axis. For example, when we did the NCAA March Madness event with CBS Sports, we found that the motion during the games was very strongly biased along the horizontal. By squeezing the video to 75 percent of the original wide (like 480 × 360 instead of 640 × 360) we were able to get better overall compression efficiency.

Square-pixel is slightly more efficient to encode, so it's the right default when there's not an obvious reason to do something else.

Bit Depth and Color Space

Modern codecs in common use for content delivery are all 8-bit per channel. While H.264 has potentially interesting 10- and 12-bit modes, neither are supported by existing CE devices and standards. 10-bit would mainly be used in making an archival or mezzanine file for later processing using a codec like Cineform, DNxHD, or ProRes.

Most delivery codecs are 4:2:0 as well, with potentially future exceptions being the same H.264 High Profile variants. Similarly, 4:2:2 would mainly be used in an intermediate/archive/mezzanine file, particularly with interlaced source.

Frame Rate

Modern formats let you specify duration per frame, changing frame rate on the fly. Some older formats like VideoCD only offer a specific frame rate, and MPEG-1 and MPEG-2 have a limited number of options (the native frame rates for PAL, NTSC, and film).

More modern formats like MPEG-4 allow variable frame rates as well, where the timebase itself changes between parts of the file. However, most tools don't support that directly.

In general, you want to use the same frame rate as the source. For anything other than motion graphics or screen recordings, less than 24 fps starts looking less like a moving image and more like a really fast slideshow. Most web and device video thus winds up in the 24-30 fps range.

However, for PC playback, we've now got PCs fast enough to do 50/60p playback well, which can deliver a sense of immediacy and vibrancy well beyond what 30p can do. So when you've got high-rate progressive content and a reasonable amount of bandwidth, consider using the full frame rate. And with interlaced sources, it's possible to use a bob technique (described in the previous chapter) to encode one frame out of each field, so taking 25i to 50p and 30i to 60p. That's a lot more effort in preprocessing, but can be worth it for sports and other content with fast motion.

Interestingly, the bigger the frame size, the more sensitive we are to frame rate. So even if an iPod could play back 60p, it wouldn't have that much of an impact on the experience. But stick 60p on an HD display, and it can be breathtaking.

Frame rate has a less linear impact on bitrate than you might imagine, and less even than frame size. When the encoded frames are closer in time, there's less time for the image to change between frames, and thus interframe encoding can be a lot more efficient. Also, the less time each frame is on the screen, the less time for any artifact or noise to be seen; they average out more, letting the underlying video shine through better.

Thus, it's only at really low bitrates where I'd even consider dropping frame rate below 24 fps; south of 300 Kbps for modern codecs. It's almost always better bang for the bit to reduce the frame size first, down to 320 × 240 or less, before reducing frame rate below 24 fps.

When reducing frame rate, the central rule is to only divide it by a full number, as discussed in preprocessing. But it's such an important point (it's second only to encoding interlaced as progressive and getting the aspect ratio wrong in things for which I will mock you in public for getting wrong), here's that table again.

Table 7.1 Acceptable Frame Rates.

Frame Rate	Full	Half	Third	Quarter	Fifth
23.976	23.976	11.988	7.992	5.994	4.795
24.000	24.000	12.000	8.000	6.000	4.800
25.000	25.000	12.500	8.333	6.250	5.000
29.970	29.970	14.985	9.990	7.493	5.994
30.000	30.000	15.000	10.000	7.500	6.000
50.000	50.000	25.000	16.667	12.500	10.000
59.940	59.940	29.970	19.980	14.985	11.988
60.000	60.000	30.000	20.000	15.000	12.000
120.000	120.000	60.000	40.000	30.000	24.000

Even the simple difference between 23.976 and 24 matters; getting the frame rate off by that 1000/1001 difference means that one of a thousand frames will be dropped or duplicated. That can absolutely be noticed if there's motion at that moment, and that'll happen about every 41 seconds.

Keyframe Rate/GOP Length

Virtually all codecs let you specify the frequency of keyframes, also called the GOP length. In most cases, the parameter is really "keyframe at least every" instead of "keyframe every." The codec will also insert natural keyframes at cuts and this resets the "counter" for keyframe every. For example, having a keyframe rate of "every 100" normally doesn't mean you'll get a keyframe at frames 1, 101, 201, 301, and so on. In this case, if you had scene changes triggering natural keyframes at 30 and 140, you'd get keyframes at 1, 30, 130, 140, 240, and so on.

This is the optimal behavior. The main use of keyframes is to ensure the bitstream has sufficient keyframes to support random access, and to keep playback from being disrupted for too long after a dropped frame or other hiccup.

The drawback of keyframes is that they don't compress as efficiently as delta frames. By sticking them at scene changes which need to be intra coded anyway, we wind up not wasting any extra bits.

One concern in having too frequent fixed frames is "keyframe strobing," where the visual quality of a keyframe is different from that of the surrounding delta frames. With a short GOP, a regular pulsing can be quite visible and annoying. Modern codecs have made great strides in reducing strobing, particularly when doing 2-pass, lookahead, or VBR encoding. Open GOP also reduces keyframe strobing.

Typical keyframe values vary wildly between formats and uses. In interframe-only authoring codecs, every frame is a keyframe. MPEG-1 and MPEG-2 typically use a keyframe every half second. This is partly because those formats are typically used in environments very sensitive to latency (set-top boxes and disc-based media), and because they have a little bit of potential drift every frame due to variability in the precision of the inverse DCT, which means you don't want to have too long a chain of reference frames.

For the web, the GOP lengths are typically 1–10 seconds. Beyond that, the cost to random access is high, but the compression efficiency gains aren't significant. However, if keyframe flashing is a significant problem, reducing the number of keyframes will make the flashes less common. Generally the GOP length goes down as the bitrate goes up, since the efficiency hit on I-frames is lower, but the cost of decoding a frame for random access is higher. Note that the random access hit is proportional to the number of reference frames per GOP, so B-frames don't count as part of that.

Inserted Keyframes

A few tools let you manually specify particular frames to be keyframes. Like natural keyframes, inserted keyframes typically reset the GOP length target.

This was critical when Cinepak was the dominant codec, as its automatic keyframe detection was so lousy. Modern codecs are a lot better at inserting keyframes automatically, so you should only do manual keyframing if there's a particular problem you're trying to address. For example, a slow dissolve might not have any single frame that's different enough than the previous to trigger a keyframe, so you can make the first full frame after the dissolve a keyframe. More commonly, manual keyframes are used to ensure easy random access to a particular frame; many DVD authoring tools set chapter marked frames as I-frames as well. The markers set in tools like Premiere and Final Cut Pro are used for that.

B-Frames

As discussed back in the compression fundamentals chapter, a B-frame is a bidirectional frame that can reference the previous and next I or P frame. These are available in the modern codecs, although off by default in some tools, particularly in older Windows Media encoders.

B-Frames normally improve quality for a few reasons:

- Bidirectional references are more efficient, reducing bits needed on average substantially.
- Since a B-frame isn't a reference frame, they can be encoded with only enough bits to look good enough, without worrying about being a good reference for later frames.
- The net effect is that bits saved on B-frames can be spent on I and P frames, raising the quality of the reference frames that the B-frames are based on in the first place.

Beyond compression efficiency, B-frames have some performance advantages as well.

- Because no frames are based on them, B-frames can be dropped on playback under CPU stress without messing up future frames.
- Because B-frames reduce the number of P-frames between I-frames, random access is faster as the number of reference frames needed to be decoded to skip to a particular frame goes down.

Open/Closed GOP

Closed GOP is one of those terms of art that seems so minor, but keeps coming up again and again.

The core idea is simple. In an Open GOP, the first frame of the GOP can actually be a B-frame, with the I-frame following. That B-frame can reference the last P-frame in the *previous* GOP, breaking the whole "each GOP is an independent set of frames" concept. But Open GOP improves compression efficiency a little, and can be a big help in reducing keyframe strobing, since that B-frame can be a mixture of the old and new GOP, and thus smoothes out the transition.

Normally Open GOP is fine, and is the default in many codecs and tools.

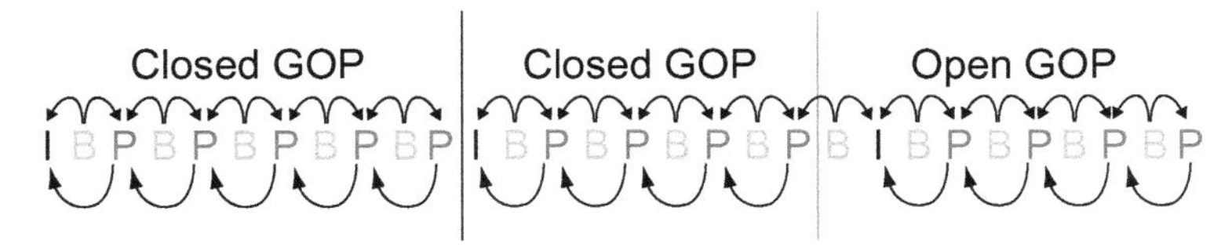

Figure 7.3 Closed and Open GOPs. An Open GOP starts with a B – frame that can reference the last P – frame of the previous GOP.

Minimum Frame Quality

Web-oriented codecs often expose a "Quality" control that is actually a spatial quality threshold. This threshold sets a maximum quantization (and thus minimum quality) allowed for any frame. However, since a bunch of big frames could overflow the VBV, the codec eventually will have to drop frames in order to maintain the minimum quality. The net effect of spatial quality threshold functions is that when the video gets more complex, the quality of each frame remains high, but the frame rate could drop, sometimes precipitously.

Most users find irregular frame rates more irritating than a lower, but steady frame rate. For that reason, I normally set spatial quality thresholds to their absolute minimum. If I find the spatial quality isn't high enough, I reduce the frame size, or in extremis, drop the frame rate in half. I'd rather watch a solid 15 fps than video bouncing between 30 fps and 7.5 fps.

Encoder Complexity

Many codecs offer options to tune for higher quality (at slower speed) or higher speed (at lower quality). The primary technique to speed up encoding is to reduce the scope and precision of the motion search. Depending on the content, the quality difference may be imperceptible or substantial; content with more motion yields more noticeable differences. Speed difference with a particular codec can vary widely; 10x or even 100x between the absolute slowest and absolute fastest modes. However, complexity only impacts the codec's portion of the compression process, not the source decode and preprocessing. So, if the bulk of the compression time is spent decoding AVCHD and doing a high quality scale down to a portable screen size, the net impact of different codec modes of the final 320x176 encode can be small.

My philosophy is that I can buy more and faster computers, but I can't buy users more bandwidth. Thus, I err on the side of slower, higher-quality modes. I certainly do comps in faster modes, to verify that the preprocessing is correct and the overall compression settings seem correct, and to test integration. But if I encode at a setting that might be leaving some pixels on the floor, I often wind up redoing it in the more complex mode "just to see if it would make a difference." That said, it's easy to get past the point of diminishing returns. Making the encode 3x slower than the default may yield an extra 15 percent efficiency. But it might take a further 5x slowdown to get that last 5 percent more efficiency. It's only worth taking more time as long as it is yielding a visible improvement in quality.

When used for real-time encoding, codecs may also make trade-offs to internally optimize themselves for speed. Speed is much more critical in real-time compression, because if the encoder isn't able to keep up, frames will be dropped on capture.

Different encoders can have wildly varying and large numbers of options, which I'll talk about in their particular chapters.

Achieving Balanced Mediocrity with Video Compression

I introduced the concept of "balanced mediocrity" in Chapter 4. Balanced mediocrity is supposed to be an amusing name for a very serious concept: getting all the compression settings in the right balance. You'll know when you've achieved the right balance when changing any parameter, in any direction, reduces the quality of the user's experience. "Quality" in this case equates with "fitness for use." A quality video compression is one that's most fit for its intended purpose—"quality" exists only in context.

For example, take a file compressed at 30 fps. If 30 fps is the optimal value, raising it to 60 fps would mean the fewer bits per frame hurt the file more than having smoother motion helped it. Conversely, reducing the frame rate to 15 would hurt the perception of motion more than any improvement in per-frame image quality would.

Choosing a Codec

Picking a codec is so basic and so important, it can be difficult to conceptualize the trade-offs. Sometimes it doesn't matter at all; only one is available for a device, or there is a clearly superior one. Other times there are several available bitstream choices or encoders for the right bitstream. There are a number of things you're looking for in a codec, but they all revolve around solving the problem posed by our three questions:

1. What is your content?
2. What are your communication goals?
3. Who's your audience?

You want the codec that can do the best job at delivering your content to your audience, while meeting your communication goals.

The three big issues you face in picking a codec are compression efficiency, playback performance, and availability.

Compression efficiency

Codecs vary radically in compression efficiency. New ones are much, much better than old ones. For example, H.264 High Profile can achieve quality at 100 Kbps that Cinepak required more than 1,000 Kbps for. Some codecs work well across a wide range of data rates; others have a minimum floor below which results aren't acceptable.

Playback performance

Playback performance determines how fast the playback computer needs to be to decode the video. The slower the decoder, the faster a computer needed for playback, or the lower resolution and frame rate acceptable. Playback performance is typically proportional to pixels per second: $height \times width \times fps$. Data rate can also have a significant effect. So, the higher the target resolution, frame rate, and data rate, the faster the playback computer will need to be. If it's a cut scene on a game that requires a Core 2 Duo processor, 1080p VC-1 is easily handled. But if the job is playing back six different video streams at once in Silverlight on a 1.8 GHz single-core P4, they better be pretty easy streams! Note that slower CPUs may allow less postprocessing in some players; hence a file may display at lower quality, but at full frame rate. B-frames may also be dropped on playback on slower machines. The worst case would be that only keyframes would be played.

Availability

Great-looking content that can't be seen doesn't count. When you think of availability, the questions you should ask yourself are: What percentage of the audience already can play that format and codec? How many of the rest are willing to download it? Are a lot of potential viewers on managed corporate desktops where they can't install anything?

CHAPTER 8

Using Audio Codecs

This chapter is about the common attributes of audio codecs and the decisions to be made when using them. It follows the structure of the previous chapter on video codecs.

While video may take 80–98 percent of the compression time and bandwidth, audio still makes up at least half the experience—audiences are more likely to stick with bad video with decent audio than decent video with bad audio. So getting the audio right is just as important as getting the video to look good. The good news is that it's also quite a bit easier; if targeting modern formats, there's really never a reason for audio to be distactingly overcompressed.

Choosing Audio Codecs

General-Purpose Codecs vs. Speech Codecs

Audio codecs fall into two main camps: general-purpose codecs and speech codecs.

General-purpose codecs are designed to do well with music, sound effects, speech, and everything else people listen to.

Speech codecs do speech well, but generally can't reproduce other kinds of content well. Commonly used speech codecs include AMR, CELP, and WMA 9 Voice. However, speech codecs can go to lower data rates than can general-purpose codecs, and generally provide better quality with speech content below 32 Kbps. They're also low complexity for easy playback on phones. Speech codecs generally only support monophonic, single-channel sound. And, of course, any nonspeech content in the audio will either be distorted or removed. Most speech codecs also have a pretty low maximum bitrate—32 Kbps on the outside.

The current generation of general-purpose audio codecs using frequency synthesis, notably HE AAC and WMA 10 Pro, can be competitive with speech at much lower bitrates than older codces, so it isn't always necessary to use a speech codec even for low-bitrate speech. Coupled with the general increase in bandwidth, speech-only codecs are increasingly relegated to mobile devices.

Sample Rate

As discussed in Chapter 3, the audio sample rate controls the range of frequencies that can be compressed. 44.1 kHz is "CD-quality" and anything much less starts demonstrating audible reductions in quality. 32 KHz is generally about as low as music can go while being at least FM quality, and 22.05 KHz is about as low as music can go and still make people want to dance. Speech can go much lower: 8 KHz is telephone quality. It's a bit low for high-pitched voices, like small kids, but fine for most adults.

As long as you've got at least 48 Kbps, you can do a decent 44.1 KHz with modern codecs and most content. Going higher than 44.1 doesn't really add any quality for the human listener, although dogs and bats would be appreciative. There's no reason not to use 48 KHz if that's your source and you've got a high enough bitrate. More than that for content delivery is just wasted bits.

Bit Depth

Thank goodness: modern audio codecs are all at least 16-bit. We had 8-bit compression in the paleolithic age of computer multimedia, and it was a lot of work to make it merely awful instead of unbearable. Once the 16-bit, 4:1 compressed IMA codec became common in 1995, 8-bit died off quickly.

Similarly to 44.1 KHz, 16-bit is a "good enough" level when correctly mastered; many fewer bits and it can sound terrible, but more are unlikely to make a perceptible improvement. That said, some codecs do support native input of 24-bit audio. Since the codec is really storing frequency data, it doesn't need a precise bit depth. If you've got greater-than-16-bit source and your codec supports that for input, go ahead and use it.

There is one 12-bit codec: DV's four-channel mode. 12-bit simply doesn't have enough bits for high-quality encoding, and shouldn't be used for professional work.

Channels

Most web video source uses mono or stereo audio, while home theater codecs like AC-3 and DTS support multichannel output from DVDs. For users that have multichannel audio out of their computers, we have good multichannel codecs in WMA Professional and AAC, both of which support up to 7.1 audio. While office workers certainly don't have 5.1 systems, they're becoming increasingly common for gamers and home theater setups. Content targeting those audiences can benefit from encoding beyond stereo. All media players will automatically "fold down" multichannel to stereo or even mono if that's all that's available for output.

Unfortunately, both Flash (as of 10) and Silverlight (as of 3) always fold down multichannel to stereo 44.1 for output. I hope we'll see this change in the future.

Older codecs allocated equal amounts of bandwidth per channel, which meant stereo required twice the bandwidth of mono. Fortunately, modern codecs use the redundancy between channels to reduce the additional bandwidth required (as described in Chapter 3). Prior to frequency synthesis, going to stereo from mono required around 20 to 50 percent more bits to achieve the same sound quality (need more bits the more stereo separation there was; identical mono audio in both channels shouldn't require an increase in bitrate).

HE AAC v2 (but not v1) and WMA 10 Pro LBR apply frequency synthesis to stereo separation as well, making stereo coding very efficient. WMA 10 Pro doesn't even offer mono for this reason.

Data Rate

Originally audio codecs didn't specify a data rate. Instead, their size was determined by number of channels and the sample rate. The math is simple: sample rate in K × channels × bits is kilobits per second for uncompressed. Divide that by the compression rate for compressed codecs. So IMA 44.1 KHz stereo is 44.1 × 2×16, divided by 4 for IMA's 4:1 compression ratio, and thus 353 Kbps.

Modern codecs offer a range of bitrates. Generally the available sample rate/channel combinations depend on the bitrate.

CBR and VBR

Traditionally, audio codecs were CBR, with each timeslice of audio (around 1/48th of a second, called a "frame" but not aligned with video frames) around the same size. But even CBR codecs have a little flexibility; MP3 and others use a "bit reservoir" where unused bandwidth from previous blocks can be used by future, more difficult to encode blocks. Like video, audio decoders use a buffer, and hence have a VBV, although that's not an encode-time configurable parameter in most audio encoders.

There are many kinds of VBR in audio codecs. There can even be big differences in implementations targeting the same bitstream.

Fixed quality

Like a quality-constrained video codec, a fixed-quality audio codec provides consistent quality with highly variable data rates. This form of codec is mainly useful for archiving content or music only playback. While the size of any one file may not be predictable, generally a given type of content will average out to a particular average compression ratio over a large number of tracks.

"Average" bitrate (ABR)

An average bitrate encode raises and lowers the data rate with quality, with a target and range. This is analogous to 1-pass video data rate-limited video codecs but with much laxer constraints. Options for maximum and minimum bitrates might also be provided. In practice, this yields a result halfway between fixed quality and true bitrate control. It's great for building out a music library where users care more about quality and total library size than bitrate control. But it's not a good fit for streaming or soundtracks in video where there are more precise requirements.

Constant bitrate (CBR)

Audio codecs can do real CBR as well, required for streaming. It's a fixed bitrate with a VBV, and prone to having challenging sections sound worse than easy sections.

Most implementations are 1-pass only. The WMA family of codecs support 2-pass CBR as well, but it's extremely rare to have an audible improvement from the second pass.

Some CBR modes will allow easy sections of the audio (like silence) to use a lower bitrate; others will insert padding bits if needed.

Variable bitrate (VBR)

There's a fair amount of confusion in terminology around VBR audio. Many tools call ABR or even fixed quality "VBR"—and they are variable. In the interest of consistency, I'm using the term in the same sense we did for video: a specified average bitrate, lower than the peak bitrate, and hence variability within the file.

This is a relatively rare mode in audio, with the WMA family the main place it has been used. But I'm a big fan of it for any time I'd be using VBR for video: bitrate is varied to keep quality more constant. Plenty of songs can encode fine at lower rates until high-frequency percussion like a hi-hat comes in, and along with it fairy bells and other artifacts. With a VBR codec, bits would be shifted to those sections for a big improvement, while still preserving a predictable file size. For highly variable content like film soundtracks, this can cut average bitrate requirements in half.

Additive VBR

An additive VBR, common to MP3, lets you set a target data rate, but it will raise the data rate above that point for difficult audio. Therefore you know the minimum data rate, but not the maximum. Still, with an output file at a given final bitrate, the additive VBR should sound better than a CBR of the same file.

Additive isn't as good as an average bit rate VBR for most uses, since neither the final data rate nor the quality can be controlled.

Subtractive VBR

In a subtractive VBR, the audio data rate can go down for easy sections, but never goes up higher for the difficult sections. Subtractive VBR can be a good mode for encoding web content, since you can assure some minimum quality, but can save bits on easy-to-encode sections.

Encoding Speed

When encoding audio as a soundtrack to video, the speed of audio is vanishingly small compared to the video. Even complex multichannel audio encoding runs many times faster than realtime on a modern machine.

The only time people really care about audio encoding time is for big batches of audio-only content, like a music library transcode. Even CD ripping is limited by the speed of the drive, not the codec these days. There are MP3 codecs that have multiple speed/quality modes, but lots of implementations have just one speed. If you have a speed control, I recommend tuning it to a slower, higher-quality option unless you're in a rush.

Tradeoffs

Generally speaking, the only tradeoff with audio is to find out how high a bitrate you need to sound awesome, and use that. Broadband and sideloaded content should always sound great; it's only streaming at low rates, particularly to phones, where there could be audible quality compromises.

In general, I'd say more compressionists err on the side of too little audio bitrate versus too high. For a 500 Kbps stream, going from 32 to 64 Kbps for audio can result in a listening experience that goes from weak to stellar, and then reducing the video bitrate from 468 to 436 may not even be noticible. So, raise your audio bitrate until a higher rate isn't appreciably better, or until the resulting reduction in video bandwidth and quality hurts more than the increase in audio helps.

Sample Rate

For broadband and CD-ROM uses, set your sample rate to at least 32 kHz. But keep in mind that higher frequencies may cause artifacts. Lower sample rates are also faster to encode and play back.

Bit Depth

This is simple. If using a 16-bit codec, great. If you've got greater-than-16-bit source and a codec that can use that, go for it, but don't go to a lot of extra work to get there.

Channels

For most listeners, stereo is less important than having a high sample rate and few artifacts. If using an older codec at low bitrates, go to stereo only if you've got enough bandwidth for the stereo to be 44.1 KHz and sound better than a mono version.

Stereo Encoding Mode

Most codecs don't offer a choice of stereo encoding modes. Of those that do, joint stereo is the best option for compression efficiency, but normal stereo is best if you need to keep accurate phase relationships for Dolby Surround–encoded content, or are doing a high bitrate archive.

Data Rate

Higher data rates make audio sound better, but they require more bits. Typically, the data rate of a file allocated to audio is much less than for video. Note that some sample rate, bit depth, and channel combinations may require particular bitrates.

CBR vs. VBR

If you're encoding a soundtrack for a VBR video file, make the audio VBR as well. Note that some codecs only offer VBR at some bitrates. For example, WMA 9.2 and 10 Pro offer VBR for their middle bitrates, but are CBR only at the high and low end.

CHAPTER 9

MPEG 1 and 2

MPEG-1

MPEG-1, released in October 1992, was the Moving Pictures Experts Group's first standard, and it offered revolutionary quality for its time. Although it has been eclipsed by later technologies, it laid the foundation for them as well as digital video for consumer electronics.

MPEG started out following the success and technical approaches of JPEG for still images and H.261 for videoconferencing, and as such is a pretty typical 8×8 DCT codec with motion compensation.

The initial hype for MPEG-1 was its application as a compression format for Video CD and interactive movies on desktop computers. The interactive movie angle never really caught on—it required users to buy a $200 card to get full-motion video, and having to sit back and watch a blocky video clip between doing stuff never amounted to compelling game play. It quickly became clear that users were much more interested in fully interactive real-time 3D games than in interactive cinema. Most CD-ROM projects made use of QuickTime or AVI, both of which offered greater flexibility and lower decode requirements. By the time computers sported CPUs fast enough for MPEG-1 playback to be ubiquitous, other codecs were emerging that offered superior compression efficiency and flexibility.

Initially, Video CD also looked to be an abysmal failure. The quality it produced was barely that of VHS, let alone the Laserdiscs then beloved by videophiles. However, once hardware costs dropped enough, Video CD became enormously popular in Asia, due to the prevalence of very cheap pirated movies and ubiquitous, inexpensive players.

MPEG-2

As soon as MPEG-1 was designed, work began on MPEG-2, focused on delivering a digital replacement for analog video tranmissions. It's called H.262 by the ITU.

Technically, MPEG-2 was a pretty straightforward enhancement of MPEG-1. The primary new feature was interlaced video, and there were many under-the-hood optimizations that improved compression efficiency.

By almost any standard, MPEG-2 is the most ubiquitous and important video standard in the world today, and will remain in wide use for years to come. While web and device video certainly get a lot of attention, MPEG-2 still gets more eyeball-hours than any other codec, through DVD, digital satellite and cable, and digital broadcasting. We are in the midst of a broad transition to H.264 from MPEG-2 in many sectors, driven by H.264 High Profile being capable of everything MPEG-2 was, but with substantially improved efficiency.

One warning: there's a huge array of stuff in the various MPEG specs that isn't used in practice (D-frames!). This chapter is going to survey MPEG-1 and MPEG-2 as they're used in practice, and doesn't attempt to be a survey of the specifications themselves.

MPEG File Formats

Elementary Stream

An MPEG elementary stream is just the video or audio track all by itself. For MPEG-2, this is generally .m2v. These are intermediate files for importing into an authoring or muxing tool, and aren't used by themselves, or even playable by many applications that can otherwise play a .mpg file.

Program Stream

A Program Stream is the MPEG file format, generally .mpg. A program stream contains multiplexed elementary streams, typically video and audio. Other kinds of data like captioning can be inserted into a program stream as "user data."

This was originally called a System Stream in MPEG-1, and later renamed; it's the same thing underneath.

Transport Stream

Transport streams are designed for real-time low-latency transmission through lossy environments, and so include a lot of error resiliency and recovery functions. They're standard in broadcast. Their extension is .ts.

The design of Transport Streams has proved to be very useful, and they continue to be the standard for broacast even when the MPEG-2 codec itself isn't used. Transport streams offer lower broadcast delay than streaming-oriented formats; we've seen some interesting use of transport stream delivery to Silverlight. Apple's new adaptive streaming technology also uses transport streams as the container file format.

Transport Streams are one of the most durable technologies to come out of MPEG, and will be used long after today's codecs have been replaced.

MPEG-1 Video

The MPEG-1 video codec took advantage of a lot of work already done in JPEG and H.261. Given the extreme limitations of the encoding and decoding horsepower available back in 1992, MPEG-1 was designed as a straight-up implementation of DCT without the advanced features modern codecs. MPEG-1 is a 4:2:0 color space codec, with 8×8 blocks and 16×16 macroblocks. It doesn't support partial macroblocks, so resolution must be divisible by 16 on both axes.

MPEG-1 is a very basic codec, designed to be playable in real time on the dedicated hardware available in the early 1990s. Thus it lacks a lot of features we've gotten used to in more modern codecs. Some of the biggies include:

- Partitioned motion vectors. All motion estimation is per macroblock. That's fine for big motion, but terrible when small things are going different directions inside a 16×16 region. Particle effects and rain are particularly problematic.
- Subpixel motion estimation. All motion vectors are aligned with the pixel grid. Slow motion isn't encoded efficiently.
- Variable block sizes. MPEG-1 is always 8×8.
- In-loop deblocking filter. If a reference frame gets quantized, all later frames carry that forward.
- Floating-point DCT. MPEG-1/2 used a classic DCT with floating-point math. This means that different hardware can vary slightly in the results of a calculation, as there's no official definition of the exact right answer with floating-point. Any single calculation will be quite close, but since the result of that calculation can be used as the input for the next, a cumulative DCT "drift" can result, making the video output less predictable over time.

The floating-point DCT has big impact, since it limits how many consecutive calcuations can be used safely. This is why MPEG-2 typcially only has a few reference frames, as discussed shortly.

In general, MPEG-1 does very well at the data rates and resolutions it was designed for: 1.5 Mbps at 352×240, 29.97 fps (NTSC) or 352×288, 25 fps (PAL). This data rate was chosen to match the transfer rate of an audio CD, so video CD players could be built around the physical mechanism of audio CD players.

Modern high-end MPEG-1 encoders with such enhancements as 2-pass encoding, better preprocessing, and more exhaustive motion search algorithms offer much higher efficiency than their predecessors available in MPEG-1's heyday. However, the lack of a in-loop deblocking filter means quality drops off sharply when the bitrate is insufficient for the quality. As you learned in Chapter 2, the human eye is much more attuned to sharp edges, so these blocky artifacts are quite distracting. While it's quite possible to do postprocessing deblocking with MPEG-1, it's not broadly implemented in players.

MPEG-2 Video

The MPEG-2 video encoder is an enhancement of MPEG-1, carrying on its features, often enhanced, and adding a few important new ones. The basic structure of macroblocks, GOPs, IBP frames, and so on, is the same. The important additions include:

- Support for interlaced video.
- Half-pel motion precision (MPEG-1 was full-pixel only; this is a big improvement).
- Allowing 8–11-bit DCT precision during the initial transform before quantization (MPEG-1 was 8-bit only) for more accurate gradients and more efficient compression at higher bitrates.
- Optional High Profile mode with 4:2:2 color.
- Defining aspect ratio as display aspect ratio, instead of the pixel aspect ratio used in MPEG-1.
- Restricted motion vector search range to simplify decoders for higher resolutions. Motion vectors can go up 128 pixels vertically and 1024 pixels horizontally. This can be an issue in HD: with 1080p24 IBBP, 128 pixels is 1/8 the frame height in an eighth of a second.
- 16×8 motion vector partitions allowed (MPEG-1 is 16×16 only).
- Some other under-the-hood tweaks that slightly improve compression; they're always on, so you don't need to worry about them.

For typical standard-definition content, MPEG-2 with a good VBR encoder is well served by average data rates between 2.5 Mbps and 6 Mbps, depending on content. MPEG-2 targeted and was tuned for higher-bitrate scenarios than MPEG-1, and can substantially outperform MPEG-1 at higher bitrates at the same frame size.

Interlaced Video

The biggest addition to MPEG-2 is native support for interlaced video, which was designed very nicely, with later codecs following the same mechanism. MPEG-2 video can be progressive or interlaced at three different levels:

- An interlaced stream can be progressive or interlaced; a progressive stream only has progressive frames.

- An interlaced stream can include both interlaced and progressive frames; progressive frames have progressive macroblocks only.
- An interlaced frame can have any macroblock and be either progressive or interlaced.

This flexibility is extremely useful. First, progressive-only content is easily detected for devices with progressive-only output. And for interlaced content, progressive frames can still be encoded as progressive. And in a frame where part of the image is moving and other parts aren't, the static sections can be encoded as progressive (for better efficiency), and the moving sections can be interlaced (which is required as the fields contain different information).

Because an interlaced macro block actually covers 32 lines of source instead of 16, the zigzag motion pattern normally used in MPEG for progressive scan must be modified. The alternate scan pattern attempts to go up and down two pixels for every one it goes left and right (see Figure 9.1). This was sometimes referred to as the "Yeltsin Walk" pattern after the famously unsteady former President of the Russian Federation.

"24p" in an interlaced stream

If we lived in a good and just world, all CE devices would support 24p MPEG-2 streams, and we'd just encode film sources with that and go home early. Alas, our world is not good and just, and the DVD and some other devices/standards (like CableLabs VOD) only support interlaced bitstreams (at least at some resolutions).

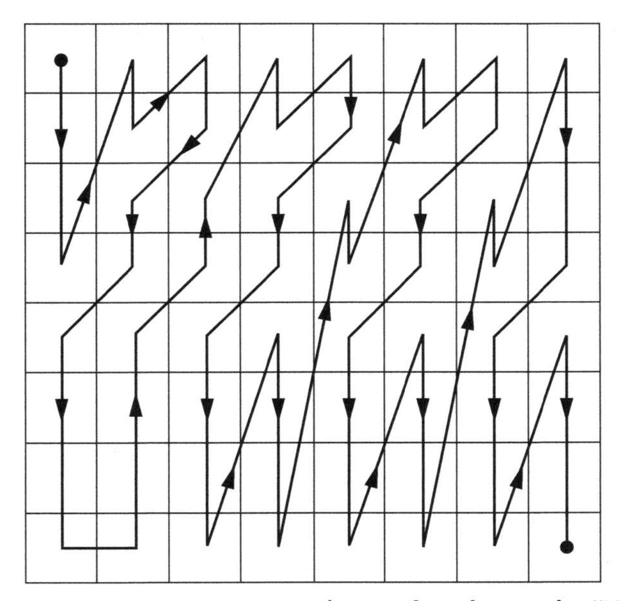

Figure 9.1 The alternate scan pattern was sometimes referred to as the "Yeltsin Walk" pattern after the famously unsteady Russian leader.

What's worse, those interlaced streams require 29.97 frame rate. So we can't even encode 23.976 progressive fps in that interlaced bitstream (although we can encode 30p that way just fine).

We actually have to encode with 3:2 pulldown, thus yielding a repeating pattern of three progressive frames and two interlaced frames. This isn't as bad as it sounds, as MPEG-2 has a field_repeat tag that can be used to mark a field as "don't waste any time on me, just reuse the same field from the previous frame." Since our 24 frames in a second get turned into 48 fields, but need to fill 60 fields, there are 12 repeated fields per second, and thus 12 field_repeat tags. This makes encoding "24p" almost as efficient as encoding real 24p.

Also, those 12 field_repeat tags come in a particular pattern that provides very good hints to a decoder for how to reassemble the original 24 frames out of the 48 unique fields, discarding the repeats. This is how "progressive" DVD players and software players on computers work. And is why movie content looks so much better from those players than 30i; no deinterlace is required and the full 480 lines of detail are preserved.

Modern NLEs and compression tools generally do a good job of hiding this for you under the covers; if you drag a 24p source into a DVD authoring app, it'll do the right thing just like it does for 4:3 and 16:9.

"24p" in 720p

The other annoying case is sticking 24p in a 720p bitstreams, since ATSC and other formats don't have a real 720p24 mode. It's the same idea as the previous example, but using frame_repeat. So, the 24p source gets turned into a repeating pattern of two and three copies of each frame to fill out the 60. It's just as efficient as a native 24p encode would be, but means that playback on the new 120 Hz displays isn't smooth. Instead of getting each frame shown five times ($5 \times 24 = 120$), we'd get a pattern of 4, 6, 4, 6 copies of each frame; classic NTSC judder again.

What Happened to MPEG-3?

MPEG-2 was originally targeted at SD resolutions, although the spec itself allows for absurdly high resolutions (the MPEG-1 spec can do 4096×4096). It wasn't obvious how well MPEG-2 could scale, so the initial plan was to produce a MPEG-3 spec for HD. However, MPEG-2 turned out to be just fine for HD without any obvious enhancements that would merit a new standard. Thus HD levels were just added to MPEG-2, and MPEG numbers skipped to MPEG-4.

Sometimes MP3 is called "MPEG-3," but that's a misnomer. It's really short for "MPEG-1 Layer III" as described next.

Table 9.1 MPEG-2 Profiles.

Abbr.	Name	Chroma Subsampling	Comment
MP	Main Profile	4:2:0	Content delivery
HP	High Profile	4:2:2	Content authoring

Table 9.2 MPEG-2 Levels.

Abbr.	Name	Max Width	Max Height	Max Framerate (fps)	Max Bitrate (Mbs)	Comment
ML	Main Level	720	576	30	15	Standard def, all DVD
H-14	High 1440	1440	1152	30	60	HDV
HL	High Level	1920	1152	720p=60 1080=30	80	ATSC, DVB, Blu-ray

MPEG-2 Profiles and Levels

These are the profiles and levels in common use for MPEG-2 (Tables 9.1 and 9.2). There's lots of others in the spec that aren't used in practice. Also, almost any CE device is going to have further constraints on encoding beyond Profile and Level. In particular, the max bitrates allowed by CE devices are almost always much lower than that of the specified level. For example, DVD is limited to a total bitrate of 9.8 Mbps, even though DVD's *Main Profile@Main Level* MPEG-2 has a maximum 15 Mbps bitrate.

Audio

MPEG-1 Audio

MPEG-1 introduced three audio codecs. All are based on similar technology with psychoacoustic modeling, with improving compression efficiency at higher levels. They're also increasingly complex on decode.

Layer I

MPEG-1 Audio Layer I sounds pretty good at higher bitrates, but offers very low compression efficiency. In practice all decoders that can do Layer 1 also do Layer 2, so that's used instead. Its main use was on the long-forgotten DAT competitor Digital Compact Cassette.

Layer II

MPEG-1 Audio Layer II audio eclipsed Layer 1 because it offers substantially better encoding efficiency and puts an insignificant hit on modern processors. Layer 2 audio is also supported in all MPEG-1 hardware playback devices, and is mandated as the audio codec for Video CD. The general rule of thumb is that Layer 2 needs about 50 percent higher bitrate to sound as good as MP3, so 192 Kbps Layer II sounds about as good as a 128 Kbps MP3. Video CD uses 224 Kbps Layer II audio, which is nearly CD audio quality.

Layer 2 is used broadly as one of the better patent-free audio codecs, as seen in HDV and DAB digital radio (although it is being replaced by HE AAC v2 in DAB+).

Layer 2 was also used in audio-only files before the MP3 explosion. MPEG-1, Layer 2 audio files use the .mp2 extension. MP2 was the high-quality format of choice on the pioneering Addicted to Noise music web site.

At the highest bitrates, Layer II can actually be more transparent than Layer III in some (albeit rare) cases.

Layer 2 data rates run from 32 to 384 Kbps, with 64 and below mono only.

Layer III

MPEG-1 Audio Layer III was never widely adopted in MPEG-1 authoring or playback tools due to patent licensing fees (otherwise not required by a MPEG-1 player). However, it became an enormously popular technology in its own right, albeit under its shortened name, MP3. Layer 3 data rates range from 32 to 320 Kbps.

You'll always want to use Layer II as a soundtrack for MPEG files, as very few players support Layer III in a MPEG file. For more details on MP3, see Chapter 10.

MPEG-2 Audio

Oddly enough, most MPEG-2 applications don't use MPEG audio codecs for audio. Instead, they use another audio bitstream that provide better efficiency or compatibility with existing players.

All DVD audio codecs default to 48 KHz, slightly higher than the 44.1 KHz of audio CD.

MPEG-2 extensions to MPEG-1 Layer II

MPEG-2 included extensions to MPEG-1 Layer II to handle new scenarios on lower and higher ends:

- Sub-32 KHz sampling for low bitrates
- MPEG Multichannel for up to 7.1 channels

Neither sees wide use. MPEG-2 isn't commonly used at bitrates low enough to take advantage of the low bitrate extensions. And while the PAL DVD spec did originally include

MPEG Multichannel, the NTSC version didn't and Dolby Digital became the universally supported multichannel codec. It was dropped as a mandatory codec for PAL DVD in 1997.

AAC

Advanced Audio Coding (AAC) was first designed for MPEG-2, although it's not used with MPEG-2 in practice. It became the default codec with MPEG-4, and is covered in chapter 13.

PCM

Consumer-level DVD authoring systems historically supported just uncompressed stereo tracks, to avoid the cost of licensing Dolby Digital. That licensing has become easier over the years, so fewer authoring tools and hence discs are PCM-only. Using uncompressed audio wastes a whole lot of bits better spent on video.

Dolby Digital (AC-3)

Most real-world MPEG-2 content is coupled with Dolby Digital (AC-3). Dolby is a mandatory codec (meaning all decoders must support it) in several important delivery formats:

- DVD
- Blu-ray
- ATSC
- DVB
- CableLabs

While AC-3 is a quite old codec this point, and not particularly efficient (AAC-LC outperforms it at lower bitrates, and WMA 10 Pro and HE AAC blow it out of the water), it has a very valuable legacy base of installed decoders. Every 5.1 reciever manufactured includes an AC-3 decoder, making AC-3 the only bistream guaranteed to work when connecting any CE devices to a receiver, be it through TOSLink, RF, or other mechanisms. Those older connection types don't have enough bandwidth for multichannel PCM, either (that was introduced with HDMI). So any device that's going to deliver multichannel audio to anything other than a system with dedicated multichannel output (like a PC with a 5.1 speaker system) will support AC-3, be it by reencoding from another codec, or just passing the bitstream on.

With sufficient bitrates, Dolby Digital provides good quality with most content, and even its maximum bitrates are quite small compared to typical MPEG-2.

Table 9.3 AC-3 Supported Channels.

Label	Channels	LFE?	Minimum Data Rate
1/0	Center	No LFE	56 Kbps
2/0	Left, Right	No LFE	96 Kbps
3/0	Left, Center, Right	LFE optional	128 Kbps
2/1	Left, Right, Surround	LFE optional	128 Kbps
3/1	Left, Center, Right, Surround	LFE optional	192 Kbps
2/2	Left, Right, Left Surround, Right Surround	LFE optional	192 Kbps
3/2	Left, Center, Right, Left Surround, Right Surround	LFE optional	224 Kbps

AC-3 provides a full range of channel options from mono to 5.1 (see Table 9.3). With a maximum data rate on DVD and ATSC of 448 Kbps, quality is very good for soundtrack content, although transients (like high-pitched percussion) can come out flattened in concert materials. Blu-ray can do AC-3 up to 640 Kbps, the maximum rate supported by existing decoders. Stereo is typically encoded at 192 or 224 Kbps.

Dolby Digital also supports Dialog Normalization, sometimes called DialNorm. This metadata should be set to the average dB of dialog, which allows the dynamic range of the audio to be compressed to allow intelligible speech while keeping explosions to a neighbor- or sleeping baby-approved level on the other. DialNorm is an art in its own right, and is something that needs to be determined by an audio engineer based on the actual mix.

There are enhanced Dolby Digital modes called Dolby Digital Plus and Dolby TrueHD used in Blu-ray, and discussed in Chapter 24. They are rarely coupled with MPEG-2.

DTS (Digital Theater Systems)

DTS, from Digital Theater Systems, is often positioned as the high-end alternative to AC-3. It is also sometimes used for theater sound, but never for broadcast. DTS's big advantage and big limitation in the marketplace is its relatively low compression. Originally designed as a theater standard with higher quality than Dolby Digital, it used CD-ROM as its data standard. The base version of DTS only does 5.1, but there's a backward-compatible 6.1 variant called DTS-ES. In DVD, DTS can use a data rate of 1,536 Kbps or 768 Kbps. The higher rate offers incredible transparency, but isn't often used on DVD due to space limitations. It's not clear whether the 768 Kbps mode is significantly better in general than AC-3 at 448 Kbps.

DTS's higher data rate requirements can mean DTS DVDs may have fewer features and cost more than AC-3 versions. However, aficionados say the superior audio makes this a small price to pay.

While most professional DVD authoring tools include AC-3, many do not include DTS. DTS is usually only used in high-end Hollywood productions.

Blu-ray makes DTS a mandatory codec, and so DTS-only titles are possible there.

MPEG Audio

Ironically, the actual MPEG-1/2 audio codecs are not commonly used with MPEG-2 in content distribution. Many ecosystems, including DVD, don't have MP2 audio as a mandatory codec. The main place where MP2 is used in in self-contained files targeting PC playback, like .mpg.

MPEG-1 for Universal Playback

Since MPEG-1 hasn't required licensing patents, it's been broadly supported for many years, and has been included with Windows since Windows 95 and every Mac since 1998. And while many codecs aren't available in Linux distributions due to patent licensing requirements, MPEG-1 is often still supported. Many devices will support it as well, including DVD players with "multiformat" options like Divx and MP3. So, if you want to have a single file that'll play on pretty much every working PC in the world, no matter how slow the CPU or ancient the OS version, MPEG-1 may be a good choice.

The drawback, of course, is that much higher bitrates are required; decent quality can require three or more times the bitrate than something using a modern codec.

For a universally supported MPEG-1, my recommendations are:

- Use square-pixel encoding. Many old media players, including Windows Media Player before version 9, always decode as square pixel so nonsquare pixels come out distorted.
- Use a maximum frame size of 640 × 480. Some ancient players don't handle anything above that.
- Use 2-pass VBR in the slowest, highest-quality mode your tool has. MPEG-1 is so simple that it'll still encode faster than real time on a laptop.
- Layer II audio needs a lot of bits. Use 192 Kbps at a minimum for decent quality. Use Joint Stereo for better efficiency at lower rates, and Normal for better quality above 224 Kbps.
- Audio should be 44.1 KHz, to improve compatibility with older players.
- Use a Program/System stream (same thing, different names) as the file format, with a .mpg extension.

- If you put it on a disc, I recommend the good old ISO 9660 format, which everything can read. That's what most ROM burners default to (some default to the superior but more recent UDF).

MPEG-2 for Authoring

Unsurprisingly, MPEG-2 is also used as a video production format, in capture, for mezzanine files, and even for editing. The consumer-focused HDV uses the same Main Profile used for content delivery. But most MPEG-2 in authoring uses High Profile, which supports 4:2:2 sampling, supported in MXF, XDCAM, and IMX.

4:2:2 is typically run with only I-frames at 50 Mbps (for SD). It's rather analogous to using Motion-JPEG at the same data rate, although it is somewhat more standardized.

For SD mezzanine files, MP@ML 15 Mbps interframe encoding can produce very nice quality, even with interlaced. And for stereo audio tracks, Layer II Normal stereo at 384 Kbps is very high quality.

One drawback to using MPEG-2 files for this is that most computers don't ship with built-in MPEG-2 decoders, so they may need to be added later. Windows 7 is the first mainstream OS to ship with out of the box MPEG-2 file playback. Most video editing and compression tools include MPEG-2 support in some fashion, but they can vary in what profiles are supported. Don't assume that any random person can play back or reencode from a MPEG-2 file.

MPEG-2 for Broadcast

MPEG-2 has been the leading codec for digital broadcasting, although it is starting to be displaced by H.264 . But the rest of the MPEG-2 technology stack, including MPEG-2 transport streams, looks to have a long future ahead of it.

The big reason for this was compression efficiency; MPEG-2 can squeeze about six channels of decent-quality SD video in the same amount of spectrum as a single analog channel. Thus it was a no-brainer for digital satellite and cable companies to adopt MPEG-2 technology, and continually push advancements in the technology, particulary for real-time encoding.

One of the key innovations here has been statistical multiplexing (statmux), which is so cool that I need to describe it briefly. Statmux is essentially an intra-band, inter-channel VBR. Using our example of six digital channels fitting into one analog channel, those six channels are combined into a single transport stream. While the total bandwidth of transport stream is fixed, any given channel can vary in its own data rate. So a statmux encoder will compress all the channels on the same band simultaneously, redistributing bits between them on the fly to achieve the best average quality at any given moment.

The unfortunate fact is that the focus of most digital broadcasting is on increasing the number of channels, with quality a much lower concern. Lots of digital content, particularly on satellite, winds up looking worse than a good analog signal would have (although certainly better than a bad analog signal).

ATSC

The U.S. format for high-definition digital television is called ATSC, from the Advanced Television Standards Committee. Like many committees, they had a lot of trouble making a definitive decision, and produced 28 different interlaced/progressive scan, resolution, and frame rate options. That was our last, best chance to kill off interlaced, but the TV manufacturers saw interlaced as a barrier to entry for the computer industry and so fought to keep it. Ironically, of course, the TV manufacturers died off or now have a big business selling computers and monitors, none of which do interlaced anymore anyway. The saga is well told in Joel Brinkley's *Defining Vision* (Harvest Books, 1998), probably the only high-drama nonfiction page-turner the compression industry shall ever have. I hope Amazon.com recommends it as a companion to this book.

Note that ATSC uses 704 × 480 instead of DVD's 720 × 480 (see Table 9.4). This is still the same aspect ratio—720 has eight pixels on either side of the frame that aren't included in the 704 broadcasts. So converting between 720 and 704 requires adding or subtracting eight pixels on the left/right.

You might think that you'll never be asked to encode ATSC, but I hope that's not true. Most broadcast content is authored well in advance, and it's silly to send it through a real-time encoder in that case. We have great software HD MPEG-2 encoders that can be used to good effect here today. That said, you definitely need an ATSC-compliant tool for ATSC, as there's lots of under the hood timing data and muxing parameters that must be followed religiously.

Personally, I still believe we only needed two HD formats: 720p60 and 1080p24. Unfortunately, all U.S. broadcasters doing 1080 are doing 1080i30.

Table 9.4 ATSC Options.

Resolution	Aspect Ratio	Interlaced fps (a.k.a. 30i)	Progressive fps (a.k.a. 24/30/60p)
640×480	4:3	29.97, 30	23.976, 24, 29.97, 30, 59.94, 60
704×480	4:3 or 16:9	29.97, 30	23.976, 24, 29.97, 30, 59.94, 60
1280×720	16:9		23.976, 24, 29.97, 30, 59.94, 60
1920×1080	16:9	29.97, 30	23.976, 24, 29.97, 30

Table 9.5 DVB/PAL ATSC Options.

Resolution	Aspect ratio	Interlaced fps (a.k.a. 25i)	Progressive fps (a.k.a. 25/50p)
352×240	4:3		25
352×576 480×576 544×576	4:3 or 16:9	25	25
704×576	4:3 or 16:9	25	25, 50
1280×720	16:9	25, 50	
1920×1080	16:9	25	25

DVB

DVB is the European/PAL equivalent of ATSC. Its biggest differences are its use of PAL frame rates, as described in Table 9.5.

DVB has branched out from just MPEG-2; some countries only do DVB with H.264. And DVB-H targeting broadcast to mobile devices is covered in the encoding for mobile devices chapter.

CableLabs

The other broadcast MPEG-2 format you may see in practice is the CableLabs VOD spec, which is used for VOD by most cable operators. The CableLabs spec is a strict subset ATSC. Fortunately, it only has three modes:

- Three-quarter screen SD: 528×480i30 with 3180 Kbps video and 192 Kbps 48 KHz stereo AC-3
- 720p: 1280×720, up to 10 Mbps video
- 1080i:1920×1080i30, up to 18.1 Mbps video

These are generally higher than used for broadcast at those frame sizes and can look quite good if encoded correctly. As they're by definition on demand, software encoders are normally used.

MPEG Compression Tips and Tricks

352 from 704 from 720

MPEG assumes the actual image area of a frame is 704 (which is generally true). So when converting to and from 352-width MPEG-1 or MPEG-2, you should scale to/from 704 wide.

If source or target is 720, add/subtract eight pixels left/right as needed. The same applies when convertiong between 720 and 704.

Slow, High-Quality Modes

MPEG-1/2 have been around forever, and are quite simple codecs for encoding and decoding; there's simply not as much work to do with them as with modern codecs. So a decent computer can encode SD MPEG-2 in faster than real time even with very complex settings. Since high QP is such a quality killer with MPEG, anything that can be done to improve efficiency can really help.

Use 2-Pass VBR

And given the risk of blocking from high QP, you should use 2-pass VBR whenever your peak bitrate can be higher than the average (often for DVD, never for broadcast). That lets the codec raise bitrate and reduce QP in order to prevent very blocky frames.

Mind Your Aspect Ratios

VCD and MPEG-2 SD is always nonsquare pixel, and HD often is as well (like 1440×1080 or 1280×1080 anamorphic compression). Make sure you're properly flagging all the aspect ratios correctly. MPEG-1 only has a few fixed-pixel aspect ratio modes, but MPEG-2 is much more flexible.

Get Field Order Straight

MPEG-2 is the only format many of us encode to in interlaced. But many of us don't have interlaced monitors to test playback on anymore. Thus, it's really important to make sure that you're encoding with the same field order in source and output. If you don't, field order will get reversed, so the display order can go from 1A 1B 2A 2B 3A 3B to 1B 1A 2B 2A 3B 3A. This results in displayed images alternating between skipping back 1/60th of a second and skipping forward 3/60th of a second! So you get a nausea-inducing strobe effect whenever there's motion. This has been the cause of many panicked emails to many compression forums over the years.

One common cause of this is DV, which is bottom field first, while many DVD encoders default to top field first. This is particularly annoying in PAL lands where everything *but* DV is top field; we in the NTSC world were never lulled into that false sense of security.

DVD and other interlaced MPEG-2 technlogies support arbitrary field order, so just set the encode to use the same as source. Some tools will automatically convert field order, but that's a visually lossy process and not needed.

If you're encoding interlaced, you should have a way to play that back as interlaced to a display with an interlaced intput. Even if it's a flat panel that bobs to 60p, it'll show this horrible behavior. A software player capable of bob can as well.

I recommend you encode a high-motion interlaced source in both correct and incorrect field order, so you're able to verify your ability to detect field order mismatch.

Progressive Best Effort

If the source is 24p/25p/30p, you can encode it in a seminative fashion. However, most MPEG-2 delivery formats are strictly PAL or NTSC only, so you may need to switch back and forth:

- 24/25p can get sped up/slowed down 4 percent for format conversion.
- 24p for 30i gets 3:2 pulldown with field_repeat tags, to make for easier progressive playback. 24p for 720p gets 3:2 frame_repeat.
- 25p to NTSC should first be slowed down to 24p (well, 23.976p) and then have 3:2 pulldown applied.
- 25p gets encoded as 25p30p gets encoded as 30p60p for non-60p formats like DVD needs to get interlaced to 30i, or progressively encoded as 30p. The choice gets down to whether temporal or spatial quality is more important for the content.

Minimize Reference Frames

Remember the issue about how different decoders can have slightly different results from DCT/iDCT? That results in a slow drift away from what the content could look like, and the drift will be different in degree and appearance on different software.

You may have noticed how few reference frames MPEG typically has. For a NTSC DVD, there's a maximum of 18 frames per GOP, and with a IBBP pattern, you only have five reference frames in a GOP: IBBPBBPBBPBBPBBPBB. So at most, one of the last B-frames would only have six frames back to the start of the GOP. This minimizes the potential for drift.

So, beware going to IBP or IP patterns, or using longer GOPs for software playback. While it might increase compression efficiency, and look fine in player with an IEEE-compliant 64-bit float DCT, it may start working quite a bit worse by the end of the GOP on a cheap hardware player with a 32-bit float engine.

Minimum Bitrate

For most codecs, we don't ever think about minimum bitrate being an issue. We want that to be as low as possible, to save bits for the hard parts. But the lack of deblocking for MPEG means that a VBR encode that targets a fixed QP can leave the easy video looking *really* bad. So most MPEG encoders include a minimum bitrate control that will keep QPs lower in the easy parts. In most cases, this doesn't do padding, so once a minimum QP is hit, bitrate can fall below the minimum QP.

Personally, this seems like laziness to me. While QP isn't a bad shorthand estimate for visual quality, it's far from perfect, and rate control algorithms should be sophisticated enough to make sure that the minimum visual quality doesn't drop too low. Modern codecs do this well, but it just hasn't happened much with MPEG-1/2.

Preprocess with a Light Hand

Since most MPEG-2 is going back out to a video device, we're not going to do a lot of the preprocessing steps we would when targeting computer or device playback:

- No deinterlacing, since we can encode interlaced.
- No cropping out safe area, since it could be played on an old CRT that needs safe area.

So, for clean source, we genenrally don't do anything. There are a few things we do want to watch for:

- Crop 720×486 to 720×480; don't scale it.
- If there's edge noise, blank it out so it's not annoying on PCs.
- Consider cropping 720 to 704 and encoding as 704 if there's horizontal blanking.
- Noise reduction can help a lot in reducing QP with MPEG-2. But if you use it with interlaced content, make sure it's a denoise filter that processes fields separately.

MPEG-2 Encoding Tools

There are zillions of MPEG-2 encoding tools available these days, and I won't try to catalog them all. And many DVD authoring packages build in MPEG-2 encoding facilities. I'm covering only MPEG-2 encoders here, and will talk about the authoring tools in Chapter 24.

Canopus ProCoder

ProCoder was long my go-to tool for MPEG-2 authoring. It uses the great Canopus MPEG-2 encoding engine, which does a very nice job of keeping QPs down and avoiding blocking in shadow areas.

ProCoder's MPEG-2 has a very cool "Mastering Mode" for good-but-slow encoding. Ignore the popup warning about being "up to 20x slower"—it's still faster–than-real-time on a modern laptop. The codec itself is single-threaded, but ProCoder will do a "grid" encode where it splits up different sections of the video to different cores on the machine for full performance. This is great for CBR, but there is some loss of efficiency with 2-pass VBR encoding using this mechanism, since it can't reallocate bits between chunks on different cores.

Some other cool features in its MPEG-2 support:

- Built in AC-3 encoding
- Full VOB generation to build a disc image from a single source file
- Automatic 480/486 and 704/352 conversion
- 24/25p speedup/slowdown
- "24p" in 30i and 60p
- Fixed quality plus VBV, allowing encoding to "good enough" quality while preserving compliance with DVD and other platforms

Rhozet Carbon Coder

Carbon is basically the big brother of ProCoder. It's the same UI and engine, but more frequently updated and with lots of high-end features. It does everything ProCoder does plus these higher-end functions:

- Integrated 5.1 AC-3 encoder
- Built-in templates for CableLabs and MPEG-2
- Presets for the High Profile production formats like XDCAM and P2
- Presets for video server formats for Grass Valley, Omenon, Quantel, MediaStream, and others

Main Concept

Main Concept is broadly licensed to many companies for MPEG-2 and other formats. Among others, it's included in Adobe's video products.

It's generally a good, fast encoder for SD. I have had trouble getting it to deliver ATSC-complaint VBV with high-motion 1080i content, however.

Apple's MPEG-2

Apple has one of the most commonly used MPEG-2 encoders out there, included in iDVD, DVD Studio Pro, and Compressor.

Early versions had rather infamous video quality issues, particularly when VBR encoding. The VBR issues were resolved several years ago, and the encoder is now reasonably fast and reaonsably good, but not a standout.

Using Apple's QMaster grid computing system, Compressor is able to farm out MPEG-2 jobs across multiple computers. However, like the Carbon implementation, each node can only redistribute bits within its own chunk of the video, so as the number of nodes goes up, efficiency can go down.

HC Encoder

HC Encoder (HcEnc) is the most popular open source/freeware MPEG-2 encoder. It's mainly focused on DVD authoring. As you'd expect, it has good integration with AVISynth and DGDecode for reencoding. I think ProCoder/Carbon do a better job in the low luma range.

CinemaCraft

CinemaCraft's MPEG-2 is best known for high-touch DVD encoding, including segment reencoding, custom quant matrices, and N-pass encoding (up to 99, although 5 is the most I've ever seen make a difference).

It also includes a good inverse telecine analysis pass that makes for good "24p" encoding from less-then-perfect telecine sources.

Tutorial: Universally Compatible MPEG-1

Most of the time, MPEG-2 encodes are used in optical disc, so tutorials for those will be in that chapter. But let's do a hands-on with that "Universally Compatible" MPEG-1 I mentioned before.

Scenario

Thumb drives are being handed out at an independent film industry event that includes marketing materials for a local digital theater chain. They'd like to take their shot-on-DV marketing piece and stick it on the thumb drive.

The theater owners got a deal on remaindered drives with their logo a few years ago, so they're tiny 512 MB models.

(*Continued*)

The Three Questions

What Is My Content?

The source is a bog-standard DV codec file in 16:9, with interlaced video (We'll use our Lady Washington source clip on the DVD as a placeholder.) Its duration is 5:48.

Who Is My Audience?

These are indie film folks, so highly variable in how technical they are, and prone to using older machines or less mainstream systems like Linux. So they need to be able to stick in the thumb drive and have a file they can double-click.

What Are My Communication Goals?

The file should be easily discovered, play on even oddball machines, and look reasonably good. Unfortunately, the thumb drives are small and will have a lot of other materials on them. We only have 120 MB left for our video file.

Tech Specs

So, what do we want to deliver? That's not a lot of bits:

120MB × 8 bits/byte × 1000 Kbits/Mbit = 960,000 Kbits
5min 48 seconds = 5 × 60min/sec + 48 sec = 348 seconds
960,000Kbits/348 seconds = 2758 Kbps

We want the audio to be decent, so will allocate 224 Kbps, leaving 2534 Kbps for video. Not terrible, but not great.

So, our tech specs will look like this:

- MPEG-1 file
- **640 × 352. 16:9 would be 640 × 360, but macroblock alignment will give us an efficiency boost. It's only a 2 percent distortion and so won't be noticeable.**
- **29.97 fps progressive.**
- **224 Kbps 44.1 KHz stereo audio.**
- **2500 Kbps video (rounding down a little from 2435 to give us some just-in-case headroom).**
- **Peak bitrate of 4000 Kbps. Those old USB keys can be a little slow sometimes. 4000 Kbps is well under the USB 1.0 10 Mbps spec.**

(*Continued*)

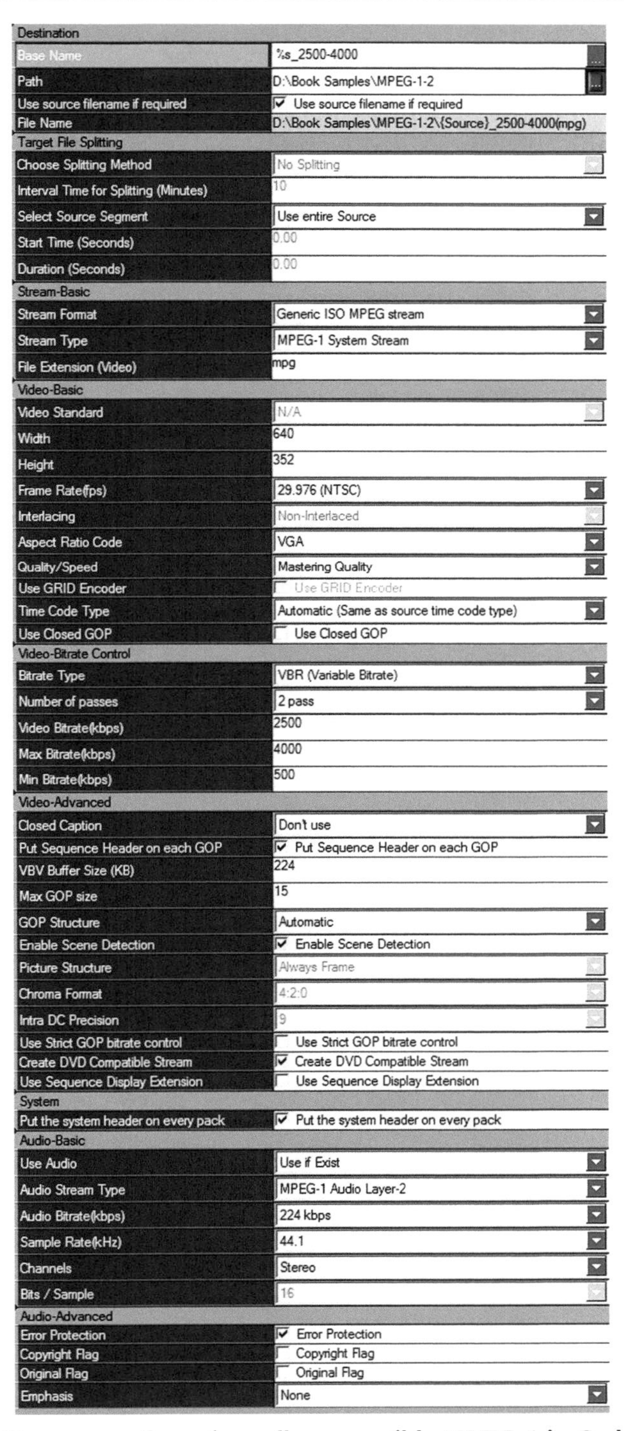

Destination	
Base Name	%s_2500-4000
Path	D:\Book Samples\MPEG-1-2
Use source filename if required	☑ Use source filename if required
File Name	D:\Book Samples\MPEG-1-2\{Source}_2500-4000(mpg)
Target File Splitting	
Choose Splitting Method	No Splitting
Interval Time for Splitting (Minutes)	10
Select Source Segment	Use entire Source
Start Time (Seconds)	0.00
Duration (Seconds)	0.00
Stream-Basic	
Stream Format	Generic ISO MPEG stream
Stream Type	MPEG-1 System Stream
File Extension (Video)	mpg
Video-Basic	
Video Standard	N/A
Width	640
Height	352
Frame Rate(fps)	29.976 (NTSC)
Interlacing	Non-Interlaced
Aspect Ratio Code	VGA
Quality/Speed	Mastering Quality
Use GRID Encoder	☐ Use GRID Encoder
Time Code Type	Automatic (Same as source time code type)
Use Closed GOP	☐ Use Closed GOP
Video-Bitrate Control	
Bitrate Type	VBR (Variable Bitrate)
Number of passes	2 pass
Video Bitrate(kbps)	2500
Max Bitrate(kbps)	4000
Min Bitrate(kbps)	500
Video-Advanced	
Closed Caption	Don't use
Put Sequence Header on each GOP	☑ Put Sequence Header on each GOP
VBV Buffer Size (KB)	224
Max GOP size	15
GOP Structure	Automatic
Enable Scene Detection	☑ Enable Scene Detection
Picture Structure	Always Frame
Chroma Format	4:2:0
Intra DC Precision	9
Use Strict GOP bitrate control	☐ Use Strict GOP bitrate control
Create DVD Compatible Stream	☑ Create DVD Compatible Stream
Use Sequence Display Extension	☐ Use Sequence Display Extension
System	
Put the system header on every pack	☑ Put the system header on every pack
Audio-Basic	
Use Audio	Use if Exist
Audio Stream Type	MPEG-1 Audio Layer-2
Audio Bitrate(kbps)	224 kbps
Sample Rate(kHz)	44.1
Channels	Stereo
Bits / Sample	16
Audio-Advanced	
Error Protection	☑ Error Protection
Copyright Flag	☐ Copyright Flag
Original Flag	☐ Original Flag
Emphasis	None

Figure 9.2 Parameters for universally compatible MPEG-1 in Carbon Coder.

(Continued)

Encoding in Canopus Procoder/Rhozet Carbon Coder

I've been a happy user of the Canopus MPEG-1/2 encoder for many years. It's not super-configurable, but does a good job of avoiding blockies at lower bitrates compared to the other encoders I've used.

Rhozet Carbon Coder (which uses the Procoder code) uses implicit preprocessing, so I don't need to do much with the source other than add it. It'll even know not to letterbox my source when going to 640×352, since that distortion falls beneath the default 5 percent aspect ratio distortion threshold.

I can just add a MPEG target and set the following parameters (also shown in Figure 9.2):

- Stream Type = MPEG-1 System Steam
- Width = 640
- Height = 352
- Aspect Ratio Code = VGA (this just means "square pixel")
- Quality/Speed = Mastering Quality (better than "Highest Quality," albeit logically questionable)
- Bitrate Type = VBR
- Number of Passes = 2 passes (much better rate allocation)
- Video Bitrate = 2500
- Max Bitrate = 4000
- Min Bitrate = 500 (make sure the easy parts don't get too blocky)
- Sample Rate = 44.1

CHAPTER 10

MP3

Although MP3 was originally designed for MPEG-1 (It's short for MPEG-1, Layer III audio), it is almost never seen in MPEG-1 files. However, it's massively popular as a standalone audio-only file format, and also used as a codec in media files, particularly "DivX" AVI files with MPEG-4 part 2 video.

MP3 was largely created by Fraunhoffer IIS. As part of an ISO standard, sample source code for MP3 encoding and decoding was released well before the licensing terms were announced. Many different software vendors created MP3 encoding and playback software while the technology's legal status was murky, and later had to pay retroactive license fees. Although it wasn't planned this way, it turned out to be an excellent way to launch the technology into ubiquity. However, this has left a certain legacy of confusion about when and where it is legal to use and distribute MP3 encoders and players, and whether or what fees need to be paid. Commercial "pay and download an MP3 file" sites definitely need to pay a fee for transactions. Commercial encoders and decoders need to do so as well.

MP3 was the first common, "good enough" format for encoding music—128 Kbps for 44.1 kHz stereo audio can be quite listenable for casual listeners, and even picky folks like me have been known to dance to a 160 Kbps MP3. Modern audio codecs like AAC and WMA Pro offer better compression efficiency, but the very broad ecosystem of MP3 players and the "good enough" compression ensure it a long life. In the end, audio bitrates are low enough that there just wasn't enough value in "just as good, but smaller" versus ubiquitous playback for anything else to really displace it.

Like other MPEG standards, file format and decoder behavior is precisely specified, but implementation of encoding is up to the vendor. Thus, many vendors have created MP3 encoders. The two most important are Fraunhoffer's own encoder (often called FhG), and the open source LAME, both of which are used by many other applications.

MP3 Rate Control Modes

MP3 files are broken up into "frames," each 26 ms long (about 38 a second). This frame rate isn't related to that of any associated video. Each frame needs to be one of a limited number of data rates from 8 to 320 Kbps (the minimum and maximum for a whole file as well). If there just isn't enough audio information in a given frame to use up its assigned number of bits, leftovers can go into a "bit reservoir," for use by later frames that may need more than they're assigned.

There are a few different encoder modes commonly seen in MP3. All commonly used MP3 encoders are 1-pass only.

CBR

Originally, all MP3 files were CBR. The nice thing about CBR files is that file size is predictable based upon its duration, as is where you need to jump into the file to start playing it at a certain point. CBR is also universally compatible on playback.

CBR MP3 today is most commonly used for audio broadcasting, à la Shoutcast, in order to give consistent bitrate.

VBR

Each frame's header specifies that frame's size. The early Xing encoder exploited this by varying data rate frame by frame in order to deliver more consistent quality at lower file sizes. These files are called VBR, and what we'd call "Quality VBR" in the video world. The file size is entirely dependent on the complexity of the audio; all you know is that it won't be more than 320 Kbps.

VBR files sound a lot better bit for bit at lower data rates. Some very old MP3 decoders didn't correctly support VBR files, but I doubt any of those are still in use. For example, QuickTime 4.0 would crash playing back VBR MP3, but that was fixed in 4.1 back in 1999.

ABR

Average Bit Rate is a flavor of VBR that offers more predictable bitrates, but still varies the bitrate within and between files to ensure minimum quality. It's the primary mode used with LAME for Audio CD ripping or other file-based playback.

Some encoders like Apple's iTunes only use VBR to raise the size of a given frame, not reduce it, so long silent sections of a file can use up a lot of bandwidth anyway.

MP3 Modes

An MP3 file can be mono, joint stereo, or normal stereo.

Mono

Mono is, of course, a single channel of audio. As always, it's generally better to maintain sample rate than number of channels, so you're generally better off encoding any MP3 less than 64 Kbps as mono.

Mid/Side Encoding

Like other Mid/Sideband solutions, instead of recording L and R as two different channels, L + R and L – R are encoded. Because most audio has a lot of information in common between the channels, L – R can be compressed a lot more aggressively, reducing data rate. Most encoders can switch between joint and M/S throughout the stream, using M/S when the two channels are similar, and joint when not. However, changing high-frequency spatial information messes up phase relationships, and thus Dolby Pro Logic. While this isn't usually a problem with music source, it can be with movie and TV soundtracks.

Mid/Side shouldn't be chosen explicitly; Joint Stereo will figure out when it's useful and when it's not.

Joint Stereo

In Joint Stereo mode, each frame can adaptively switch between Mid/Side and Normal Stereo per frame based on what's most efficient. Normally, this helps compression efficiency and hence quality, and should be used for 128 Kbps and lower bitrates. However, it can cause audible degradation in audio with flanging/phasing effects—most famously, the guitar from "Mrs. Robinson."

Normal Stereo

In Normal Stereo, the two audio channels are encoded as separate tracks, with no information shared between them. However, the encoder is able to use different frame sizes for each, if one is a lot more complex than the other. This is the optimum mode for high-bitrate archiving, or when you need to preserve Dolby Pro Logic information.

FhG

FhG is Fraunhoffer's licensable, professional MP3 encoder. For a long time it was by far the best MP3 encoder, although LAME's continual improvements have brought it into the ballpark with FhG. Most FhG-based encoders offer very few controls, providing choices of bitrate, sample rate, channel mode, and maybe a speed-versus-quality option—enough to suit the needs of most projects.

LAME

LAME, originally the recursive acronym for "LAME Ain't an MP3 Encoder," started as an open source patch to the available reference encoder source code. This reference encoder worked, but wasn't optimized for speed or quality, and so started out enormously worse than the FhG encoder. However, lots of hackers cared about MP3, and so have progressively been enhancing it for years, focusing on features most useful for CD archiving.

LAME, which is freely distributed, doesn't license the MP3 patent. Commercial products that want to include LAME need to pay the MP3 license fee.

The official LAME is distributed as source code, which you can compile yourself for a command-line version, or a third party can incorporate it into their own tools. Whichever way it's used, the standard parameters of LAME are available in most encoders.

While I'm generally a fan of knob-tweaking with encoders, the LAME presets are quite good and have seen a lot of tuning. If there's one that matches your scenario, it's unlikely you'd be able to come up with something dramatically better. A few of note:

–abr (Average Bit Rate)

This is a 1-pass VBR, where the file shoots for an average data rate while letting the size of each frame to go up or down as much as needed. Actual final file size can vary quite a decent amount (perhaps +/− 5 percent), but not as much with the VBR mode.

–c Constant Bit Rate

CBR is good old CBR. It's appropriate for fixed-buffer streaming or when you need to encode to an exact file size, but otherwise ABR is higher quality.

–v (Variable Bit Rate)

This is the fixed-quality VBR. It's useful for archiving and personal music libraries, but very unpredictable file size makes it suboptimal for distribution.

–q (Quality)

The quality mode is a speed-versus-quality control. It defaults to –q 4, but I normally use –q 2, which is a little higher quality (–h is the same as –q 2). And while it's slower, it'll still be many times faster than real time on modern hardware. Q 0 and 1 are much slower yet and rarely product a perceptible improvement.

Using –q with –v or –preset results in a smaller file, as it makes encoding more efficient—perhaps 10 percent from –q 9 to –q 0, with a 3–4x increase in encoding time.

--preset

LAME also includes a number of high-quality presets for particular tasks. These include tuned psychoacoustics models for the particular tasks, like telephony, "CD quality," and voice. The voice preset is especially useful for encoding intelligible low-bitrate speech. Preset is mainly used for quality-limited encoding. In order of quality and file size, they are:

- –preset medium
- –preset standard
- –preset extreme
- –preset insane

Even medium sounds pretty good for soundtracks, and most people find standard high enough quality for headphone listening. I've never been disappointed by extreme. Insane seems aptly named—it's really a 320 Kbps CBR.

MP3 Encoding Examples

When you encode with LAME, you get a histogram like the following one, indicating the relative size of the different frames. Note the data rates of frames used run from 32 Kbps to 320 Kbps.

Here's an example of a vocal rock song encoded with LAME, using the "standard" preset meant to offer transparent compression to most listeners. The "–h" flag tells it to use the high-quality mode; I always use it.

```
LAME>lame ForgiveMe.wav ForgiveMe_standard.mp3 --preset standard -h
Using polyphase lowpass filter, transition band: 18671 Hz-19205 Hz
Encoding as 44.1 kHz j-stereo MPEG-1 Layer III VBR(q=2) qval=2
Frame | CPU time/estim | REAL time/estim | play/CPU | ETA
9718/9718 (100%)| 0:13/ 0:13| 0:13/ 0:13| 18.458x| 0:00
32 [ 89] %*
40 [ 1] *
48 [ 5] %
56 [ 15] *
64 [ 16] *
80 [ 4] *
96 [ 16 *
112 [ 23] %
128 [ 47] %
160 [1148] %%%%**************
192 [4368] %%%%%%%%%%%%%%%%%%%%%%%%%%%%%%**************************************
224 [3025] %%%%%%%%%%%%%%%%%%%%%%%%%%%**********************
256 [ 880] %%%%**********
320 [ 81] %*
-----------------------------------------------------------------------
kbps LR MS % long switch short %
202.4 40.3 59.7 97.1 1.8 1.1
```

Even though our average here wound up at a potentially reasonable 202 Kbps, we got a lot of frames at higher bitrates, some up to the max of 320. Conversely, quite a few frames didn't need the full average. So our audio quality is a lot better than it would have been otherwise. But we didn't have any rate control. If we wanted to be smaller and still very good, --preset medium is a good compromise (it gave me 160 Kbps with the same song). But if we needed to be around an Internet-friendly 128 Kbps but still get the quality benefits of VBR, we could have used this:

```
LAME>lame ForgiveMe.wav ForgiveMe_abr-128.mp3 --abr 128 -h
Using polyphase lowpass filter, transition band: 16538 Hz-17071 Hz
Encoding as 44.1 kHz j-stereo MPEG-1 Layer III (11x) average 128 kbps
qval=2
 Frame | CPU time/estim | REAL time/estim | play/CPU | ETA
9718/9718 (100%)| 0:14/ 0:14| 0:14/ 0:14| 17.805x| 0:00
32 [ 80] %
40 [ 0]
48 [ 0]
56 [ 0]
64 [ 0]
80 [ 12] *
96 [ 405] %****
112 [5541] %%%%%%%%%%%*********************************************************
128 [3220] %%%%%%%%%%%%*****************************
160 [ 378] %%***
192 [ 66] %
224 [ 12] %
256 [ 4] %
320 [ 0]
----------------------------------------------------------------------------------
kbps LR MS % long switch short %
118.5 20.1 79.9 97.5 1.5 0.9
```

We get a pretty similar distribution to "standard," but shifted to lower bitrates as we'd expect. Still, quality will be quite a bit better than with a true CBR encode, particularly wherever there are transients like percussion. And we actually came in almost 10 Kbps below our target.

mp3Pro

The mp3PRO codec was an enhanced version of MP3 with Spectral Band Replication. Like WMA 10 Pro and HE AAC, it combined a baseband (of MP3 in MP3 Pro's case) at a lowered sample rate with frequency synthesis hints to recreate the higher sample rates. For example, a 64 Kbps stereo MP3 would normally encode at 22.05 kHz. When encoded with mp3PRO, perhaps 60 Kbps would be spent on conventional MP3 data and the remaining four kilobits on the SBR data. An MP3 Pro decoder can use this extra information to synthesize the missing frequencies all the way up to 44.1 kHz, while a conventional MP3 decoder would ignore the SBR and play back at 22.05 kHz.

Since it offered poor quality on existing MP3 players, which is the main reason to use MP3, users just stuck with classic MP3 or used the more advanced SBR codecs, and so mp3PRO decoders and encoders never became common.

CHAPTER 11

MPEG-4

After the clear resounding success of MPEG-2, the Moving Picture Experts Group (MPEG) started work on MPEG-4 in 1993. Version 1 of the specification was approved in 1998, version 2 followed in 2000, and the all-important H.264 (MPEG-4 part 10) was unveiled in 2003.

In general terms, MPEG-4 is a standard way to define, encode, and play back time-based media. In practice, MPEG-4 can be used in radically different ways in radically different situations. MPEG-4 was designed to be used for all sorts of things, including delivering 2D still images, animated 3D facial models with lip sync, two-way video conferencing, and streaming video on the web. Many of those features have never been used in practice, and the particular features supported in any given industry or player vary widely.

Like MPEG-1 before it, MPEG-4 fell far short of the initial expectations many had for it. In particular, ambiguity around licensing terms slowed adoption around the turn of the millennium, the original MPEG-4 Part 2 video codec had practical limitations that kept it from becoming a viable replacement for MPEG-2, and by the time it was widely available it was outperformed by the available proprietary codecs. And some of the most-hyped features—in particular, the Binary Format for Scene (BIFS) rich presentation layer—never made it past cool demos into real deployments.

But as we saw with MPEG-1 and MPEG-2, different markets found different early uses for the MPEG-4 technologies:

- "DivX" files commonly used for pirated content were the biggest area of use for MPEG-4 Part 2, and led to lots of development for (largely noncommercial) encoders and decoders.
- The post-MP3 audio format for Apple's iTunes and iPods, .m4a, is AAC-LC in an MPEG-4 file.
- Early, pre-H.264 podcasting.
- QuickTime 6-era video files.
- 3GP/3GPP video on cell phones.

But the real explosion in MPEG-4 has happened since H.264 came out:

- The first broadly supported media format since MPEG-1! All major players and devices—QuickTime 7+, Windows Media Player 12, iPod, Zune, Xbox 360, PlayStation 3, Flash, Silverlight, etc.—support H.264.
- The broadcast industry is finally shifting from MPEG-2 to H.264.
- HE AAC streams are becoming the new digital radio and audio webcasting standard.

The momentum of MPEG-4 and its component technologies is inarguable.

MPEG-4 Architecture

MPEG-4's official designation as a standard is ISO/IEC-14496. MPEG-4 picks up where MPEG-1 and MPEG-2 left off, defining methods for encoding, storing, transporting, and decoding multimedia objects on a variety of playback devices. Whereas MPEG-1 and -2 were all about compressing and decompressing video and audio, MPEG-4's original aims were "low bitrate audio/video coding" and defining "audio-visual or media objects" that can have very complex behaviors.

There is no such thing as a generic MPEG-4 player or encoder. Because the specification is so huge and flexible, subsets of it have been defined for various industries. A file intended for delivery to a digital film projector isn't meant to work on a cell phone! Thus MPEG-4 has a more complex set of Profiles and Levels than MPEG-2. And it also includes "Parts," independent sections of the spec, to enable new aspects of the format. H.264 (MPEG-4 part 10) is the biggest of these.

MPEG-4 File Format

The base MPEG-4 file format was closely based on the QuickTime file format, and will feel very familiar to old QuickTime hands. They're so close that the same parser is often used to read both .mov and .mp4 files, as in Flash, Silverlight, and Windows Media Player 12.

The MPEG-4 file format is extensible. There are many mappings for codes not formally part of MPEG-4 to the file format, including MP3 and VC-1.

Boxes

A box is the basic unit of a MPEG-4 file, inherited from QuickTime (which called them atoms). A box can contain other boxes, which themselves contain other boxes. Each has a 4CC. The most important is the movie header or "moov" (see Figure 11.4).

Name	Start Time	Duration	Data Size	Data Rate	Format
Bad Habit QVGA.mp4	00:00:00.00	00:02:13.33	7.86 MB	494.50 kbits/sec	-NA-
☑ Sound Track	00:00:00.00	00:02:13.33	1.31 MB	82.50 kbits/sec	AAC
☑ Video Track	00:00:00.00	00:02:13.33	6.22 MB	391.02 kbits/sec	MPEG-4 Video
☑ Hinted Video Track	00:00:00.00	00:02:13.33	233.40 KB	14.32 kbits/sec	Hint
☑ Hinted Sound Track	00:00:00.00	00:02:13.33	108.54 KB	6.66 kbits/sec	Hint

Figure 11.1 Tracks in a hinted file. They increase the file size a little more than 3 percent in this case.

Tracks

As in QuickTime, media in an MPEG-4 file is in tracks, normally a sequence of video or audio. A single .m4a file will have a single track of AAC-LC audio. A simple podcasting file would have a track of AAC-LC audio and a track of H.264 Baseline video.

But more complex combinations are possible. A file could have multiple audio tracks for different langauges, with the player picking which one to play.

Hint tracks

MPEG-4 files meant for real-time streaming with standard Real Time Streaming Protocol (RTSP) streaming require hint tracks. This is an extra track (one per track to be streamed) containing an index showing how to packetize the media for real-time delivery. Hint tracks are easy to make during encoding or afterwards. They shouldn't be included in files meant for local playback only, as the extra data increases file size. Figures 11.1, 11.2, and 11.3 demonstrate the characteristics of hint tracks in various implementations.

Flash Media Server (FMS) doesn't require hint tracks for its proprietary RTMP (Real Time Messaging Protocol) protocol, but there's no downside to including them either, as they aren't transmitted by FMS even if they're there.

Text tracks

Text tracks are how MPEG-4 stores subtitles, synced URLs, and other time-based data. This has significant use only recently, with different player implementations expecting somewhat different versions of subtitle presentations.

Fast-Start

Traditional MPEG-4 content, and all .mp4/.m4a files, have been nonfragmented, following the classic QuickTime model. In a nonfragmented file, the video and audio tracks are continuous, and to jump to a particular frame, the player needs to know where that frame and its associated audio are in memory. That information is stored in an index called the header, which is a "moov" box.

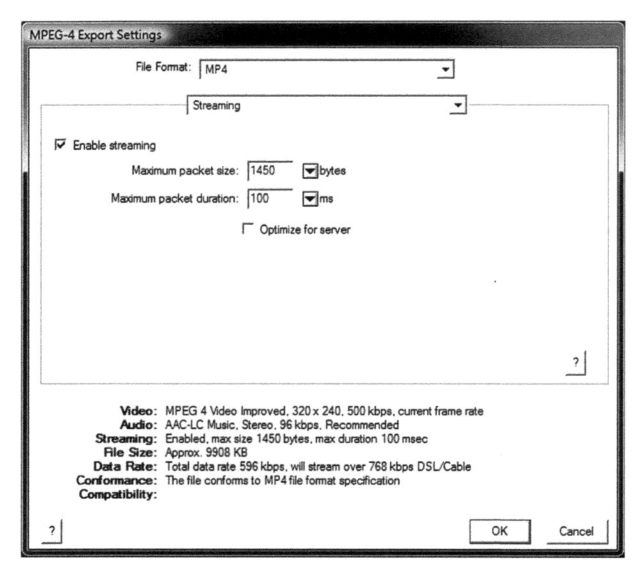

Figure 11.2 QuickTIme has pretty basic hinting options. The defaults are fine unless you're targeting a specific network that is best with a different packet size. The "optimize for server" mode hasn't been useful for ages; it just makes the file much bigger without actually helping anything.

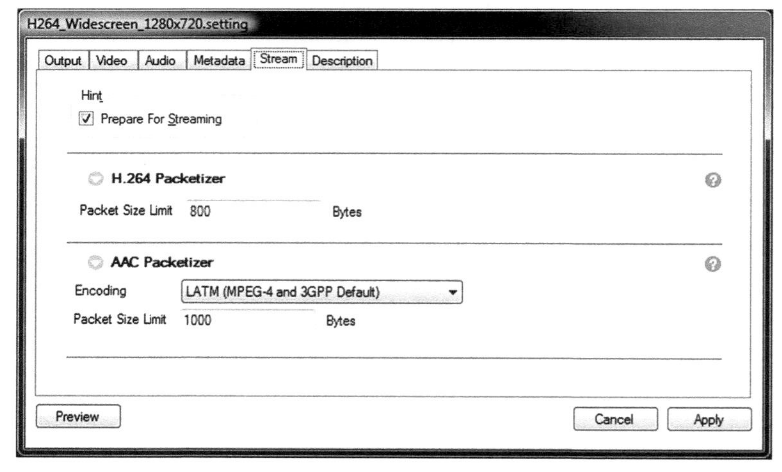

Figure 11.3 Episode Pro started as an encoder for mobile devices, and continues to have a very deep feature set for that market. It has much more configurability for hinting than most applications. This allows encoding to be tuned for particular devices or networks.

The header needs to know the location of all the samples in the media tracks, so it isn't created until after encoding is done. The simplest way to do this is to append the "header" at the end, with an offset value at the front of the file indicating where it is. That works great when random access is possible. But it's not a good experience when the content is being

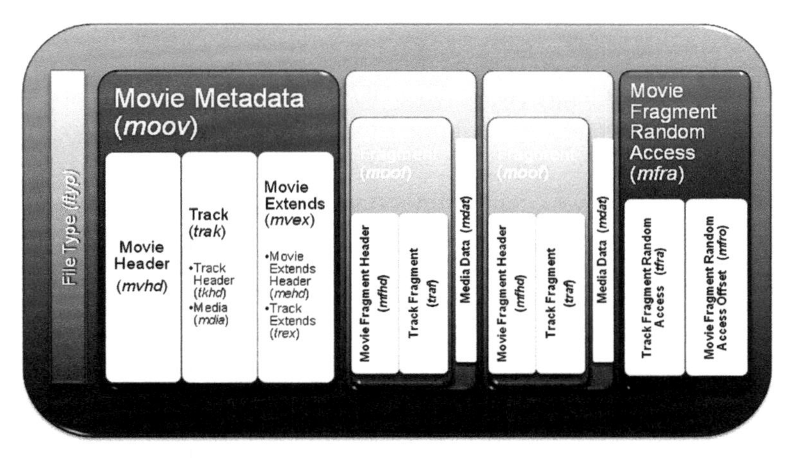

Figure 11.4 The structure of a fragmented MPEG-4 file (fMP4). Each fragment is self-contained, with its own video, audio, and index needed for playback.

accessed serially, like classic progressive download without byterange access; all the media has to be downloaded before anything can be played.

Hence the "Fast Start" file, which moves the movie header to the beginning of the file. This requires that the file be resaved after the initial write, but that's plenty fast on modern machines. Most MPEG-4 tools automatically make the files Fast Start.

Fragmented MPEG-4 files

The other way to handle random access into a MPEG-4 file is to use the fragmented subtype. In fragmented MPEG-4 (fMP4), instead of having a big movie header for all the samples, there are periodic partial headers that cover one portion of the content. This type of header is a "moof" box, and along with the associated video and audio content is called a fragment.

Historically, fMP4 was mainly used for digital video recorders (losing power during recording leaves the file playable except for the final partial fragment). However, it's now used as the file format for Smooth Streaming, and in other major implementations not yet announced.

The tragedy of BIFS

The big thing that MPEG-4 promised that went beyond the QuickTime file format and then-existing technologies was BIFS—the Binary Format for Scenes. It was a derivative of VRML (Virtual Reality Markup Langauge, another mid-1990s rich media disappointment), and provided control over the video presentation, including synchronized text, multiple video windows, integrated user interfaces with audio feedback, and so on.

It could have changed everything.

Except it didn't. It wasn't ever supported by QuickTime/RealPlayer/Windows Media Player. And while there were some interesting authoring tools, from iVast and Envivio, it never got enough traction in players to drive content, or enough traction in content to drive adoption by players. MPEG-4 in the end was mainly important as a way to get much better standards-based video codecs. But the video format itself didn't really enable anything MPEG-1 couldn't in interactivity.

In the end, the promise of BIFS was finally realized by Flash, and later, by Silverlight. But in those cases it was about embedding media in a rich player instead of having the media itself host the interactivity. HTML5's <video> tag has been proposed as another way to address this market, but that so far lacks any codec or format common denominator.

MPEG-4 Streaming

MPEG-4 uses RTSP for streaming. As a protocol, RTSP is broadly used, with RealPlayer and Windows Media using implementations as well. RTSP uses UDP and RTSP, as we'll discuss in more detail in Chapter 23.

RTSP is a pretty canonical streaming protocol with all the features one would expect, like:

1. Multicast support
2. Random access in on-demand content
3. Live support
4. Session Descrption Protocol (SDP) as metafile format

That said, pure MPEG-4 RTSP hasn't been that widely used for PC delivery; RealMedia, Windows Media, and Flash have long dominated that market. Apple keynotes and presentations have probably been the most prominent. RTSP has been much more used in device streaming.

MPEG-4 Players

While there is a reference player available, MPEG-4 player implementation is left open to interpretation by the many different vendors. Different players can support different profiles. They can also support different levels of decompression quality within a single profile, which can result in lower-quality playback on lower-speed devices. For example, postprocessing (deblocking and deringing) may not be applied—these tools are not required in the standard.

This implementation dependence is the source of much confusion among end users and developers alike. There's no way to tell in advance whether any given .mp4 file is playable in a given player, and clear documentation can be hard to come by, particularly for devices.

MPEG-4 Profiles and Levels

When someone asks if a piece of technology is "MPEG-4-compatible," the real question is, "Which MPEG-4 profiles and levels does it support?" Like in MPEG-2, a profile is a subset of the MPEG-4 standard that defines a set of features needed to support a particular delivery medium, like particular codec features. And level specifies particular constraints within a profile, such as maximum resolution or bitrate. Combined, these are described as *profile@level*, e.g., Advanced Simple Visual@Level 3. Compliance with profiles is an all-or-nothing proposition for a decoder—if you have a device that can handle the Advanced Simple Visual Profile Level 3, it has to be able to play any file that meets that specification. However, many players don't conform to the official descriptions. Apple made up their own "H.264 Low Profile" for the initial video iPods.

MPEG-4 Video Codecs

So far, there are two primary video codecs as part of MPEG-4, with other codecs that can be used in MPEG-4 but aren't part of the formal standard.

MPEG-4 Part 2

The original MPEG-4 codec was MPEG-4 part 2. It's also known as Divx and Xvid by its most common implementations, and as ASP after its most popular profile (Advanced Simple Profile). It's a big topic, and hence the subject of the next chapter.

H.264

MPEG-4 part 2 was a relative disappointment in the market, but H.264 has become the real deal, and is rapidly becoming the leading codec in the industry. It's the topic of Chapter 14.

VC-1

VC-1 is the SMPTE standardized version of the Windows Media Video 9 codec. A mapping for VC-1 into MPEG-4 has existed for a while, but it has only come into common use with Smooth Streaming. The default .ismv file format for Smooth Streaming is a fragmented MPEG-4 file containing either VC-1 or H.264 as the video codec. Smooth Streaming itself is covered in Chapter 27, and VC-1 in Chapter 17.

MPEG-4 Audio Codecs

MPEG-4 has a rich set of audio features. Like most platforms, it provides separate codecs for low bitrate speech and general-purpose audio.

Advanced Audio Coding (AAC)

Following in the footsteps of MP3 (a.k.a. MPEG-1, Layer III audio), born of MPEG-1, AAC, born of MPEG-2, has became a dominant audio codec in the MP4 era.

It's covered in Chapter 13.

Code-Excited Linear Prediction (CELP)

Code-Excited Linear Prediction (CELP) is a low-bitrate speech technology (related to the old ACELP.net codec in Windows Media) used in the 3GPP2 MPEG-4 implementation for mobile devices. It supports bitrates from 4 to 24 Kbps at 8 kHz or 16 kHz mono by spec, but most implementations just offer 6.8 and 14 Kbps versions. If available, you'd use 12 Kbps HE AAC over 14 Kbps CELP. QuickTime includes encode and decode.

Adaptive Multi-Rate (AMR)

Adaptive Multi-Rate (AMR) is another low-bitrate speech codec also used in 3GPP and 3GPP2 devices. It offers from 1.8 to 12.20 Kbps, depending on implementation. QuickTime and Episode can both encode, and QuickTime can play back.

CHAPTER 12

MPEG-4 part 2 Video Codec

This chapter is about the original MPEG-4 part 2 video codec (also known as just "MPEG-4," "DivX," "Xvid," or "ASP"). The MPEG-4 part 10 codec (better known as H.264) is covered in Chapter 14. As influential as MPEG-4 has been as a format, we have to admit that its first attempt at a video codec was something of a failure, something of a speed bump between the towering successes of MPEG-2 and H.264. So, what happened?

In the end, I'd suggest it was a mix of technical limitations and timing. On the technical side, part 2 simply wasn't enough of an improvement over MPEG-2 to justify the switching costs. While software players on a PC are easy to update, replacing hundreds of millions of cable boxes and other media players with new models containing updated hardware decoders is an enormously expensive, challenging, and long process. And MPEG-4 part 2 wound up not being as efficient as hoped, while MPEG-2 wound up getting better than its creators imagined. Part of that was due to the huge market for MPEG-2 that made even small improvements in efficiency valuable. Part 2 never developed that kind of market. But there were also real limitations to part 2 that capped how good it could really get. In the end, I don't think there ever was a moment where part 2 was more than 30 percent better than MPEG-2 for important scenarios, which kept it from being worth the trouble of switching for the big guys.

From a timing perspective, part 2 straddled three eras:

- It launched as MPEG-2 was hitting its stride with DVD and digital broadcasting.
- Part 2 implementations hit the market during the early web video days, but licensing ambiguities kept big content companies wary, while the proprietary formats and codecs offered simple licensing and then better real-world compression efficiency.
- By the time the licensing issues were all worked out and the Streaming Wars were ending, H.264 was already on the horizon and clearly poised to offer big efficiency advantages over both MPEG-2 and part 2.

In the end, piracy and phones were the primary markets for part 2.

The DivX/Xvid Saga

The piracy tale is an interesting part of the bigger story. Microsoft had been an early participant in the MPEG-4 process, and had created an early reference encoder and decoder.

And Windows Media actually shipped several generations of MPEG-4 codecs—first Microsoft MPEG-4 v1, v2, and v3 based on standards drafts, and then an ISO MPEG-4 codec based on the final standard. These were decent codecs of their era, but locked to the ASF file format. An enterprising hacker took the MS MPEG-4 v3 .dll and changed some bits in order to let it work as an AVI codec, which was a lot easier to work with in video authoring tools of the day. And then the hacker broadly distributed the hacked version, under the name "DivX ;-)"—annoying emoticon and all (and a reference to the failed DivX competitor to DVD). There are people at Microsoft still irritated by this. But it became a pretty popular way to create and distribute higher-quality video files via download and peer-to-peer than was possible with real-time streaming of the era.

Somehow this hacked codec led to financing, and DivX Networks was born to commercialize the technology. However, said technology consisted of the MPEG-4 part 2 video codec, the AVI file format, and, typically, MP3 audio. The company dabbled in open source for a while with OpenDivX, then reverted to a closed source model. Participants in OpenDivX forked the code from the last open version and created their own version, the palindromically named "Xvid."

At that point DivX Networks tried to move the ecosystem to their new DivX Media Format (DMF), which was still based on AVI for backward compatibility. But most content targeting "DivX" was still just vanilla AVI, and increasingly encoded using Xvid.

DivX Inc's (as it is now named) biggest moneymaker appears to have been certifying DVD players as also supporting DivX, so that users could burn their DivX files to disc and play them back easily. The market for that has been declining, however, as optical media is increasingly uncompetitive with hard drive and flash-based storage. And since there wasn't any actual DivX Networks technology required to play the files back, many devices simply shipped compatibility for the files without paying for the logo.

Today, many media players can handle the AVI + MPEG-4 part 2 + MP3 combination, including VLC, Windows 7's WMP, Xbox 360, and PlayStation 3.

DivX Networks itself is now focused on DivX 7, which uses H.264 and the open source Matroska container format (MKV).

Encoding for DivX-compatible devices is covered in Chapter 24.

Why MPEG-4 Part 2?

Honestly, there are fewer and fewer reasons to use part 2 these days; the main one remaining is compatibility with existing players. Every class of players that supported part 2 is rapidly adopting H.264. There are a few places where legacy part 2 only players might matter, but they're shrinking.

Consumer Electronics

There was a burst of "DivX-certified" DVD players a few years ago, all capable of playing the DivX-style part 2 + MP3 in an AVI file. While specifics specs varied, lots of them are capable of doing at least 720p. Xbox 360 and PS3 are also able to play back these files.

Mobile

Various flavors of part 2, particularly Short Header and Simple Profile, were used in phones before H.264 was common. I haven't seen any recent breakdowns of which phones are capable of decoding what, but given the rapid rate of phone replacement, H.264 Baseline will soon be a safe choice if it isn't already.

Low Power PC playback

Part 2 is a lot simpler than H.264 to decode. So older computers can play back higher resolutions using part 2.

The challenge there is only Macs have out-of-the-box part 2 decode, but it's Simple Profile .mp4 only. So this only works if you're able to install a decoder along with your media, or give up B-frames and hence a bunch of compressor efficiency.

Why Not Part 2?

There are some pretty big reasons not to use part 2 by default anymore.

H.264 or VC-1 Is Already There

Most of the phones, PCs, and consoles that do part 2 can also do H.264, which is a much more efficient and broadly supported codec. And Windows PCs already have WMV, which is more efficient and offers better performance on playback on most machines today.

So lack of good out-of-the-box support is a challenge. Even older Macs support at least H.264 Main Profile. Windows 7 will be the first OS to ship with full ASP support out of the box, and it includes H.264 as well.

Lower Efficiency

Both VC-1 and H.264 offer better compression efficiency than part 2, making them better choices for anywhere that bandwidth is a premium.

The efficiency gets even worse if QuickTime is being targeted for playback, due to the Simple Profile limitation.

What's Unique About MPEG-4 Part 2

So what's part 2 is like as a codec? First off, it was based on the ITU H.263 videoconferencing codec, with all profiles in significant use being supersets of that. Unfortunately, it was based on the original H.263, not including many of the useful features of H.263+ (also released in 1998). This was not uncommon; H.263 is more advanced than MPEG-1 but doesn't have patent licensing fees associated with it.

MPEG-4 part 2 was perhaps the last 8 × 8 floating-point DCT. Very broadly, VC-1 can be thought of as the culmination of that course, keeping the core structure while addressing its weaknesses, while H.264 represents a decisive break from that past and adds fundamental changes.

Custom Quantization Tables

Part 2 supports custom quantization tables that can be tuned to content and bitrate (like MPEG-2). There are quite a few available for download with different tools, but it can require a lot of trial and error to figure out which ones are well-suited to different kinds of content. In general, the H.263 matrix is better for lower bitrates (errs on the side of softness) while the MPEG-4 matrix is better for higher bitrates with film/video source (retains more detail). Most part 2 encoders default to the H.263 matrix.

B-Frames

Part 2 has B-frames, but they can't contain intra-coded blocks, and so BI-frames like VC-1 uses for flash frames aren't possible. Still, B-frames are highly useful in compression, and are by far the most valuable tool in Advanced Simple Profile not in Simple Profile.

Quarter-Pixel Motion Compensation

Part 2 goes beyond MPEG-1's full-pixel and MPEG-2's half-pixel precision by supporting quarter-pixel precision. However, the implementation wasn't as efficient as in later codecs, providing a slight improvement in efficiency at best, and is often not used by encoders. DivX's own recent profiles don't include it.

Global Motion Compensation

Global motion compensation (GMC) sounded like a great idea—do a global motion search and provide an overall description of the motion in the frame. And not just pans, but zooming! But it's very expensive to calculate, provides a very slight improvement at best, and thus is very rarely used.

Interlaced Support

Part 2 has good MPEG-2 style interlaced support, but it is rarely used in practice; 720p is generally the highest resolution decoded.

Last Floating-Point DCT

MPEG-4 was the last significant interframe codec to use a MPEG-1/2 style floating-point DCT without an ironclad definition of the right answer. So different decoders can potentially drift over time from what the encoder assumed due to a succession of slightly different rounding errors, although in practice this problem is less severe than in MPEG-2.

No In-Loop Deblocking Filter

There's no in-loop deblocking in part 2; if you get a reference frame with high quantization, that'll get propagated forward to the next frame.

MPEG-4 Part 2 Profiles

Short Header

MPEG-4 short header is simply canonical, original flavor H.263. It was commonly used for phone playback due to its low decoder requirements. However, its compression efficiency is similarly low. Modern phones with better ASICs use more efficient codecs to preserve bandwidth.

Simple Profile

Simple Profile was the most commonly used for web video, being the best mode QuickTime could export and play. The main difference is that Simple Profile adds optional error resilience to reduce the impact of corrupt/missing data.

Advanced Simple Profile

Advanced Simple Profile (ASP) is the one everyone expected to dominate. It added four main features to Simple Profile:

- B-frames
- Global motion compensation
- Quarter-pixel motion estimation
- MPEG-4 and custom quant matrices (Simple just used H.263)

B-frames turned out to be by far the most useful feature of the bunch, and many ASP encoders default to doing SP + B-frames. And while there isn't a formal profile for that, there are plenty of devices, including the Zune HD and many DivX certified hardware players, that support SP + B-frames as a de facto profile, while lacking support for the other ASP features.

Probably the biggest barrier to ASP's broad adoption was that QuickTime has never supported it. Due to an architectural limitation prior to QuickTime 7, QuickTime codecs couldn't natively support B-frames. The QuickTime 6 betas could play ASP content by dropping B-frames, but that was removed for the released version. Even though QT7 fixed the general B-frame issue in 2005, the part 2 decoder has never been updated to play ASP.

In general, all the other software players, such as VLC and ffmpeg, support all of ASP. But as QuickTime has long been the architecture most likely to try to play a double-clicked .mp4 file, ASP files will fail to play on many out-of-the-box systems.

Windows 7 includes full ASP support in Windows Media Player 12.

Studio Profile

The MPEG-4 part 2 Studio Profile was intended for high quality content acquisition and archiving. It goes far beyond 8-bit 4:2:0. Its only significant use so far is in Sony's excellent HDCAM-SR tape format, which does both 10-bit 4:2:2 and the glorious 12-bit 4:4:4.

MPEG-4 Part 2 Levels

MPEG defines resolution with such terms as QCIF, CIF, 2CIF, and 4CIF, derived from the MPEG "Common Intermediate (meaning between PAL and NTSC) Format." Q = Quarter, 2 = double width, and 4 = double width and double height. Each has a canonical resolution (Table 12.1). However, the actual limitation in the spec is the maximum number of 16×16 macroblocks allowed. So, in theory you can redistribute the blocks into any shape you like. So, for QCIF, instead of 176 × 144, you could deliver 256 × 96 (for streaming *Ben-Hur*, perhaps). The one exception to this redistribution rule is Simple Level 0, which is limited

Table 12.1 CIF Family of Frame Sizes.

Name	Canonical resolutions	16 × 16 blocks
QCIF	176 × 144	99
CIF	352 × 288	396
2CIF	352 × 576	792
4CIF	704 × 576	1620

Table 12.2 Simple Profile Levels.

Level	Max size	Max data rate (Kbps)
0	QCIF (fixed to 176 × 144)	64
1	Equivalent to QCIF	64
2	Equivalent to CIF	128
3	Equivalent to CIF	384

Table 12.3 Advanced Simple Profile Levels.

Level	Max size	Max data rate (Kbps)	Interlace
0	QCIF	128	No
1	QCIF	128	No
2	CIF	384	No
3	CIF	768	No
3b	2CIF	3000	No
4	2CIF	3000	Yes
5	4CIF	8000	Yes

to 176 × 144 actual pixels. However, in practice device decoders may also have a fixed maximum height and width.

Simple

The Simple Profile is mainly seen in QuickTime and phones. It may also be used for recording video with low-power devices. Level 0 also has a limit of 15 fps (Table 12.2).

Advanced Simple

Advanced Simple (Table 12.3) is a superset of Simple, so it can play all Simple content. This profile adds a number of enhancements to support better visual quality. Interlaced content is supported in levels 4–5. Advanced Simple Level 3b is used in ISMA Profile 1, and support for higher levels are planned for later ISMA Profiles.

MPEG-4 Part 2 Implementations

DivX

DivX, Inc., was really the popularizer of MPEG-4 part 2. It's often called "DivX" even by those who don't use their tools.

While DivX dabbled in commercial encoding tools for several years, their Dr. DivX product (Figure 12.1) is now free and open source. They also have a commercial DivX Pro implementation (although that's much more focused on H.264 these days).

Figure 12.1 The Dr. DivX OSS interface. Basic and simple, aimed at consumers.

Xvid

Xvid (Figure 12.2) was created as a fork from the OpenDivX project. It has had years of sustained development from dedicated codec hackers who have pushed part 2 about as far as it's likely to go.

As an open source technology, Xvid is widely used in many free encoders, including ffmpeg, Handbrake, and MeGUI.

Sorenson Media

Sorenson Media is best known today for their Squeeze product and online services, but their heritage was in codecs, notably the H.263-derived Sorenson Video and the H.263 Spark implementation for FLV (Figure 12.3).

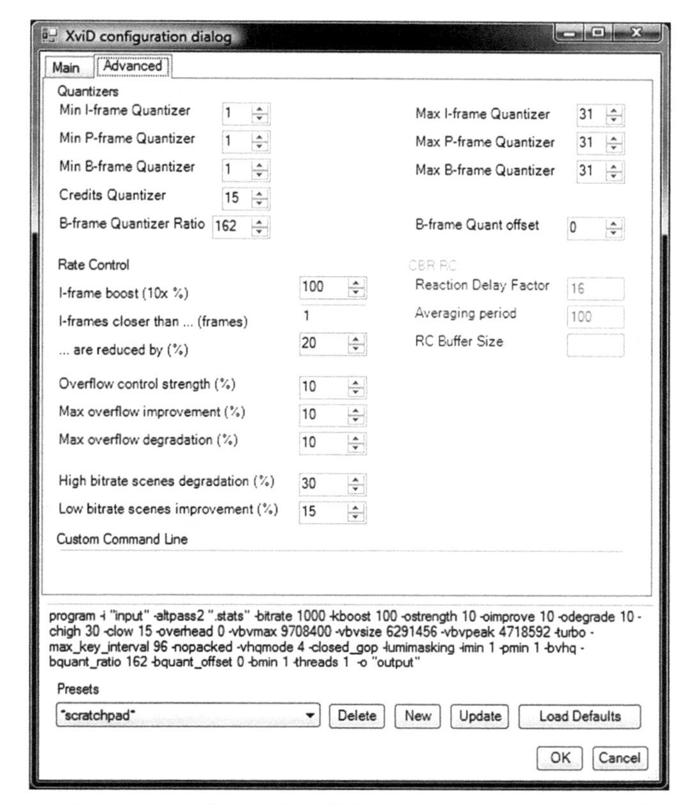

Figure 12.2 The Xvid advanced configuration dialog in MeGUI. It's rather imposing; fortunately, the presets are quite good for most content.

Telestream

Telestream's purchase of PopWire and its Compression Master product (now Episode, Figure 12.4) gained them encoding technology very focused on mobile encoding, which includes excellent part 2.

QuickTime

QuickTime had one of the most widely used part 2 implementations, which was a shame, as it was always one of the weakest, and hasn't been updated in ages. Beyond being Simple Profile only, it also only has 1-pass CBR with mediocre efficiency at best (Figure 12.5). I can't think of a reason to use it today.

A number of compression products encode part 2 via the QuickTime API, and they'll do no better in quality.

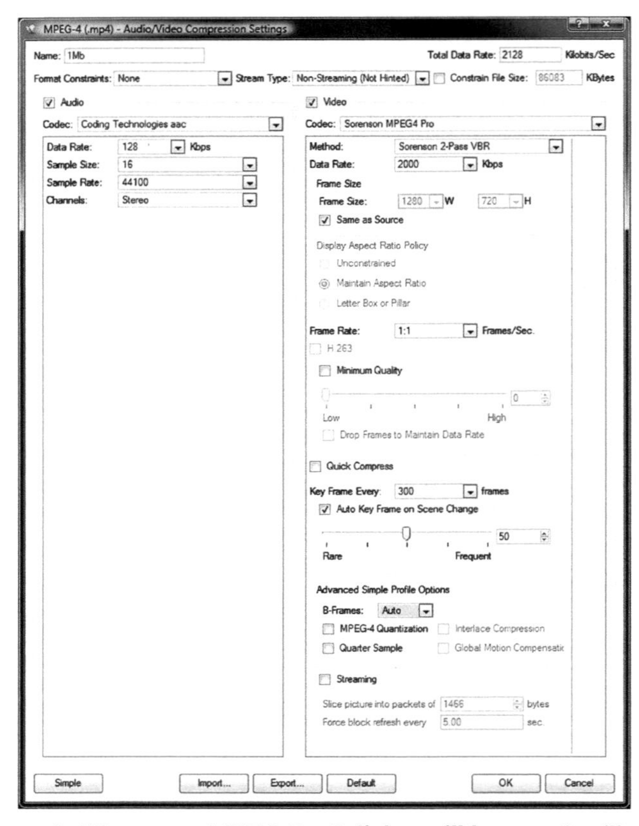

Figure 12.3 Squeeze's "Sorenson MPEG Pro" dialog will be very familiar to Spark users. It supports .mp4 files, but not .avi or other containers.

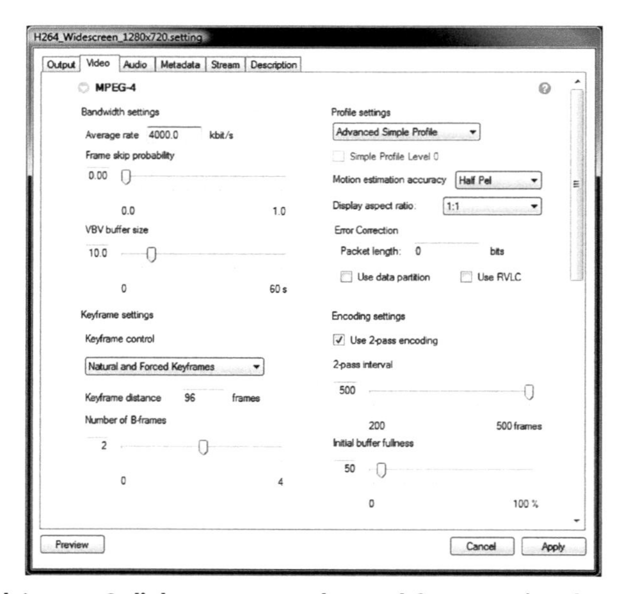

Figure 12.4 Episode's part 2 dialog exposes advanced features aimed at the mobile market.

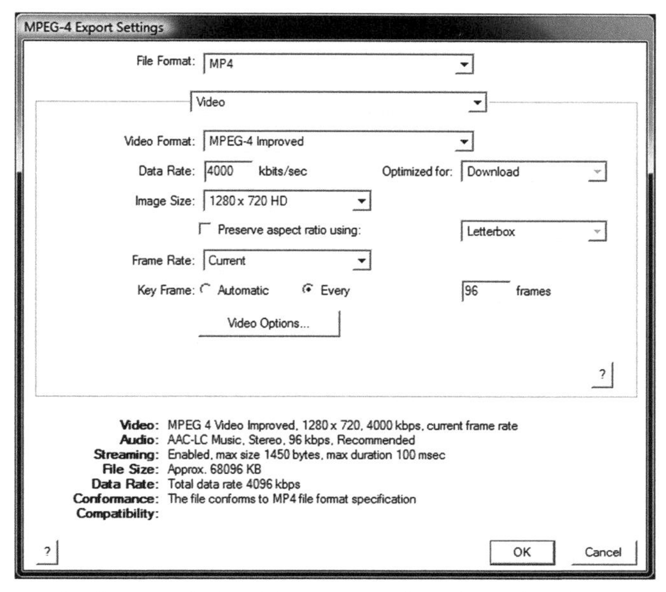

Figure 12.5 Unfortunately the "Optimized for" option is only available for QuickTime H.264, not part 2. Apple's part 2 is 1-pass CBR only.

MPEG-4 Part 2 Tutorial

So, what would we actually use MPEG-4 part 2 for?

Scenario

We and our friends have made a cool animated short, which we want to make available to the world. It's in HD, and we figure people might want to watch it in home theaters as well as PCs.

Three Questions

What Is My Content?

Our source is 1920 × 1080p24 animation, currently living as a sequence of PNG files. Audio is a simple stereo mix.

Who Is My Audience?

People inclined to download amusing videos from the Internet and watch them on their PCs, consoles, and DVD players: they're younger, tech-savvy, often but not always with the latest gear. And probably with time on their hands; they're prefer a download for higher quality over streaming.

(*Continued*)

What Are My Communications Goals?

We want our content to be awesome; we'd rather it take longer to download than look bad. And we want it to be in HD, at least 720p. We're in it for the glory and adulation of our peers; our grandmothers aren't going to watch it, and that's probably how we want it.

Technical Specs

We're going to go for 1280 × 720 24p, at a data rate that'll deliver great quality. We want the audio to sound good too.

For broadest compatibility, we're going to do "DivX"-style part 2: MPEG-4 part 2 Simple Profile + B-frames with MP3 audio in an AVI wrapper.

We're not going to sweat bitrate; we want to be as small as it can and still be awesome.

Encoding in MeGUI

We're going to encode with the open source MeGUI, which is free and a nice front-end to Xvid and other codec implementations (Figure 12.6).

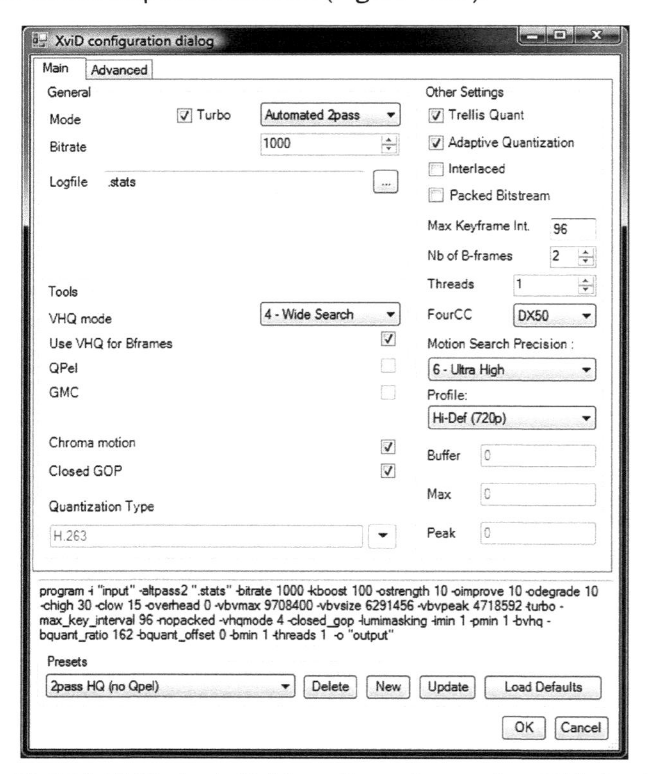

Figure 12.6 MeGUI's basic Xvid configuration option. This is where most real-world tuning will be done.

(Continued)

First off, we have to turn our PNG sequence and WAV files into a source. MeGUI uses AVISynth for video input, and can use AVS or other audio formats for audio. Here's a simple script that will do all of the following:

- Turn the PNG sequence into video
- Add the audio
- Merge them together
- Resize the video to 1280 × 720
- Convert to the required YV12 (4:2:0) color space ready for the codec:

```
a = imageSource("D:Dream108005d.png",1,15691,24)
b = wavsource("D:DreamCM-St-16bit.wav")
audiodub(a,b)
BilinearResize(1280,720)
ConvertToYV12()
```

Then we need to configure our settings.

MeGUI comes with a number of good presets. We'll start with "2pass HQ (no Qpel)," which will gives us the broadest compatibility for 720p capable devices. For audio, we'll start with "LAME MP3-128 ABR" (Figure 12.7). We'll want to make a few modifications, though:

- Raise Bitrate to 4000. We want our stuff to look good!
- Change Profile to Hi-Def (720p), which adds a few further constrains to make sure it's compatible with DivX-certified 720p players.

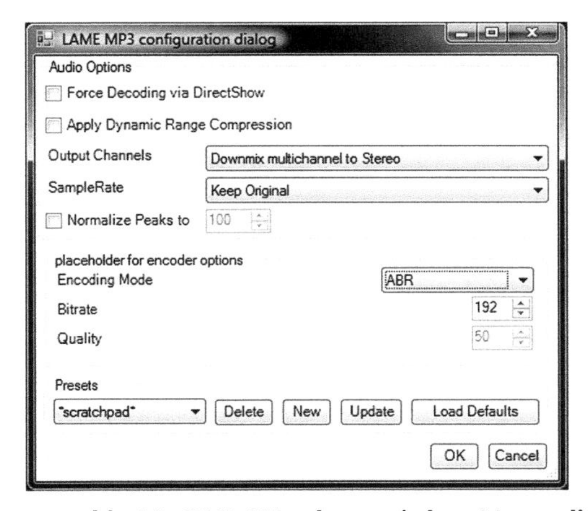

Figure 12.7 LAME as exposed in MeGUI. Watch out; it has Normalize Peaks on by default.

(Continued)

- Reduce Max Keyframe Int from 250 to 96, keeping the max GOP length to 4 seconds. We figure people will want to scrub through the video looking for cool stuff, so want random access to be smoother.
- Raise the MP3 ABR from 128 to 192 for better audio quality.

And there we have it. A very nice-looking presentation of our masterpiece to share with the world.

CHAPTER 13

Advanced Audio Coding (AAC) and M4A

M4A File Format

The M4A file format is simply an MPEG-4 file with only a single track of AAC audio. It's not ". mp4" simply to make it easy to tell it's an audio-only file. M4A was made famous by Apple with iTunes, and has become broadly supported in media players and consumer electronics devices.

The main advantage of M4A over MP3 is better compression efficiency. In its typical AAC-LC profile, it provides the same quality in a third less bitrate.

An iTunes .m4p file is a FairPlay DRM-encrypted M4A and as such isn't interoperable outside of the Apple ecosystem. It's something only Apple can make, and only Apple products can play it back.

AAC Profiles

AAC Low Complexity

AAC Low Complexity (AAC LC) was originally the low-complexity implementation of AAC, a simplified version of the forgotten AAC Main. But in real-world listening tests with optimized encoders, AAC LC quality at 128 Kbps was essentially as good as AAC Main, but with lower encode and decode complexity. Thus it has become the default version of AAC, with enhancements building off LC instead of Main.

AAC-LC is the only version supported in QuickTime as of 7.6 and Silverlight as of 3, although the High Efficiency variants provide lower-fidelity backward compatibility to LC.

AAC-LC is a very flexible codec, going from 8 KHz mono to 7.1 96 KHz in common implementations. While many encoders are CBR-only, various VBR modes also exist.

Beyond its use for audio in .mp4 files, AAC-LC has also been used in billions of .m4a CD rips in iTunes, where it has long been the default codec.

AAC-LC was a good efficiency improvement over the previous music codec standard of MP3, offering around a 50 percent efficiency improvement (CBR AAC-LC at 128 Kbps sounds about as good as CBR MP3 at 192 Kbps). However, the broader ecosystem of MP3 players have kept MP3 in broad use; AAC delivers the same quality at the same size, but rarely better quality.

But if you're in an environment where you can count on AAC-LC being there, use it.

High efficiency AAC v1

High Efficiency AAC (HE AAC) uses Spectral Band Replication (SBR) for improved efficiency. The key insight is that while high frequencies are necessary for audio to sound rich, most of what we hear in high frequencies that matters is an overtone of a lower frequency. Since an overtone is a multiple of a lower frequency, HE AAC reduces the frequency range that gets a full DCT encode (typically by half, so 22.05 out of 44.1), and hence the bitrate of the recoding. Then it includes some "hint" data that shows how to synthesize the missing overtones based on the retained lower-frequency data. The net effect is that bandwidth requirements get cut nearly in half without a big loss in perceptual quality.

The base band in HE AAC is AAC-LC, which remains backward compatible. So while software that only decodes AAC-LC will have lower fidelity, listeners will still hear something, just at half the sample rate. This can sound pretty weak for music, but fine for intelligible speech.

HE AAC is sometimes called "AAC+" after the trademark Coding Technologies used for their implementation. Coding Technologies has since been bought by Dolby, which now calls its implementation "Dolby Pulse." Note that this is an entirely different codec than "Dolby Plus," which is the evolution of Dolby Digital (AC-3). Yes, this can get very confusing on conference calls!

While HE AAC supports multichannel, many decoders that can do HE AAC in stereo don't handle HE AAC 5.1.

HE AAC v2

HE AAC v2 extends the SBR concept to stereo channels. Instead of encoding separate channels using L + R/L − R, it encodes all the frequencies together, with more "hint" data explaining which channel to steer each frequency to. That makes stereo encoding nearly free compared to mono.

Again, the core is AAC-LC, so AAC-LC decoders can still play back HE AACv2, albeit in half sample rate mono.

While extending SBR to more than stereo makes sense, it's not something enabled in encoders and decoders yet.

HE AAC v1 is supported by many mobile devices, and HE AAC v1 and v2 are supported in Flash since 9 Update 3. QuickTime X on Mac OS X 10.6 also can decode v1 and v2, but that hasn't been enabled in QuickTime for Windows or older Mac OS versions.

AAC Encoders

Apple (QuickTime and iTunes)

QuickTime was the first major architecture to support AAC-LC (introduced with QuickTime 6.0 in 2002). Since it's a free API that includes output, the QuickTime implementation has been broadly used in other tools. And of course, Apple's AAC-LC encoder is famously used in iTunes, where it has ripped billions of .m4a files. Even though QuickTime X introduced HE AAC decode, it doesn't include an HE AAC encoder.

MPEG-4 Export vs. QuickTime Export

One highly annoying aspect of QuickTime's AAC-LC is that most of the cool advanced options aren't available if you're exporting to a .mp4/.m4a file (see Figure 13.1). You need to export as a QuickTime Movie for the full dialog to show up (Figure 13.2). It's not a compatibility thing; it's trivial to reopen the .mov, and then export to MPEG-4 with audio set to "Pass through" (Figure 13.3). Hopefully, this will change with QuickTime X.

The Mac does include a command-line encoder that will do this in a single step. You can find it by typing the following in Terminal:

```
/usr/bin/afconvert
```

It has a variety of promising sounding parameters that appear to be entirely undocumented, none of which are exposed in any GUI compression tools, and there are cases where afconvert produces seemingly valid output that isn't QuickTime-compatible.

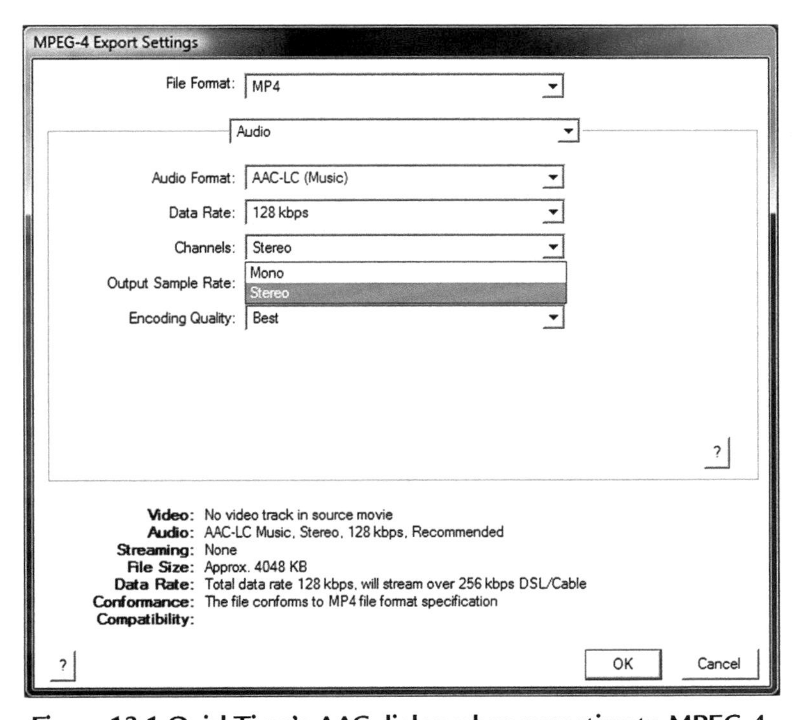

Figure 13.1 QuickTime's AAC dialog when exporting to MPEG-4.

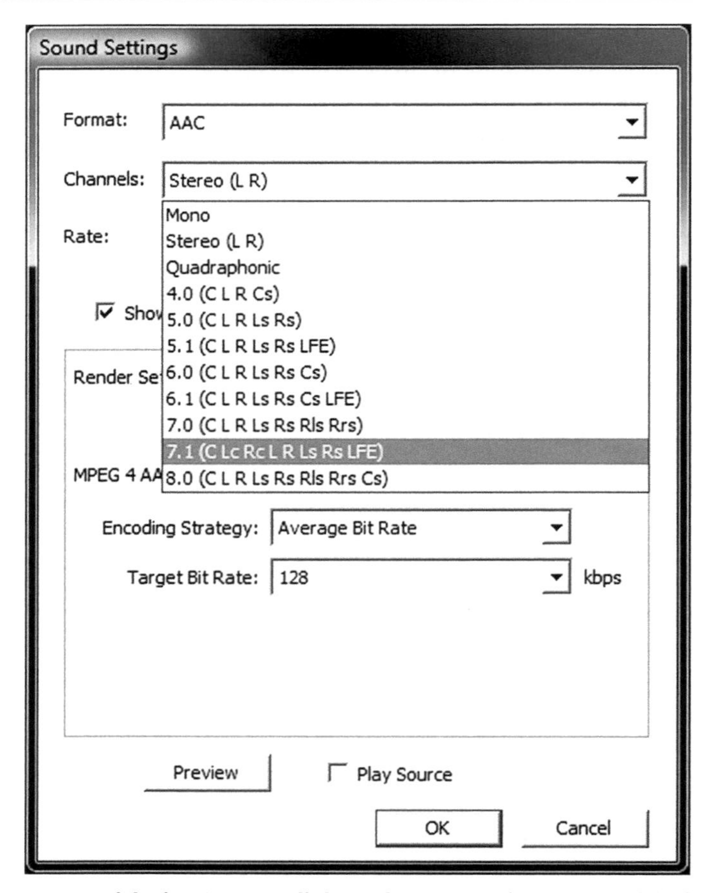

Figure 13.2 QuickTime's AAC dialog when exporting to a QuickTime file.

Quality

The classic Quality versus Speed control, going Faster-Fast-Normal-Better-Best. This is more than just the codec; it also impacts sample rate and bitrate conversion; Best is definitely recommended if your source is 24-bit so that you get high-quality rounding/dithering to 16-bit. If you're encoding audio-for-video, just stick it at Best and forget about it – the increase in total encoding time would be a rounding error.

Encoding strategies

Constant Bit Rate

The default Constant Bit Rate mode is a classic CBR encoder. It's the least efficient overall, of course, and is mainly useful when doing real-time encoding, as it has the lowest and most predictable broadcast delay.

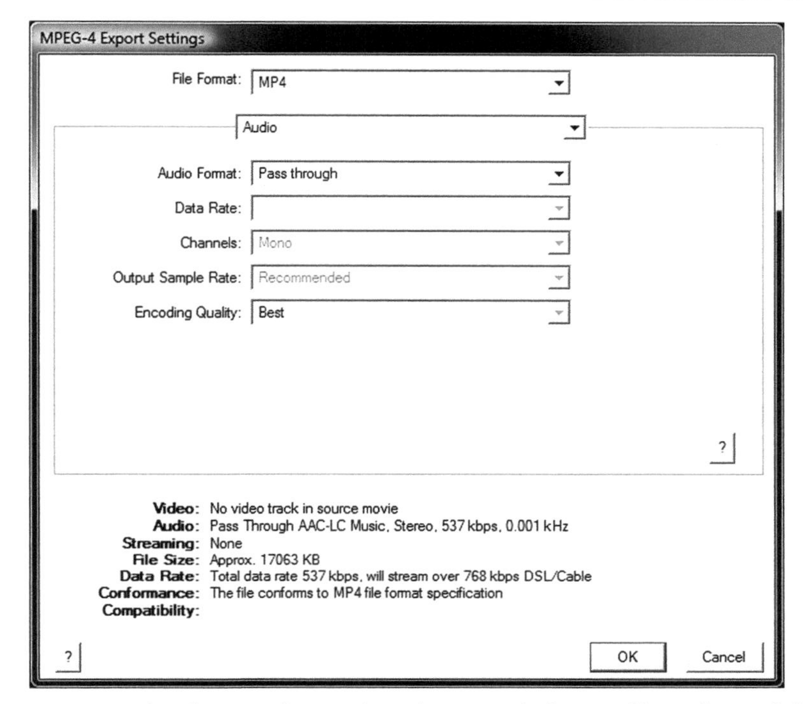

Figure 13.3 But you can load your advanced setting encoded .mov file and pass it back through to MPEG-4 by using "Pass through."

Average Bit Rate

Average Bit Rate is a ABR-style encode that sticks closely to the target bitrate—it hits the average over several seconds, so total file size will be with a percent or two for content of at least a minute. This should be your go-to mode for on-demand encoding. It's fine for streaming as well, given the relatively small variability.

Variable Bit Rate

Variable Bit Rate asks for a target bit rate, but it's otherwise pretty similar to other quality-based VBR audio codecs. The output size can easily be 20 percent higher or lower than the target. VBR mode is good for music collections where the actual bitrate will average out. Its unpredictability can be a problem for web distribution.

VBR has a smaller range of available bitrates than ABR/CBR, but as it can undershoot/overshoot its target, the practical range is essentially the same.

Variable Bit Rate constrained

VBR Constrained is a hybrid of ABR and VBR, keeping the final bitrate within a range and thus offering a lot more adaptability. It's a good compromise for progressive download, as it offers quite a bit better efficiency than ABR.

Table 13.1 shows the allowed data rates for the different modes in QuickTime. I'm keeping to reasonable sample rate/channel combos; even though it allows 8 KHz 7.1, you wouldn't ever want to do that.

Coding Technologies (Dolby)

Coding Technologies were the creators of HE AAC (called AAC+ in their implementation). They were then purchased by Dolby, with the implementation renamed Dolby Pulse. But it's still the same thing, and lots of applications, such as Squeeze and Episode, are still using the old name (see Figures 13.4 and 13.5).

Table 13.1 Allowable Data Rates for Different QuickTime Modes.

Channels	Sample rate	CBR/ABR Bitrate range	VBR range 44.1/48
Mono	8	8–24	8–24
Mono	11.025	8–32	9–36
Mono	12	12–32	9–36
Mono	16	12–48	12–48
Mono	22.05/24	16–64	18–72
Mono	32	24–96	24–96
Mono	44.1/48	32–256	24–96
Stereo	22.05/24	40–128	36–144
Stereo	32	48–192	48–192
Stereo	44.1/48	64–320	48–192
4.0	44.1/48	128–640	96–384
5.0/5.1	44.1/48	160–768	120–480
6.0	44.1/48	192–960	144–576
7.0/7.1	44.1/48	224–960	168–672
8.0	44.1/48	256–1280	192–768

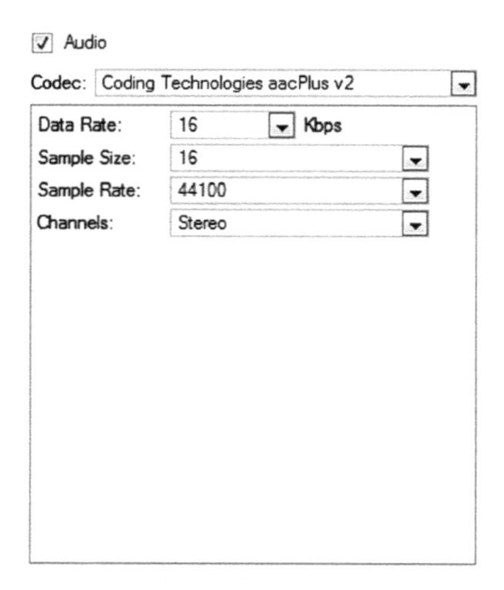

Figure 13.4 Squeeze calls out HE AAC by the Coding Technologies brand. It doesn't offer a way to turn off v2; if you want v1, you have to use mono.

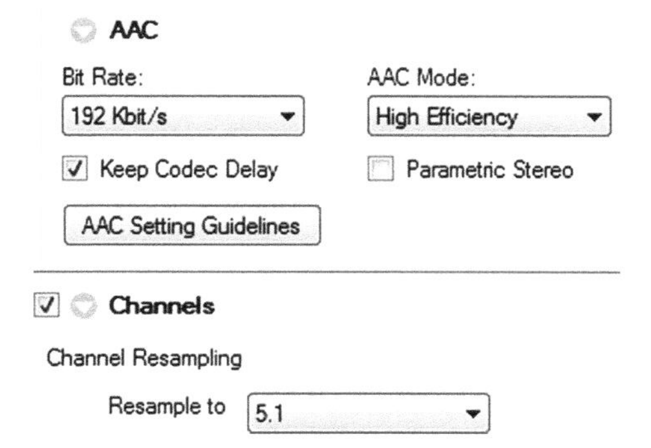

Figure 13.5 Episode gives explicit options for High Efficiency (v1) and Parametric stereo (v2). It's one of the few tools that can encode multichannel in HE v1. However, it doesn't constrain your choices: you can ask for 96 KHz 7.1 HE AAC v2 at 8 Kbps, and you won't find out that it doesn't work until the encode errors out.

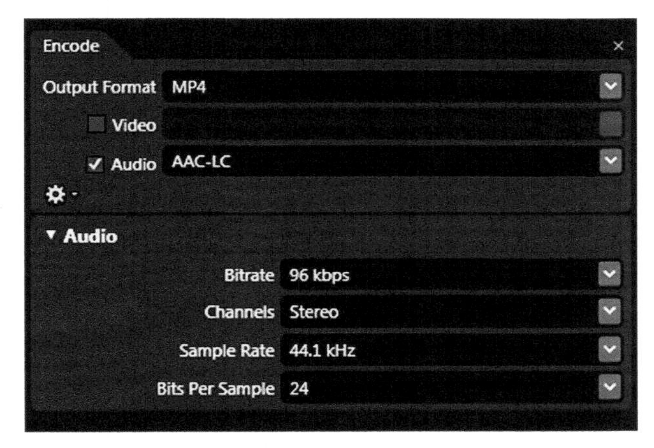

Figure 13.6 The Microsoft AAC-LC encoder from Expression Encoder 3. It's not very flexible, but offers good quality in the available bitrates.

These implementations offer good quality, and are generally CBR only.

Microsoft

The Microsoft AAC-LC implementation in Expression Encoder 3 and Windows 7 doesn't have very complex options, but provides good quality within those (Figure 13.6). It is mono and stereo only with 96, 128, 160, and 192 Kbps CBR options. It does support both 16-bit and 24-bit audio input.

CHAPTER 14

H.264

H.264 has entered into the public consciousness only in the last few years, but the effort actually started back in 1997, as H.263+ was finishing up. At that point, it was becoming clear to some that adding new layers on top of H.263 was running out of steam, and yielding bigger increases in complexity than efficiency. What was then called "H.26L" was formally launched in the ITU's Video Quality Expert's Group (VCEG) in 1999, chaired by my now-colleague Gary Sullivan.

By eschewing the "let's enhance H.263 again!" trend and building a new codec from the ground up, all kinds of proposed tools to improve compression efficiency were considered, even if they broke the traditional models of codec design. Their goal was to at least double the efficiency of other standard codecs.

As it became clear that MPEG-4 Part 2 wasn't scaling as well as hoped, MPEG put out a call for proposals for new codecs. H.26L was proposed, and the ITU and MPEG agreed to collaborate on its development, forming the Joint Video Team (JVT).

The H.264 spec was delivered in 2003, with the Fidelity Range Extensions (FRExt), including the broadly used High Profile, coming in 2004.

The effort paid off, and H.264 is clearly the most efficient codec on the market, particularly as the bits per pixel drops down to very low levels. However, there's a price to pay: Decoding the more complex profiles of H.264 requires more MIPS in software or more silicon in hardware to decode than past codecs. However, even the Baseline profile offered notable efficiency improvements over older complexity-constrained codecs, and has quickly taken over the market for lower-power media players.

The Many Names of H.264

The codec now called H.264 has been known by many other names. There are people who have political feelings about one name or another. The names you might hear or have heard it called include the following:

- H.264. This is the ITU designation, following H.261, H.262 (aka MPEG-2), and H.263 (aka MPEG-4 part 2 short header). It has clearly emerged as the most commonly used name for the codec, probably because that's what Apple called it in QuickTime.
- MPEG-4 part 10. This is the MPEG designation, from the section of the MPEG-4 specification where it is defined.
- Advanced Video Coding (AVC). This has been proposed as a politically neutral alternative to H.264 and MPEG-4 part 10. However, it has disambiguation issues with other AVC acronyms.
- H.26L: This was the working name for H.264.
- JVT: The codec was designed by the Joint Video Team, composed of codec experts from MPEG and ITU.

Why H.264?

Compression Efficiency

H.264's biggest draw has been its industry-leading compression efficiency. Its novel design enabled the bitrate requirements for common delivery scenarios to drop substantially compared to past codecs as H.264 encoders matured.

While we'll certainly have even better codecs in the next few years, right now H.264 High Profile is the undisputed king of low-bitrate quality, with plenty of room for further optimization in current profiles.

Ubiquity

Nothing begets success like success. By combining a big technical advantage over MPEG-2 and MPEG-4 part 2, H.264 has quickly been added to new products that used the older technologies. It's an ongoing process, but H.264 is certainly becoming a given in any consumer electronics device that does any form of video playback.

And since it's an international standard, everyone has equal access to the specification and licensing, and can compete on implementation while participating in a big interoperable ecosystem.

Why Not H.264?

Decoder Performance

The biggest downside to H.264, particularly when using the more advanced settings with more advanced profiles, is decoder performance. It's no big deal when GPU acceleration or an ASIC decoder is available, but in pure software, H.264 can require 2–4x the MIPS/pixel as simpler codecs.

Of course, there are optional features like CABAC (defined later in this chapter) that can be turned off to improve decoder performance, albeit with a reduction in compression efficiency.

Older Windows Out of the Box

Windows only started shipping with H.264 in-box with Windows 7, and even then not in the very low end Starter and Home Basic editions. For content targeting Windows machines, WMV is the only format that absolutely will play on every one, particularly on slow-to-update corporate desktops. Flash or Silverlight are likely to be installed as well, but they can't play a double-clicked media file.

Profile Support

While H.264 High Profile is a clear efficiency winner, that's not the profile supported everywhere. In particular many devices support both H.264 Baseline and VC-1 Main, and VC-1 can offer superior detail retention at moderate bitrates in that case.

Licensing Costs

Lastly, there has been some concern about MPEG-LA licensing fees. I am not a lawyer, but my reading of the published fees is that anyone with a business big enough to be impacted would have a business big enough to pay the fees. And these are nothing new; MPEG-2 and MPEG-4 part 2 had licensing fees, and VC-1 has similar ones today. However, this is the first intersection of the web world with standards-based codecs, so it's new to a lot of folks. Concerns about H.264 patent licensing have been the major source of interest in the Theora and Dirac codecs.

What follows is my personal summary of a PDF from MPEG-LA's web site; if you have legal questions about this, seek real legal advice.

The agreements renew every five years, with prices rising no more than 10 percent each.

Branded encoders and decoders (Built-in decoders are covered by Apple and Microsoft)

- Up to 100 K/year: No fee
- $0.20 per unit more than 100 K/year up to 5 M

- $0.10 per unit over 5 M
- Maximum fee per company per year: $5 M

Video content and service providers

End user pays directly on title-by title basis

- No royalty for titles of 12 minutes or less
- The lesser of $0.02 or 2 percent of price paid to licensee per title over 12 minutes

End user pays on subscription basis (yearly fee)

- Fewer than 100 K subscribers per year: No fee
- 100 K to 250 K subscribers per year: $25 K
- 250 K to 500 K subscribers per year: $50 K
- 500 K to 1 M subscribers per year: $75 K
- More than 1 M subscribers per year: $100 K

Remuneration from other sources, like free TV not paid by end user (yearly fee); may apply to internet broadcasting in 2010+

- Choice of $2500 K per broadcast encoder, *or*
- Broadcast market under 100 K: No fee
- Broadcast market between 100 K and 500 K: $2500
- Broadcast market between 500 K and 1 M: $5000
- Broadcast market more than 1 M: $10 K

Annual cap for service provider enterprises

- $4.25 M in 2009
- $5 M in 2010

What's Unique About H.264?

If VC-1 can be thought of as the culmination of the H.261/MPEG-1/MPEG-2/H.263 lineage, keeping what worked (IBP structure, 8 × 8 as the "default" block size, field/frame adaptive interlace coding) and addressing what didn't (using an integer transform, adding in-loop deblocking, offering more flexibility and simplicity in block types per frame type), H.264

changed many of the fundamentals of how codecs work, from DCT and blocks all the way up to I-, B-, and P-frames.

4×4 blocks

Perhaps the most startling thing about H.264 was replacing the long-standard 8 × 8 block size for the much smaller 4 × 4 (although 8 × 8 was added later as an option in High Profile). With smaller block sizes, ringing and blocking artifacts from high QP blocks are smaller and hence less noticible. Plus, edges go through fewer blocks, and so fewer pixels need higher bitrates to encode well.

4 × 4 blocks were designed along with the in-loop deblocking filter, so that the loss of efficiency from having only 16 samples instead of 64 samples per block had less of an impact, reducing the visual and predictive impact of higher quantization.

The return of 8×8 blocks

While 4 × 4 worked well for the lower resolutions that H.264 was originally tested at, they didn't scale as well to larger frame sizes, particularly with film content where there can be large areas that are relatively flat except for grain texture. In the initial DVD Forum testing to select a HD codec for HD DVD, several MPEG-2 and early H.264 encoders were tested along with VC-1, and to broad surprise, VC-1 won the quality shootout, followed by MPEG-2, and with the H.264 encoders in last place. That result likely had more to do with the relative refinement of the implementations of each codec. Before long, the High Profiles were launched, including the optional 8 × 8 block type, and H.264 was a more competitive HD codec.

Strong In-Loop Deblocking

Probably the single most valuable feature added in H.264 is its strong, adaptive in-loop deblocking filter. The goal of in-loop deblocking is obvious; soften the edges between highly quantized blocks in order to remove the blocking artifacts. And not just in postprocessing, but on reference frames, keeping block artifacts from propagating forward.

But H.264 was the first to allow the filter to be so strong (in terms of how much it softened) and broad (in terms of how many pixels it touches) as to really be able to eliminate blocking artifacts outright. As you crank up the compression with H.264, the image gets softer and softer, but doesn't get the classic DCT sharp-edged blocks. Some H.264 encoders rely on this too much, in my opinion; it allows some sloppiness in encodes, throwing away more detail than necessary. Even if viewers don't realize it's due to compression, not a soft source, it's still throwing away important visual information. But done correctly, this makes H.264 less vulnerable to QP spikes and distractingly poor quality, particularly with 1-pass CBR encoding.

The downside of in-loop deblocking is that it can also soften the image somewhat at even moderate QPs. Fortunately, it can be tuned down or even turned off in many encoders. Typically Blu-ray H.264 encodes won't use in-loop deblocking for most scenes, reserving it for those for those high-motion/complexity scenes that would otherwise have artifacts (see Figures 14.1 and 14.2).

Figure 14.1 A H.264 encode with in–loop deblocking off.

Figure 14.2 Same source with in–loop deblocking on. Deblocking not only reduces blocking, but improves image data by improving reference frames.

Variable Block-Size Motion Compensation

H.264 supports a wide variety of different block sizes for motion compensation compared to the simple 16 × 16 and 16 × 8 options in MPEG-1/2. Each macroblock can use 16 × 16, 16 × 8, 8 × 16, 8 × 8, 8 × 4, 4 × 8, or 4 × 4 partitions.

Quarter-Pixel Motion Precision

H.264 has a more efficient implementation of quarter-pixel motion estimation than part 2's, delivering a good efficiency improvement.

Multiple Reference Frames

In the classic codecs, there would be at most two reference frames at a time: A B-frame would reference the previous and next I- or P-frame. H.264 radically expanded the concept of a reference frame by allowing any frame to have up to sixteen reference frames it could access. And that goes for P-frames, not just B-frames. Essentially, the decoder can be required to cache up to 16 past frames, any or all of which can referenced in a motion vector to construct a future frame.

H.264 doesn't require that decoded and displayed order match either. It's perfectly legal for a P-frame to reference a later P-frame in display order, although it'd have to come first in encode order.

And frames don't even need to be all P or B. Each frame can be made up of I-, P-, and B- slices (see Figure 14.3). What we think of as an I-frame in most codecs is called an IDR

Figure 14.3 One possible structure of slices in a four–slice encode.

in H.264—Instantaneous Decoder Refresh. All frames that reference each other must all reference the same single IDR frame. Non-IDR I-frames are also allowed in H.264 (like a Open GOP to an IDR's Closed GOP).

While awesome-sounding, and awesomely complex-sounding, it's not as dramatically hard to implement or as dramatically useful as it appears on first blush. The main value of multiple reference frames is when a part of the image is occluded (covered) in the previous reference frame, but visible in a frame before that. Imagine the classic shot through a spinning fan. At any given point, the fan's blades are covering a third of the screen. So frame 100 would include a part of the image obscured in frame 99 and frame 98, but visible in frame 97. If the encoder was using three reference frames, frame 100 could pull a motion vector from frame 97 to predict frame 100, instead of having to use intrablocks to encode that as a new part of the image.

The number of allowed reference frames is constrained by a maximum number of macroblocks for the given *profile@level*. Thus smaller frame sizes can support more reference frames than ones near the max capacity of the level. That said, it's rare for the full 16 to be materially better than 3–4 for most film/video content. Cel animation or motion graphics can take better advantage of lots of reference frames.

All of this information can seem conceptually confusing, but the details aren't really exposed to a compressionist. Normally, the choices are about how many reference frames to use. The rest follows from that.

While software decoders typically don't enforce limits on maximum reference frames, hardware decoders very often will. This includes GPU decoder technologies like DXVA in Windows. So stick to the constraints of your target profile@level.

Pyramid B-Frames

Pyramid B-frames are another concept that can blow the mind of a classical compressionist.

It's long been a matter of definition that a B-frame is never a reference frame, and is discarded as soon as displayed. Thus a B-frame only needs as many bits as it takes to look decent, without any concern for it being a reference frames.

But with pyramid B-frames, a B-frame can be a reference frame for another layer of B-frames. And that layer itself could be used as a reference for yet another layer of B-frames.

Typically the first tier of B-frames is written as "B" while a second tier that could reference those is written as b. So instead of the classic IBBPBBPBBP pattern, you could get IbBbPbBbPbBbP, with the "B" frames based on I- and P-frames, and the "b" frames referencing the I-, P-, *and* B-frames (see Figure 14.4).

Figure 14.4 Pyramid B−frames strucuture.

These complex structures, when coupled with an encoder that can dynamically adjust its pattern to the content, can further improve efficiency of encoding with motion graphics, cel animation, rapid or complex editing patterns, and lots of flash frames. It doesn't make that much of a difference with regular content, but is still worthwhile.

Weighted Prediction

Weighted prediction allows for whole-frame adjustments in prediction. This is mainly useful for efficient encoding of fades, both to/from a flat color and cross-fades. Coupling weighted prediction with multiple reference frames and B-frames can finally make that kind of challenging content encode well.

Logarithmic Quantization Scale

The scaling factor in H.264's quantizer isn't linear as in other codecs, but logarithmic. So going up one QP has less visual impact on the lower end of the scale. So, while QP 10 is typically pretty ugly in MPEG-2 or VC-1, QP 10 in H.264 is around the threshold of visually lossless.

Flexible Interlaced Coding

I have mixed feelings about H.264's interlaced support. I'm impressed it was done so well. But I worry that gives the industry one more excuse to keep the interlaced millstone around our necks another decade.

There are two basic levels of interlaced in H.264.

Macroblock adaptive field-frame

Macroblock Adaptive Field-Frame (MBAFF) is conceptually similar to classic MPEG-2 and VC-1 interlaced coding, where individual macroblocks can be either interlaced or progressive, with some clever tweaks around the edges for in-loop deblocking and the like.

Picture adaptive field-frame

Picture Adaptive Field-Frame (PAFF) allows individual frames to be MBAFF (good for interlaced frames with mixed areas of motion and static elements), or fields separately encoded (for when everything is moving).

So far, interlaced H.264 has been mainly the domain of broadcast, and hasn't been supported in QuickTime, Silverlight, Flash, or many devices. (Well, Flash can display interlaced H.264 bitstreams, but it leaves them as interlaced, so you should deinterlace on encode anyway.)

CABAC Entropy Coding

Entropy coding has been pretty similar for most codecs, with some flavor of Huffman coding with variable length tables. And H.264 has a variant of that with its CAVLC mode—Context Adaptive Variable Length Coding. It's a fine, fast implementation of entropy coding.

What's new in H.264 is CABAC—Context Adaptive Binary Arithmetic Coding. As you may remember from the section on data compression, Huffman compression can't be more efficient than one-bit-per-symbol, but arithmetic compression can use fractional bits and thus can get below one. Thus, CABAC can be quite a bit more efficient than CAVLC; up to a 10–20 percent improvement in efficiency with high compression, decreasing with less compression. With visually lossless coding it is more like 3 percent more efficient. CABAC is probably second only to in-loop deblocking in importance to H.264's efficiency advantage. But it comes with some limitations as well:

- CABAC is a lot slower, and not parallelizable within a slice. CABAC can increase decoder requirements up to 40%. That speed hit is proportional to bitrate, so it's less of a problem at the lower bitrates where CABAC is most useful. Keeping peak bitrate as low as is feasable provides adequate quality and can help significantly. Each slice in the video is independently decodable, so using more slices can increase decoder performance on multicore machines, and is required in some cases, like Blu-ray at Level 4.1.
- CABAC is also slower to encode, and requires a feedback mechanism that can add some latency to the encoding process.
- CABAC's speed and complexity excludes it from the Baseline profile, and hence it's not available for portable media device content.

Differential Quantization

H.264 can specify the quantization parameter per macroblock, enabling lots of perceptual optimization. Improved use of DQuant has been a big area of H.264 advancement in recent encoders.

Quantization Weighting Matricies

This is a High Profile–only feature brought over from MPEG-2. Custom quantization matricies allow tuning of what frequencies are retained and discarded.

The popular x264 recommends a flat matrix and then aggressively uses Differential Quantization.

Modes Beyond 8-bit 4:2:0

Relatively unique for an interframe codec, H.264 has profiles that support greater-than-8-bit-per-channel luma, and both 4:2:2 and 4:4:4 sampling. H.264 also supports using a full 0-255 luma range instead of the classic 16-235, making round-trip conversion to/from RGB much easier.

None of those modes are widely supported in decoders yet, though. H.264 is still mainly used at good old 16-235 8-bit 4:2:0.

H.264 Profiles

Baseline

The Baseline profile is a highly simplified mode that discards many of the computationally expensive features of H.264 for easy implementation on low-power devices. Generally, if it runs on a battery or fits in your hand (like an iPod, Zune, or phone), it likely will only have Baseline support. Baseline leaves out several prominent features:

- B-frames
- CABAC
- Interlaced coding
- 8 × 8 blocks

So, Baseline requires quite a few more bits to deliver the same quality as the more advanced profiles, particularly High. But it works on cheap, low-power decoder ASICs.

While Baseline is certainly easier on software decoders than a full-bore High encode, that absolutely doesn't mean that Baseline should be used for software decodes even when decode complexity is a constraint. In particular, B-frames and 8 × 8 blocks offer big improvements in efficiency, and minor increases in decoder complexity. In software, they'll outperform Baseline when comparing the same quality.

Constrained baseline

There's an effort ongoing to add a new "Constrained Baseline" profile to document what's emerged as the de facto Baseline standard. It turns out that there are some required features in Baseline not used by Baseline encoders, and hence not tested and not working in existing "Baseline" decoders. Constrained Baseline would just codify what is currently implemented as Baseline.

Extended

The Extended profile targets streaming applications, with error resiliency and other features useful in that market. However, it hasn't been implemented widely in encoders or decoders. Its most prominent appearance has been as a mysteriously grayed-out checkbox in QuickTime since H.264's addition in 2005.

Main

Main Profile was intended to be the primary H.264 codec for general use, and includes all the useful features for 8-bit 4:2:0 except for 8 × 8 block sizes. However, adding 8 × 8 turns out to help compression a lot without adding significant complexity, so any new decoder that does Main also does the superior High.

The primary reason to use Main today is if you're targeting legacy players that do it but not High. Silverlight, Flash, and WMP all included High in their first H.264 implementations. QuickTime was Baseline/Main only in QuickTime 7.0 and 7.1, with High decode added in 7.2.

Many Main decoders can't decode interlaced content, although that's a required feature of the Main Profile spec.

High

High Profile adds three primary features to Main:

- 8 × 8 block support!
- Monochrome video support, so black-and-white content can be encoded as just Y′ without CbCr.
- Adpative Quantization Tables.

8 × 8 blocks are quite useful in terms of improving efficiency, particularly with film content and higher resolutions.

Monochrome is used much less in practice (and it's not like coding big matrices where Cb and Cr = 128 took a lot of bits), and decoder support is hit-and-miss.

Adaptive Quantization Tables have been used to good success in some recent encoders, and are becoming another important High advantage.

High 10

After the initial launch of H.264, attention quickly turned to "Fidelity Range Extensions"—FRExt—to get beyond 8-bit luma coding. However, it was the addition of 8 × 8 blocks to the baseline High 8-bit codec that was broadly adopted. It's great, but so far not used for content distribution. Having 10-bit would require less dithering on authoring. However, as the end-to-end ecosystem is all based around 8-bit sampling, it would be a slow transition as best.

The only widely supported player that decodes High 10 is Flash. Adobe doesn't support it in their compression tools, however, and Flash itself has an 8-bit rendering model that doesn't take advantage of the additional precision.

High 4:2:2

High 4:2:2 simply adds 4:2:2 chroma sampling with 10-bit luma. It would be most useful for interlaced content.

High 4:4:4 Predictive

High 4:4:4 Predictive (normally just called high 4:4:4) goes further to 14-bit 4:4:4. Again, points for awesome, but it is much more likely to be used as an intermediate codec than for content delivery for many years to come.

Intra Profiles

H.264 supports intra-only implementations for production use (like an updated MPEG-2 High Profile I-frame only). These are subsets of the following:

- High 10 Intra
- High 4:2:2 Intra
- High 4:4:4 Intra
- CAVLC 4:4:4 Intra

Really, intra codecs should use CAVLC anyway. CABAC is worth around only 3 percent extra efficiency at these high bitrates, while these same high bitrates make the decode complexity of CABAC much more painful.

Scalable Video Coding profiles

I can't decide if Scalable Video Coding (SVC) is going to be the Next Big Thing in IP-based video delivery, or just something neat used in video conferencing.

The idea of scalable coding has been around for quite a while. The concept is to use prediction not just for future frames, but to enhance a base stream. B-frames are actually scalable video already, and could be thought of as a temporal enhancement layer. Since no frames are based on B-frames (putting pyramid B aside for a moment), a file could have all its

B-frames stripped out and still be playable, or have new B-frames inserted to increase frame rate without changing the existing frames.

Temporal scalability

For example, take a typical 30-fps encode at IBPBPBP. A lower-bitrate, lower-decode complexity version could be made by taking out B-frames and leaving a 15 fps IPPP. Or extra B-frames (pyramid b, typically) could be added making it IbBbPbBbPbBbP and 60 fps (see Table 14.1).

It's conceptually simple. However, only the original 30 fps encode would have the best coding efficiency. If you knew you wanted 15 fps or 60 fps all along, you could have used the optimal bitrate and frame structure to encode that band. But where this really shines is when you don't know how much bitrate is going to be available; instead the video can be encoded once (as IbBbP in the example) and depending on bitrate and/or processing power, just the IP, IP+B, or IP+B+b frames could be delivered on the fly. Each layer of extra data that can be added is called an enhancement layer. For our example here, the I- and P-frames are the base layer, the B-frames are the first enhancement layer, and the b-frames are the second enhancement layer.

And thus the basic concept of scalable video: trading some compression efficiency for increased flexibility.

Spatial scalability

Another place where scalable coding has long been used is "progressive" JPEG and GIF files, where a lower resolution of the image is transmitted and displayed first, with a higher resolution version then predicted from that. This is of course the basic mode of operation of wavelet codecs, and wavelet codecs can be thought of as being natively scalable, with each band serving as an enhancement layer.

H.264 SVC goes well beyond that simplified example in spatial scalability, but the concept is the same—predicting a higher-resolution frame from a lower-resolution frame. The key innovation in H.264 prediction is adaptive mixing of both spatial prediction (of the same frame in a lower band) and a temporal prediction (using another frame of the same resolution).

Table 14.1 Temporal Scalabilty à la Pyramid B-Frames. The reference frames are at 15 fps, the first set of B-frames is at 30p, and then the second layer of B-pyramids.

Sec.	1/60	2/60	3/60	4/60	5/60	6/60	7/60	8/60	9/60	10/60	11/60	12/60	13/60
15p	I				P				P				P
30p	I		B		P		B		P		B		P
60p	I	b	B	b	P	b	B	b	P	b	B	b	P

Quality scalability

The last form of scalability in SVC is quality scalability, which adds extra data to improve the quality of a frame. Essentially, the enhancement layers add additional coefficients back to the matrix. Imagine them as the difference between the coefficients of the base stream's matrix QP and a less coarse QP.

SVC profiles

There are three profiles for SVC today, mapping to the existing profiles in use:

- Scalable Baseline
- Scalable High
- Scalable High Intra

Note the lack of Main; there's no point in using it instead of High, since SVC requires a new decoder. High is getting by far the most attention, as it as the best compression efficiency. High Intra could be of use for workflow project to enable remote proxy editing of HD sources.

Why SVC?

Looking at SVC, we can see that it has the potential to offer better scalability than classic stream switching or adaptive streaming. This is because stream switching needs to make sure that there's some video to play, and so it can't be very aggressive at grabbing higher bitrates with low latencies. With stream switching, if a 600 Kbps stream only gets 500 Kbps of bandwidth, there's a risk of running out of buffered frames to decode and the video pausing. But with SVC, each enhancement layer can be pulled down one-by-one—a 100 Kbps base layer can be delivered, and then additional 100 Kbps layers could be downloaded. If a player tried to pull down the 600 Kbps layer but didn't get it all, the layers at 500 Kbps and below are still there, and there's no risk of a loss of video playback.

Buffering could be prioritized per stream, with the base stream always buffering ahead several minutes, with relatively less buffer the higher the enhancement layer. And with this buffering comes greater responsiveness. With stream switching, it takes a while to request a new stream; if background CPU activity causes dropped frames, it may be a few seconds before a new stream can be selected. But SVC can turn layers on and off on the fly without having to request anything new from the server.

Lastly, SVC can save a lot of server storage space. 4 Mbps across 10 layers is a lot smaller than a 4 Mbps stream + 3 Mbps stream + 2 Mbps + 1.5 Mbps…

Why not SVC?

There are a few drawbacks for SVC, of course. First, decoders aren't widely deployed, nor are encoders widely available. It's essentially a new codec that hasn't solved the chicken-and-egg problem of being used enough to drive decoders, or being in enough decoders to drive use. This is particularly an issue with hardware devices.

SVC also has more decoder overhead; 720p24 4 Mbps with eight bands requires more power to play back than 720p24 4 Mbps encoded as a single layer.

There's also an efficiency hit, particulary when spatial scalability is used. It might take 10–15 percent higher bitrate by the highest layer of a low bitrate to HD layer set than to encode a single-layer HD stream. So, to get the quality of a 3000 Kbps single stream, SVC with eight layers may need 3300–3450 Kbps.

When SVC?

SVC is more interesting the less predictable the delivery environment is. It makes little sense for file- and physical-based playback, since the available bandwidth is already known. And for IPTV-like applications with fixed provisioned bandwidth, it's more efficient to just encode at the desired bitrate.

But when bandwidth is highly variable, like with consumer Internet use, SVC could shine. And the lower the broadcast delay required, the harder it is to use stream switching.

Thus, the first place we're seeing SVC catch on is in videoconferencing, where the broadcast delay is a fraction of a second.

Where H.264 Is Used

QuickTime

QuickTime was the first major platform to support H.264, introduced with QuickTime 7 in 2005. Initially only Baseline and Main were supported for decode. QuickTime 7.2 added High Profile decoding in 2007; it's safe to assume nearly all Mac and iTunes users would have upgraded by now. QuickTime doesn't support interlaced decode at all.

QuickTime's decoder is decent, albeit software-only and not very fast.

Flash

Flash added H.264 decode in version 9.115. They have one of the deeper implementations (based on the Main Concept decoder), and can decode three profiles:

- Baseline
- Main
- High (including 10 and 422)

Table 14.2 H.264 Levels.

Level number	Max macro-blocks/second	Max macro-blocks/frame	Max bitrate Baseline/Main	Max bitrate High	Max 4:3	Max 16:9	Max 4:3 24p	Max 16:9 24p	Max 4:3 30p	Max 16:9 30p
1	1485	99	64 Kbps	80 Kbps	176 × 144	208 × 128	144 × 112	160 × 96	128 × 96	144 × 96
1b	1485	99	128 Kbps	160 Kbps	176 × 144	208 × 128	144 × 112	160 × 96	128 × 96	144 × 96
1.1	3000	396	192 Kbps	240 Kbps	368 × 272	432 × 240	208 × 160	240 × 128	192 × 128	208 × 128
1.2	6000	396	384 Kbps	480 Kbps	368 × 272	432 × 240	288 × 224	336 × 192	256 × 208	304 × 176
1.3	11880	396	768 Kbps	960 Kbps	368 × 272	432 × 240	416 × 304	480 × 272	368 × 272	432 × 240
2	11880	396	2 Mbps	2.5 Mbps	368 × 272	432 × 240	416 × 304	480 × 272	368 × 272	432 × 240
2.1	19800	792	4 Mbps	5 Mbps	512 × 400	608 × 336	528 × 400	608 × 352	480 × 352	544 × 304
2.2	20250	1620	4 Mbps	5 Mbps	736 × 560	864 × 480	528 × 416	624 × 352	480 × 368	560 × 304
3	40500	1620	10 Mbps	12.5 Mbps	736 × 560	864 × 480	752 × 576	880 × 496	672 × 512	784 × 448
3.1	108000	3600	14 Mbps	14 Mbps	1104 × 832	1280 × 720	1232 × 928	1424 × 816	1104 × 832	1280 × 720
3.2	216000	5120	20 Mbps	25 Mbps	1328 × 992	1520 × 864	1744 × 1328	2016 × 1136	1568 × 1168	1808 × 1024
4	245760	8192	20 Mbps	25 Mbps	1664 × 1264	1936 × 1088	1872 × 1408	2160 × 1216	1664 × 1264	1936 × 1088
4.1	245760	8192	50 Mbps	62.5 Mbps	1664 × 1264	1936 × 1088	1872 × 1408	2160 × 1216	1664 × 1264	1936 × 1088
4.2	522240	8704	50 Mbps	62.5 Mbps	1728 × 1296	1984 × 1120	2720 × 2048	3152 × 1760	2432 × 1840	2816 × 1584
5	589824	22080	135 Mbps	168.75 Mbps	2736 × 2064	3168 × 1792	2896 × 2176	3344 × 1888	2592 × 1936	2992 × 1680
5.1	983040	36864	240 Mbps	300 Mbps	3536 × 2672	4096 × 2304	3728 × 2816	4096 × 2304	3344 × 2512	3856 × 2176

Flash can decode interlaced H.264, unlike QuickTime and Silverlight. However, it doesn't have any deinterlacing support, so both fields will be painfully visible on playback.

Silverlight

Silverlight 3 introduced MPEG-4 and H.264 support, and it supports the same H.264 profiles as QuickTime:

- Baseline
- Main
- High

Smooth streaming with H.264

With Silverlight 3, H.264 is now supported in Smooth Streaming. While tool support for that combination was just emerging as this book was being written, support was forthcoming in Expression Encoder 3, plus products from Inlet, Telestream, Envivo, and others.

The Smooth Streaming Encoder SDK used for VC-1 Smooth Streaming encoding provides resolution switching to maintain minimum quality. There's less need for resolution switching with H.264 due to its strong in-loop deblocking.

One nice feature of Smooth Streaming is that different versions of the content can be encoded at the same bitrate. So codecs can be mixed and matched; for example, a single player could link to different 300 Kbps streams containing:

- VC-1 Advanced Profile stream for desktop playback on computers
- H.264 Baseline for mobile devices
- H.264 High Profile for consumer electronics devices

Windows 7

While there have been third-party DirectShow H.264 decoders for years, and hardware acceleration support for H.264 since Vista, Windows 7 is the first version with out of the box support for the MPEG-4 formats, including H.264.

Windows 7 has a deep implementation, including interlaced support and hardware acceleration via the Windows 7 Media Foundation API, including HD H.264 playback on the next generation of netbooks.

Portable Media Players

H.264 Baseline has become the dominant codec for portable media players and mobile phones. We'll give specific examples in Chapter 25.

Consoles

Both the Xbox 360 and PlayStation 3 can play back H.264 files and streams from media services. There is not an equivalent for the Wii (which has much more limited hardware capabilities).

Settings for H.264 Encoding

Profile

Profile is your first question, and largely dependent on the playback platform. This will almost always be Baseline or High. Even if you're concerned about decode complexity with a software player, use High if you can, while turning off the expensive-to-decode features. Features like 8 × 8 blocks improve compression without hurting decode complexity.

Level

Level defines the constraints of the encode to handle particular decoders. The level is stored in the bitstream, and decoders that can't play back that level may not even try. It's important to always have the level set and specified, even if implicitly (some encoders figure out the lowest Level compatible with your encoding settings). It's bad practice to do 640 × 360 encoding with Level left at 5.1; use the lowest Level your content needs in order to deliver broader compatibility. QVGA 30p encodes never need more than 2.1, SD encodes never need more than 3.1, and 1080 p24 encodes never need more than 4.1.

Bitrate

Compared to older codecs, a well-tuned H.264 encode can deliver great quality at lower bitrates. However, that comes at the cost of extra decoder complexity; encoding H.264 at MPEG-2 bitrates is a waste of bits and a waste of MIPS. Make sure that your average (and, if VBR, peak) bitrates aren't higher than needed for high quality. There's no need to spend bits beyond the point where the video stops looking better.

Conversely, H.264 isn't magic, and if you're used to just watching out for blocking artifacts when compressing, you'll need to take a more detailed look with H.264 since in-loop deblocking yields softness instead of blockiness when overcompressed. Use enough bits to get an image that looks good, not just one that's not obviously bad.

Entropy Coding

The two entropy coding modes in H.264 Main and High are CAVLC and CABAC. CAVLC is faster and less efficient, CABAC is slower to encode and decode, but more efficient.

Most of the time, you'll want to use CABAC; CAVLC only makes sense when you're worried about decode complexity.

Since CABAC's decode hit is proportional to bitrate, watch out for performance with VBR and high peak bitrates. Make sure the hardest part of the video plays well enough on the target platform. A peak of 1.5x the average bitrate is a good starting point; don't use a higher peak than one that significantly contributes to quality.

Slices

One way to allow entropy decoding to parallelize is to encode with slices, breaking the frame into (typically) horizontal stripes. Each slice is treated somewhat like an independent frame, with separate entropy coding.

This can improve playback on some multicore software decoders, and is required by the Blu-ray spec for encodes using Level 4.1. There's a slight loss in efficiency with slices (perhaps 1 percent per slice as a ballpark). The hit is bigger the smaller the frame; a good rule of thumb is for each slice to be at least 64 pixels tall.

It's generally a lot more bitrate efficient to encode CABAC with slices than CAVLC without slices.

The Main Concept and Microsoft encoders, among others, also use slices for threaded encoding; the default is one slice per physical core in the encoding machine. However the trend in H.264 is definitely single-slice encoding for maximum efficiency.

Number of B-frames

Two B-frames is a good default for most non-pyramid H.264 encodes. Most H.264 encoders treat B-frame number as a maximum value when B-frame adaption is turned on, with Microsoft's an exception.

Pyramid B-frames

Pyramid B-frames can be a welcome, if not huge, boost to encoding efficiency. You'll need to be using at least 2 B-frames to use pyramid, obviously. Pyramid should be used with an adaptive B-frame placement mode for best results.

Number of Reference Frames

As mentioned above, large numbers of reference frames rarely pay off for film/video content, although they can be quite useful with cel animation and motion graphics. Four is generally sufficient. However, using the full 16 often has little impact on decode performance for playback, but can slow down random access and delay the start of playback.

Strength of In-Loop Deblocking

The default in-loop deblocking filter in H.264 is arguably tuned for video and lower bitrates more than for film grain and higher bitrates, and may soften detail more than is required. For tools that offer finer-grained control over it, turning it down can increase detail. Beyond on/off, there are two parameters that control behavior of in-loop deblocking:

- Alpha determines the strength of deblocking; higher values produce more blurring of edges.
- Beta modifies that strength for blocks that are more uniform.

It'd be rare to increase the strength of the filters. The range for each is −6 to +6, with 0 being the default. If adjusted, the values almost always are negative, typically −1 or −2 is sufficient.

Some Blu-ray encoders default to having in-loop deblocking off entirely in order to preserve maximum detail. It seems unlikely to me that the job couldn't be accomplished by just using lower settings.

H.264 Encoders

Main Concept

There's a huge number of H.264 encoders out there, but just a few that are used across a variety of products you'll see again and again. Main Concept is the most broadly licensed H.264 encoder SDK, and provides support in video products from companies including the following:

- Adobe
- Sorenson
- Rhozet
- Inlet

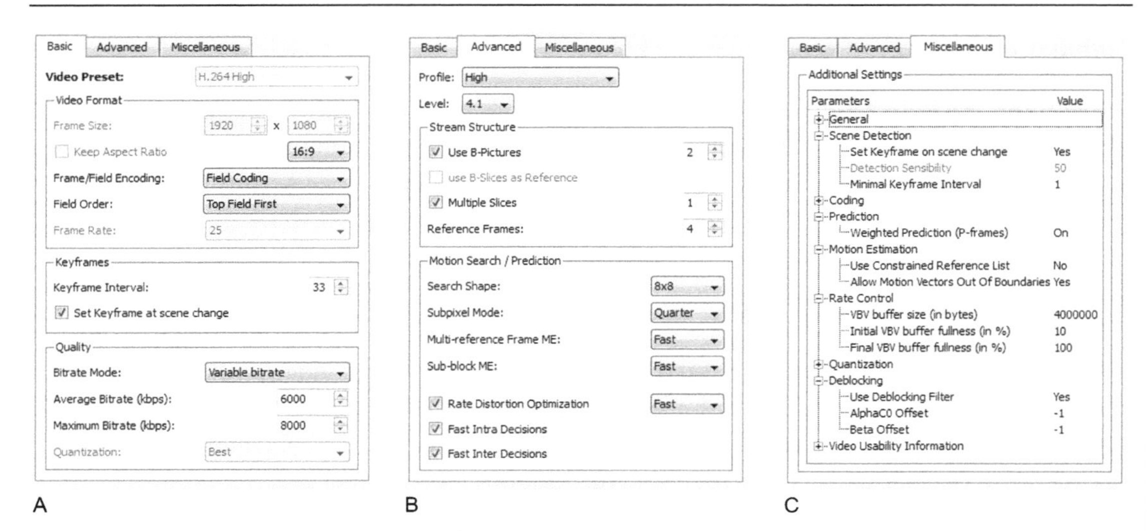

Figure 14.5 The Main Concept encoding controls as presented in their own Reference product. I just wish they had more documentation on the impact of these settings.

The Main Concept SDK has seen progressive improvement in speed, quality, and configurability over the years. At any given time, most products will be using the same implementation; differences are mainly in what controls are exposed and what defaults are used.

In general, Main Concept provides the same settings as other encoders (Figure 14.5). One difference is that it doesn't have an explicit Complexity control, but rather a series of options with a Fast/Complex switch. Generally the Fast modes are faster than they are bad, but the Complex modes will eke out a little extra quality if needed. Unfortunately Main Concept's documentation has infamously lacked useful settings guidelines to users or licensees, so we're often in the dark as to the impact of any given option. However, a few comments:

- Differential Quantization can be a big help in preserving background detail. I find using Complexity: -50 is a good default and really improves many encodes.
- Search shape of 8 × 8 is more precise than 16 × 16. This presumably specifies partition size.
- 2-pass VBR is way better than 1-pass VBR, particularly for long content.
- Hadamard Transform is also a slower/higher quality setting.

All that said, the general defaults are pretty good in most products these days. Perhaps the biggest criticism of Main Concept is that it's less tuned for detail preservation than other implementations, particularly x264.

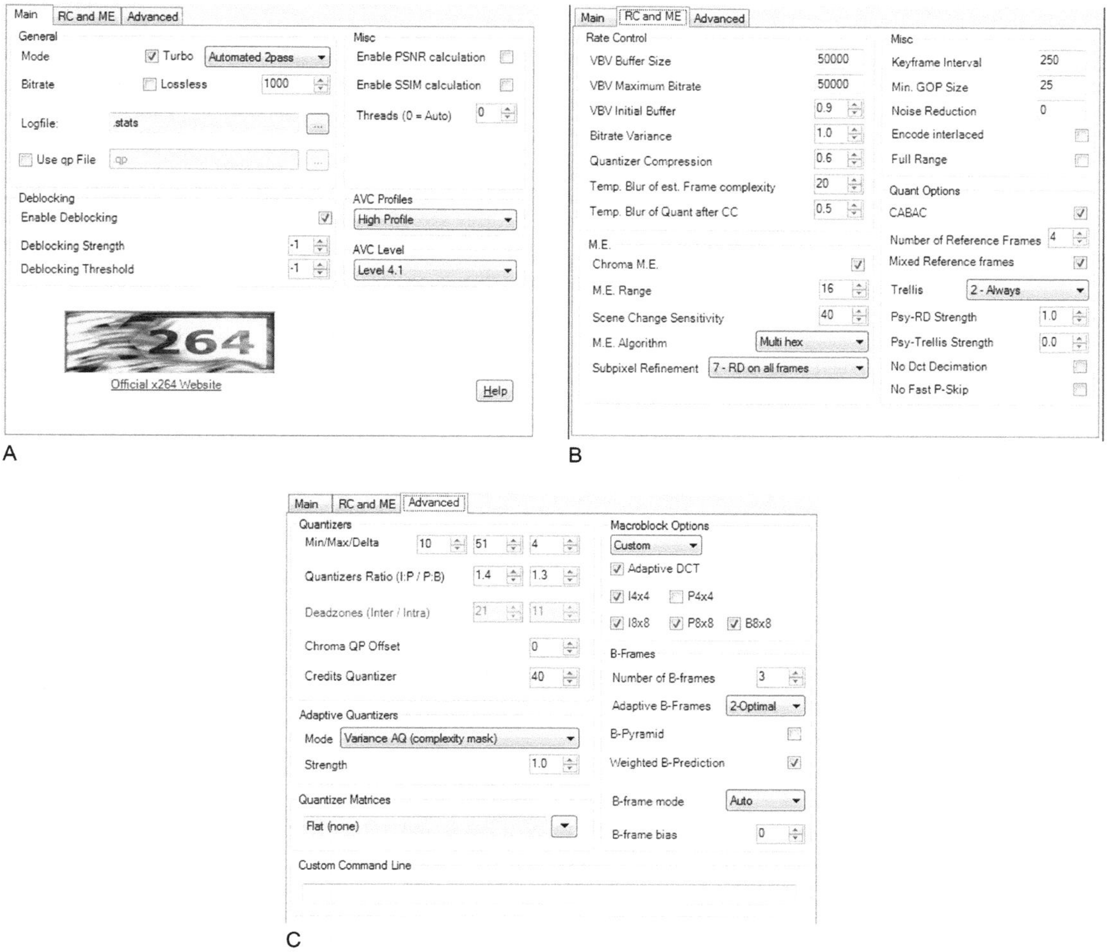

Main | RC and ME | Advanced
General
Mode | Turbo | Automated 2pass
Bitrate | Lossless | 1000
Logfile: | .stats
Use qp File | qp
Deblocking
Enable Deblocking
Deblocking Strength | -1
Deblocking Threshold | -1
Misc
Enable PSNR calculation
Enable SSIM calculation
Threads (0 = Auto) | 0
AVC Profiles
High Profile
AVC Level
Level 4.1
264
Official x264 Website
Help
A

Main | RC and ME | Advanced
Rate Control
VBV Buffer Size | 50000
VBV Maximum Bitrate | 50000
VBV Initial Buffer | 0.9
Bitrate Variance | 1.0
Quantizer Compression | 0.6
Temp. Blur of est. Frame complexity | 20
Temp. Blur of Quant after CC | 0.5
M.E.
Chroma M.E.
M.E. Range | 16
Scene Change Sensitivity | 40
M.E. Algorithm | Multi hex
Subpixel Refinement | 7 - RD on all frames
Misc
Keyframe Interval | 250
Min. GOP Size | 25
Noise Reduction | 0
Encode interlaced
Full Range
Quant Options
CABAC
Number of Reference Frames | 4
Mixed Reference frames
Trellis | 2 - Always
Psy-RD Strength | 1.0
Psy-Trellis Strength | 0.0
No Dct Decimation
No Fast P-Skip
B

Main | RC and ME | Advanced
Quantizers
Min/Max/Delta | 10 | 51 | 4
Quantizers Ratio (I:P / P:B) | 1.4 | 1.3
Deadzones (Inter / Intra) | 21 | 11
Chroma QP Offset | 0
Credits Quantizer | 40
Adaptive Quantizers
Mode | Variance AQ (complexity mask)
Strength | 1.0
Quantizer Matrices
Flat (none)
Custom Command Line
Macroblock Options
Custom
Adaptive DCT
I4x4 | P4x4
I8x8 | P8x8 | B8x8
B-Frames
Number of B-frames | 3
Adaptive B-Frames | 2-Optimal
B-Pyramid
Weighted B-Prediction
B-frame mode | Auto
B-frame bias | 0
C

Figure 14.6 MeGUI settings. A good front-end to the great x264, but I can't say it does a very good job of guiding users to setting that are tweakable and the ones that should really be left alone. But I can always find out *exactly* what each option does.

x264

x264 (Figure 14.6) is an open source H.264 implementation following the successful model of LAME for MP3 and Xvid for MPEG-4 Part 2. It's absolutely competitive with commercial implementations, has an active and innovative developer community, and is used in a wide variety of non-commercial products. Some of its cool features:

Great performance

Using defaults, it can easily encode 720p in real-time on a decent machine, and is still quite fast even with every reasonable quality-over-speed option selected.

Great quality

x264 has done great work with classic rate distortion optimizations, and is able to perform rate distortion down to the macroblock level. But perhaps even more impressive is its perceptual optimizations. Its Psy-RDO mode does a pretty incredible job of delivering more psychovisually consistent encodes, reducing blurring, retaining detail, and maintaining perceptual sharpness and texture. It's my go-to encoder these days when I need to get a good looking image at crazy-low bitrates.

MB-tree

The new Macroblock rate control tree mode optimizes quality based on how future frames reference different macroblocks of the current frame. That lets it shift bits to parts of the image that have a long-term impact on quality away from transient details that may last for just a frame. This yields a massive efficiency improvement in CGI and cel animation, and a useful improvement for film/video content.

Single-slice multithreading

It can use 16 cores with a single-slice encode for maximum efficiency. This is a significant advantage over the Main Concept and Microsoft implementations.

CRF mode

This is a perceptually tuned variant of constant quantizer encoding, allowing a target quality and maximum bitrate to be set, and the encode will be as small as it can be while hitting the visual quality target and maintaining VBV. It's a great feature for file-based delivery, where we care more about "good enough and small enough" than any particular bitrate.

Rapid development

There can be new builds with new features or tweaks quite often. On the other hand, these rapid changes can be difficult to stay on top of.

Telestream

Telestream's H.264 codec (Figure 14.7) is derived from PopWire's Compression Master, and so has many advanced features of use in the mobile and broadcast industries, a la Main Concept. It includes some unique features:

- Can set buffer as both VBV bytes and bitrate peak.
- Lookahead-based 2-pass mode makes for faster encoding but somewhat weaker rate control for long content.

It had some quality limitations in older versions, but as of Episode 5.1 is competitive.

A

B

C

Figure 14.7 (a) Episode has all the right basic controls, including a choice of specifying buffer as VBV or bitrate. Oddly, its multiple reference frames maxes out at 10. (b) Episode uses its usual Lookahead–based 2–pass mode for H.264 as well. It can also do High 4:2:2, unlike many other encoders. (c) Episode also has some more unusual settings that allow streams to meet some unusual requirements, like easy switching between bitrates. It uses multithreaded slicing.

QuickTime

Apple's encoder is relatively weak, particularly beyond Baseline. It is Baseline/Main only and its implementation lacks three key features:

- CABAC
- Multiple reference frames
- 8 × 8 blocks

In fact, Apple's Main is pretty much Baseline + B-frames.

There's very little control in QuickTime's export dialog (Figure 14.8), but here are a few tips:

- Set Size and Frame Rate to current if you're not doing any preprocessing.
- If you've got some time on your hands, use the "Best quality (Multi-pass)" Encoding mode. It'll take as many passes as the codec thinks are necessary per section of the video. I've counted up to seven passes in some encodes.

MPEG-4 Export Settings

File Format: MP4

Video

Video Format: H.264

Data Rate: 256 kbits/sec Optimized for: Download

CD/DVD-ROM

Download

Streaming

Image Size: Current

Preserve aspect ratio using:

Frame Rate: Current

Key Frame: Automatic Every 150 frames

Video Options...

Video: H.264 Video, 720 x 480 (Current), 256 kbps, current frame rate

Audio: AAC-LC Music, Stereo, 96 kbps, Recommended

Streaming: Enabled, max size 1450 bytes, max duration 100 msec

File Size: Approx. 4224 KB

Data Rate: Total data rate 352 kbps, will stream over 512 kbps DSL/Cable

Conformance: The file conforms to MP4 file format specification

Compatibility:

OK Cancel

A

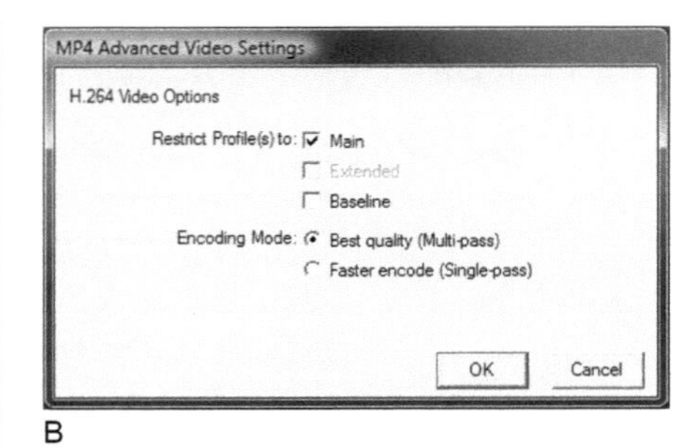

B

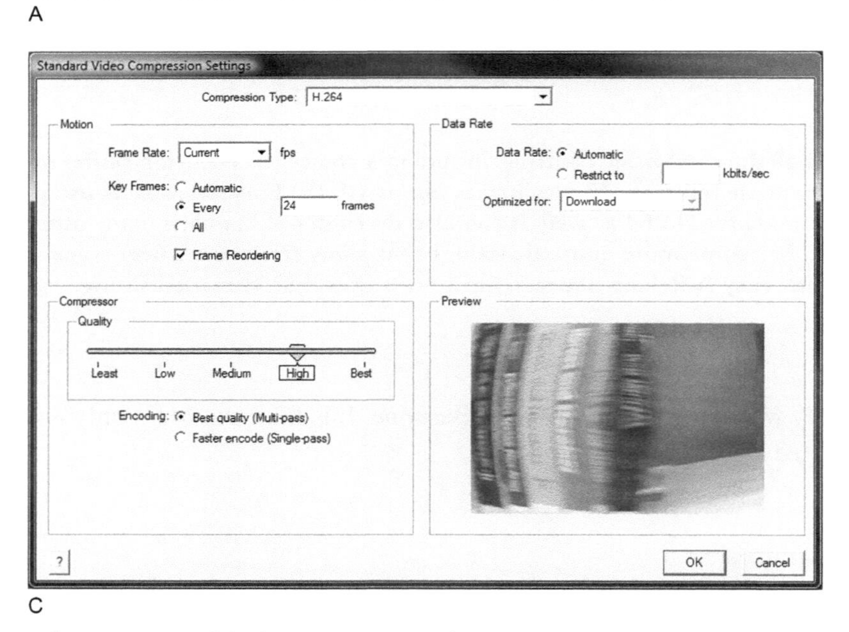

C

Figure 14.8 QuickTime's export settings. It's the same underlying codec, but you get 14.6a and 14.6b when exporting to a MPEG−4 file (I hate modal dialogs), but a nice unified one in 14.6c when exporting to .mov. Note the CD−ROM/Streaming/Download control is only available in multi−pass mode. We lack good documentation on the difference between CD − ROM and Download; I presume Streaming is CBR.

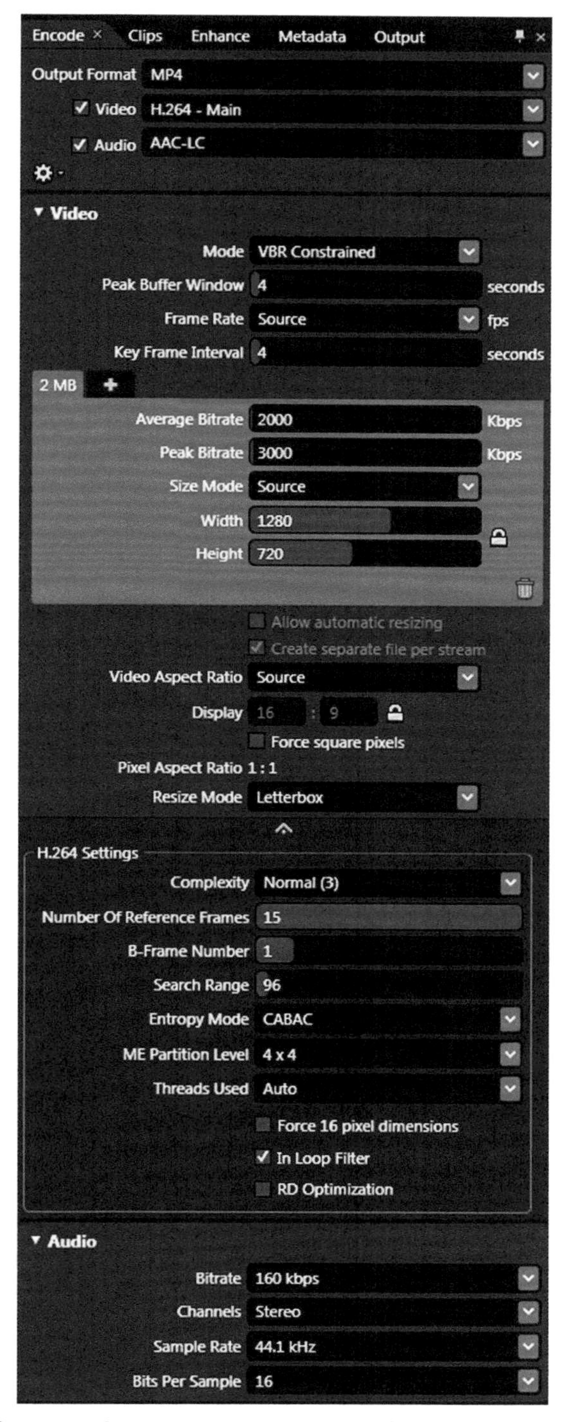

Figure 14.9 Expression Encoder 3's H.264 controls. I like having a single–pane nonmodal control like this.

- Don't drive yourself crazy trying to activate the "Extended Profile" checkbox. This has been grayed in every version of QuickTime with H.264. Extended is a streaming-tuned mode that hasn't been used significantly in real-world applications.

In general, QuickTime's exporter is a fine consumer-grade solution, but I can't imagine using it for quality-critical content.

Microsoft

Microsoft first shipped an H.264 encoder with Expression Encoder 2 SP1, which was a quite limited implementation just for devices. Windows 7 adds much improved H.264 support, although the implementation in Windows itself is focused on device support. Windows 7 also supports hardware encoding, which can make transcoding for devices as fast as a file copy.

A fuller software implementation with a bevy of controls is included in Expression Encoder 3, for both .mp4 and Smooth Streaming .ismv files.

EEv3's H.264 (Figure 14.9) is quite a bit better than QuickTime's, although it's also missing a few quite useful features:

- 8 × 8 blocks (Baseline/Main Only)
- Pyramid B-frames
- Adaptive frame type decisions

However, unlike QuickTime, it does support multiple reference frames, peak buffer size, and CABAC. It also has higher complexity modes that can deliver quite good quality within those limitations. I look forward to a future version with High Profile, of course.

Tutorial: Broadly Compatible Podcast File

Scenario

We're a local community organization that wants to put out a regular video magazine piece about industry events. Our audience is a broad swath of folks, many of whom are not very technical. So we want to make something that'll work well on a variety of systems and devices with minimal additional software installs required.

We want to make a simple H.264 file that'll play back on a wide variety of devices.

The content is a standard def 4:3 DV file. It's a mix of content, some handheld, some tripod, some shot well, some less so.

(*Continued*)

The Three Questions

What Is My Content?

A reasonably well-produced 4:3 480i DV clip. We'll yell at them next time about shooting progressive, but what we've got is what we've got.

Who Is My Audience?

The people in the community, everyone from tech-savvy teenagers to retired senior citizens. We want something that'll be very easy for them to use without installing anything. We want a single file that can work on whatever they've got, and will offer multiple ways to consume it from our web site.

- Via a podcast subscription to iTunes, Zune client, or any other compatible podcasting application.
- Embedded in the web page via a Flash player.
- As a download for desktop, or portable media player playback.

Users will have a wide range of computer configurations and performance. While most will have broadband, we'll still have some potential modem users out there who mainly use their computers for email; the file shouldn't be too big a download for them.

What Are My Communications Goals?

We want to share all the cool things we're doing with our community. We want the content to look good, but that's just a part of the whole experience: it must be easy to find, play back well, and leave them excited about downloading the next episode. We want viewers to feel more connected to their community and more excited about engaging in it.

Tech Specs

We'll try both QuickTime and Sorenson Squeeze for this encode. Squeeze is one of the most approachable compression tools, and a good first choice for someone more interested in doing some compression without becoming a compressionist. Squeeze includes good deinterlacing and a good Main Concept implementation.

We've got a lot of constraints to hit simultaneously. In general, we can aim for the iPod 5G specs; other popular media players do at least that or better. Thus, we can use the following parameters:

- H.264 Baseline Profile Level 3.0
- Up to 640 × 480, 30 fps, 1.5 Mbps
- Up to160 KHz 16-bit stereo 48 KHz audio

(*Continued*)

Reasonably straightforward. The one thing we'll want to keep our eyes on is that bitrate; that'll have to be our peak.

Encoding Settings

QuickTime iPod encoding is dead simple (Figure 14.10). You just export the file as "iPod" and you're done. No options, no tuning, and it'll look great on the iPod's small screen. However, it'll also use more bits than needed for its quality, and can have some quality issues when scaled up to full screen (although Apple's Baseline encoding is much more competitive than their Main).

Alas, sometimes with great simplicity comes great lameness. QuickTime exports the 480i DV assuming it's progressive square-pixel, and so we get horrible fields and the image scaled down to 640 × 426:720 × 480 proportionally scaled to 640 width. This will not stand!

Squeeze comes with an "iPod_Lg" preset that sounds exactly like what we're looking for. Applying it to the source file, the automatic preset modes of Auto Crop and Auto Deinterlace are applied. Those filters are much improved from previous versions of Squeeze in quality.

Look carefully at the settings, though; "iPod_Lg" is Sorenson MPEG4 Pro—actually their part 2 implementation. It gives us the choice of Sorenson, Main Concept, or Apple's H.264 encoders. Main Concept is the only one which allows a 2-pass VBR with both average and peak bitrates, and it's also the highest-quality one; we'll pick that.

The default settings aren't bad, but we'll make a few changes, also shown in Figure 14.11:

- **Switch Method from 1-Pass VBR to 2-pass VBR; important to get the bits where we need them for optimal file size.**

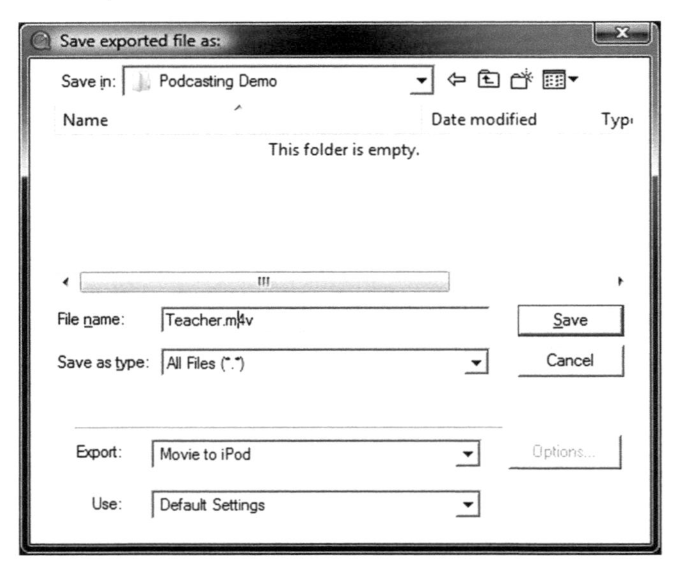

Figure 14.10 The optionless iPod export in QuickTime.

(Continued)

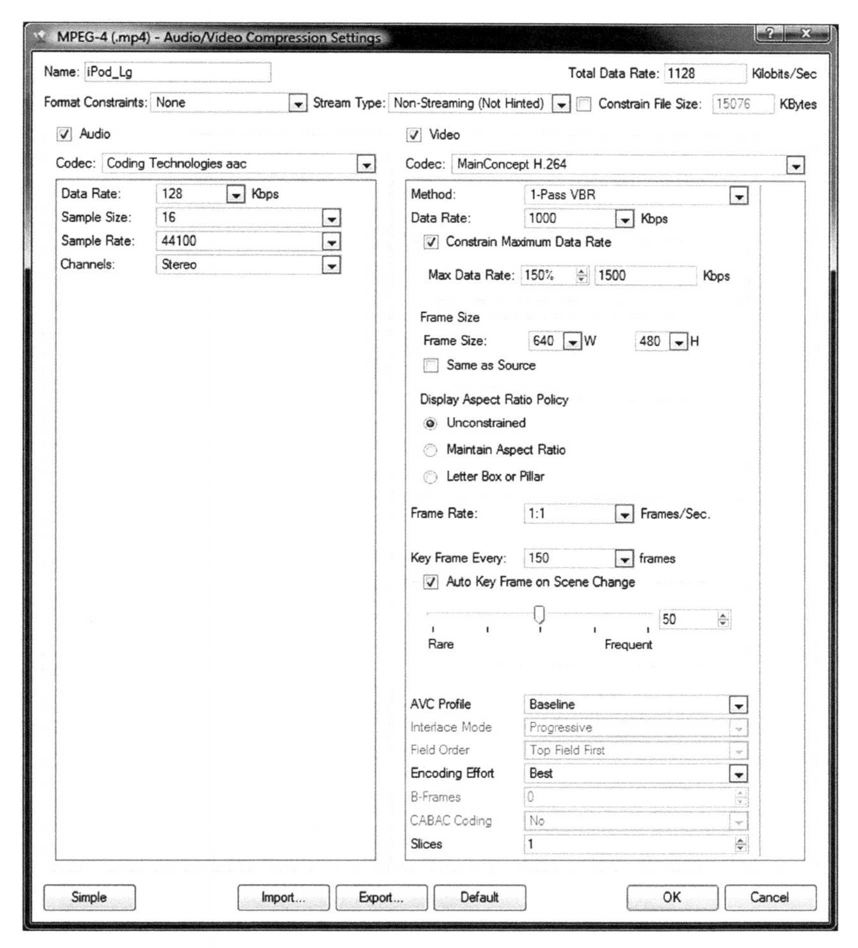

Figure 14.11 Our final Squeeze settings. It also gets points for a single-pane interface, although there are plenty of options I wish were exposed.

- Lower data rate to 1000 Kbps to reduce our download size for modem users.
- Reduce sample rate to 44.1. That's slightly more compatible on very old computers, and will sound just as good.
- Set Max Data Rate to 150 percent, giving us our 1500 Kbps peak for compatibility.
- Encoding Effort (Complexity) is already on Best, so we'll leave it there.
- Slices defaults to 0. That's probably fine; can't use CABAC in baseline.

And the result are pretty decent. If we weren't worried about download time, we would have been a little better off using the full 1500 Kbps. But the hard parts of the

(*Continued*)

Figure 14.12 The Squeeze encode is well preprocessed, which makes for much better quality even at a third lower bitrate.

Figure 14.13 QuickTime's output with this source is a good refresher of what not to do from the Preprocessing chapter. It has interlaced coded as progressive, anamorphic coded as square–pixel, and a reduced luma range, due to a wrong 16–235 to 0–255 conversion.

video were already going to be at the 1500 Kbps peak, so it wouldn't have helped all that much. At 1000 Kbps, the Squeeze output (Figure 14.12) looks a lot better than the QuickTime (Figure 14.13) at 1500 Kbps.

H.265 and Next – Generation Video Codec

Of course, H.264 isn't the end of codec development by any means. Its replacement is already in early stages. Recognizing that many of H.264's most computationally expensive features weren't the source of the biggest coding efficiency gains, a new codec could provide a mix of better compression and easier implementation. The VQEG has ambitious targets:

- Allow a 25 percent reduction in bitrate for a given quality level with only 50 percent of H.264 decode complexity
- A 50 percent reduction in bitrate without decode complexity constraints

It's too early to say if either or both will be met, but getting close to either should be as disruptive to H.264 as H.264 was to MPEG-2. These improvements may be addressed via an extension to H.264 or an all-new H.265. Both are very much research projects at this point. A H.264 extension could potentially emerge in 2010, but a full H.265 is unlikely before 2012 at the earliest.

CHAPTER 15

FLV

Flash Video—FLV—was where Flash came of age as a media platform, not just an animation plug-in. Flash started as FutureSplash Animator back in 1996, just doing simple vector graphics animation. Its potential for media was seen early on—I first saw it in the keynote of the RealNetworks conference that year. After several years of kludgy third-party technologies for video-in-Flash, Macromedia licensed the Spark codec from Sorenson Media, introduced the FLV format, and built them into Flash 6 player in 2002. This worked well as a way to add video to Flash presentations, but compression efficiency wasn't sufficient for video-centric projects. The much-improved VP6 video codec was added in Flash 8 in 2005, and found widespread use in many high-profile sites.

Adobe added support for standards-based MPEG-4 files with H.264 and AAC audio in a Flash 9 update, and much of the Flash ecosystem has been moving to that for broader encoder support, increased efficiency, and interoperability. See Chapters 11 and 26 for more details.

Still, FLV offers some unique features and advantages, like alpha channels, easier decoding, and backwards compatibility, and remains widely used. Adobe has helped this effort by releasing a free specification for the FLV file format itself.

Why FLV?

FLV is increasingly becoming a legacy format, but it does have its places. The "default" encoding for Flash is now MPEG-4 H.264 ("F4V"), but there are still good reasons to go to FLV in some cases, as described in the following sections.

Compatibility with Older Versions of Flash

While most consumers upgrade to new versions of Flash pretty quickly, corporate users require IT assistance to update their Flash players, and so they lag behind. If your market contains a significant number of users behind the firewall, FLV will offer greater backwards compatibility.

Decoder Performance

Both Spark and VP6 require less CPU per pixel than H.264, allowing higher frame sizes on older machines. Spark in particular is very simple and thus very fast. And VP6 can dynamically reduce or turn off its postprocessing to offer stable frame rates on older hardware (with reduced video quality, of course). There's also the VP6-S simplified mode, which reduces decode complexity another 20 percent or so, albeit with a slight reduction in compression efficiency.

That said, there's also flexibility in H.264 to tune for simpler decoding. If decoder performance is a primary concern, you can try making H.264 with the following settings:

- CAVLC instead of CABAC.
- Only 2 reference frames.
- And at an extreme, perhaps even turning off in-loop deblocking.

The first is automatic with Baseline Profile, so that's easy to try. These wouldn't be as efficient as a full-bore High Profile, but certainly is still more efficient than VP6 and especially Spark. Even without in-loop deblocking, a good H.264 Baseline would dramatically outperform Spark in coding efficiency, but may not outperform VP6 with its rather advanced postprocessing filter.

Alpha Channels

FLV is the only way so far to get alpha channel video in Flash, supported in both Spark and VP6. While there are alpha channel modes for H.264 and MPEG-4, Adobe doesn't support them as of Flash 10.

Why Not FLV?

Flash Only

The biggest drawback to FLV is that it's a Flash-only file. An .f4v is just an .mp4 (and can be renamed as such), and thus is already compatible with QuickTime, Silverlight, and many other media players.

There are third-party players that include FLV playback, like VLC, but out of the box QuickTime and Windows Media Player do not. And both OSes have had better in-box codecs for many years.

Lower Compression Efficiency

VP6 and especially Spark need more bits for the same quality level as H.264. If you're more limited by bandwidth than decoder performance, a well-tuned H.264 can be a dramatic improvement.

Fewer and More Expensive Professional Tools for VP6

Spark is just H.263, so it's cheap and readily available, even in open source tools like ffmpeg.

VP6, however, is expensive to license, particularly for the full-featured versions and incredibly so for enterprise licenses. This is why some high-end tools like Carbon just ship with the very limited 1-pass only QuickTime Export Component encoder from Adobe. Telestream is the only vendor to bundle the full VP6 across most of their product line.

Sorenson Spark (H.263)

The Spark video codec is an implementation of the standard H.263 videoconferencing codec. It was introduced with Flash 6, along with FLV. H.263 is a lightweight codec, supported on all kinds of devices, and fast to decode.

Given all of this, Spark can offer decent quality at higher bitrates, particularly when using 2-pass VBR encoding.

Not all Spark versions are the same. The implementation in Squeeze has many options (Figure 15.1), while the one in Adobe Media Encoder (Figure 15.2) really just does bitrate and Key Frame Distance. Telestream's implementation in Episode is based on their own MPEG-4 part 2 implementation (Figure 15.3).

Quick Compress

Quick Compress is a faster, lower-quality mode. Leave it off unless you're in a huge rush; Spark is very fast regardless.

Minimum Quality

Minimum Quality sets a minimum visual quality (presumably mapping to maximum QP), below which the codec won't drop. Instead, it raises the bitrate of each frame until it hits that target. With Drop Frames turned off, the Minimum Quality feature can cause the data rate to overshoot the target substantially. Turn Drop Frames on if you must use Minimum Quality and don't want to overshoot the target. Minimum Quality should generally be off unless legibility of elements on the screen is critically important.

Drop frames

If this option is on and you're encoding in CBR mode, the frame rate will drop to maintain data rate. This option doesn't do anything in 2-pass VBR encoding. You should have it on in CBR mode, even if you turn Minimum Quality down to 0. All CBR encodes should at least have Minimum Quality = 0 and Drop Frames on to ensure output really has a constant bitrate.

Figure 15.1 Sorenson invented Spark, and still has the deepest implementation of H.263 for FLV.

Automatic Keyframes

This feature lets the codec start a new GOP at scene changes and high motion. It should always be on. The slider controls how sensitive the codec is to keyframes. The default of 50 is good for most content.

Image Smoothing

Image Smoothing is a flag telling the decoder to apply a deblocking postprocessing filter. This helps improve quality of compressed frames. It does increase decode complexity somewhat, so shouldn't be used when targeting very slow playback systems with higher data rates and resolutions, but most of the time should be used. Spark is a very fast decoder in general.

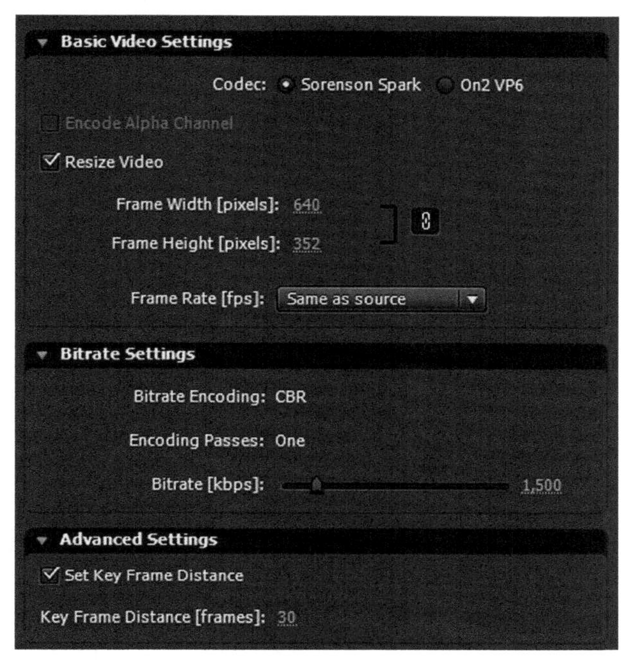

Figure 15.2 Adobe Media Encoder's H.263 support is a very limited 1-pass CBR only.

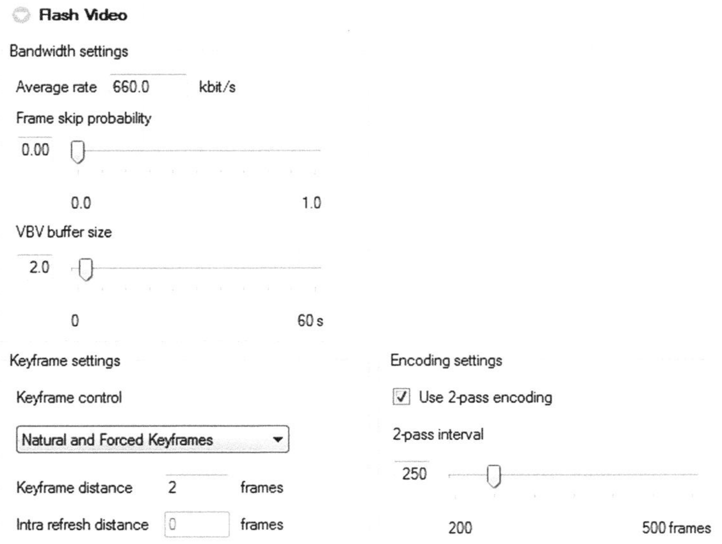

Figure 15.3 Episode's H.263 is a subset of its MPEG-4 part 2, including Telestream's unique Lookahead-based 2-pass mode. It has no control over peak bitrate, just buffer duration.

Playback Scalability

With the playback scalability option, the video can drop the frame rate in half instead of presenting uneven stuttering on slower machines. Playback scalability uses the scheme from the ancient Sorenson Video 2, with even and odd video frames encoded as two parallel separate streams. Scalability is achieved by simply not playing one of the tracks. This reduces compression efficiency, because it doubles the temporal distance between frames, making motion estimation much less efficient. It also makes decoding both streams somewhat slower than just doing a single stream.

Because of these limitations, Playback scalability hurts more than it helps and shouldn't be used.

On2 VP6

The big video upgrade in Flash 8 was On2's VP6 codec. On2 (formerly the Duck Corporation) has been engineering codecs for well over a decade now, and VP6 offered a good combination of compression efficiency and easy decoding. VP6 is also used in JavaFX, but not in a FLV wrapper. On2's VP7 and VP8 are better codecs yet, but Adobe has since switched to H.264 as its primary high-efficiency codec for Flash, and hasn't added support for the more recent VPx decoders. On2 supports MPEG-4 in their Flix products now as well. Google announced their aquisition of On2 in 2009. It remains to be seen what if any impact this would have on further development of VP series of codecs.

VP6 offers a pretty massive improvement in compression efficiency over Spark—you can get away with half the bitrate in some cases. Decode is somewhat slower, although VP6 varies postprocessing quality based on available CPU power, so it can vary quite a bit.

The biggest drawback to VP6 is that it's pretty slow compared to other modern codecs; it's still single-threaded in its most common implementation (although that's changing; see "New VP6 Implementation" later in this chapter).

In some ways, postprocessing is the most impressive aspect of VP6. Beyond normal deblocking (which it does quite well), VP6 even will add grain texture to the video to mask the loss of detail from compression, making it seem quite a bit more detailed than the retained details would otherwise provide.

Alpha Channel

VP6 also: supports alpha channels. If you have alpha-channeled source (like from After Effects), you can encode that alpha in the movie file for real-time compositing on playback.

Flash doesn't support alpha channels in F4V, so FLV is the only way to provide that today.

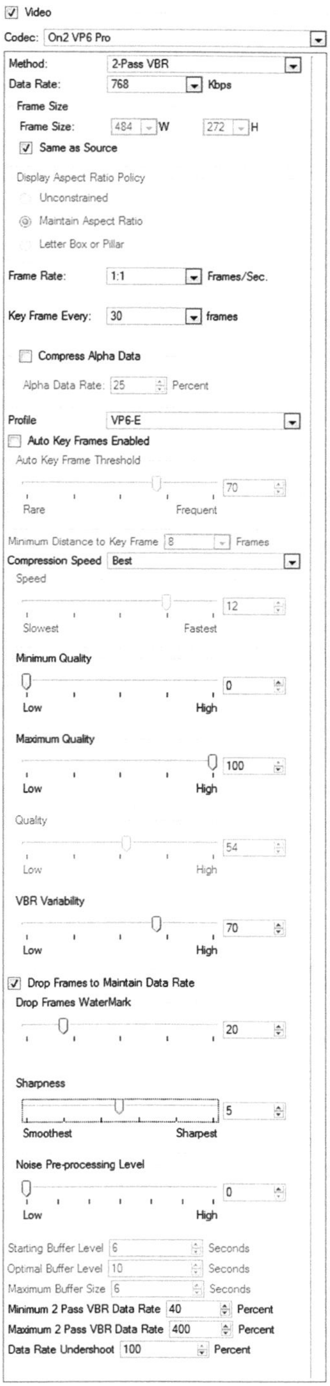

Video
Codec: On2 VP6 Pro
Method: 2-Pass VBR
Data Rate: 768 Kbps
Frame Size
Frame Size: 484 W 272 H
Same as Source
Display Aspect Ratio Policy
Unconstrained
Maintain Aspect Ratio
Letter Box or Pillar
Frame Rate: 1:1 Frames/Sec.
Key Frame Every: 30 frames
Compress Alpha Data
Alpha Data Rate: 25 Percent
Profile VP6-E
Auto Key Frames Enabled
Auto Key Frame Threshold
70
Rare Frequent
Minimum Distance to Key Frame 8 Frames
Compression Speed Best
Speed
12
Slowest Fastest
Minimum Quality
0
Low High
Maximum Quality
100
Low High
Quality
54
Low High
VBR Variability
70
Low High
Drop Frames to Maintain Data Rate
Drop Frames WaterMark
20
Sharpness
5
Smoothest Sharpest
Noise Pre-processing Level
0
Low High
Starting Buffer Level 6 Seconds
Optimal Buffer Level 10 Seconds
Maximum Buffer Size 6 Seconds
Minimum 2 Pass VBR Data Rate 40 Percent
Maximum 2 Pass VBR Data Rate 400 Percent
Data Rate Undershoot 100 Percent

Figure 15.4 The incredibly complete and long Squeeze VP6 dialog. I love having all my parameters in one pane, but should it really require a 2560 × 1600 display?

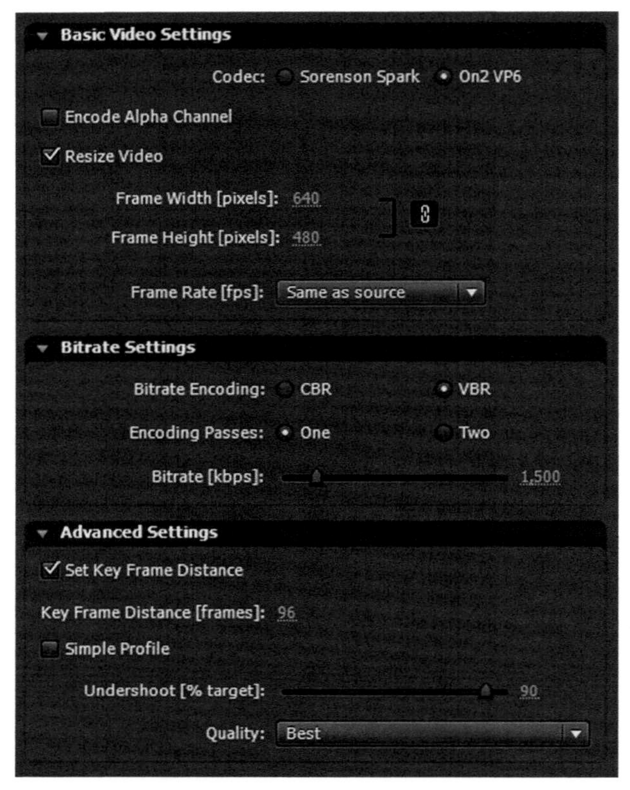

Figure 15.5 Adobe Media Encoder's VP6 support has 2-pass and VBR, but not a lot of other parameters.

VP6-S

VP6-S is a newer variant of VP6 that constrains decode complexity somewhat while remaining backward-compatible. It does lose some encoding efficiency as well, but the playback improvement is bigger, so it can increase maximum quality when CPU is the limit, not the network.

New VP6 Implementation

On2 has been working on an updated VP6, promising improved quality and (finally!) multi-threading for much improved encoding performance on modern hardware. These are initially available in their Enterprise products, but should eventually migrate down to their Pro products.

VP6 Options

VP6 offers a wide variety of options in its Pro version, as supported in Sorenson Squeeze (with installation of the "On2 VP6 for Adobe Flash" add-on), in On2's own Flix products, and any product using QuickTime Export Components with the "Flix Exporter" installed.

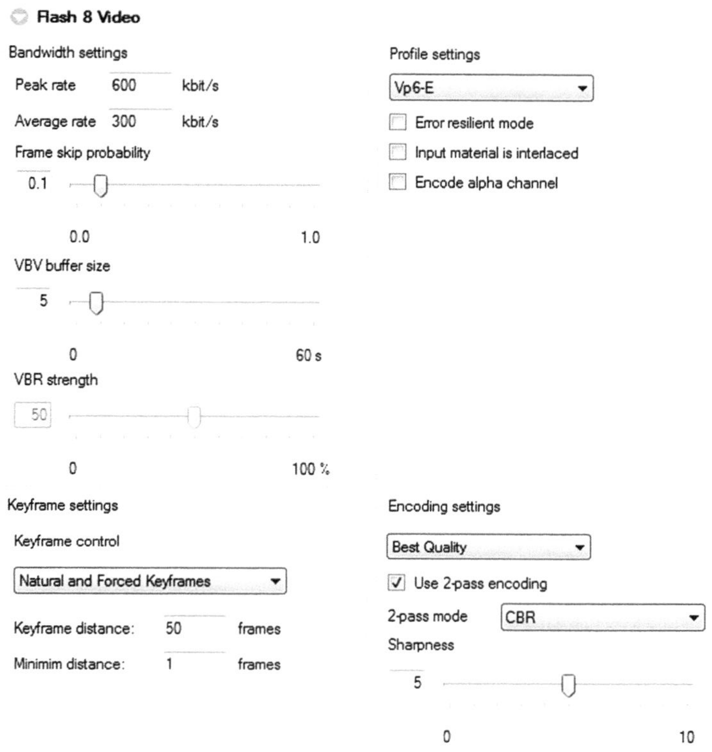

Flash 8 Video

Bandwidth settings

Peak rate 600 kbit/s

Average rate 300 kbit/s

Frame skip probability

0.1

0.0 1.0

VBV buffer size

5

0 60 s

VBR strength

50

0 100 %

Keyframe settings

Keyframe control

Natural and Forced Keyframes

Keyframe distance: 50 frames

Minimim distance: 1 frames

Profile settings

Vp6-E

Error resilient mode

Input material is interlaced

Encode alpha channel

Encoding settings

Best Quality

Use 2-pass encoding

2-pass mode CBR

Sharpness

5

0 10

Figure 15.6 Episode's VP6 setting. Tuned for mobile, it also exposes error resiliency.

There's also the much more basic Adobe version, implemented in Adobe Media Encoder and the "Flash Video" export component that's installed along with the Flash products. That's 1-pass CBR only and lacks all of the following advanced settings. Use Adobe Media Encoder at least, or if you need QuickTime export integration, buy the Flix export component.

The advanced modes definitely pay off in improved quality and efficiency for almost every project. If you've got a good reason to use FLV, you should get access to the full version.

The Sorenson and Flix products expose a slightly different set of controls, so I'll discuss both here.

CBR vs. VBR

In general, you should use CBR when you'll be delivering video using Flash Media Server, and VBR the rest of the time. The 2-pass modes for both CBR and VBR give higher quality than the 1-pass modes, albeit at twice the encoding time. VBR in particular benefits from 2-pass encoding.

Compress alpha data

This applies some mild lossy compression to the alpha channels. The default is generally okay, but reduce this compression if you're having trouble getting a clean edge.

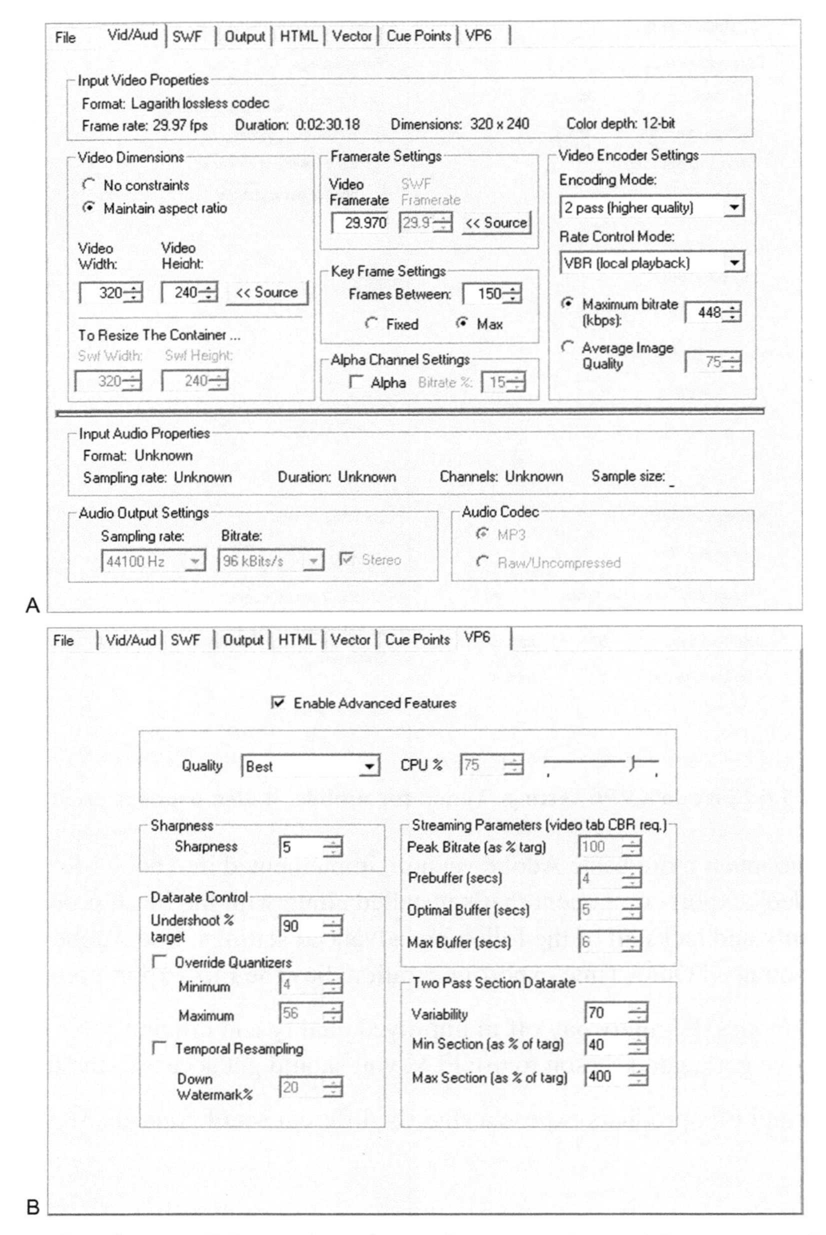

Figure 15.7 The Flix VP6 dialogs. The Advanced Features (15.7B) have some of the most esoteric terminology in the industry.

Auto key frames enabled/key frame settings

In Squeeze, this turns on automatic keyframes. That's quite important for quality, and should always be used for any content with scene changes.

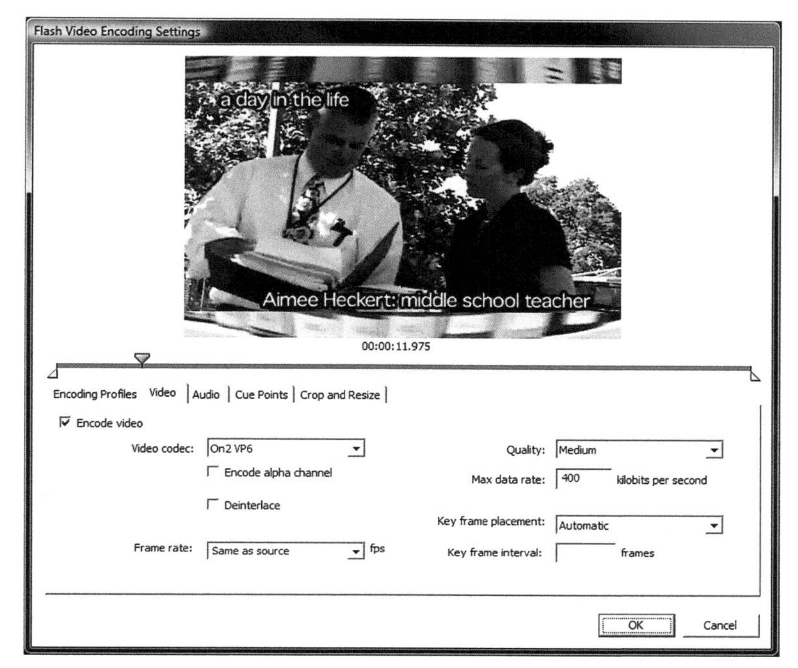

Figure 15.8 Adobe's QuickTime exporter. 1-pass only.

In Flix, turning Key Frame Settings to Fixed turns off automatic keyframes.

Auto key frame threshold

This specifies how different two frames need to be to trigger a keyframe. The default is almost always fine; you'd only change if you were seeing keyframe flashing.

Minimum distance to keyframe

Sets the maximum number of frames before the next keyframe. 1 is keyframe-only.

Compression speed

Best is the highest quality, but also the slowest. Good is faster, but a little lower quality. Speed is the fastest, but lowest quality. Use the highest quality you have time for, and be thankful if you have the multithreaded version.

Minimum quality/override quantizers: maximum

This is the maximum quantizer the encoder can use, and hence represents how much compression it can apply to a given frame. Higher values will lead to more dropped frames, but with a higher minimum quality for frames. I generally set to 0 (Squeeze) or 56 (VP6) to avoid dropped frames, and use other optimizations to achieve target quality.

Maximum quality/override quantizers: minimum

Maximum quality specifies the minimum quantizer for a particular frame, and hence the maximum quality and least compression. Using a lower value can preserve more bits for the more difficult portions of the video. The default maximum of 4 is fine for most web work, but a lower value may improve detail a bit if you've got bits to burn.

Drop frames to maintain data rate

Lets the encoder drop frames if there aren't enough bits to do all frames at the minimum quality. If off, the data rate will just go up when there aren't enough bits, and so this should be used with CBR.

Drop frames watermark/temporal resampling

This oddly named parameter has nothing to do with watermarks. It sets the percentage that the buffer goes down to before frames start getting dropped. It's much more applicable to live and 1-pass CBR encoding. Default of 20 is a fine start.

Sharpness

Sharpness tells the encoder to emphasize a smooth low-artifact image rather than a sharper but potentially artifacted image. It won't add sharpness beyond what's in the source. Generally, the highest value that doesn't introduce artifacts should be used; the more challenging the encode, the lower Sharpness will be optimal.

Noise pre-processing level

This activates the codec's own internal noise reduction. Normally noise reduction should be done in preprocessing; if that's not available, this mode seems to work reasonably well for sources with visible noise. Higher levels can make for a soft image, but can dramatically improve compression efficiency.

CBR settings

The next four options apply to CBR encoding only.

Maximum data rate/peak bitrate

This specifies how much of the available bandwidth the codec will try to use. While 100 would be the obvious choice, VP6's rate control has historically not been very accurate, so using 90–95 leaves some headroom. 100 can be used with 2-pass CBR.

Starting buffer level/prebuffer

This specifies how many seconds of buffer are used at the start of the file. Higher values help quality, but delay start of playback.

Optimal buffer level
This denotes the buffer level the encoder tries to maintain throughout the file after it starts. It is generally higher than the Starting Buffer Level.

Maximum buffer size
The maximum length allowed for the buffer if a complexity spike pushes past the Optimal Buffer value. The codec will drop frames to keep within Maximum Buffer.

VBR settings

The following settings are VBR-only. Flix calls this "Two-Pass Section Datarate" even though it applies to 1-pass VBR as well.

Maximum 2-pass VBR data rate/max section (as % of target)
This determines how high the data rate can go. Higher peaks make for higher decode requirements, so the lowest value that provides adequate quality in the hardest sections should be used. 200 percent is a good target for HD encodes; the lower the bitrate, the higher the variability can be without risking decoder performance issues.

Minimum 2-pass VBR data rate/min section (as % of target)
This specifies the minimum bitrate to be used. Like MPEG-2, VP6 can require a minimum bitrate ensuring VBR algorithms don't bit-starve very easy sections of the video. The default of 40 percent seems to work, but lower values can preserve more bits for other sections of the video.

VBR variability
This specifies how much the data rate can go up and down in a VBR encode. 0 is the same as a CBR. It seems like this should be bound by the Min/Max VBR bitrates, but it's exposed as a parameter. Lacking any good data as to what it's doing internally, I stick with the default of 70.

Data rate undershoot/undershoot % target
This sets a percentage of the target data rate to shoot for, to leave a little headroom in case you really can't afford to go over. This is most useful in 1-pass encoding, to leave a bit reserve in case of very difficult sections.

FLV Audio Codecs

Almost all FLV files use MP3, but there are some other options that used to be available.

MP3

MP3 is by far the best general-purpose audio codec in FLV, and by far the most used.

Note that some very old versions of Flash had trouble playing back MP3 above 160 Kbps. If you're encoding Spark FLV for backward compatibility, you may want to stick to that limit.

Remember to encode MP3 in mono at lower bitrates; it's more important to keep the sample rate high than to have two channels. Generally, content delivered at 64 Kbps and below should be mono.

Nellymoser/Speech The

Flash encoder used to include a licensed version of the Nellymoser Asao codec, called "Speech" or "Nellymoser" depending on the version. Like the other examples of its genre, Speech is suited to low-bitrate, low-fidelity speech content. MP3 does a fine job with speech and can provide better compression efficiency, so Nellymoser Speech was only useful at very low bitrates. Encoding it isn't supported in the current tools, for which my largest regret is not being able to say "Nellymoser" during classes.

ADPCM

ADPCM is a very old 4:1 compressed audio codec that's supported in old versions of Flash. You'd never encode with it now, but it may be found in some very old files.

PCM

The PCM audio codec is straight-up uncompressed audio. It would be very unusual to use it in a FLV. Authoring tools include it because older versions of the Flash application would automatically re-encode all the audio imported in the project.

FLV Tools

A variety of tools offer FLV support, although they can vary widely.

Some tools are dropping built-in support for FLV now that Flash supports H.264. But any tool that supports QuickTime Export Components can encode to FLV using either the Adobe or (much better) On2 implementations.

Adobe Media Encoder CS4

Media Encoder CS4 has basic FLV encoding, tuned for simplicity over configurability, as is the case for most of its settings. It has added 2-pass and VBR modes, a good improvement over previous versions.

QuickTime Export Component

Adobe used to bundle a QuickTime Export Component with Flash that lets QuickTime-based apps encode to FLV. It's very limited, unfortunately, providing 1-pass CBR only and few other configuration options. It is no longer installed as of CS4.

Flix

Wildform pioneered the video-in-Flash market with Flix, a system that provided limited video functionality to Flash prior to Flash 6 and FLV. On2 purchased Wildform's Flix products and has turned them into pretty rich encoding products for FLV and MPEG-4, with good integrated skin support.

There are three Flix products: the flagship Flix Pro (Mac and Windows), the more limited Flix Standard (Windows-only), and the Flix Exporter (Mac and Windows) QuickTime Export Component. Note that the $199 desktop version of the Exporter is limited to 1500 files encoded per month; more than that requires the $3,000 "Server Edition." They're otherwise identical. $3,000 is an unprecedentedly high price for a codec, particularly one that isn't multithreaded. That pricing has driven a lot of F4V adoption.

Most professionals would use either Pro or Exporter—Standard lacks basic pro features like 2-pass VBR and batch encoding, and really targets the amateur market.

There are also SDK versions of Flix used to build custom workflows and implement enterprise encoding.

Telestream Flip4Factory and Episode

Telestream's FlipFactory 6 is the only enterprise encoding tool that comes with the full Server Edition for VP6, offering all the parameters in Flix and Squeeze. And their desktop Episode Pro (but not the non-Pro version) includes a full implementation as well. Unlike most of the other Episode codecs, it implements On2's 2-pass rate control mode instead of Telestream's unique Lookahead-based multipass.

H.263 FLV support in both products uses Telestream's own implementation. They are the only products other than Squeeze to support the error resiliency feature of H.263 for improved quality over lossy networks, via the Intra Refresh Distance option.

Sorenson Squeeze

Sorenson Media was the creator of the original Spark codec for Flash, and also provides deep support for VP6 with a $199 codec pack (and, of course, makes great F4V files).

Squeeze exposes all the parameters Flix does; it's your preference as to which encoder you want to use.

ffmpeg

The open source ffmpeg includes FLV support, based on its existing support for MPEG-4 Part 2 (H.263 is MPEG-4 part 2 Short Header). It doesn't have VP6 encoding support.

I bet ffmpeg is responsible for more eyeball-hours of FLV encoding than anything else; that's what YouTube used for its FLV files.

While ffmpeg front-ends don't generally include FLV support, it's not that hard to achieve via the command line. The example you'll find by searching "ffmepg FLV encoding" is pretty much identical everywhere, cloning the old YouTube settings:

```
ffmpeg -I "foo.avi" -acodec mp3 -ar 22050 -ab 32 -f flv -s 320 × 240
"foo.flv"
```

Alas, this replicates the horror of old YouTube audio at 22.05 kHz, and doesn't even set a bitrate.

Much better sounding would be:

```
ffmpeg -I "foo.avi" -acodec mp3 -ar 44100 -ab 64 -f flv -b 400-s 320 × 240
"foo.flv"
```

Which improves audio and sets the video bitrate to 400 kbps. There's plenty of further tweaking that can be done to improve quality; with enough effort, this would probably match the highest-quality Spark files, but it won't match VP6 Pro.

FLV Tutorial

Scenario

We've been hired to make a short video clip using transparency for an online banner ad using Flash. We have specific technical requirements from the advertising agency:

- Frame size of no more than 320 × 240 – that's how big the SWF will be
- File size of no more than 350 KB, with a peak bitrate no more than 300 Kbps
- No audio—that will be included in the SWF by the agency

Three Questions

What Is My Content?

The content is a 15-second 640 × 480 PNG codec .mov file with an alpha channel.

(*Continued*)

Who Is My Audience?

The agency. We're not exposed to the whole project, just this clip, so we're just trying to make the agency happy with something they think will make the audience happy.

What Are My Communication Goals?

Make the agency happy the first time, so we continue to get work from them.

Tech Specs

It's always nice to get specific specs for this kind of project. There aren't many for this one, but they're tight.

When making alpha video, there's no reason to include blank pixels, so we'll crop down to the active image area to the maximum of 320 × 240.

We have a measly 350 KB to spend on our video; fortunately we've only got 15 seconds of it. So:

- 350 Kbytes × 8 bits/byte = 2800 Kbits
- 2800 Kbits/15 seconds = 186 Kbits/sec

Not a lot, but it's mainly black pixels as well; it should encode okay.

Settings

We'll use Squeeze for this one. It has fine support for alpha input.

For preprocessing, we'll turn off deinterlacing (progressive source, thank goodness). In looking at the image, we can see that it doesn't use the full horizontal motion range. Our maximum frame size is 320 × 240, but there's no reason we can't reduce one axis. By cropping to the region of motion, and then rounding up to a mod32 width in the 640 × 480 source, we can get an image that will scale down to a mod16 within the 320 × 240 allocation. A crop of 28 on the left and 100 on the right gives us 640–100–28 = 512 and 512/2 = 256, so our output is 256 × 240, a nice savings in pixels from the maximum 320 × 240. This is particularly useful when doing alpha channels, as each pixel of each frame needs to get blended, which can really increase CPU requirements. See Figure 15.9. And it'll look the exact same, since the part we cropped out was all alpha anyway.

For settings, we'll start with the "F8_256 K" preset, and make these modifications:

- Turn off audio—not needed by spec.
- Make Method 2-pass VBR. We want to nail rate control at this bitrate and short duration.
- Data Rate = 185 Kbps (always good to round down a bit for some headroom). Squeeze even lets us type in 350 for "Constrain File Size" and will then figure out the data rate for us.

(Continued)

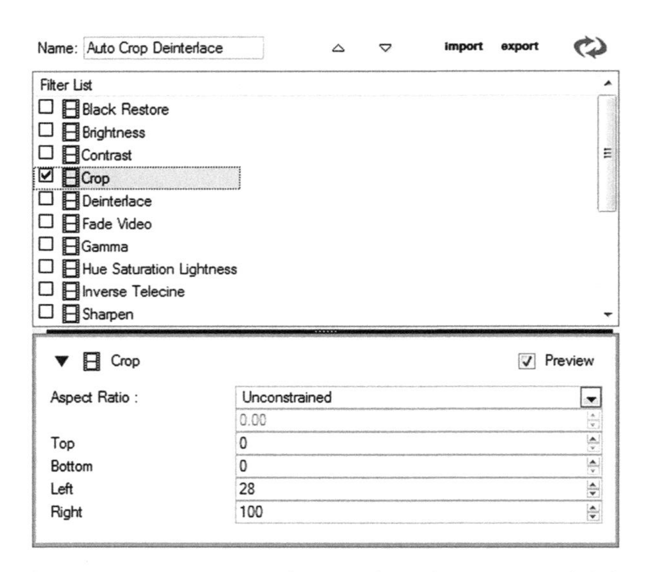

Figure 15.9 The Squeeze preprocessing settings for our tutorial; just cropping.

- Frame size = 256 × 240. It actually types in the "240" for us as we key in 256, as "Maintain Aspect Ratio" correctly is based on the aspect ratio post-crop.
- Frame Rate: 1:1 for same as source.
- Key Frame Every: 60 frames, to get a little extra efficiency.
- Compress Alpha Data: ON! That's the whole point of the exercise!
- We can leave Alpha Data Rate on the default, but will definitely need to make sure there are clean edges after compression.
- Since this is a low bitrate, and the content doesn't need a lot of detail, we can take the Sharpness control down to 0 and hopefully get fewer artifacts.
- Our average is 185 Kbps and our peak is 300 Kbps. That means peaks are 300/185 = 1.62 = 162 percent value for "Maximum 2 Pass VBR Data Rate." We'll round down to 160 percent.

And it all looks good. But as we double-check the file size—whoops! It's 431 KB, not 350! It's critical to double check the output against the customer's constraints. Remember that 25 percent value for "Compress Alpha Data"—that actually is how much extra bits get spent on alpha on top of the specified bitrate. 25 percent more than 350 is 438.

(*Continued*)

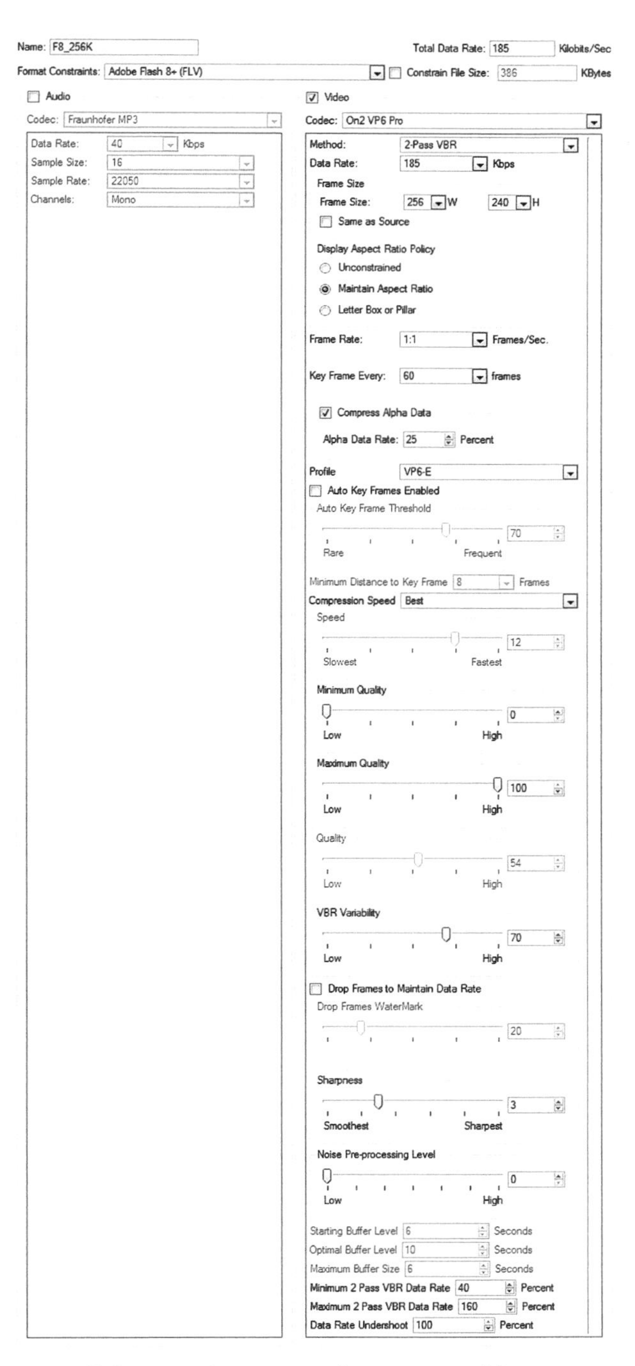

Figure 15.10 Parameters all the way down. It took some tweaking to get around the VP6 rate control challenges, but we're getting a good quality result.

(Continued)

So, if we want to keep 25 percent for the alpha, we need to take our 185 Kbps target and account for the extra 25 percent being on top of the 100 percent of the normal bitrate. So, 125 percent = 1.25, and 185/1.25 = 148. However, that lets us now raise the Maximum 2 Pass VBR Data rate to 200 percent.

We reencode, but we're still only down to 377 KB. VP6's rate control isn't particularly accurate. 350/377 = 0.93, or 93 percent. Let's try a Data Rate Undershoot of 90 percent.

And that gets us down to 372. This is where it starts to suck to be a compressionist.

Okay, time to pull out the big guns and drop frame rate in half. The specs didn't say anything about frame rate, so we're going to change that before risking busting the bit budget. See Figure 15.10.

And that gets us down to 320 KB. Bump undershoot back up to 100 percent, and we're at 328 KB. Good enough. And we've got a client-pleasing FLV!

I wish I could say that type of noodling at the end was rare, but VP6's rate control has been problematic for years now. Be prepared to confirm that your file size is what you need.

Green/Blue Screen vs. Chroma-Keyed

I'm not even going to pretend to teach you how to pull a good key out of content shot on a blue or green screen. That's a postproduction skill, not part of compression.

But as a compressionist, you can be asked to do post by people who don't know the difference. Great if you've got the chops (don't forget to charge extra!), but if it's not your job and doesn't match your skills, be up front about that.

Squeeze, Flix, and so on take content that has an alpha channel and can pass the alpha channel through. That requires a file with an alpha channel exported from a tool like After Effects. And it can be quite a lot of work to get from camera footage of something shot in front of a green screen to a nice clean alpha channel.

Also, if you're shooting for compositing, use green instead of blue. Green gets less chroma subsampling, because Y′ is mainly green. See Color Figure C.23. ■

CHAPTER 16

Windows Media

This chapter is about Windows Media the format, not about media on Windows the platform. Windows playback of other media types is covered in Chapter 28.

Windows Media is Microsoft's first-party digital media platform. Like many a Microsoft technology, the first few versions (known as NetShow) were weak. But over time, it matured into a true powerhouse and the eventual victor of the turn-of-the-millennium QuickTime—RealMedia—Windows Media streaming wars. And needless to say, it's something I've worked with quite a bit myself, both before and during my own time at Microsoft. But I was a fan well before I worked on it, and I think it remains the most complete end-to-end media delivery architecture. Windows Media does a great job of handling live broadcast and on-demand video, as progressive download and streaming files on PC (and not just Windows) and Macintosh computers as well as mobile devices. Windows Media was the first architecture with integrated digital rights management (DRM), and remains the most broadly used and deployed DRM platform.

Microsoft's strategy is much less Windows Media—centric these days. Windows Media certainly is still supported and enhanced, but MPEG-4 and H.264 are also supported in Silverlight, Windows 7, Xbox, Zune, and beyond. The Windows Media Video 9 codec has been standardized as VC-1 and is used outside of Windows Media, including in Smooth Streaming .ismv files and on Blu-ray optical discs.

Windows Media is most dominant in the enterprise market, since it is preinstalled on the Windows desktops prevalent in corporations, with streaming supported from their Windows Server 2003 and 2008 installations. Windows Media is also very broadly used for downloadable and streaming video, particularly of premium content in Europe and Asia. In the United States, Netflix, Blockbuster, CinemaNow, and the Xbox and Zune marketplaces all use Windows Media.

The biggest historic weakness of Windows Media has been relatively weak cross-platform and cross-browser support; if your content ran on a Microsoft operating system and Internet Explorer, everything was generally great, but otherwise you could run into many limitations. This was fine in corporate environments, but was a problem when targeting a broad consumer audience. Addressing that limitation was a big reason for the development of Silverlight, and Silverlight's cross-platform and cross-browser compatibility has been central to its own success.

Microsoft also makes underlying technologies for Windows available for license as porting kits including source code for both encoding and playback. Third-party products like Flip4Mac and Episode incorporate these.

While Windows has had a Media Player for nearly two decades, the current era of Windows Media really started in 2003 with the release of Windows Media 9 Series, which included dramatically improved versions of the codecs, player, server, and file format. The player, server, and codecs have been updated multiple times since then, but still maintain broad compatibility with the 9 series launch. For typical web streaming use, it's easy to maintain perfect compatibility with WMP 9 and later while taking advantage of improved authoring tools and servers.

Why Windows Media

There are a variety of reasons why Windows media is so broadly used, as discussed in the following sections.

Windows Playback

If you've got content targeting the Windows platform (which is about 90 percent of all PCs in the world) Windows Media is universally supported with a high level of functionality. As long as users are on at least XP Service Pack 2, they'll have Windows Media Player 9 at a minimum, and so will be able to handle the 9 Series technologies.

Enterprise Video

Windows Media, like Windows itself, has placed a large emphasis on the enterprise, and provides unique functionality for large organizations. Being preinstalled on most corporate and government desktops makes deployment easy, and it supports multicast, critical for being able to deliver live corporate events across a WAN.

Interoperable DRM

Windows Media DRM is the most broadly deployed content protection technology, and the only one so far approved by all the major film studios. Beyond WMP, it's also supported by a wide variety of devices, and on Mac via Silverlight in many markets, Windows Media is required for Hollywood-licensed content for DRM.

Why Not Windows Media

Not Supported on Target Platform

The biggest reason not to use Windows Media is if your users don't have support on their target platform. Out of the box, Macs and Linux PCs don't, nor do many CE and portable media devices.

For web video, Silverlight is changing this dynamic quickly by supporting WMV cross-platform, cross-browser. But as of this writing, Silverlight was still catching up with Flash's ubiquity.

The Advanced System Format

The file format of Windows Media is called Advanced Systems Format, or ASF, and it's quite a bit more complex than the older AVI format. The principal design goal of ASF was for it to be highly efficient for a "bit-pump" server, minimizing the file parsing required and thus CPU requirement per stream. This has been a major part of Windows Media strength in server scalability, allowing a single box to deliver more content simultaneously than for competing formats.

ASF mainly contains video and audio, plus file- and time-based metadata. ASF files can also include markers and scripts mainly used for closed captions and to trigger web browser or Silverlight events. Some enterprising vendors have successfully implemented other data types in ASF; Accordent embeds JPEG slides inside ASF streams with their presentation products.

ASF can also contain multiple video and audio tracks. This enables bitrate switching (called "Intelligent Streaming") and multiple audio languages.

An ASF file requires indexing to enable random access. In most cases, indices are dynamically generated by the Windows Media Format SDK as the file is created, but if a file doesn't have an index (for example, a live broadcast currently in progress), it will need to be indexed later (the free Windows Media File Editor can do this).

The original file extension for ASF files was, unsurprisingly, .asf. However, because Windows uses file extensions to discriminate among file types, there was no way to assign different icons or default actions to video files (which couldn't play on music devices back then) and audio files (which could). So, with the introduction of Windows Media v7, .wmv (Windows Media Video) and .wma (Windows Media Audio) extensions were created. These don't represent any change to the underlying file type, just to the extension, and so you can generally mix and match extensions with no fundamental problem. Windows Media can also use metafiles that reside on web servers for server-side playlists and redirection. These metafiles link to the actual media. The original extension for those metafiles was .asx, but the more recent (although less often used) .wax (audio) and .wvx (video) discriminate between audio and video metafiles.

Windows Media Player

The canonical player for Windows Media content is good old Windows Media Player (WMP). It has a long heritage going back to 1991 and the Media Player application included with the Multimedia Extensions to Windows 3.0. There've been plenty of versions since then, with the oldest version still seen widely being WMP 9, preinstalled with Windows XP Service Pack 2, and an optional download back to Windows 98 and 2000. WMP 9 was part of the huge "Windows Media 9 Series" wave of products, and remains a fine baseline to target for broad compatibility.

The current versions are WMP 11, preinstalled with Vista and a recommended download for XP, and WMP 12 in Windows 7. (WMP 12 isn't available for older versions of Windows), and contains much broader format support, including MPEG-4 and H.264.

Windows Media Player support of formats beyond Windows Media is covered in Chapter 28.

Windows Media Video Codecs

Since the Windows Media Video 9 codec was standardized as VC-1, detailed information on using it is in Chapter 17. This section will discuss the particulars of VC-1 in Windows Media.

Windows Media Video 9 ("WMV3")

Windows Media Video 9 (WMV 9) has been the mainstream choice for most Windows Media content since the 9 Series launch. There have been five generations of backward-compatible encoder enhancements that provide quality and encoding speed improvements while maintaining full compatibility.

In general, the only time you'd use a codec other than WMV 9 for web video use is if you were trying to stream native interlaced content or have a single bitstream compatible with Windows Media Player and Smooth Streaming (and hence would use WMV 9 Advanced Profile, described shortly) or screen capture content (potentially using WMV 9.2 Screen, also described in this chapter).

Profiles

For computer playback, WMV 9 Main Profile is the broadly compatible high quality choice. Simple Profile is a simplified version targeting low-power devices like mobile phones. This simplicity means it requires less horsepower to play back, but it also requires more bits to provide equivalent quality to Main Profile since Main has some additional tools to improve compression efficiency (particularly B-frames and the in-loop deblocking filter). That said, most devices from Windows Mobile 2003 forward support Main Profile, although potentially requiring a lower bitrate. And mobile devices with hardware decoding like more advanced phones and the Zune handle Main Profile with aplomb. See Figure 16.1.

Encoding tools that don't provide a profile control are Main Profile only.

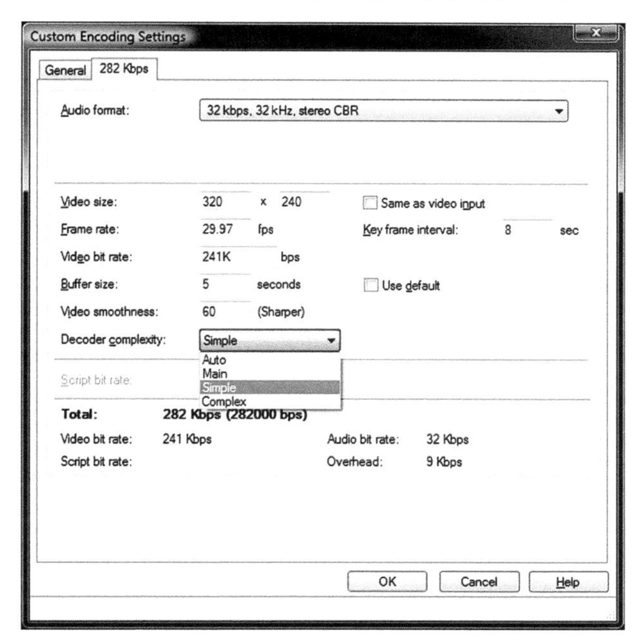

Figure 16.1 The Simple/Main profile switch in Windows Media Encoder.

WMV 9 = "WMV3"?

There are two profiles supported in WMV 9—Main Profile and Simple Profile (Advanced Profile is a separate codec). Both use the somewhat unfortunate 4CC of "WMV3"—Windows Media Video 7 was "WMV1," thus WMV 8 was "WMV2" and WMV 9 is "WMV3." It doesn't make sense, but that's the way it is.

Complex Profile

There was also a Complex Profile introduced with WMV 9, but that has since been deprecated in favor of the superior Advanced Profile. As Complex Profile didn't receive further development, the current Main Profile implementation outperforms Complex Profile anyway. While Microsoft decoders handle Complex Profile, it isn't part of the VC-1 spec, and so is not supported by other players based on the VC-1 spec, particularly hardware decoders.

Bottom line: don't use Complex Profile even if you have a tool that supports it. The results will be lower-quality and less compatible than Main Profile. Playback of Complex Profile-encoded content is also quite a bit more CPU-intensive than Main or Advanced.

Windows Media Video 9 Advanced Profile ("WVC1")

Windows Media Video 9 Advanced Profile is a non-backward-compatible enhancement to WMV 9 focused on professional video applications. The two most important additions are native interlaced support and transport independence enabling use in formats other than ASF. For traditional WMP playback, neither is needed. WMP 11 on both XP and Vista don't automatically convert interlaced WMV9 AP to progressive on playback unless a registry key is set, so interlaced playback is really only of use in customized apps and in dedicated players (software Blu-ray players being the most common). The only addition that significantly helps quality is the ability to use Differential Quantization on I-frames, discussed in the next chapter.

"WMVA" and "WVC1"

WMV9 AP was introduced with WMP 10 and the Format SDK 9.5 back in 2004. However, there were a few format tweaks in the v11 implementation, to make it compatible with the final SMPTE VC-1 spec meaning that WMP 9 and 10 will require a codec download to play back v11 encoded WMV 9 AP. The original, non-SMPTE AP implementation had the 4CC of "WMVA," with the current SMPTE compatible using a 4CC of "WVC1." WMVA, like Complex Profile, is supported by Microsoft decoders but not decoders based on the SMPTE VC-1 spec. Thus it's less compatible and lower quality, and should not be used for any new encoding.

VC-1 and Windows Media Video 9

Thee relationship between the Windows Media Video codecs and the SMPTE (Society of Motion Picture and Television Engineers) VC-1 spec can be a little confusing. VC-1 is the SMPTE designation for their standardization of WMV 9. In essence, you can now think of Windows Media Video 9 as Microsoft's brand for its implementation of VC-1, as implemented for ASF files. Microsoft's Expression Encoder product uses the VC-1 name, but it's the same compatible bitstream.

The VC-1 Simple and Main Profiles are progressive scan only, and hence part of the existing WMV 9 profile. Advanced Profile requires the new WMV 9 AP implementation.

Table 16.1 Windows Media Video and VC-1 Equivalents.

Windows Media Codec Name	WMV 9 Codec Profile	VC-1 Profile Name
Windows Media Video 9	Simple Profile	VC-1 Simple Profile
Windows Media Video 9	Main Profile	VC-1 Main Profile
Windows Media Video 9	Complex Profile (deprecated)	n/a
Windows Media Video 9 Advanced Profile	Always Advanced Profile	VC-1 Advanced Profile

Windows Media Video 9 Screen

The WMV 9 Screen codec was designed for efficient compression of screen recordings. It's a special-use codec—very efficient for this task, but not for normal video. Screen is most efficient with simpler, flat graphics like Windows 2000 or the "Windows Classic" themes in XP, Vista, and Windows 7. However, richer graphical environments like Vista and Win 7's Aero Glass include lots of gradients and transparencies, and normally encode more efficiently using standard WMV 9.

For efficiency, WMV Screen needs access the uncompressed RGB source video of the screen shots—it doesn't work well when compressing from screen shots that have had any lossy encoding already applied to them, or even simple resizing. It's typically used in conjunction with lossless screen recording products like TechSmith's Camtasia. On a fast computer, it's possible to use Windows Media Encoder to record or even broadcast live screen activity with the screen codec, but not recommended. As discussed in Chapter 5, newer DWM-based screen capture technologies like that in Expression Encoder 3 provide much better performance. Compared to other screen capture codecs, WMV 9 Screen is unique among screen codecs providing full support for 2-pass VBR and CBR encoding, making it possible to use for streaming (live and on-demand). Still, for most modern screen recordings, VC-1 is a better choice. Expression Encoder's screen recording targets VC-1 only.

The Screen codec is Windows-only, and not supported in Silverlight, Flip4Mac, and other porting kit–based players.

Windows Media Video 9.1 Image

Also known as PhotoStory, the WMV 9.1 Image codec isn't a video codec per se. Instead, it stores still images and metadata describing how they can be animated for a slideshow, including transitions. For this special class of content, it can be much more efficient than a typical video codec. It's not widely supported outside of Windows; Silverlight and Flip4Mac don't play it back.

Legacy Windows Media Video Codecs

Given Windows Media's longevity, it's had many different codecs over the years. While WMV 9 is all that need be authored these days, I'll list the details of the legacy codecs in case you stumble across them. In general, Microsoft-derived WMV 9 decoders like those in Silverlight and Flip4Mac can also decode WMV 8 and WMV 7, but not earlier codecs.

Windows Media Video 8 ("WMV2")

Windows Media Video v8 was the predecessor to WMV 9, naturally. It was about 15–20 percent less efficient than the original version of WMV 9, and is much further behind current implementations.

It was notable for having a very fast encoder compared to the original releases of WMV 9, but on modern hardware, that's very rarely a concern anymore.

Windows Media Video 7 ("WMV1")

Like WMV 8, WMV 7 was yet faster to encode and yet less efficient to compress than its successor. Again, there no reason to make these files anymore. And this is where the 4CC version number offset started; they could have saved some confusion if it had been "WMV7"

MS MPEG-4 v3

This was the last Microsoft codec based on MPEG-4 Part 2 Simple Profile. The codec had three versions (v1, v2, and v3). The latter was on the market the longest, when it became clear that MPEG-4 part 2 was hitting a wall for how far it could be optimized, and thus the superior WMV codec series was begun. MS MPEG-4 v3 was the subject of the original "DivX ;-)" hack to enable its use in AVI files.

The encoder hasn't been included in Windows since XP, although some video tools may install it as part of an older Format SDK.

MS MPEG-4 v1 and v2

These older codecs were the initial versions of Microsoft's MPEG-4–derived codecs, and predate the final MPEG-4 standard. They offered substantially lower quality than MS MPEG-4 v3. Encoders for these haven't been in the Windows Media SDK since 1999, so only very old tools can encode to them. MS MPEG-4 v2 was the last version of the codec that also functioned as an AVI codec, although the DivX hack makes MS MPEG-4 v3 work in AVI files.

ISO MPEG-4

MS MPEG-4v3 implemented a draft of MPEG-4 part 2, and wasn't quite compatible with the final spec, so this ISO-compatible version was introduced alongside the initial WMV 7 codec. It didn't see significant use, as Windows Media didn't otherwise support the file format, stream, or audio codecs from MPEG-4, and WMV 7 outperformed it handily.

Windows Media Screen 7

Windows Media Screen 7 was an earlier, weaker version of Screen 9. It's no more broadly supported, and thus shouldn't be used anymore.

Windows Media Audio Codecs

The Windows Media Audio codecs are covered in detail in the next chapter.

Encoding Options in Windows Media

Data Rate Modes

Windows Media supports the full range of rate control modes discussed back in Video Codecs. Unusually, Windows Media supports 2-pass modes, including 2-pass VBR, for audio as well as video.

Constant Bitrate (CBR)

Windows Media tools normally define CBR buffers by duration of the buffer, but some VC-1 implementations use VBV in bytes.

Normally CBR is used for streaming, although Windows Media Services can stream VBR content as well, treating it as a CBR file at the peak bitrate. CBR is also required for live encoding for streaming.

1-pass

1-pass CBR is required for live streaming, of course. It's also the fastest bitrate-limited mode available in Windows Media (the bitrate-limited VBR modes are 2-pass only).

Note that 1-pass CBR encoding can be significantly improved by the use of the Lookahead and Lookahead Rate Control features. In current Windows Media implementations, 1-pass CBR with those modes normally outperforms 2-pass CBR in quality as well as speed.

2-pass

2-pass encoding lets the encoder see into the future, ramping the bitrate up and down as needed to account for future changes. In theory, it should be better than 1-pass CBR, but 1-pass CBR recieved tremendous tuning for live streaming and now generally surpasses 2-pass CBR. The only reason you would get much use out of 2-pass CBR is with a very long buffer (more than eight seconds) or when only stock settings are used. This may change with future implementations.

The first pass runs much faster than the second pass or a single pass would be, so going to 2-pass doesn't double encode time—it's more typically a 20 percent increase.

Variable Bitrate (VBR)

The essential difference between CBR and VBR is that VBR files have a peak bitrate higher than the average bitrate. This lets VBR file be more efficient for a given file size, since it can redistribute bits throughout the file to provide optimal quality.

You can also think of the difference as "CBR maintains bitrate by varying quality, and VBR as maintaining quality by varying bitrate."

1-pass Quality-limited VBR

1-pass (quality-limited) VBR uses a fixed quantization for I- and P-frames (with B-frames varying some by amount of motion). Each frame using as many bits as needed. File size varies tremendously depending on content, and there's no limit on peak bitrate at all. Obviously, this makes quality-limited VBR a poor choice for content distribution. It's mainly used for mezzanine files, and since it is 1-pass, it can be captured in real time on a sufficiently powerful box.

2-pass Bitrate VBR (Constrained)

Bitrate VBR (Constrained) is the optimal encoding mode for progressive download and file based playback where the peak bitrate can be higher than the average bitrate. Almost every web video (WMV) file not streaming from Windows Media Services should be encoded using Constrained VBR.

With older versions of the codec, 1-pass CBR was recommended for HD content due to a problem with occasional dropped frames with 2-pass at high bitrates. This was fixed in the v11 codecs, and 2-pass VBR can be safely used for HD content. This can result in big savings in file size at a given quality level.

As a rule of thumb, VBR becomes significantly better than CBR when the peak is around 1.5x or more the average bitrate.

2-pass Bitrate VBR

The unconstrained 2-pass Bitrate VBR mode is the same as Constrained VBR without any peak constraint. This means that playability isn't predictable; the constrained mode ought to be used instead. While many tools expose this mode, it's never the best choice.

Where Windows Media Is Used

Windows Media for ROM Discs and Other Local Playback

Windows Media is widely used for the video portions of Windows-only ROM discs, although web delivery is rapidly shrinking that market.

One common use of WMV for local playback is embedded video in PowerPoint presentations. This was frustratingly not cross-platform for over a decade, but current versions of Office for Mac 2008 are now able to play an embedded WMV using Flip4Mac without any trouble.

Windows Media for Progressive Download

Although Windows Media was originally targeted at streaming, it has become a great technology for progressive delivery.

There aren't any particular tricks to encode for local playback of WM files, except to use a single video track instead of the Intelligent Streaming multiple bitrate mode, and to use 2-pass VBR Constrained for video and audio.

Windows Media for Streaming

Windows Media adopted the RTSP protocol for streaming many years back, dropping the older, proprietary Microsoft Media Server (MMS) protocol. Still, a URL for a streaming ASF file typically still starts with mms://. This serves a hint to Windows Media–incompatible players that they probably can't play back the content, and to Silverlight and WMP to stream the file instead of trying progressive download. Windows Media Services (WMS) was built into Windows Server 2003 and is a free download for Windows Server 2008. RealNetworks' Helix Server includes some Windows Media support, but it hasn't been actively developed for some years and can have trouble with newer content and players.

Windows Media Services is a full-featured streaming server, familiar to administrators of Microsoft's Internet Information Services (IIS). It offers all the appropriate features of a modern server architecture. By default, it will fall back to TCP delivery if UDP is blocked by a firewall, and it can be configured to fall back to HTTP. The Intelligent Streaming MBR mode will correctly switch data rates even when being used in this mode.

On-demand streaming

The ideal encoding mode for on-demand streaming is CBR. The larger the buffer, the higher the average quality, but the greater the latency for both startup time and random access. If the server is using the Advanced Fast Start and Fast Cache features, random access will be reduced as the user's connection speed goes up, so you can get away with bigger buffers.

Live streaming

Windows Media provides a good, mature system for doing live streams. Most folks have used Windows Media Encoder for capture, but other tools are much better today. Expression Encoder 3 offers a better experience at the WME price point. On the higher end, there are great real-time encoders from companies like Viewcast, Digital Rapids, Inlet, Envivio, and Telestream.

WMV can use up to eight processors for video compression on Windows 7 and Expression Encoder 3 (4 in older versions of Windows without EEv3), so if you're doing live broadcasting, a hefty multiprocessor machine can substantially increase the maximum resolution, frame rate, and data rate you can target. 720p30 is realistic on a fast 4-core, and 720p60 and 1080p30 are possible with an Intel i7 8-core.

WME can also simultaneously capture and serve video from the same computer. However, unless you're going to have more than a handful of viewers, it's best to send a single copy of the stream to a dedicated server, so CPU cycles don't get gobbled up by serving instead of being used for encoding.

Intelligent streaming

Intelligent Streaming is Windows Media's Multiple Bitrate (MBR) technology. It bundles different video and audio bitstreams of the content into a single WMV file, with the server delivering just the video and audio the player requests based on available bandwidth. Different audio streams can also be used for different languages

The streams can vary in bitrate, frame size, frame rate, and buffer, but must share the same codec.

Intelligent Streaming was fine on the server side, but like all classic streaming MBR technologies, it was hard for the player to know which bitrate to use. While it was easy to figure out when the requested bitrate is too high (buffering), there wasn't any good way to figure out if there was extra bandwidth to increase bitrate. Also, any bitrate switch would yield several seconds of buffering, which isn't ideal.

Silverlight's Smooth Streaming is a much more robust solution to the problems that Intelligent Streaming targeted, and is discussed in Chapter 27.

Windows Media for Portable Devices

Windows Media has long been supported on Microsoft's portable devices like Windows Mobile phones and Zunes, plus devices from Nokia and other vendors.

Of course, many mobile devices are limited in processing power, memory, storage, and bandwidth compared to desktop computers.

All Windows Mobile devices since 2003 have supported WMV 9 Simple and Main Profile plus WMA Standard, Voice, Pro, and Lossless. However Advanced Profile and WMA 10 Pro's LBR modes are not as widely supported.

For a device that supports WMV and H.264, a good WMV 9 Main Profile encode can sometimes retain more detail at moderate bitrates than H.264 Baseline Profile,

Embedding Windows Media in a Web Page

Windows Media Player is available as an OCX (Object Linking and Embedding [OLE] Control EXtension component; glad you asked?) for embedding in Internet Explorer, the primary method Windows Media files have been embedded in web pages. The WMP OCX works well on Windows, offering full hardware acceleration for playback, DRM, etc. It's commonly used that way inside IE-only corporate environments. But that hasn't addressed consumers with Macs, Linux, or other browsers.

For Firefox users on Windows, Microsoft provides a plug-in that embeds the WMP OCX in a Firefox-style plug-in. Its permanent home is:
`http://port25.technet.com/pages/windows-media-player-firefox-plugin-download.aspx`

Mac users can use the free version of Telestream's Flip4Mac from Flip4Mac.com. It's based on the Windows Media porting kit, and offers increasingly good compatibility with existing sites, although complex JavaScript can sometimes cause trouble.

And if course Silverlight provides a cross-platform, cross-browser mechanism to embed Windows Media content. However, Silverlight requires a Silverlight app to control the media playback; it won't used on WMP OCX-HTML targeted web pages.

Windows Media and PlayReady DRM

Digital rights management is required by many content owners to protect against piracy. Windows Media DRM was released back in 1999, and ran through WMDRM 10. Its successor technology PlayReady, which supports other formats including MPEG-4 and H.264.

The basic model of WM DRM is similar to other DRM solutions, like Apple's FairPlay. The content is encrypted, and can't be decrypted without a license key. The license key itself can be delivered by a server at play time, securely stored on the computer, or even stored in the file itself.

Different players can support more or fewer rules on how the content is used. Silverlight 2 and 3 really only support "Play Now" and don't offer a way to store the local key. But Windows Media Player offers the full functionality. Some of these rules include:

- Expiration date
- Output protection (whether analog or unencrypted digital outputs can be used)
- Whether the file can be transferred or transcoded to another secure device

From a compression workflow perspective, DRM can either be applied as part of the encoding process (if the encoding tool integrates DRM) or an unencrypted file can then be encrypted later in the workflow.

And from a business perspective, Windows Media DRM is the only broadly supported DRM available for license and approved by the Hollywood movie studios. In markets not large enough for publishers to negotiate one-off deals with each studio, WMDRM is often the only option to publish premium content.

Windows Media Encoding Tools

Windows Media and the VC-1 codec are supported by a vary wide array of tools, but most of them incorporate one of a few underlying SDKs. That'll determine the features you have available. And in some cases, the same tool may offer different quality and performance depending on the version of Windows or Windows Media Player is installed.

Products using the VC-1 Encoder SDK don't have any dependence on the OS for video, but still do for audio.

VC-1 Encoder SDK

The VC-1 Encoder SDK is a professional-grade implementation of VC-1. It supports encoding for Blu-ray and Smooth Streaming as well as WMV. The SDK, its features, and the tools that use it are covered in detail in the following chapter. But the important takeaway is the VC-1 Encoder SDK encodes faster and with higher quality than the Format SDK, and doesn't require registry keys for advanced parameters.

Windows Media Format SDK

Windows Media Encoder and other older WMV encoders are graphical front-end to Microsoft's Windows Media Format SDK. The actual codec version you're running is determined by the Windows Media .dll (Dynamic Link Library) files on your system (Figure 16.2). Most people get those updates bundled with Windows Media Player, so as long as you have the most recent WMP version for your platform (that's WMP 11 for XP and Vista) and the encoding hotfixes listed next—or, better yet, Windows 7—you're good to go.

For vendors of compression tools, it's always good to include the redistributable installer with your product, to ensure that users have the current version.

I've see far too many production encoders are still running WMP 9 and thus the 9.0 SDK. That's a codec implementation from 2003, and will offer much lower quality and performance than the current versions deliver.

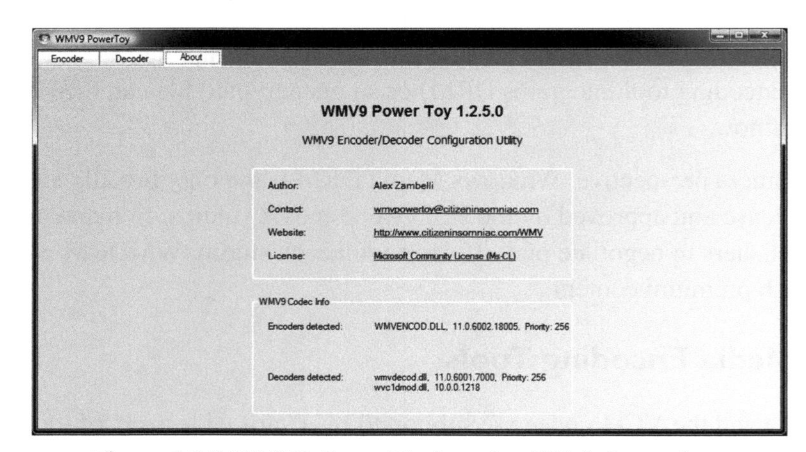

Figure 16.2 WMV9 PowerToy's codec DLL information.

Windows XP, Vista, or Server 2008: Format SDK 11

Windows XP shipped with Windows Media Player XP, a flavor of WMP 7. However, the almost universally installed Service Pack 2 included WMP 9. WMP 11 is also available for XP, and is the minimum version seen in the wild today.

Windows Vista shipped with WMP 11 preinstalled, so no update is required beyond hotfixes listed here.

On Windows Server 2008, Windows Media Player, and hence the Format SDK (FSDK), is installed with the Desktop Experience feature; that's installed through the management interface like all other Server 2008 features. You'll want to enable Desktop Experience on any encoder box (and in general for any 2K8 box you'll be using as a workstation).

Windows Server 2003: Format SDK 9.5

The most recent version for Windows Server 2003 is the Format SDK 9.5 (still better than 9.0), released with Windows Media Player 10.

Running on Server 2003 means no access via the SDK to the SMPTE-compliant "WVC1" flavor of Windows Media Video 9 Advanced Profile, nor WMA 10 Pro LBR modes (the 32–96 Kbps range). The video codec is also only 2-way threaded instead of 4-way threaded, slower in general, and lacks the advanced registry keys.

If you need to encode on a Server OS, you should use a VC-1 Encoder SDK-based product or upgrade to Server 2008 (ideally both!). However, WMA 10 Pro will still be unavailable.

There have been reports of using the WMFDist11-WindowsXP-X86-ENU.exe installer from FSDK 11 set to XP compatibility mode to install the FSDK 11 .dll files onto Server 2003. While there aren't known issues with this, this isn't a supported configuration for Server 2003, and Microsoft wouldn't support issues when running with mismatched .dll files.

Windows 7

Windows 7 comes with updated Windows Media .dlls, but not a new Format SDK; they're only available in Windows 7 and the Windows 7 based Windows Server 2008 R2. It doesn't have any new registry keys or configurations, but it has two big performance improvements:

- 8-way threading (automatically activated when frame size is at least 480 lines)
- 40 percent speedup when encoding with B-frames

Windows 7 also has a quality improvement for noisy textures, particularly in shadows, reducing cases where noisy textures wind up getting turned into motion vectors. This reduces the appearance of "swirling" noise.

Figure 16.3 (A) A challenging frame from Windows Media in 2003 using the first-generation VC-1. (B) And in 2009 with the fifth-generation implementation.

The same improvements are included in Expression Encoder 3, and hopefully soon in VC-1 Encoder SDK. See Figure 16.3.

Format SDK Hotfixes

WME is ancient code at this point, and all updates since the original Windows Media 9 Series launch have been done as hotfixes. You need to have these installed for security, stability, and performance if running the specified software. None apply to Windows 7, which you should be using for FSDK encoding anyway.

Hotfixes can be downloaded from http://support.microsoft.com.

Vista Compatibility Hotfix (119591)

If you're on Vista, you absolutely need to install this one.

This hotfix addresses three issues:

1. When you run the Windows Media Encoder command-line script WMCmd.vbs, the script host Cscript.exe may crash.
2. The icons that appear on the encoder toolbar and in the encoder dialog boxes are displayed by using a low bit depth. Therefore, the icons appear to have a low resolution.
3. When you configure the encoding profile or start an encoding session, the encoder may crash.

Multithreaded Multiple Bitrate Encoding Hotfix (945170)

This one fixes a pretty embarrassing bug – Format SDK 11 wouldn't use multiple processors when encoding in multiple bitrates (a.k.a. "Intelligent Streaming"). So you could have eight simultaneous bitrates all stuffed into a single thread, while other cores waited unused.

Critical Security Fix for Windows Media Encoder (954156)

This is a security fix for a critical vulnerability that could allow remote code execution on a machine running WME. Install it before you launch WME! Note that the vulnerability requires two things that shouldn't be happening anyway on a production encoder:

- WME is running logged in as an administrator.
- The machine is being used to browse the web.

Low-Latency Webcasting

While Windows Media offers 15–20-second end-to-end delay by default, it's possible to drive it down to 2–3 seconds with best practices on a good network. The critical thing is to tune the encoder, server, and player latency together.

Latency is the delay between when video enters the encoder and is seen on the user's display. It's sometimes called "broadcast delay." Latency is something that doesn't matter at all in some markets, and matters a lot in others.

The primary reason we have latency is buffering for quality and reliability. By having the server wait several seconds after a video packet is received before sending it out, it can smooth out peaks and valleys in the data rate, and can detect and resend dropped packets before their display time. Buffering in the player averages out data rates and allows recovery of dropped packets as well.

Large buffers were extremely important in the modem era, and the defaults can deliver high-quality content over a variety of networks. But when minimizing the latency is important and reliable networks are available, the delay can be reduced substantially.

Since total latency is the sum of encoder latency, server latency, and player latency, plus packet transmission time, improving latency requires tweaks to the encoder, server, and player in parallel.

Encoder Latency

WME enables video encoding latencies down to one second, and EEv3 can go down to 0.1 second. Using Lookahead or Lookahead Rate Control (LRC) will increase latency beyond the buffer value (by the GOP duration for LRC).

Windows Media includes the "Low Delay" audio codecs for both WMA and WMA Pro. This can be recognized by having bitrates 1 less than the typical (like 127 instead of 128); actual bitrates are the normal (so 127 really uses 128Kbps). The specific low-latency codec modes are listed for each codec in Chapter 18.

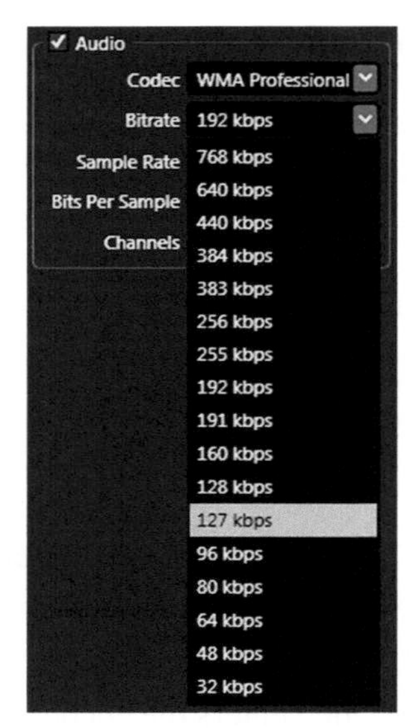

Figure 16.4 Expression Encoder's audio codec list. The "–1" settings like 127 are Low Latency, and minimize encoder-side buffering.

Server Latency

WMS features like Advanced Fast Start can dramatically reduce latency for on-demand content, but don't apply to live streaming. Server buffering can be turned off entirely in WMS for Windows Server 2003 and 2008 (Figure 16.5), yielding a significant drop in latency (and a drop in robustness to dropped packets).

Player Latency

Normally the player settings aren't under control of the encoder or server. By default, WMP picks a buffer size based on measurements of network and stream performance. However, it's possible for the user to lower the buffer size in the player's options. This can help reduce latency when watching streams with a good connection, but could produce pauses in the video when watching video from the general Internet.

Silverlight makes the player buffer size a controllable parameter, so an optimal setting for a given stream can be specified. This is controlled by the BufferingTime parameter in a Silverlight MediaElement. See Figure 16.6.

Encoders for Windows Media

There are lots of tools available for Windows Media. The technology has been available for years, and SDKs have been free of charge for Windows all along. Essentially every

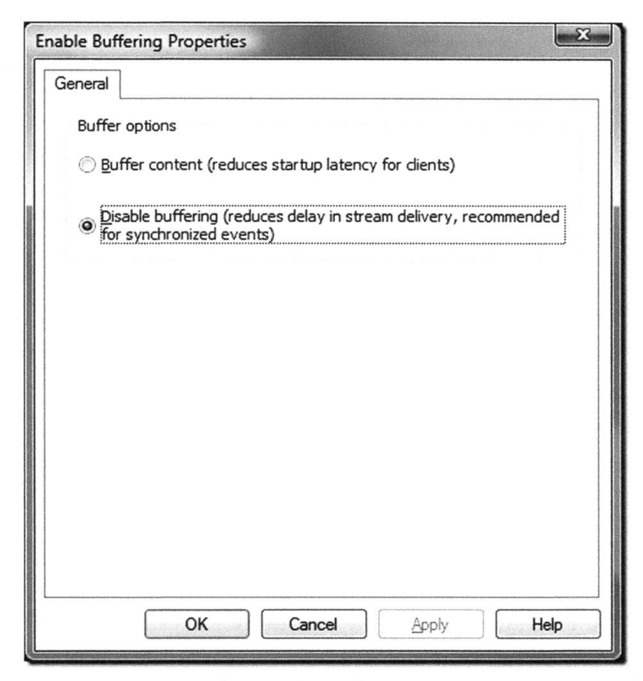

Figure 16.5 The Windows Media Services dialog to turn off server-side buffering.

general-purpose compression tool on Windows supports Windows Media, as do many on other platforms.

I'm covering WMV-centric products in this chapter. More advanced encoders with deeper VC-1 support like Fathom, Rhozet, and CineVision PSE are covered in Chapter 17.

Expression Encoder

Expression Encoder is Microsoft's current first party compression tool for Windows Media. With version 3, it's largely replacing Windows Media Encoder. EE also supports Smooth Streaming and H.264 output, as described in the relevant chapters.

Expression Encoder 3 is now available in a free version that supports the WMV features of WME. There's also a $49 upgrade with additional codecs and format support, including:

- MPEG-4/H.264 export
- Smooth Streaming export (VC-1 and WMV)
- MPEG-2, MPEG-4, and AC-3 source decoders

```
1  <UserControl
2      xmlns="http://schemas.microsoft.com/winfx/2006/xaml/presentation"
3      xmlns:x="http://schemas.microsoft.com/winfx/2006/xaml"
4      x:Class="Low_latency_media_player.Page"
5      Width="1280" Height="720">
6
7      <Grid x:Name="LayoutRoot" Background="White" >
8          <MediaElement Source="http://foo.com/lowlatency.asx" BufferingTime="00:00:00.000"/>
9      </Grid>
10 </UserControl>
```

Figure 16.6 XAML code for turning off client-side buffering.

EEv3 includes all the Windows 7 and VC-1 Encoder SDK improvements, so its WMV encoding is top notch in quality and performance. Better yet, it provides the settings in a humane manner, offering drill-down to features as needed, but also offering a simpler gradient of Faster to Better presets.

It's also the first lower-cost live encoder to include the full VC-1 Encoder SDK, including dynamic complexity and lookahead rate control (full details in the VC-1 chapter). But that basically means it'll automatically tune to the best quality your hardware can deliver, and will offer a lot better quality WME at the same bitrate.

Expression encoder notes

- You can use Expression Encoder (Figure 16.7) to trim files, edit or add metadata, apply Silverlight templates, and so on without having to re-encode. Just pick "Source" for video and/or audio profile.
- Expression Encoder treats the frame rate as a maximum. So if you combine sources that have different frame rates, or set a higher frame rate than the content uses, it won't encode any repeated frames.
- If you open a file that has metadata, it'll carry that metadata through, with or without editing. Metadata can also be exported to XML, or imported from XML, SAMI, and dxfp files.
- EEv3 can do multiple-bitrate encoding, encoding all streams in parallel to an Intelligent Streaming bundle or separate files. This makes managing multiple versions of the same source easier, and encoding faster. Each output stream can use up to 8 cores, so this is a great excuse for a new workstation.
- There is a .NET object model SDK for EE that enables automated workflows to be built around it, like the old WME SDK, but offering full control over live encoding and advanced codec settings.

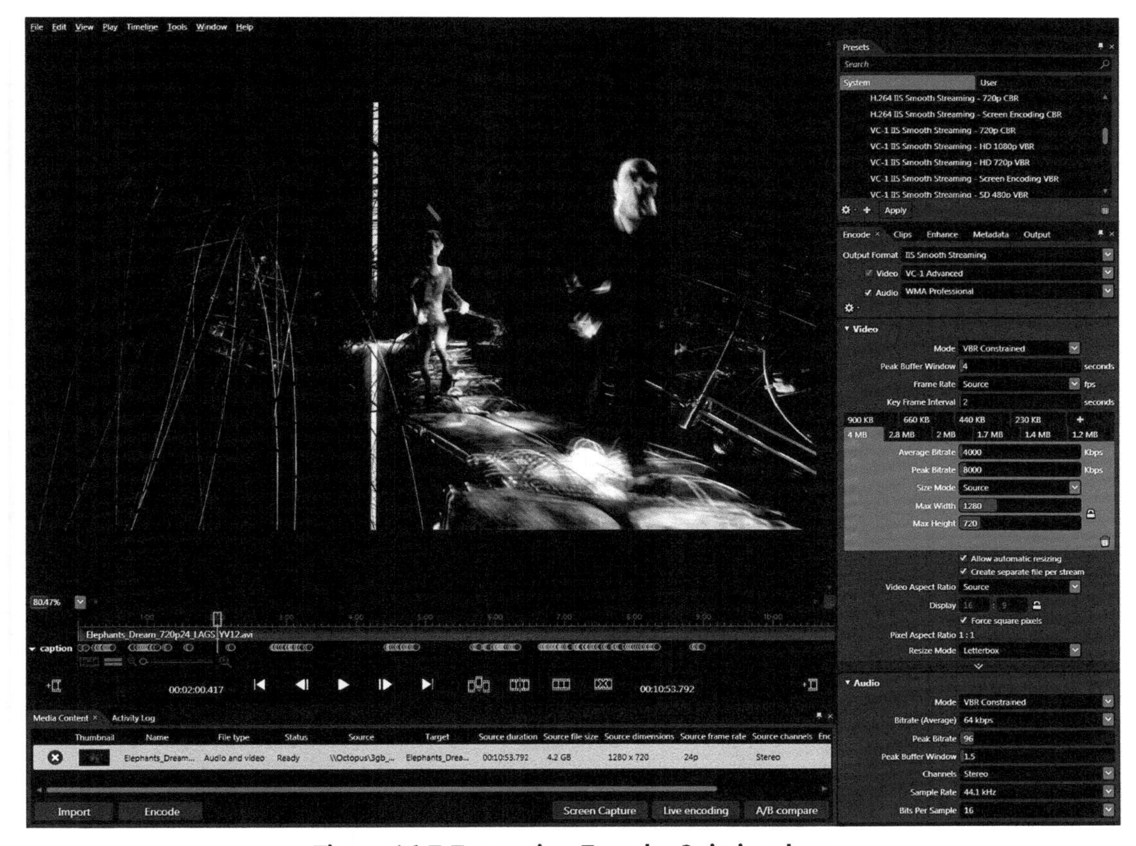

Figure 16.7 Expression Encoder 3, in its glory.

- Markers can be used to specify particular frames as keyframes, which are made Closed-GOP I-frames on encode.
- EE can automatically wrap a WMV in a Silverlight player template and publish it to a web server without recompression.

All that said, Expression Encoder is really focused on being a high-touch replacement for WME. There are many industrial-grade WMV encoders that use the same underlying VC-1 Encoder SDKs for mission-critical live encoding or high-volume on-demand encoding.

Windows Media Encoder

Windows Media Encoder has long been the most used encoder in the industry. Part of this was due to its deep integration with the Format SDK and implementation of the broad sweep of Windows Media features. And, perhaps as importantly, it was free. WME also has an "Encoder SDK" that enables simpler automation of live and on-demand encoding than the more complex Format and VC-1 Encoder SDKs.

However, the era of Windows Media Encoder is coming to a close, for a few reasons:

- WME's last release was the 9 Series launch in 2003.
- It's now aged out of Microsoft's "Mainstream Support" category, and so doesn't have formal support on Windows 7.
- It's stuck using the old Format SDK .dll files, and thus lacks advanced features of the newer SDKs, and requires registry keys for advanced parameters.
- The free version of Expression Encoder 3 does most of what WME does, but with better performance, quality, control, and user interface.

Now, the lack of mainstream support on Windows 7 doesn't mean it doesn't work; it seems to work just fine. But Microsoft isn't obligated to fix any nonsecurity bugs going forward. Companies with existing WME workflows can continue to use it, but really should be looking to migrate to EE or commercial encoder product.

WMCmd.vbs

The ungainly named WMCmd.vbs is a CScript script that automates the Windows Media Encoder SDK from the command line, and it's installed by default in `C:\Program Files\Windows Media Components\Encoder\WMCmd.vbs`. However, if you plan to use it, download the enhanced version from `http://www.alexzambelli.com`/wmv that includes full programmatic support of the various registry key options. For example, this is my old script for turning a 16:9 .avs from a DVD into a high-quality file playable on a Zune 2 and pretty much anywhere else. Needless to say, I switched to using Expression Encoder as soon as it was available (where I could just pick the "Zune 2 A/V Dock" preset):

```
cscript "C:\Program Files\Windows Media Components\Encoder\WMCmd.vbs" -input
MyDVD.avs -output MyDVD_Zune2.wmv -a_codec WMASTD -a_mode 4 -a_setting 128_
44_2 -a_peak 192000 -v_codec WMV9 -v_mode 4 -v_keydist 4 -v_bitrate 2000000 -
v_peakbitrate 3000000 -v_peakbuffer 4000 -v_performance 80 -v_profile MP -pix-
elratio 40 33 -v_msrange 0 -v_mslevel 2 -v_mmatch 0 -v_adz 2 -v_bframedist
1 -v_dquantoption 2 -v_mvcost 0 -v_loopfilter 1 -v_overlap 1
```

Flip4Mac

Flip4Mac WMV is a component for QuickTime on the Mac that adds WMV support to QuickTime. The free version of Flip4Mac is playback only, but the higher-end versions support import of WMV files to tools like Final Cut Pro, and export to WMV. The available versions are:

- Free: Playback only
- Pro: Import only
- Studio Preset-based encoding only
- Studio Pro: SD 1-pass only
- Studio Pro HD: Full support for HD, 2-pass, and multichannel audio

Beyond import and export, Flip4Mac (seen in Figure 16.8) will emulate the WMP OCX embedding options so that those web pages work on the Mac. This support has evolved over the years and now supported many more web sites.

Windows Media in Flip4Mac notes

- For whatever reason, Flip4Mac calls Advanced Profile "Broadcast" in the Advanced dialog and "WMV 9 Advanced" in the main dialog.

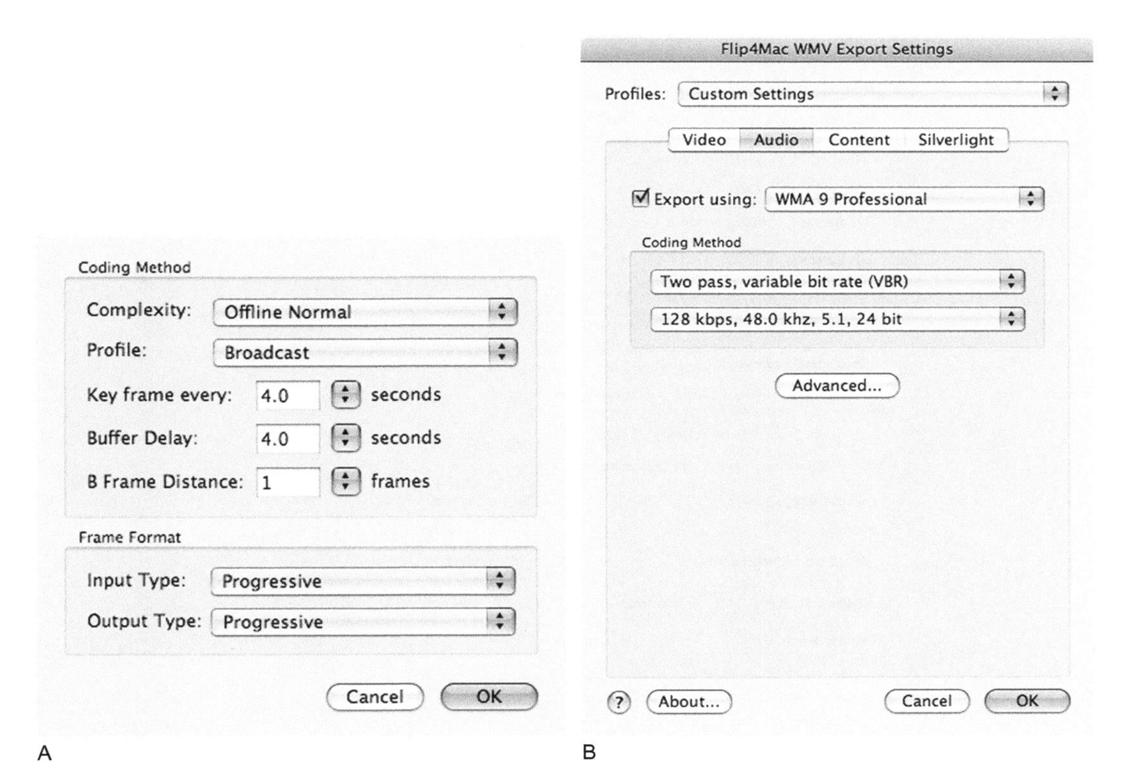

Figure 16.8 (A) Flip4Mac supports some advanced encoding modes, including interlaced VC-1. (B) Unlike Episode, Flip4Mac supports 2-pass audio.

- Flip4Mac includes built-in support for applying Silverlight templates to encoded files. The output is a directory including the media files, Default.html, and Silverlight player, just like Expression Encoder.
- Make sure to turn up B-frames to at least 1; they're off by default.
- Flip4Mac has built-in deinterlacing and interlaced pass-through modes. As QuickTime lacks a good interframe compressed interlaced codec, Flip4Mac can export interlaced WMV files from Final Cut laptops for news gathering. Using 1-pass VBR for video and audio at around 85–90 quality will be around 10 percent the size of DV with similar 720 × 480i quality, making for 10x faster uploads.

Episode

Telestream's Episode compression tools began life as Compression Master from PopWire. Originally a Mac-only product, they now have Mac and Windows versions. Episode for Windows is unique in that it uses Telestream's Porting– Kit derived encoder instead of either the Format SDK or VC-1 Encoder SDK, so quality and performance may vary from that of other tools on the same machine. See Figure 16.9.

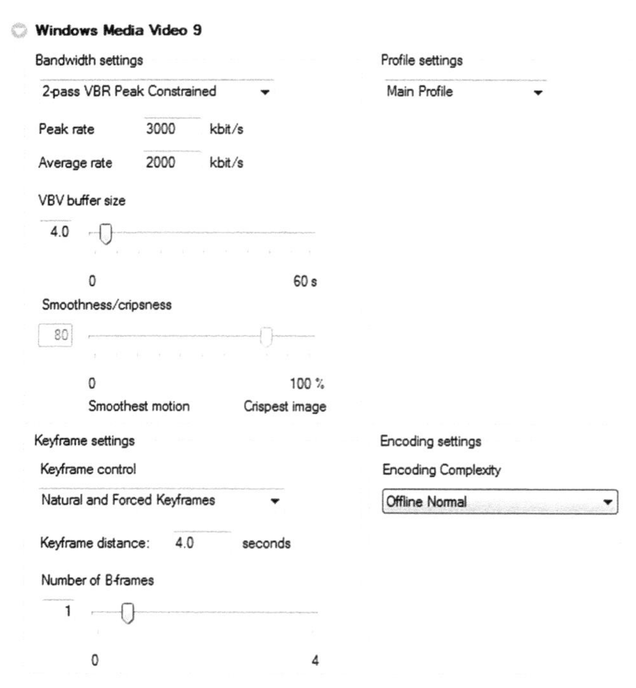

Figure 16.9 The Episode "WMV 9" dialog. Unlike FSDK-based tools, it exposes the important B-frame parameter.

Windows Media in Episode notes

A few other notes on Flip4Mac:

- Episode's VC-1 implementation, like its other codecs other than VP6, does a pipelined 2-pass, essentially using a Lookahead of up to 500 frames. Thus, even though it's only a 2-way threaded implementation, it can saturate a 4-core machine as both passes are running at the same time. However, this limits its ability to redistribute bits over longer, highly variable content; that 500 frames is just 22 seconds for 24p content.
- Episode and Flip4Mac were the first products that supported B-frames and lower-bitrate WMA modes later introduced in FSDK 11.
- Because Episode is based on the porting kit, it only supports WMA Pro to v9, and thus none of the LBR bitrates below 128 Kbps.
- Episode's Windows Media Video 9 is simple/main profile. Advanced Profile is called Windows Media VC-1.
- Episode supports the full range of video 1-pass and 2-pass modes, but its audio output is 1-pass CBR and 1-pass quality VBR only.
- Complexity modes have names, not numbers, in Episode. Their mappings to the typical values are:
  - Live Fast — Complexity 0
  - Live Normal — Complexity 20
  - Offline Fast — Complexity 40
  - Offline Normal — Complexity 68
  - Offline Slow — Complexity 80
  - Offline High Quality — Complexity 100

WMSnoop

WMSnoop's not really an encoding utility, but it's cool enough I wanted to mention it. It's a free WMV analysis tool from Sliq Media Technologies that provides useful information about WMV files, including a cool data rate graph. See Figure 16.10.

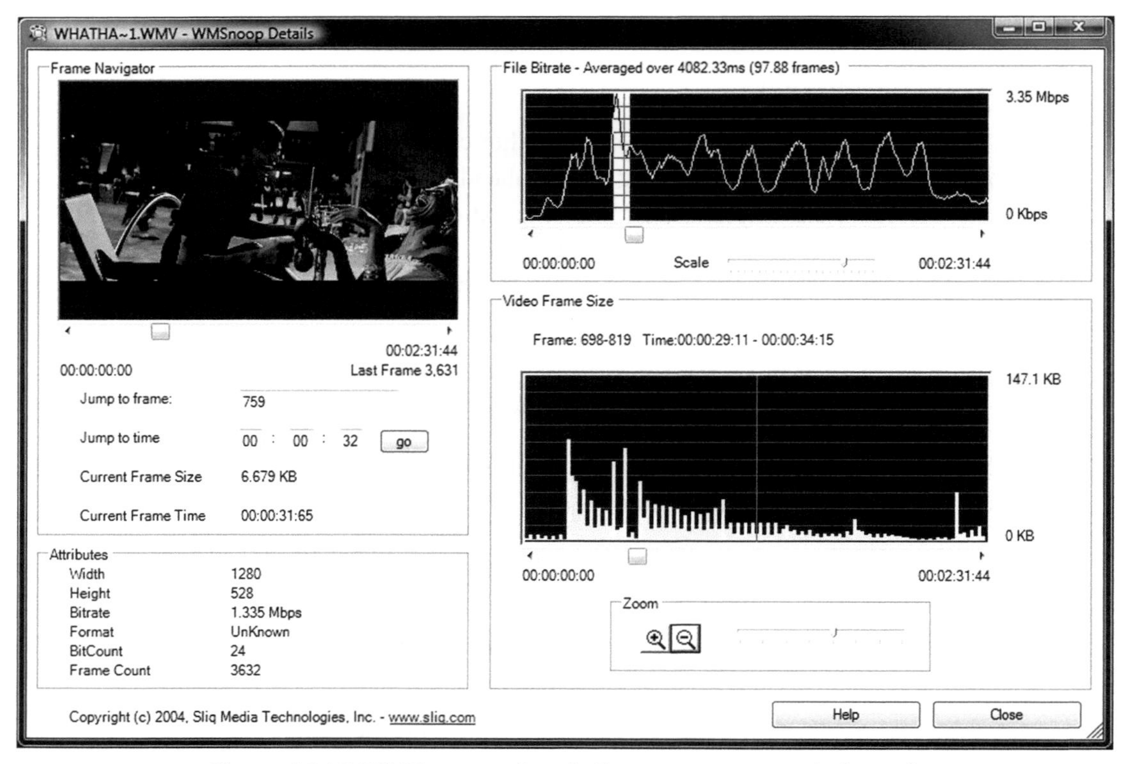

Figure 16.10 WMSnoop, a handy freeware WMV analysis tool.

WMV Tutorial: WMV File for Embedding in Cross-Platform PowerPoint

We'll be tackling a detailed VC-1 compression tutorial in the next chapter, so we'll do a simple one here.

Scenario

We work in the in-house A/V department of a mid-sized company. A VP has asked us to put a corporate branding video into PowerPoints she can take around to customer sites. She wants the file to be able to play on a presentation computer at an event, and to leave the client a copy. She's also got some customers who use Macs at home, and wants to make sure the presentation would work for them as well.

The total file has to be under 50 MB.

Three Questions

What Is My Content?

The source is a nicely produced 1080p24 HD corporate branding piece provided as a QuickTime file using the PNG codec. It's 2 minutes long with 5.1 audio.

(*Continued*)

Who Is My Audience?

The primary audience is our VP; we work for her and want her to see us as a successful problem solver. Her audiences are the people who view the presentation. It has to work and look good.

What Are My Communication Goals?

The content looks and sounds great on any machine the VP has to play it back on. We don't want dropped frames or stuttering audio, and definitely not a blank screen.

If she leaves a copy with someone, they should be able to play it back with minimum hassle.

Tech Specs

We've got 50 MB to provide the final presentation, but the .ppt file provided is already 31 MB! What to do?

First, we can convert from .ppt to the newer .pptx format (built into Office 2007/2008, and a free download for 2003/2004). .pptx gets an automatic Deflate (.zip) compression, leaving more room for the video. That takes the file size down to 20 MB; 30 MB is a lot more to work with than 19. With our 2-minute clip, that gives us 2000 Kbps total bitrate.

Historically, the most reliable format for cross-platform PowerPoint has been MPEG-1, since that's playable in both QuickTime on Mac and DirectShow on Windows. However, we need to make HD; 2000 Kbps isn't really enough for even SD with MPEG-1.

Fortunately, current versions of PowerPoint 2008 on Mac can use Flip4Mac to play embedded WMV in PowerPoint. And all Windows Machines can play the embedded WMV already.

We want the video to be playable on reasonably low-end machines, but look good on the big screen. Most presentation projectors are sadly still just 1024 × 768; we can encode our 16:9 source at 1024 × 576 at its original 24p. With 5.1 audio, we can give it 256 Kbps WMA Pro, leaving (less 9 Kbps for WMV overhead) 1735 Kbps for video. 1024 × 576 isn't a huge frame size, but we'll definitely want to use VBR encoding to get maximum bang for the bit. We can use peaks of 1.5x the average to keep decode complexity reasonable, and 3 Mbps peaks should still play fine off USB drives. Video will be VC-1 Main Profile, for XP SP2 compatibility.

We'll also make sure to include a link to Flip4Mac in the presentation so Mac users can find it if they don't have it installed already.

Settings

We'll use Expression Encoder 3, of course. It makes this project pretty easy.

We can start with a preset: Encoding for Silverlight > VC-1 > Variable Bit Rate > VC-1 HD 720p VBR. And then we can apply a higher-quality, slower preset by then double-clicking on Encoding Quality > Best Quality.

(*Continued*)

We can then make a few modifications:

- Video Bitrate to 1735 average and 2604 peak
- Width to 1024 (it automatically updates height to 576 to maintain aspect ratio)
- Audio Bitrate Average to 256 and peak to 384
- Channels to 5.1

I've always appreciated how Expression Encoder tries to do the right thing by default so that I don't have to mess with specifying frame rate and aspect ratio like Windows Media Encoder would have required. Figure 16.11 shows the final settings.

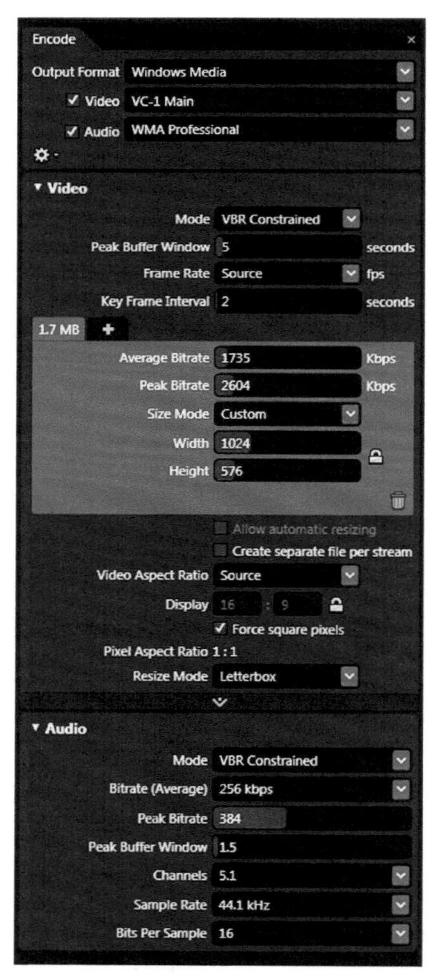

Figure 16.11 Final settings for our tutorial. It took enormous restraint for me not to dive into the advanced codec settings, but the defaults were already fine.

CHAPTER 17

VC-1

VC-1 (Video Codec 1) is the SMPTE designation for its standard 421M based on the Windows Media Video 9 codec. As an international standard, a full description of the bitstream is available, as well as conformance test files, and source code for reference encoders and decoders.

VC-1 is one of the Big Three codecs offered as an international specification and available for license from the MPEG Licensing Authority (MPEG-LA). While different codecs dominate in different industries, in general modern ASIC decoding hardware supports all of MPEG-2, H.264, and VC-1.

Why VC-1?

Windows Media Compatibility

The biggest reason VC-1 is used is that it's the best codec in Windows Media, and hence the best way to deliver video to out-of-the-box Windows PCs.

Quality@Perf

VC-1 was designed to offer excellent efficiency along with good decoder performance, particularly on the x86 processors that run Windows. Thus, assuming equally good decoders, maximum complexity VC-1 has about only half the CPU requirements of H.264 High Profile.

VC-1 performance is a lot more predictable; knowing the *profile@level*, frame size, and bitrate, it is possible to make a pretty accurate estimate of playback for a particular decoder. Of course, H.264 has a lot of latitude for tuning between compression efficiency and decode complexity.

For devices, VC-1 Main Profile can often retain better detail than H.264 baseline for mobile device encodes.

Smooth Streaming

The Smooth Streaming Encoder SDK is a VC-1 implementation highly and specifically tuned for Smooth Streaming that provides better encoding time, visual quality, and decoder performance. The techniques it uses (described later) could certainly be applied to H.264, but until a similar H.264 implementation is created, SSE SDK's VC-1 output will generally outperform H.264 for Smooth Streaming.

CineVision PSE

CineVision PSE is Sonic Solutions' high-end Blu-ray encoding product, developed with Microsoft. Beyond including a great VC-1 implementation, it also offers industry-leading dithering quality and a compressionist-tuned workflow for segment re-encoding. Many Blu-ray compression facilities prefer CineVision PSE and so encode VC-1.

Why Not VC-1?

Compression Efficiency Paramount

When both H.264 High Profile and VC-1 Advanced Profile decoding is available and decode complexity isn't a concern, the best H.264 encoders can outperform the best VC-1 encoders.

Licensing Costs

VC-1 is licensed by MPEG-LA, and has very similar terms—see H.264 page 227 for details. So, any concerns with H.264 fees would apply to VC-1 as well. The only significant difference is the current agreement is in force until 2012 instead of 2010.

What's Unique About VC-1?

VC-1 can be thought of as a continuation of the H.261, H.263, MPEG-1, MPEG-2, and the MPEG-4 part 2 codec philosophies, with the goal of adding the maximum improvements in coding efficiency without unduly increasing decoder complexity. To that end, the new techniques it adds not in older codecs are focused on improving compression efficiency in ways minimizing increases in decoder complexity. In general, VC-1 requires about twice as much CPU power per pixel per second as MPEG-2, while requiring half or fewer the bits to achieve the same quality level in many typical uses. The spec runs to hundreds of pages, so I'm just going to focus on a few of the more unique or user adjustable aspects of VC-1.

Adaptive block size

VC-1 was the first codec to introduce variable block size, and still has the deepest implementation of it. VC-1 uses the standard 8 × 8 DCT for intra-coded macroblocks, and adds 4 × 4, 4 × 8, and 8 × 4 blocks for motion vectors. By using smaller blocks on areas of rapid change, less ringing and blocking is produced than when only 8 × 8 blocks are used, and by using bigger blocks on smoother areas, more efficient encoding is delivered.

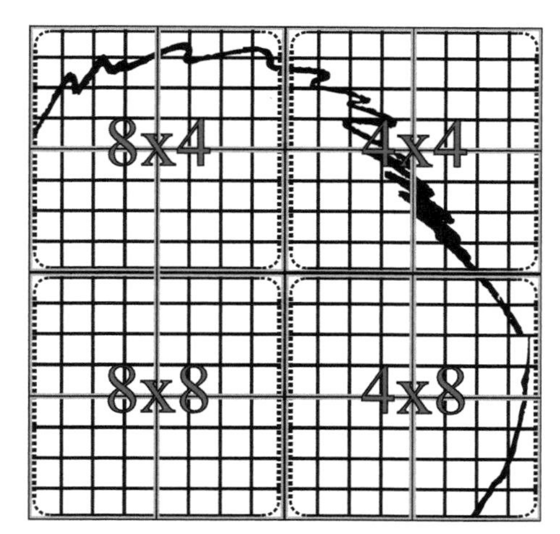

Figure 17.1 The variable motion vector block structure of VC - 1. A single macroblock can mix sections of four different block types.

16-bit integer iDCT

One big problem with MPEG-1/2 and MPEG-4 part 2 is their use of floating-point math in its DCT. Because different floating-point implementations can have slightly different results, different decoders could reconstruct a particular frame with slight differences. But that slight difference would get exaggerated each time a slightly off result was used as the reference for further frames resulting in noticeable image drift. Also, floating-point calculations are more expensive in silicon, and thus are ill-suited to implementation on very low-power devices, particularly in software.

Instead, VC-1 uses a 16-bit integer variant of DCT. While it serves the same function, there is a single correct result for each calculation, so no decoders have drift. It's also much easier to implement on processors without floating-point support.

Quarter-pixel motion estimation

Like MPEG-4 ASP, VC-1 added quarter-pixel motion estimation, to more accurately code subtle and slow motion between frames. The MPEG-4 implementation wasn't particularly efficient and is often not used, but the VC-1 implementation is improved, and is almost always used.

Overlap transform

The Overlap transform reduces blocking artifacts by smoothing the borders between adjoining intra coded blocks with high quantization. This offers some of the value of in-loop deblocking at very little computational cost (it's even available in Simple Profile). Overlap isn't used for motion-predicted blocks, and is automatically turned off at low QP where it's not needed. VC-1 is the only major interframe codec that supports Overlap transform; the other place it's seen is the experimental Dirac wavelet-based codec.

Overlap can soften the image slightly, and so isn't used in very high bits-per-pixel encodes like for Blu-ray.

In-Loop Deblocking Filter

VC-1's in-loop deblocking filter softens the border between two motion-estimated blocks proportional to their quantization. So it has little effect on lightly compressed content, but can help reduce blocking quite a bit for more compressed frames. It is called "in-loop" deblocking because future frames are predicted from the deblocked version, reducing the propagation of compression artifacts forward. It's more complex than Overlap, and is therefore off in Simple Profile. Overlap and In-Loop are complementary, and are used together in most Main and Advanced encodes.

One of the biggest differences between H.264 and VC-1 is in the strength of the in-loop deblocking filter. H.264's is stronger, and so it does a better job of reducing blocking, but is more expensive to process and can also smooth out detail more at moderate bitrates, particularly in Baseline profiles.

The same highly-compressed frame encoded with (A) In-loop deblocking and Overlap off, and (B) on. These deblocking features can for a healthy reduction in blocking artifacts.

Figure: In-Loop On/Off

Intensity compensation

Intensity compensation modify the overall brightness level of a frame, which makes compression of fades to or from black or white much more efficient than with past codecs.

Dynamic resolution change

VC-1 offers the ability to dynamically change frame size in the bitstream, so harder-to-encode scenes can be encoded at a smaller frame size to avoid strong artifacting. In Main Profile only 2x scaling is allowed, so a 640 × 480 stream could only throttle down to 320 × 240 if needed. In Advanced Profile, any legal resolution can be chosen per GOP; even pixel aspect ratio can be changed. This is a key feature in the new Smooth Streaming Encoding SDK.

Unfortunately, some hardware decoders using DXVA aren't capable of smoothly handling dynamic resolution changes midstream, leading to some flashing or other visual distortions. The feature shouldn't be used in generic WMV files, since they may be played on an incompatible system.

WMV and VC-1 FAQ

Is VC-1 The Same as Windows Media Video 9?

Yes: VC-1 is the SMPTE standardized version of Windows Media Video 9. The output of of a VC-1 encoder is bitstream compatible with a Windows Media Video 9 decoder that supports the profile, level, and features used.

Is VC-1 The Same Thing as WMV?

The VC-1 spec just defines the video bitstream; it doesn't cover the ASF file format (used in .wmv and .wma) or any WMA codec. VC-1 Advanced Profile can be stored in other file format wrappers, including MPEG-4 (standard or fragmented, the latter used for Smooth Streaming) and MPEG-2 transport stream (as used on Blu-ray or IPTV). Only VC-1 Advanced Profile has format independence; VC-1 Simple and Main profiles are only supported in WMV files.

Can I Use Expression Encoder to make Files That Play in Windows Media Player 9?

Yes, absolutely. VC-1 Main Profile is identical to the old Windows Media Video 9. A Windows Media file you make using VC-1 Simple and Main will work perfectly in WMP 9 out of the box. A stock XP Service Pack 2 machine that hasn't been updated in years would play it without issue.

Will Users Have to Install An Update If I Encode with VC-1?

No. Everyone with Windows Media Player 9 or higher can play VC-1 Simple and Main Profile already. Windows Media Player 11 can play Advanced Profile as well. Users of

WMP 9 or 10 will be prompted to install a VC-1 Advanced Profile decoder if they try to play a WMV file using that. All Vista and Windows 7 systems shipped with full VC-1 Advanced Profile support, so the only issue is XP users who haven't installed WMP 11 or the updated decoder.

Can I Encode for Blu-ray from Windows Media Encoder or Expression Encoder?

Since Windows Media Video 9 Advanced Profile is the same as VC-1 Advanced Profile, and VC-1 AP is supported on Blu-ray, one might think you could encode Blu-ray compatible stream in Windows Media Encoder. Alas, no. There are specific technical constraints for Blu-ray VC-1 that WME and the Windows Media Format SDK weren't designed to support. The VC-1 Encoder SDK was designed with optical discs in mind, and can produce Blu-ray-compatible VC-1 elementary stream with products that expose it. Expression Encoder 3 doesn't do elementary streams, but Rhozet's Carbon Coder 3.12 and higher do. The primary product used for making VC-1 on Blu-ray is CineVision PSE.

How do Windows Media Video 9 and VC-1 Profiles relate?

The details are in Table 17.1.

What was Complex Profile?

Windows Media Video 9 originally had three profiles: Simple, Main, and Complex. Simple and Main are the same as the VC-1 Simple and Main Profiles. Complex was deprecated, and replaced by Advanced Profile. As Complex isn't part of the VC-1 spec, it will normally only work on decoders from Microsoft, or based on Microsoft's porting kit implementations. Even if you have a product that can encode Complex Profile, don't use

Table 17.1 VC-1 Profiles and Windows Media Video 9.

VC-1 Profile	Windows Media codec	WME Decoder Complexity	Four Character Code (4CC)	Introduced with...	Downlevel to...
VC-1 Simple Profile	Windows Media Video 9	Simple	WMV3	WMP 9	WMP 6.4
VC-1 Main Profile	Windows Media Video 9	Main	WMV3	WMP 9	WMP 6.4
	Windows Media Video 9	Complex	WMV3	WMP 9	WMP 6.4
	Windows Media Video 9 Advanced Profile (original)		WMVA	WMP 10	WMP 9
VC-1 Advanced Profile	Windows Media Video 9 Advanced Profile (SMPTE compliant)		WVC1	WMP 11	WMP 9

it. Beyond the lesser compatibility, it hasn't been tuned in ages, and will offer lower-quality and lower decoder performance than modern Main and Advanced Profile implementations.

What's the Difference between WMVA and WVC1?

There were two flavors of Windows Media Video 9 Advanced Profile. The first, with a "WMVA" 4CC, was introduced with WMP 10 and used as the basis for VC-1 AP. But some changes were made in the final specification that kept it from being entirely compatible. An updated SMPTE VC-1 compliant version of Windows Media Video 9 Advanced Profile was introduced with WMP 11, using a "WVC1" 4CC was introduced with WMP 11, and that is what should be used.

Anyone wanting to make the real "WVC1" WMV 9 Advanced Profile needs to have Format SDK 11 installed (the easiest way is to install WMP 11), use a VC-1 Encoder SDK based product like Expression Encoder 2, or use another updated app like Flip4Mac or Episode. This is mainly an issue with running Windows Media Encoder on Windows Server 2003, which only supports up to Format SDK 9.5.

If it's the Same Codec, Why are There Different Windows Media Video 9 and VC-1 Encoders?

All that said, there is one last practical difference between WMV 9 and VC-1: encoders that call it VC-1 are normally based on the newer, faster, higher-quality, and more flexible VC-1 Encoder SDK. This isn't about a change in the codec standard, just the usual improvement of encoder implementations over time, like we saw with the Format SDK 9, 9,5, and 11 releases, and will be seeing again in future releases.

■

VC-1 Profiles

Main Profile

We start with Main Profile because it's the mainstream choice for WMV files. If you're looking for classic WMV playback on the web or desktop, it's really all you need. There are three main features available in Advanced missing from Main Profile (described in detail in the "Advanced Profile" section):

- Interlaced coding
- Transport independence
- Differential quantization on I-Frames (P and B DQuant are supported in Main)

Only I-frame DQuant matters at all for typical WMV use.

Most tools default to Main Profile over Simple Profile, as they should; 99 percent of "Windows Media Video 9" content is Main Profile.

Simple Profile

VC-1 Simple Profile is a pared-down verison that leaves out some computationally expensive features to decode. The significant features it leaves out from Main Profile are:

- In-loop deblocking filter
- B-frames
- Motion search range beyond 64 pixels horizontally and 32 pixels vertically
- Differential quantization
- Intensity compensation

Honestly, unless you're targeting mobile phones that only have software decoding, Simple Profile is probably not needed. There aren't any Simple Profile–only decoders in the wild, and the simplification of Simple Profile means higher bitrates are needed to achieve the same quality. A phone that can only do Main Profile at 200 Kbps may be able to do Simple at 300 Kbps, but Simple is unlikely to look that much better.

Advanced Profile

Advanced Profile (AP) VC-1 adds two major and one minor feature over Main Profile, as mentioned earlier.

Transport independence

What's turned out to be the most-used feature of AP is transport independence. Simple and Main Profile were created to live inside the ASF format, and therefore work natively only inside a WMV. There have been various hacks around that in special cases, but nothing that's been either satisfying or broadly supported.

With AP, the codec was designed to support arbitrary file types, and there are official VC-1 mapping standards for a variety of common file formats. The notable non-ASF formats include the following:

- VC-1 elementary stream. This is a .vc1 file that is just the binary video stream, equivalent to an .m2v file. This is what is produced by Blu-ray VC-1 compression tools and imported by authoring/mux tools.

- MPEG-2 Transport Stream, as used in digital broadcasting (IPTV, ATSC, DVB, etc.) and Blu-ray discs.
- MPEG-2 Program Stream, as was used in HD DVD.
- MPEG-4 Program, as used in typical .mp4 files.
- MPEG-4 Program in fragmented mode, as used in Smooth Streaming.

Interlaced coding

AP added a native interlaced mode for VC-1. Similar to the MPEG-2 implementation, each frame can be either progressive or interlaced, and each 16 × 16 block can be either progressive or interlaced.

While not widely used outside of Blu-ray, the VC-1 interlaced format offers good efficiency. For example, encoding 480i30 content as interlaced and then performing a bob deinterlace on playback to 480p60 can be done at about 30 percent lower bitrate with the same quality as bobbing source to 480p60 and encoding as that.

The Windows Media [Pipeline supports this bob deinterlacing, but on XP and Vista it's off by default for Windows Media Player and needs to be turned on with a registry key (Figure 17.2). Interlaced video simply doesn't play in Silverlight. Thus interlaced VC-1 is really only useful with Windows 7, custom applications, and hardware devices with interlaced support, like Blu-ray.

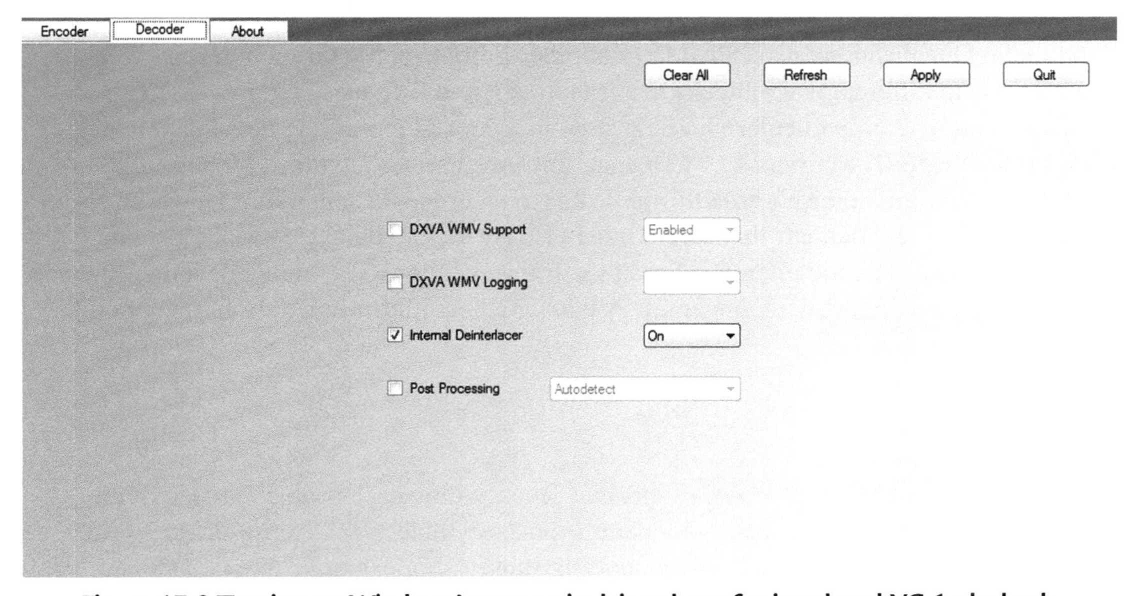

Figure 17.2 Turning on Windows' automatic deinterlacer for interlaced VC-1 playback.

Table 17.2 Profiles and Levels in VC-1.

Profile	Level	Maximum Bit Rate	Max Resolution/Frame Rate Combos
Simple	Low	96 Kbps	176×144p15
Simple	Medium	384 Kbps	240×176p30 352×288p15
Main	Low	2 Mbps	320×240p24
Main	Medium	10 Mbps	720×480p30 720×576p25
Main	High	20 Mbps	1920×1080p30
Advanced	L0	2 Mbps	352×288p30
Advanced	L1	10 Mbps	720×480p/i30 720×576p/i25
Advanced	L2	20 Mbps	720×480p60 1280×720p30
Advanced	L3	45 Mbps	1920×1080p24 1920×1080p/i30 1280×720p60
Advanced	L4	135 Mbps	1920×1080p60 2048×1536p24

Differential quantization on I-frames

One of the VC-1 developers cited this as his most regretted omission from Main Profile, simply because there wasn't any technical reason not to have it. Main Profile can apply Differential Quantization (also called DQuant; the ability to apply different levels of compression to different macroblocks in a frame) to P and B frames, but not I-frames. DQuant can be useful, particularly in reducing blocking and preserving detail in smoother areas of the image. The current VC-1 DQuant implementations are tuned for very high bitrates, so this omission isn't broadly applicable to web-rate content today. However, for scenes with smooth gradients that would otherwise get visibly blocky, DQuant just on I-frames can provide a good reference for all the frames of the GOP. And while it does increase the bitrate of the I-frame, with longer GOPs that won't significantly reduce bitrate of other frames.

Levels in VC-1

Like other standardized codecs, VC-1 specifies levels within each profile that determine maximum bitrate, frame rate, and other parameters (see Table 17.2). Actual devices rarely map to these profiles directly, instead supporting subsets. For example, the second- and third-generation Zune players support Main Profile at Medium Level resolutions, but with a lower maximum bitrate of 3 Mbps (more than sufficient for progressive standard def content).

Where VC-1 Is Used

Windows Media

VC-1, as Windows Media Video 9 or Windows Media Video 9 Advanced Profile, has been the primary video codec in Windows Media since the 9 Series launch in 2003. Main Profile dominates Windows Media use, mainly because the additional AP features aren't all that useful, nor are the performance improvements of SP meaningful to most scenarios.

Smooth Streaming

Smooth Streaming is a new media delivery technology for Silverlight. There are lots more details in Chapter 27, but from a content creation perspective, the critical fact is that Smooth Streaming can dynamically switch between different video streams at different data rates. Seamless switching is done by requesting a new chunk of video every few seconds, assembling video and audio streams from those, and then feeding a constant bitstream into the decoder. There are a few requirements to enable this seamless switching:

- Each chunk of video needs to start with a sequence header indicating frame size and frame rate.
- They need to start with a closed GOP, so there are no references outside the chunk.
- All streams must start each chunks on the same frame.

Beyond that, there's a huge amount of potential variability with chunks (whether in the same stream or different streams) being able to vary by the following:

- Frame size
- Frame rate (as integer multiples so the switching frames align)
- Data rate
- Aspect ratio

For products using the VC-1 Encoder SDK, Smooth Streaming compatibility will be delivered by following these settings:

- Closed GOP: On
- Adaptive GOP: Off

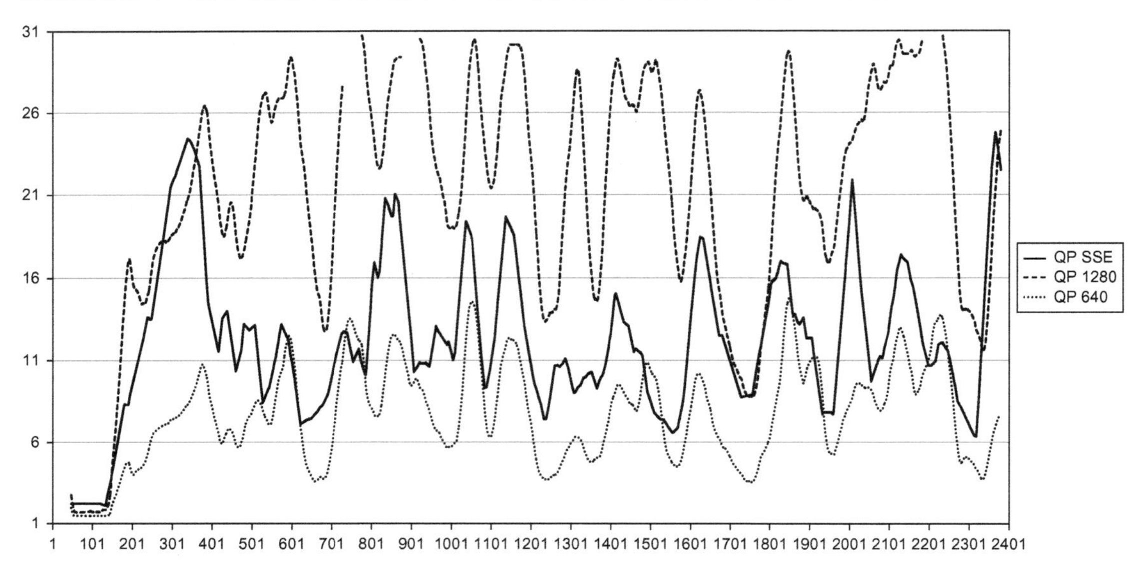

Figure 17.3 How the Smooth Streaming Encoder SDK keeps QP more consistent by varying frame size, enabling easy sections to match the detail of 1280x, and the hard sections to match the low artifacting of 640x.

- Insert Dropped Frames
- Elementary Stream with Sequence Header

These settings ensure that all streams will follow the fixed cadence of the GOP length, not resetting based on scene changes. Expression Encoder 2 Service Pack 1, the first Smooth Streaming encoder, used this mechanism.

The Smooth Streaming Encoder SDK (SSESDK) produces Smooth Streaming–compatible content with dramatically better quality than VC1ESDK (Figure 17.3). In particular, the SSESDK will:

- Vary the GOP length to align with edits, instead of using a fixed cadence (typically an average of every 2 seconds with a maximum of 4 seconds)
- Use 2-pass VBR for each chunk for much better rate control and consistent quality
- Dynamically adjust encoded frame size per chunk based on image complexity, reducing frame size as needed to minimize objectionable artifacts. This includes aspect ratio changes, compressing the image along the primary axis of motion where motion blur would reduce detail anyway

Variable Resolution: How It Works

One of the cool new features of Smooth Streaming is its abilty to vary the frame size used in a chunk based on the complexity of the video in that chunk. Variable Resolution takes advantage of Smooth Stream's ability to have independent chunks, with no references to frames outside of that chunk. Since they're independent, each can have its own sequence header, which defines the frame size, aspect ratio, frame rate, and other parameters. That was originally used to switch between streams at different data rates, but we later realized it could be used to switch parameters within a particular bitrate. Thus we could get out of normal encoding's "one size fits all" approach of picking a frame size that's neither too big for the hardest parts of the video nor too small for the easiest portion.

There are a lot of approaches that could be taken to determine what the resolution of a chunk should be, and I'm sure we and others will continue to tune it over time. But the basic concept is to keep quantization inside a given chunk from getting high enough for distracting artifacts, and instead picking the frame size that offers optimal quality.

Since only VC-1 Advanced Profile supports sequence headers, Main or Simple Profile output won't use this feature. Lacking transport independence, they aren't compatible with Smooth Streaming anyway.

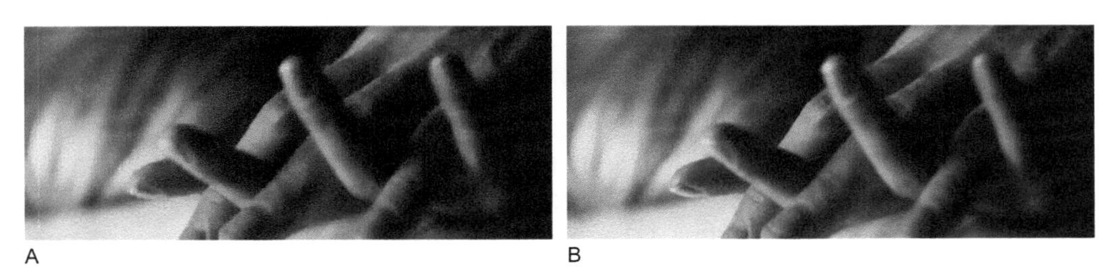

Figure 17.4 A frame from the encode used in Figure 17.3, showing how lower resolution can produce better quality by preventing artifacts. Figure 17.4A is the full 1280 × 528, while 17.4B was down - rezed to 704 × 304.

Blu-Ray

VC-1 is one of the mandatory codecs in Blu-ray, along with MPEG-2 and H.264, with all players required to support it. VC-1 dominated the rival HD DVD format; and several studios used the same VC-1 encode for both formats.

One theoretical advantage of VC-1 on Blu-ray is that its better decoder performance would make playback in software easier. PC-based Blu-ray playback hasn't caught on broadly, however, and most PCs that play Blu-ray are high-end models with GPUs capable of hardware acceleration of all three codecs.

The biggest advantage to VC-1 for Blu-ray has been the CineVision PSE encoding product available for it.

IPTV

VC-1 has been used in a number of IPTV deployments. There's long been a spec for mapping VC-1 to the industry standard MPEG-2 transport stream wrapper. However, since any IPTV set-top box with VC-1 also includes High Profile H.264 decoders, the decoder complexity advantage of VC-1 isn't meaningful. IPTV streams proper are largely converging on H.264 and the big market of broadcast-grade H.264 live encoders there.

VC-1 is more commonly used in the "over-the-top" content, delivered via cable broadband, but using more PC-like delivery technologies like Windows Media and soon Smooth Streaming.

Basic Settings for VC-1 Encoding

Most VC-1 implementations in the market are created by Microsoft or based on code licensed from Microsoft, so there's more commonality in settings than with MPEG-2 or H.264, where different vendors can provide wildly different encoder modes and presentations thereof. So I'm able to offer more specific advice for VC-1 best practices than for the other major codecs.

Complexity

Complexity often gets forgotten and left at the product's default, but it's quite important. Complexity controls the tradeoff between speed and quality of the encode—each value is roughly 2x slower than the one below it, but can provide better quality at a given data rate. By default, Windows Media Encoder and many Format SDK apps default to a complexity of 4 for offline encoding and 1 for live encoding, a reasonable default when it shipped. But today's dual-core laptops are many times more powerful than the biggest encoding box available back then. So a typical machine today should be able to do Complexity 3 for live 1 for 320 × 240 29.97 fps. And for offline, tuning of Complexity 3 in recent versions have given it nearly the same compression efficiency as 4, at about twice the speed.

As frame size and frame rate go up, complexity needs to come down on the same hardware. 720p30 at complexity 1 is quite doable on a quad-core Intel i7 system.

One key addition in the VC-1 Encoder SDK is Dynamic Complexity for live encoding, which automatically adjusts complexity in real-time to the highest the hardware can handle at any given moment, ensuring that quality is optimized without any frames dropped. This was introduced in Inlet's Spinnaker and is supported in professional-grade live encoders from

A

B

Figure 17.5 The motion search ranges of VC - 1. For inside out: 64 × 32, 128 × 64, 512 × 128, and 1024 × 256. 512 × 128 might sounds like a huge area, but it gets pretty small when comparing two 1080p frames separated by two B - frames.

ViewCast, Winnov, Digital Rapids, Envivio, and Telestream. It is also included in Expression Encoder.

Recommendation: 3 for general use, 4 for high-quality offline encodes. As high as doesn't drop frames for live encodes. See Figure 17.5.

Buffer Size

The buffer size is the duration over which the data rate is averaged. For Windows Media, this is described in bitrate over time. For example, encoding with a peak of 200 Kbps with a 2-second buffer means that any 2-second chunk out of video needs to be 400 Kbits or less. Conversely, using a 10-second buffer with the same 200 Kbps bitrate means any 10-second chunk needs to be 2000 Kbits or less. But within that 2000 Kbits, there's much more flexibility to reduce bitrate in easy parts and increase it in complex parts, delivering more consistent quality. The big drawback to large buffers is that they can increase the start-up and random access times. A 4-second buffer is a reasonable default for most content where latency isn't a concern. While WME doesn't expose it, buffer size can go down below 1 second in EE and some other tools for when very low latency is required.

Recommendation: 4–seconds if no other constraints, potentially lower for very low bitrates.

Keyframe Rate

Typically, your keyframe rate is going to be around your buffer size, since they both control latency and random access timing. Too frequent keyframes hurt compression efficiency, and can result in keyframe "pulsing" where there's a visible change or "jump" in the image at every keyframe. A keyframe every 4–10 seconds is typical for streaming, with higher data rates using more frequent keyframes. Blu-ray has a maximum GOP duration of 1 second (2 for 15 Mbps or lower peak bitrates), and Smooth Streaming uses 2 on average and 4 as a typical max.

Recommendation: 4 seconds if no other constraints.

Advanced Settings for VC-1 Encoding

The various VC-1 SDKs have many options available for fine tuning. Some offer quality/ speed tradeoffs beyond Complexity, others are only applicable for certain types of content and scenarios.

Setting these with the Format SDK requires setting registry keys, which is a pain, particularly for automation as each encode uses the state of the registry keys at the time its encode starts.

VC-1 Encoder SDK offers parameters via the API, making it easier to offer full control in a user interface.

The initial release of the Smooth Streaming Encoder SDK locks in most of the advanced settings, using the optimum choice for Smooth Streaming automatically.

GOP Settings

Number of B-frames

As discussed in Chapter 3, a B-frame is a bidirectional frame that can reference both the previous and next I- or P-frames. They are therefore more efficient to encode, and can be dropped on playback for scalability.

Turning on B-frames in VC-1 also allows flash compensation, where single-frame images (like those caused by strobe lights and camera flashes) get turned into intracoded B-frames (BI-frames; see Figure 17.6), functionally a form of I-frame (or keyframe) that will allow the following frame to reference the previous frame. Thus, a keyframe doesn't have to get inserted after every flash. This dramatically improves quality for those parts of video, without any drawbacks. To get flash compensation, you'll need to be using:

- 2-pass encoding, or 1-pass encoding with Lookahead on
- Number of B-frames at least 1

I recommend setting at least 1 B-frame to enable those features. 1 is also the optimum value for the vast majority of content (contrasting with MPEG-2, where two B-frames are typically best). Rarely animation content without any noise is better with two B-frames, and two B is often better for 50/60p, and I've used up to four B-frames for screen recordings.

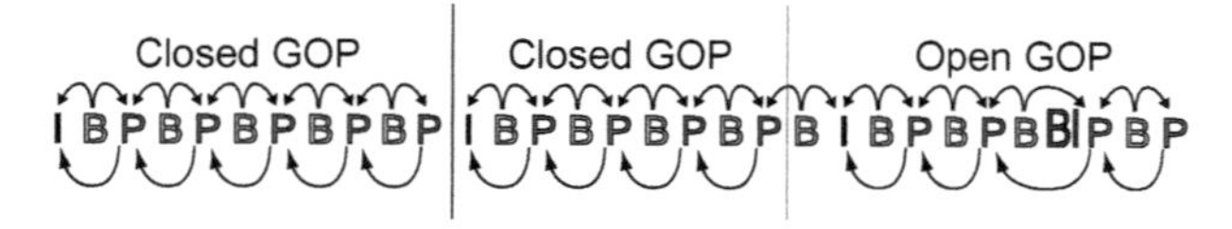

Figure 17.6 The GOP structure of VC-1 with a BI-frame.

Random access times are proportional to the number of P-frames per GOP, so using more B-frames means that I-frames can be further apart while preserving the same random access performance.

Recommendation: 1.

Lookahead

The Lookahead parameter tells the codec to analyze and buffer a specific number of frames before deciding how to process them. This enables the codec to peek into the future and optimize frame type accordingly. Specifically, it enables some important 2-pass encoding features in 1-pass mode:

- Improves I-frame detection and selection in 1-pass encoding, reducing chances of getting a bunch of I-frames or no I-frames with a funky transition
- Enables flash compensation by setting a flash frame to be a BI-frame
- Enables fade compensation in 1-pass encoding by turning off B-frames during fades

The full Lookahead of 16 is recommended unless very low latency is needed. Even for low-latency encodes, four frames of Lookahead can be quite useful.

Recommendation: On for 1-pass CBR or VBR, unless Lookahead Rate Control available.

Lookahead rate control

Lookahead Rate Control (LRC) is a mode in the VC1 Enterprise SDK that buffers ahead about half a second before allocating bitrate. This lets the codec provide much more consistent quality with content that varies a lot in complexity. It generally offers a quality boost beyond Lookahead, and should be used instead when available.

In the 2009 VC-1 implementations, LRC in 1-pass CBR typically produces better quality than 2-pass CBR (and is obviously faster to encode as well).

LRC also adds about a half-second of latency to the encode, limiting its utility for extremely low latency scenarios.

Recommendation: On if available for 1-pass CBR encoding.

LRC is turned on in Expression Encoder 3 if

- 1-pass CBR is being used
- GOP duration is less than 5 seconds
- Clossed GOP is On

Filter Settings

Noise reduction

Noise Reduction is just a simple filter to reduce video noise in the source. Ideally clean source footage will be used, but shooting in low light can get pretty noisy. Five levels of reduction are supported in FDSK—use the lowest one that makes the video clean enough; too high a level will make the video distractingly soft. VC1ESDK and thus EEv3 only has on/off for noise reduction, but it does a better job of preserving edges than the FSDK version. While upstream noise reduction can offer a lot more control and choice of mechanisms, the VC1ESDK version rarely degrades content, and is helpful when noise is pushing QP up to where visible artifacts are seen.

Recommendation: Off unless noisy source with no other noise reduction available.

Edge noise removal

The Edge Noise Removal filter attempts to blank out noise in the image typically seen around the edges of an analog video source. In particular, it'll blank out the Line 21 (subtitle) dashed lines at the top of the screen. Again, it's much better to have clean sources in the first place (a simple crop would be much better), but this feature is there in a pinch.

Recommendation: Only if enocoding a live source with no other way to crop.

Perceptual Options

In-Loop filter

The In-Loop deblocking filter should always be used for MP and AP, and isn't available in SP. While it does have a slight impact on decode performance (perhaps 5 percent on modern players), it helps efficiency more than that, so you can get better quality at the same performance by using a slightly lower bitrate with In-Loop on than a higher bitrate with In-Loop off.

Recommendation: On.

Overlap filter

The Overlap transform isn't as complex as the Loop filter, as it only applies to intra-coded blocks, and so is available in Simple Profile. However, it can sometimes reduce detail. Overlap is nearly always useful at web rates, but should be left off for encodes aiming for quality irrespective of bitrate. Overlap is automatically turned off below QP 4 regardless.

Recommendation: On.

Adaptive deadzone (ADZ)

This fearsome sounding feature allows the codec to sacrifice some detail in order to reduce artifacts. The Conservative mode is generally helpful for most content, and in particular can reduce blocking artifacts in shadows and other smooth areas. The Aggressive mode is theoretically useful to further reduce banding at challenging bitrates, but in practice almost always hurts the image more than it helps. Turning off ADZ can yield a more mathematically accurate compression, but rarely helps quality; as it's an adaptive mode, the times it reduces detail are times where artifacts would otherwise likely intrude.

Recommendation: Conservative.

B-Frame Delta QP

B-Frame Delta QP specifies how much more to compress B-frames versus reference frames. As B-frames are discarded after playback, increasing B-Frame QP saves bits for I- and P-frames.

This can improve quality of the B-frames, as the saved bits are spent to improve the quality of the reference frames B-frames are based on. Beyond a certain point, there can be a visible change in image quality between frames, yielding a distracting "strobing."

However, all current VC-1 implementations adaptively apply B-frame Delta QP from 0–2 based on the amount of motion in the frame. It's only in special cases (again, like animation or screen recordings) that a different value would be any better. And with low-motion content, B-frames are already quite small.

Recommendation: Default/unspecified.

DQuant

DQuant is a form of perceptual optimization that varies QP across a frame, giving smoother parts of the image lower QP than more complex parts. This can help video quality, especially where there might be visible blockiness in smoother areas like gradients. This is particularly an issue with dark content as viewed on LCD displays.

The drawback to DQuant is that the bits it applies to the smooth areas are taken from the other parts of the image, increasing their QP. While the FDSK version is tuned for lower bitrates, VC1ESDK's implementation was tuned for HD DVD and Blu-ray. While very effective at reducing blocking in flat areas, it uses so many bits doing so that the textured areas can look a lot worse. For 1080p24, it's generally only worthwhile at average bitrates around 15 Mbps or higher. There was a much better implementation of DQuant introduced in CineVision PSE 2.1; hopefully it will make it into future VC-1 implementations. For web rate, if it is useful at all, it's I-frame only for longer GOP content. That helps static parts of the image look good for the whole GOP referencing that I-frame, while only increasing bitrate on a single frame.

Table 17.3 DQuant Options.

Options
Off
Apply to I-frame only
Apply to I- and P-frames
Apply to I-, P-, and B-frames

Table 17.4 DQuant Method Options.

Options	Description
Off	No DQuant
Regular	Standard mode, most useful for lower-moderate bitrates
Fixed QP - Smooth	QP 2 for smooth areas
Fixed QP - Very Smooth	QP 1 for smooth areas (only mode in VC1ESDK)
Fixed QP - Dark	QP1 for dark areas
Regular - Dark	Regular DQuant applied only to dark areas
Adaptive	Newer mode only in CineVision PSE to date

The default is Off (Table 17.3). Even in FSDK it's best on be I- and P-frame only. This is a good compromise for much content, since the B-frames can be based on the DQuant'ed frames, but more bits don't get spent on non-reference frames that are immediately discarded.

DQuant method

The FSDK DQuant has a variety of modes (Table 17.4). If you're using DQuant, most of the time you're just fine using Regular mode. The Fixed QP modes are very bit-hungry, and not really suitable for typical Internet use. The Dark modes apply DQuant only to low luma levels, like shadow detail. The VC-1 Encoder SDK just has a simple on/off.

Recommendation: Off or I-frame only for VC-1 Encoder SDK, I + P Regular for FSDK Advanced Profile.

Motion Estimation Settings

Motion search level

Motion Search Level (Table 17.5) controls whether and how color is included with motion search. Most previous codecs have done motion search purely in the luma (black-and-white) channel, just assuming that the chroma (color) channels follow along. Motion Search Level activates a separate chroma search, which can provide big improvements in quality and efficiency with highly colorful content, like motion graphics. The True modes apply chroma search to the whole frame; the Adaptive modes use it on the 50 percent of the macroblocks estimated most likely to benefit.

Table 17.5 Motion Search Level Values.

Usage
Luma only
Luma with nearest-integer chroma
Luma with true chroma
MB-adaptive with true chroma
MB-adaptive with nearest integer chroma

Table 17.6 Motion Search Range Values.

Usage
+63.75/−64.0 H, +31.75/−32.0 V
+127.75/−128.0 H, +63.75/−64.0 V
+511.75/−512.0 H, +127.75/−128.0 V
+1023.75/−1024.0 H, +255.75/−256.0 V
Macroblock-adaptive

For absolute best quality, the True Chroma option delivers, but encoding can be up to 50 percent slower than the default luma-only mode. For decent speed and improved quality, Adaptive True delivers most of the value of chroma search, with only half the performance hit. If there isn't enough horsepower for even that, adaptive-integer chroma can still help with a more modest performance hit yet.

Recommendation: Luma Only for speed, Luma with True Chroma for quality, particularly with animation or motion graphics.

Motion Search Range

Motion Search Range (Table 17.6) controls how big an area the codec looks in for motion vectors in its reference frame(s). A larger search window will help find faster motion, but has a big impact on performance—CPU requirements roughly double each level. This is not available in Simple Profile, which always uses the lowest 64/32 range.

For FSDK live encoding, your best bet is to pick the highest value you can encode without dropping frames. For on-demand, or for when using an encoder with Dynamic Complexity, the Macroblock Adaptive mode is generally the best—it'll pick the most efficient mode dynamically. When coupled with Dynamic Complexity, complexity may go down as motion search goes up, making for a coarser motion search over a broader range for high-motion scenes, but reverting to a more precise search for lower-motion scenes. This can provide a big boost in the quality of high-motion frames.

Recommendation: Adaptive.

Table 17.7 Motion Match Method.

Usage
SAD
Adaptive
Hadamard

Table 17.8 VideoType Values.

Usage
Progressive frame
Interlace frame
Interlace field
Auto Interlace frame/field detection
Auto Progressive/Interlace frame/Interlace field detection

Motion Match Method

Motion Match Method (Table 17.8) chooses between the fast Sum of Absolute Differences and the better but slower Hadamard methods to measure the accuracy of motion matches. Hadamard is a pretty expensive feature, so set this only once you've already got at least Complexity 4 and Chroma Search.

The Adaptive mode uses Hadamard for the 50 percent of macroblocks with the most motion, and so is between SAD and Hadamard in performance. Interestingly, slightly outperforms Hadamard in quality more often than not.

Recommendation: SAD for speed, Adaptive for quality.

VideoType

The VideoType parameter (Table 17.9) is applicable only when doing interlaced encoding with Advanced Profile. The default of Field interlaced does well for most frames. Some are more efficiently encoded as interlaced frame, but that varies from scene to scene, so Auto Field/Frame is a good default although somewhat slower. If the content is likely to contain truly progressive frames, Auto Field/Frame/Progressive is the most efficient mode yet, albeit the slowest.

Recommendation: Interlaced Field for speed, Auto Frame/Field for general use, Auto Field/Frame/Progressive for best quality or if content contains many truly progressive frames.

Number of Threads

This parameter controls the number of threads that are used in a particular instance of the VC-1 encoder. The maximum number of threads was 4 in FSDK 11 and the initial version of the VC1ESDK, and 8 in EEv3, SSE SDK, and Windows 7. In the current implementation, the frame is sliced into as many horizontal bands as the number of threads, and motion vectors don't cross between those bands. So small frames with a lot of vertical motion lose a bit of efficiency as the number of threads goes up. For high quality encoding, I recommend using at least 64 pixels high per slice, thus:

- < 128 pixels high: 1 thread
- 128–255 pixels high: 2 threads
- 256–511 pixels high: 4 threads
- 512+ pixels high: 8 threads

Options per Profile

Not all of the codec features are available in each profile. Table 17.9 lists what is or isn't available with each. Using a registry key to set a mode that isn't supported in that profile can yield a noncompliant stream and so should be avoided. For example, Setting NumBFrames to anything other than 0 shouldn't be done when encoding to Simple Profile.

Table 17.9 Options per Profile.

Setting	SP	MP	AP
DenoiseOption	Yes	Yes	Yes
Dquant Option	Yes	Yes	Yes
Dquant Strength	Yes	Yes	Yes
Force B Frame Delta QP	No	Yes	Yes
Force Encoding Height / Force Encoding Width	No	No	Yes
Force LoopFilter	No	Yes	Yes
Force NoiseEdgeRemoval	Yes	Yes	Yes
Force NumThreads	Yes	Yes	Yes
Force Overlap	Yes	Yes	Yes
Force Video Scaling	Yes	Yes	Yes
Lookahead	Yes	Yes	Yes
Macroblock Mode Cost Method	Yes	Yes	Yes
Motion Match Method	Yes	Yes	Yes
Motion Search Level	Yes	Yes	Yes
Motion Search Range	No	Yes	Yes
Motion Vector Coding Method	No	No	Yes
Motion Vector Cost Method	Yes	Yes	Yes
NumBFrames	No	Yes	Yes
Adaptive Deadzone	Yes	Yes	Yes
VideoType	No	No	Yes

Registry keys – a guide and an apology

Yes, registry keys can be fearsome and a pain, but what's the best way to go about them?

1. Your best bet is to use a product that uses the VC1ESDK and hence offers all controls integrated into the application, so registry keys aren't needed. This includes EEv3 and most enterprise-grade WMV compression products. But, if not, see option 2.
2. Regedit.exe has been built into Windows forever, and allows registry keys to be input. But it's also quite terrifying, particularly when a negative number is needed.
3. If you must use registry keys, download Alex Zambelli's WMV9 PowerToy from http://www.alexzambelli.com/wmv (Figure 17.7). This app offers a convenient way to set registry keys, with integrated tooltip help. It also includes useful decoder options and diagnostics.

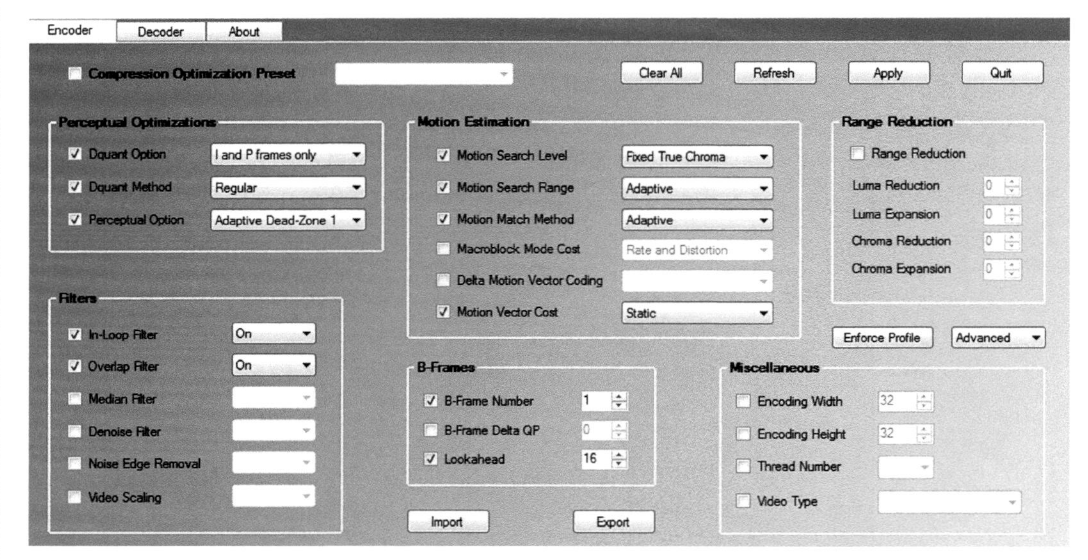

Figure 17.7 WMV9 PowerToy's encode settings.

4. If you want to batch-encode using different registry key values, Alex also has an updated WMECmd.vbs script that can batch encode using the Format SDK, with full control over registry keys.

Now that you've gotten this far, option 1 is probably looking better than ever.

Encoding Mode Recommendations

"Always-On" choices

This is a good set of defaults with no significant performance downside that almost always helps quality.

Perceptional option: Adaptive Deadzone

This maps to the "Adaptive Deadzone: Conservative" option from Expression Encoder. This lets the codec reduce detail before introducing artifacts, and generally improves quality at lower bitrates.

In-loop filter: On

This turns on the In-Loop deblocking filter, which softens the edges of block artifacts. This improves the current frame, and also future frames based on it.

Overlap filter: On

This filter further smoothes the edges of blocks. It can reduce detail a little at high bitrates, but is almost always helpful at typical web bitrates.

B-frame number: 1

Turns on B-frames, and hence enables flash/fade detection when using Lookahead or 2-pass encoding, and also improves compression efficiency.

Lookahead: 16

Tells the codec to buffer ahead 16 frames in 1-pass (CBR or VBR) encoding, letting the codec detect flash frames and fades and switch the frame type based on it. Always on in Expression Encoder. It will increase end-to-end latency by that many frames in live encoding, but is generally well worth it due to quality improvements.

Complexity: 3

Complexity 3 is the sweet spot; higher values are a lot slower and don't look that much better, while lower values get a significant quality degradation. Note the WME/FSDK default is 4, not 3, and it's controlled by the app, not by registry key.

Graphic:
WMV9 PowerToy and EEv3 settings for Always On choices

High-quality live settings

Assuming you have a machine fast enough to run these settings in at least Complexity 3, they will improve the live experience.

Motion Search Range: Adaptive

This tells the encoder to switch to a bigger motion search range for frames with high motion, and then go back to a smaller range when motion dies down. This dramatically improves quality with higher motion at bigger frame sizes, while keeping performance high when there is less motion. The default range is 64 pixels left/right and 32 pixels up/down, so if any objects move more then that between any two P-frames, then this feature will help.

Lookahead Rate Control: On

LRC offers a great quality boost. It's only available in VC-1 Encoder Enterprise SDK–based products.

Dynamic Complexity: On

Dynamic Complexity lets your system use all its horsepower without dropping frames. Also Enterprise Only.

High-quality offline

These settings offer maximum quality for offline encoding, and are slower yet. Use them when you've got the time.

Complexity: 4

A little better, and quite a bit slower than Complexity 3.

Motion Search Level: Fixed True Chroma

This is a full-precision motion search for chroma. It never hurts, and can help quality a lot with motion graphics and animation.

Motion Match Method: Adaptive

This switches between the Sum of Absolute Differences (SAD) and the Hadamard method to compare motion between frames as appropriate for each macroblock. Full Hadamard can be higher quality for some very complex content, but the Adaptive mode is faster and produces better results most of the time.

Insane offline

This is for those cases where you don't want to worry that it's not a smidge worse than it could possibly be, and don't care if it takes 10x longer to encode than something essentially as good. If nothing else, no one will complain you were lazy! Most of the time this is just a placebo with a lot of waste heat.

Complexity: 5 (VC1ESDK) or 4 (FSDK)

Complexity 5 is slower and ever so slightly better than 4 in VC1ESDK. Complexity 5 disables the specified chroma search and motion match modes in FSDK and so generally produces worse quality than a tuned Complexity 4.

Threads: 1

In FSDK and VC1ESDK, motion vectors don't cross slice boundaries. So for content with a lot of vertical motion (which is rare; we largely live on the horizontal), you can get a slight efficiency improvement for that Handycam footage of the International Pogo Stick Finals.

Tools for VC-1

This section focuses on tools with specialized depth for VC-1, all of which are based on the VC-1 Encoder SDK. Windows Media–only tools are also covered in the previous chapter.

Expression Encoder 3

Expression Encoder 3 is the only free encoder with relatively complete support for VC-1 Encoder SDK functions. It's a bit of a hybrid, combining features from both Windows 7 (8-way threading, speedup, "swirling" fix) and the VC-1 Encoder SDK.

EEv3 also includes the first public release of the Smooth Streaming Encoder SDK VC-1 implementation, which is used in the "Smooth Streaming VBR" mode.

EEv3 doesn't expose all modes from the VC-1 Encoder SDK (Figure 17.8). Specifically, it doesn't:

- Support interlaced VC-1
- Create elementary streams

Inlet Fathom

Inlet's Fathom was the first professional-grade VC-1 encoder, originally introduced in 2005 as a front-end to a hardware accelerated capture and compression board. Its current 3.7 version mainly is used as a software encoder, with either live or file-based input, including live streaming to WMS. Its standout VC-1 feature is segment reencoding; it'll allow tweaking of compression parameters for particular sections of a video.

Fathom has the deepest implementation of VC-1 Encoder SDK, with the SSE SDK coming soon as well. See Figure 17.9.

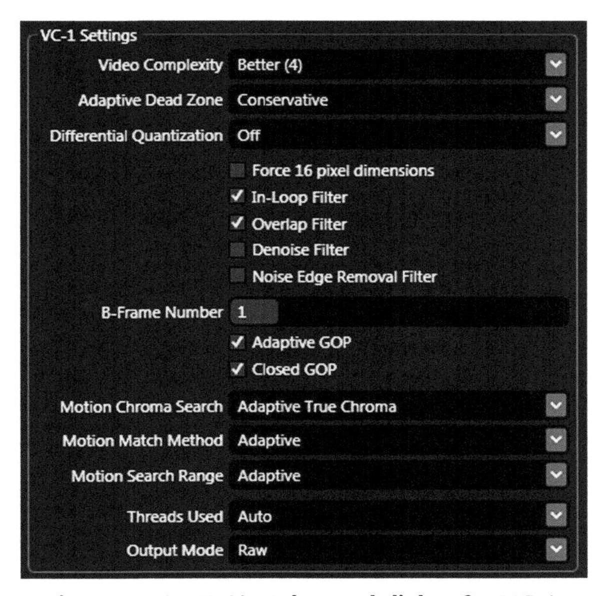

Figure 17.8 EEv3's Advanced dialog for VC-1.

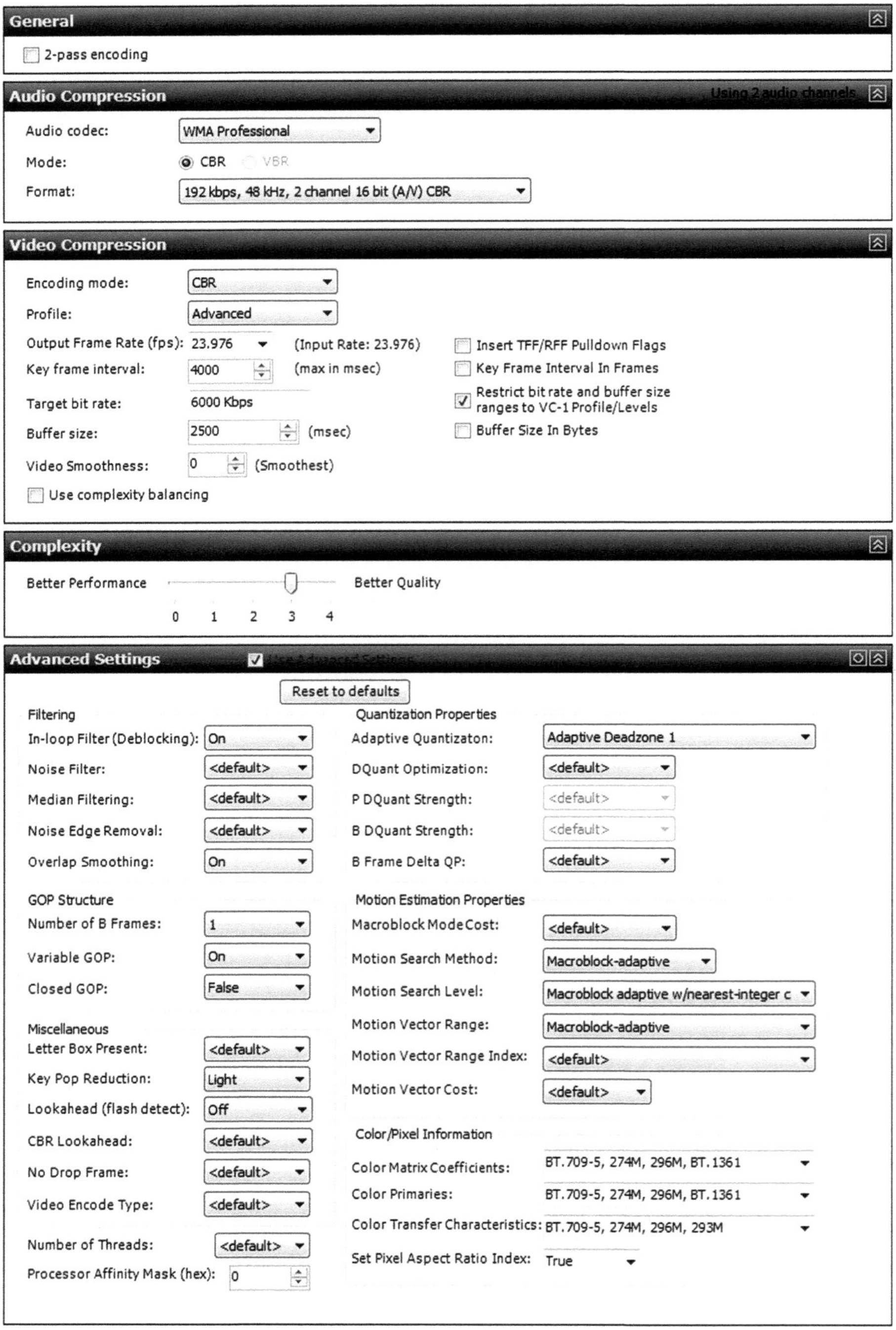

General
2-pass encoding

Audio Compression
Using 2 audio channels
Audio codec: WMA Professional
Mode: CBR VBR
Format: 192 kbps, 48 kHz, 2 channel 16 bit (A/V) CBR

Video Compression
Encoding mode: CBR
Profile: Advanced
Output Frame Rate (fps): 23.976 (Input Rate: 23.976)
Key frame interval: 4000 (max in msec)
Target bit rate: 6000 Kbps
Buffer size: 2500 (msec)
Video Smoothness: 0 (Smoothest)
Use complexity balancing
Insert TFF/RFF Pulldown Flags
Key Frame Interval In Frames
Restrict bit rate and buffer size ranges to VC-1 Profile/Levels
Buffer Size In Bytes

Complexity
Better Performance 0 1 2 3 4 Better Quality

Advanced Settings
Reset to defaults

Filtering
In-loop Filter (Deblocking): On
Noise Filter: <default>
Median Filtering: <default>
Noise Edge Removal: <default>
Overlap Smoothing: On

GOP Structure
Number of B Frames: 1
Variable GOP: On
Closed GOP: False

Miscellaneous
Letter Box Present: <default>
Key Pop Reduction: Light
Lookahead (flash detect): Off
CBR Lookahead: <default>
No Drop Frame: <default>
Video Encode Type: <default>
Number of Threads: <default>
Processor Affinity Mask (hex): 0

Quantization Properties
Adaptive Quantizaton: Adaptive Deadzone 1
DQuant Optimization: <default>
P DQuant Strength: <default>
B DQuant Strength: <default>
B Frame Delta QP: <default>

Motion Estimation Properties
Macroblock Mode Cost: <default>
Motion Search Method: Macroblock-adaptive
Motion Search Level: Macroblock adaptive w/nearest-integer c
Motion Vector Range: Macroblock-adaptive
Motion Vector Range Index: <default>
Motion Vector Cost: <default>

Color/Pixel Information
Color Matrix Coefficients: BT.709-5, 274M, 296M, BT.1361
Color Primaries: BT.709-5, 274M, 296M, BT.1361
Color Transfer Characteristics: BT.709-5, 274M, 296M, 293M
Set Pixel Aspect Ratio Index: True

Figure 17.9 Fathom's main VC - 1/WMV pane, with a potentially overwhelming array of options.

Rhozet Carbon

Carbon offers both Format SDK ("Windows Media File") and VC-1 Encoder SDK ("VC-1 Exporter")–based encoding modes. Carbon's pioneering implicit preprocessing mode makes it a good choice to do many-to-many transcoding operations, with multiple targets rendering simultaneously:

- Carbon supports interlaced VC-1 encoding, not available in EEv3.
- Elementary stream output and byte-based buffer control allows for Blu-ray-compatible .vc1 exports.
- The VC-1 Exporter supports 1-pass CBR audio only, alas.
- Rhozet has announced Smooth Streaming Encoder SDK support will be available in 2009, likely before this book arrived in your hands.

CineVision PSE

Last but not least, we have the grand-daddy of VC-1, Sonic's CineVision PSE (Figure 17.10). This is a very high-end commercial Blu-ray encoder born out of Microsoft work on HD DVD. It has been responsible for around 90 percent of HD DVD titles and about a quarter of Blu-ray titles to date. It includes some great features that haven't been adopted by other VC-1s yet:

- Grid encoding: the encode can be automatically split over multiple machines for fast turnaround.
- Studio-grade Segment re-encode: individual scenes, shots, or frames can be re-encoded with tweaked settings.
- Region of interest encoding: change parameters for specific areas of the frame.
- Adaptive DQuant: a Differential Quantization mode that works better at lower bits per pixel.

VC-1 Tutorial

This project is publishing a HD movie trailer to the web for a trailer site, playable in Silverlight, WMP, and a variety of other media players.

Scenario

We're on the postproduction team of the marketing department of a Hollywood studio. We're creating a destination site for one of our summer blockbusters, and want to provide a high-end HD experience to get the fans juiced.

(*Continued*)

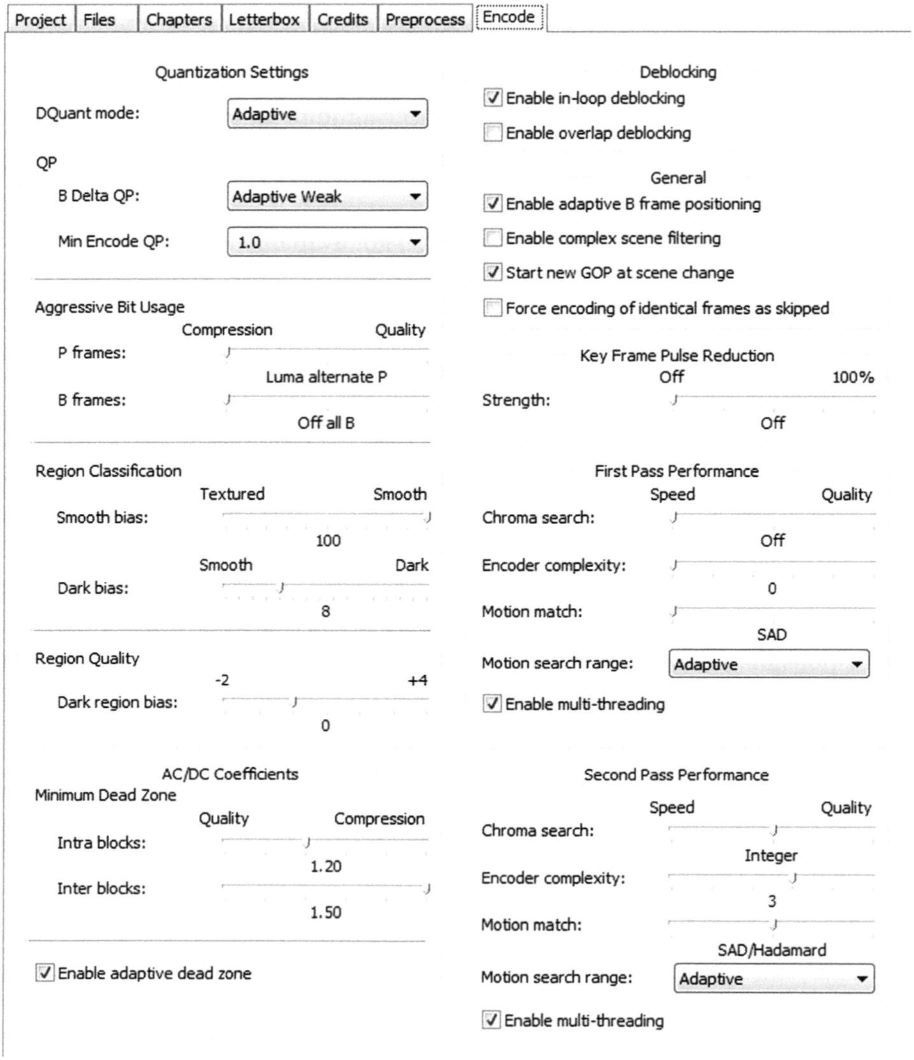

Project | Files | Chapters | Letterbox | Credits | Preprocess | Encode

Quantization Settings
DQuant mode: Adaptive
QP
B Delta QP: Adaptive Weak
Min Encode QP: 1.0

Aggressive Bit Usage
Compression Quality
P frames: Luma alternate P
B frames: Off all B

Region Classification
Textured Smooth
Smooth bias: 100
Smooth Dark
Dark bias: 8

Region Quality
-2 +4
Dark region bias: 0

AC/DC Coefficients
Minimum Dead Zone
Quality Compression
Intra blocks: 1.20
Inter blocks: 1.50

Enable adaptive dead zone

Deblocking
Enable in-loop deblocking
Enable overlap deblocking

General
Enable adaptive B frame positioning
Enable complex scene filtering
Start new GOP at scene change
Force encoding of identical frames as skipped

Key Frame Pulse Reduction
Off 100%
Strength: Off

First Pass Performance
Speed Quality
Chroma search: Off
Encoder complexity: 0
Motion match: SAD
Motion search range: Adaptive
Enable multi-threading

Second Pass Performance
Speed Quality
Chroma search: Integer
Encoder complexity: 3
Motion match: SAD/Hadamard
Motion search range: Adaptive
Enable multi-threading

Figure 17.10 CineVision PSE's VC - 1 dialog, with options not yet seen in any other tool.

The web site uses a Silverlight web player, and we want to make the same files also available for download for local playback, as well as through media extenders like the Xbox 360 and PlayStation 3.

Following typical convention, we've been asked for 480p, 720p, and 1080p versions so that users of slow computers and connections can also get a good experience. The decision has been made not to use Smooth Streaming, given the short duration of the content and high-quality goals; for marketing purposes, we'd rather have viewers buffer then see low-bitrate video.

(*Continued*)

Three Questions

What Is My Content?

Our source file is a 1920 × 1080 AVI file using the 10-bit 4:2:2 Cineform HD codec. The film itself is a letterboxed 2.39:1. The audio mix had some last-minute changes, and we have an updated 5.1-channel WAV file.

The content was short on film, and has nearly constant, fast cutting, and a fair amount of grain. It looks like a hard encode.

Who Is My Audience?

The audience is going to be summer movie fans, typically younger males, but with crossover appeal to young women.

We also know that film critics and bloggers are going to review the trailer, so we want to make sure they have a great, thrilling experience.

We're counting on them to be patient for a little buffering; they can pick the 480p clip if they're in a hurry. As movie buffs, they'll be familiar with being given three bitrate choices. They include a disproportionate number of Mac users (about 25%), so we want to make sure our Flip4Mac playback experience is good as well.

What Are My Communication Goals?

To provide an awesome HD experience for potential fans of the movie. It's more important for quality to be great than to have fast startup time, but we also don't want to use any more bits than we need to to hit our quality targets.

Tech Specs

We know we want to play back in:

- Silverlight
- Windows Media Player 9 or higher
- Flip4Mac
- Xbox 360
- PlayStation 3
- Zune HD

All of these support at least VC-1 Main Profile (most do Advanced) and WMA 9 Pro (but not the WMA 10 Pro LBR modes covered in the next chapter).

(*Continued*)

Because the source is letterboxed, HD "1080p" really means "1920 wide." So, given our 2.39:1 aspect ratio, and rounding to Mod 16, we get:

- "1080p" = 1920/2.39 = 803 = 1920 × 800 = 1536000 pixels/frame
- "720p" = 1280/2.39 = 536 = 1280 × 528 = 675840 pixels/frame
- "480p" = 640/2.39 = 268 = 640 × 272 = 174080 pixels/frame

We're not exactly sure what bitrates we should be using yet. We'll start by trying a few encodes at the top bitrate, and then based on the "power of 0.75 rule" extrapolate from that to the lower bitrates. We'll keep our maximum bitrate for 1080p at 8 Mbps, though, based on published recommendations for Silverlight playback.

We'll encode with 5.1 audio in all cases, using 192 Mbps at our lowest band, 256 Mbps at the medium band, and 384 Mbps at the high band. This version of Carbon didn't support 2-pass VBR audio with VC-1 exports, but that will be supported in Carbon 3.14 by the time you're reading this. Since audio peaks aren't hard to decode, I would have just used 2-pass VBR Unconstrained with the same bitrates.

Settings

We'll do this with Rhozet Carbon Coder, a popular compression tool in postproduction facilities.

First, we load our source video as a source, and then use the "Audio Stream Selection" to attach our updated audio file to the video.

Going into Video Filter, we use a Relative Crop to specify the active video region. Movie trailers are often a little bit off, and can have a little black on the left and right, but this trailer is a perfect 140 crop top/bottom to make 1080–140 × 2 = 800 lines.

The audio sounded a bit quiet, so we add a Normalize to Mean RMS filter, which will raise the volume while keeping dynamic range accurate.

Carbon lets us apply multiple targets to the same source, making for more efficient encoding. We'll start with a basic VC-1 target for 1080p, and add some quality-over-bitrate tuning choices. The default settings are already very tuned for quality over speed, but we'll set:

- Width and Height to 1920 × 800
- Frame rate to 23.976
- Bitrate Mode to 2-pass constrained VBR
- Profile: Main
- Video Bitrate at 6000 (the max we'll try)
- Closed GOP Off (slight efficiency improvement)
- Maximum Bitrate at 8000

(*Continued*)

- GOP Size 96 frames (4 seconds)
- VBV Buffer Size: 4000000 (8000 Kbps = 1000000 Bytes/sec × 4 sec = 4000000)
- Overlap Filter On (reduces artifacts)
- Audio Codec: WMA 10 Pro: 6 Channel 48 KHz 24-bit 384 Kbps

And we'll then duplicate that with additional average bitrates at 5000, 4000, and 3000 Kbps and encode.

Based on that test encode, the 5000 Kbps version looks like our winner; it's obviously of higher quality than the 4000 Kbps, but almost as good as the 6000. When content varies, sometimes it's just trial and error.

Using the power of 0.75 rule and our pixels/frame counts, we can estimate:

- 720p: (675840/1536000)^0.75 = 0.54 × 5000 = 2700 Kbps
- 480p: (174080 /1536000)^0.75 = 0.20 × 5000 = 1000 Kbps

And thus we can take tune in final settings.

1080p:

- Raise Complexity to 5 (this is our summer blockbuster, after all)

720p:

- Width and Height to 1280 × 528
- Video Bitrate: 2700
- Maximum Bitrate: 4000
- VBV Buffer: 2000000
- Audio Bitrate: 256 Kbps

480p:

- Width and Height to 640 × 272
- Video Bitrate: 1000
- Maximum Bitrate: 1500
- VBV Buffer: 750000
- Audio Bitrate: 192 Kbps 44.1 KHz

And there we go (Figure 17.11). We hit Convert, and Carbon outputs our three final files in high quality. We chose the slowest, highest-quality modes so that even on an 8-core machine, it might take an hour or two to output.

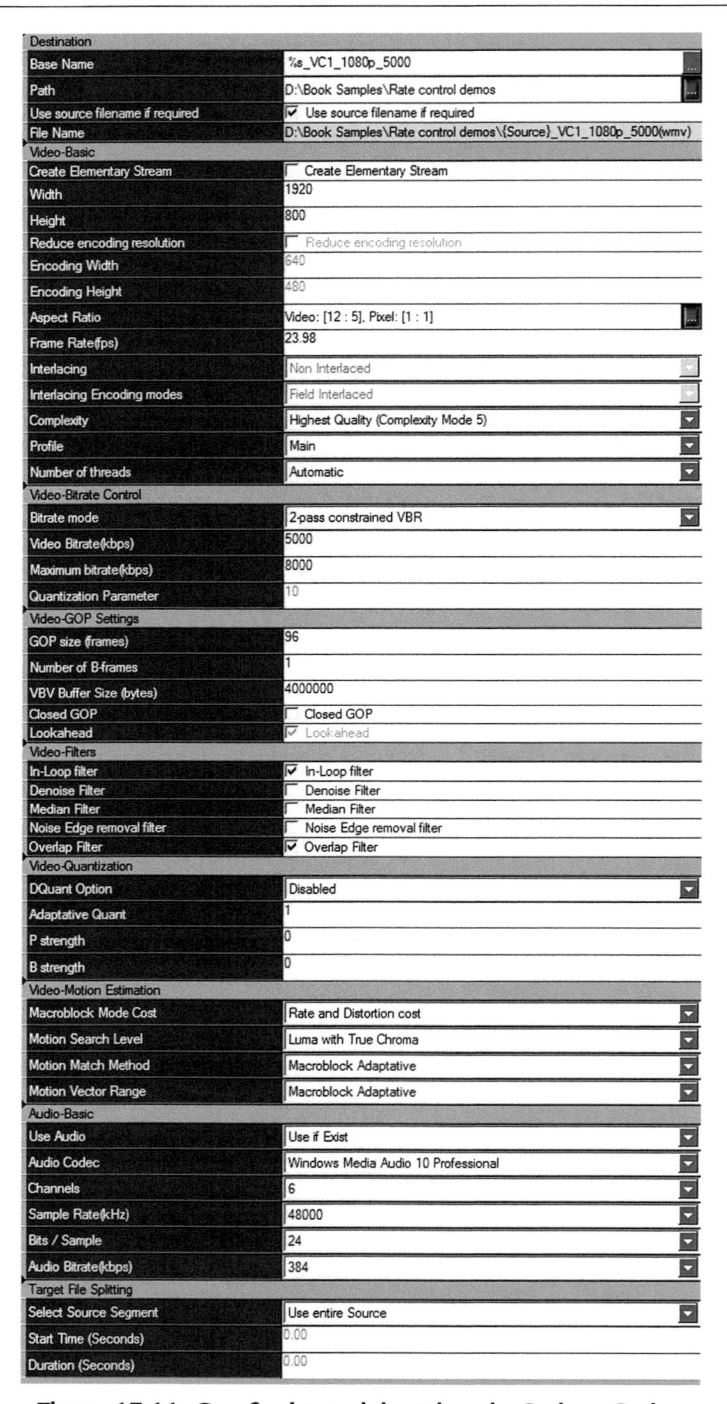

Destination	
Base Name	%s_VC1_1080p_5000
Path	D:\Book Samples\Rate control demos
Use source filename if required	☑ Use source filename if required
File Name	D:\Book Samples\Rate control demos\{Source}_VC1_1080p_5000(wmv)
Video-Basic	
Create Elementary Stream	☐ Create Elementary Stream
Width	1920
Height	800
Reduce encoding resolution	☐ Reduce encoding resolution
Encoding Width	640
Encoding Height	480
Aspect Ratio	Video: [12 : 5], Pixel: [1 : 1]
Frame Rate(fps)	23.98
Interlacing	Non Interlaced
Interlacing Encoding modes	Field Interlaced
Complexity	Highest Quality (Complexity Mode 5)
Profile	Main
Number of threads	Automatic
Video-Bitrate Control	
Bitrate mode	2-pass constrained VBR
Video Bitrate(kbps)	5000
Maximum bitrate(kbps)	8000
Quantization Parameter	10
Video-GOP Settings	
GOP size (frames)	96
Number of B-frames	1
VBV Buffer Size (bytes)	4000000
Closed GOP	☐ Closed GOP
Lookahead	☑ Lookahead
Video-Filters	
In-Loop filter	☑ In-Loop filter
Denoise Filter	☐ Denoise Filter
Median Filter	☐ Median Filter
Noise Edge removal filter	☐ Noise Edge removal filter
Overlap Filter	☑ Overlap Filter
Video-Quantization	
DQuant Option	Disabled
Adaptative Quant	1
P strength	0
B strength	0
Video-Motion Estimation	
Macroblock Mode Cost	Rate and Distortion cost
Motion Search Level	Luma with True Chroma
Motion Match Method	Macroblock Adaptative
Motion Vector Range	Macroblock Adaptative
Audio-Basic	
Use Audio	Use if Exist
Audio Codec	Windows Media Audio 10 Professional
Channels	6
Sample Rate(kHz)	48000
Bits / Sample	24
Audio Bitrate(kbps)	384
Target File Splitting	
Select Source Segment	Use entire Source
Start Time (Seconds)	0.00
Duration (Seconds)	0.00

Figure 17.11 Our final tutorial settings in Carbon Coder.

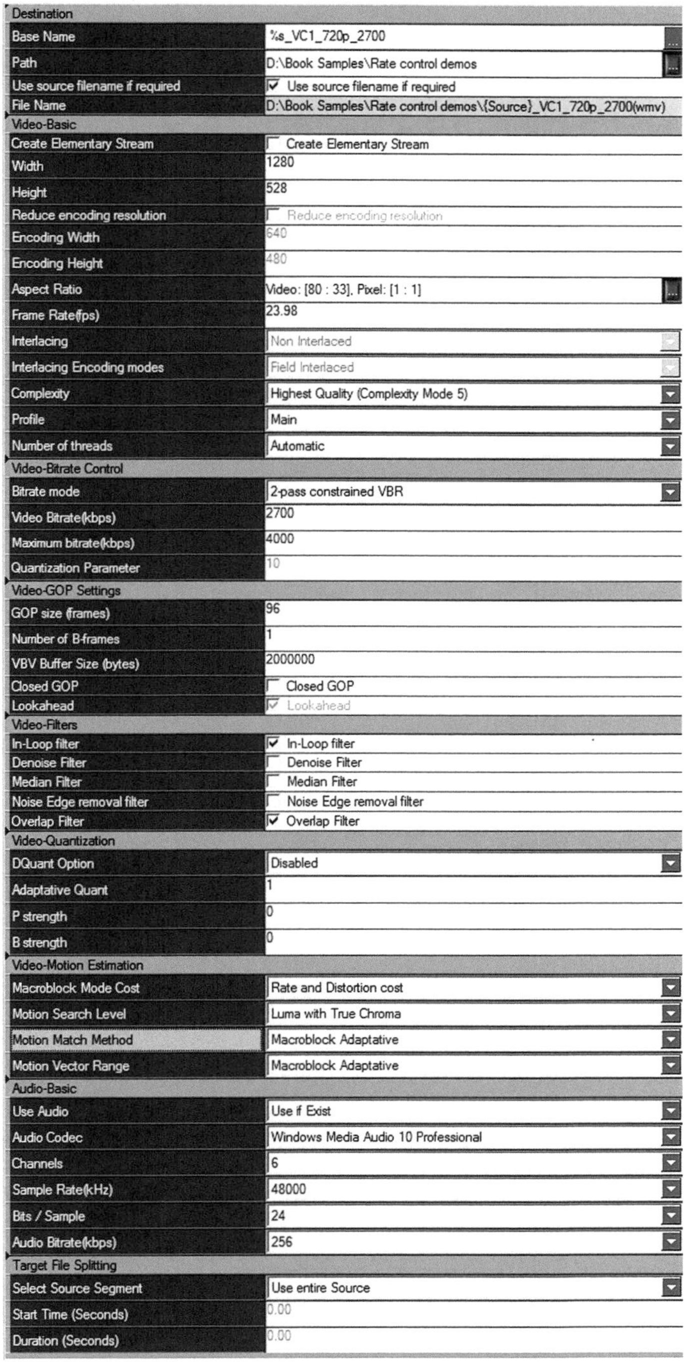

Destination	
Base Name	%s_VC1_720p_2700
Path	D:\Book Samples\Rate control demos
Use source filename if required	☑ Use source filename if required
File Name	D:\Book Samples\Rate control demos\{Source}_VC1_720p_2700(wmv)
Video-Basic	
Create Elementary Stream	☐ Create Elementary Stream
Width	1280
Height	528
Reduce encoding resolution	☐ Reduce encoding resolution
Encoding Width	640
Encoding Height	480
Aspect Ratio	Video: [80 : 33], Pixel: [1 : 1]
Frame Rate(fps)	23.98
Interlacing	Non Interlaced
Interlacing Encoding modes	Field Interlaced
Complexity	Highest Quality (Complexity Mode 5)
Profile	Main
Number of threads	Automatic
Video-Bitrate Control	
Bitrate mode	2-pass constrained VBR
Video Bitrate(kbps)	2700
Maximum bitrate(kbps)	4000
Quantization Parameter	10
Video-GOP Settings	
GOP size (frames)	96
Number of B-frames	1
VBV Buffer Size (bytes)	2000000
Closed GOP	☐ Closed GOP
Lookahead	☑ Lookahead
Video-Filters	
In-Loop filter	☑ In-Loop filter
Denoise Filter	☐ Denoise Filter
Median Filter	☐ Median Filter
Noise Edge removal filter	☐ Noise Edge removal filter
Overlap Filter	☑ Overlap Filter
Video-Quantization	
DQuant Option	Disabled
Adaptative Quant	1
P strength	0
B strength	0
Video-Motion Estimation	
Macroblock Mode Cost	Rate and Distortion cost
Motion Search Level	Luma with True Chroma
Motion Match Method	Macroblock Adaptative
Motion Vector Range	Macroblock Adaptative
Audio-Basic	
Use Audio	Use if Exist
Audio Codec	Windows Media Audio 10 Professional
Channels	6
Sample Rate(kHz)	48000
Bits / Sample	24
Audio Bitrate(kbps)	256
Target File Splitting	
Select Source Segment	Use entire Source
Start Time (Seconds)	0.00
Duration (Seconds)	0.00

Figure 17.11 (Continued)

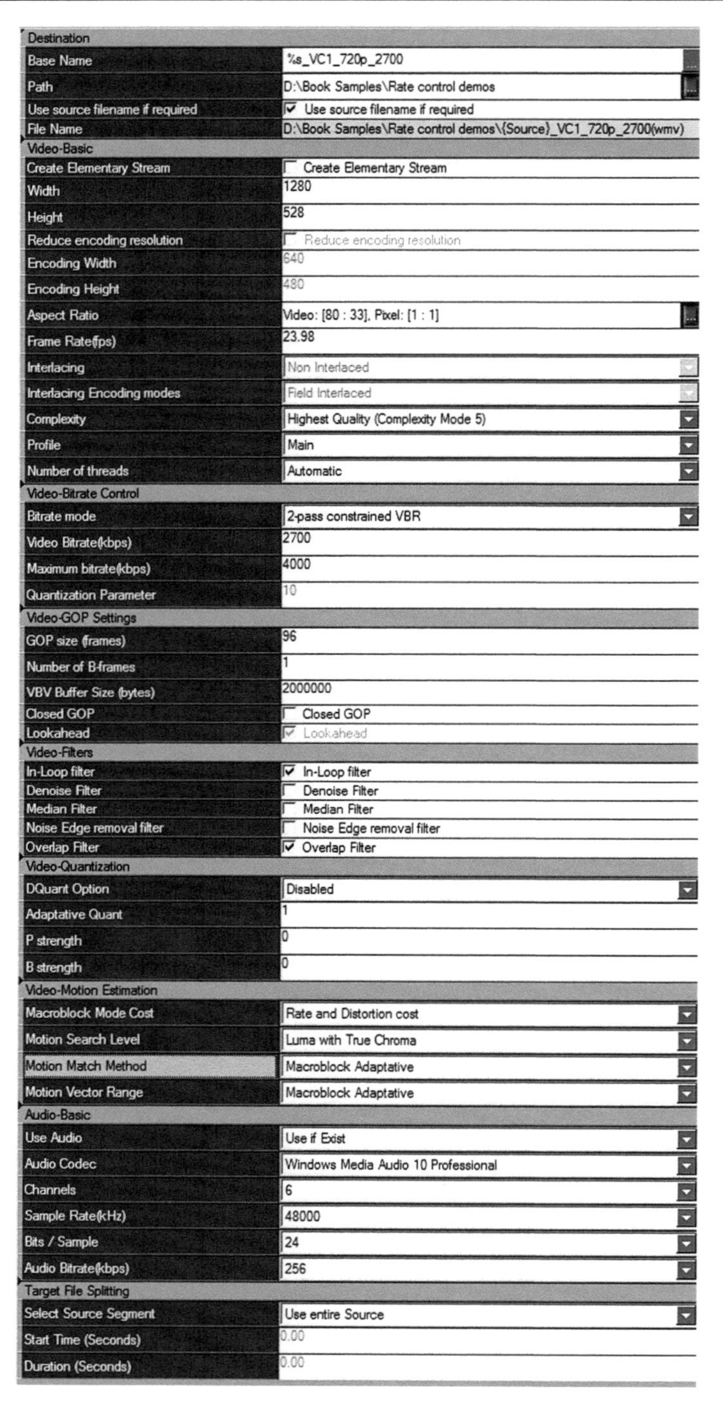

Destination	
Base Name	%s_VC1_720p_2700
Path	D:\Book Samples\Rate control demos
Use source filename if required	☑ Use source filename if required
File Name	D:\Book Samples\Rate control demos\{Source}_VC1_720p_2700(wmv)
Video-Basic	
Create Elementary Stream	☐ Create Elementary Stream
Width	1280
Height	528
Reduce encoding resolution	☐ Reduce encoding resolution
Encoding Width	640
Encoding Height	480
Aspect Ratio	Video: [80 : 33], Pixel: [1 : 1]
Frame Rate(fps)	23.98
Interlacing	Non Interlaced
Interlacing Encoding modes	Field Interlaced
Complexity	Highest Quality (Complexity Mode 5)
Profile	Main
Number of threads	Automatic
Video-Bitrate Control	
Bitrate mode	2-pass constrained VBR
Video Bitrate(kbps)	2700
Maximum bitrate(kbps)	4000
Quantization Parameter	10
Video-GOP Settings	
GOP size (frames)	96
Number of B-frames	1
VBV Buffer Size (bytes)	2000000
Closed GOP	☐ Closed GOP
Lookahead	☑ Lookahead
Video-Filters	
In-Loop filter	☑ In-Loop filter
Denoise Filter	☐ Denoise Filter
Median Filter	☐ Median Filter
Noise Edge removal filter	☐ Noise Edge removal filter
Overlap Filter	☑ Overlap Filter
Video-Quantization	
DQuant Option	Disabled
Adaptative Quant	1
P strength	0
B strength	0
Video-Motion Estimation	
Macroblock Mode Cost	Rate and Distortion cost
Motion Search Level	Luma with True Chroma
Motion Match Method	Macroblock Adaptative
Motion Vector Range	Macroblock Adaptative
Audio-Basic	
Use Audio	Use if Exist
Audio Codec	Windows Media Audio 10 Professional
Channels	6
Sample Rate(kHz)	48000
Bits / Sample	24
Audio Bitrate(kbps)	256
Target File Splitting	
Select Source Segment	Use entire Source
Start Time (Seconds)	0.00
Duration (Seconds)	0.00

Figure 17.11 (Continued)

CHAPTER 18

Windows Media Audio

WMA File Format

Windows Media Audio can refer to both a file format and a family of codecs. This chapter is mainly about the codecs. The .wma file format is just good old ASF, as described in Chapter 17; a WMA is just a WMV without video.

Rate Control in Windows Media Audio Codecs

One feature of the Windows Media Audio family rarely seen in other codecs is support for 2-pass encoding; most other audio codecs are 1-pass CBR or VBR. While 2-pass CBR rarely makes for a practical improvement over 1-pass CBR, 2-pass VBR can be a big improvement, particularly for soundtracks where audio complexity varies a lot. Progressive download content should use the VBR modes. However, VBR modes aren't always available at every bitrate CBR is. Even then, VBR can outperform a CBR at a higher bitrate—a VBR file with an average of 64 Kbps and a peak of 96 Kbps can sound better than a CBR at a fixed 80 Kbps data rate.

Similarly to MP3 and AAC, 1-pass quality-limited VBR is a good choice for personal music libraries where the size of any given file isn't as important as the average size and guaranteed quality.

Windows Media Audio 9.2 "Standard"

Windows Media Audio is a venerable codec format, compatible back to WMP from the mid-1990s (predating .wma files!), and can be played back on a very wide variety of hardware and software players. In this book, I refer to it as WMA Standard to distinguish it from the Voice, Pro, and Lossless versions of WMA. WMA is from the post-MP3 generation of audio codecs, and is generally capable of producing better quality than MP3 at 128 Kbps and below, although it's not competitive with the low bitrate performance of frequency synthesis codecs like High Efficiency AAC and WMA 10 Pro.

While the bitstream of WMA was locked down years ago, the encoder has seen many generations of tweaking to improve quality and performance. The current version is WMA 9.2 and includes minor but welcome performance and quality enhancements compared to previous versions. It also added some new bitrate modes, like 48 Kbps 44.1 kHz mono.

WMA is the safe codec choice for any Windows Media file. It's flexible, offers high quality with sufficient bitrate, and is playable by anything that can play a .wma file. However, it's dramatically outperformed by WMA 10 Pro at 96 Kbps and below (see Table 18.1).

Windows Media Audio 9 Voice

WMA 9 Voice is designed for low-bitrate applications below 32 Kbps, in which WMA and WMA Pro don't perform as well. Despite its name, it does a credible job with music and other nonvoice content. You're not going to dance to WMA Voice at low bitrates, but it can intelligibly compress music interludes and sound effects in otherwise voice-centric content (see Table 18.2).

WMA Voice 9 is supported in WMP 9 and higher. Note that Voice is not currently supported in all non-Windows players. Most notably, it isn't supported in Silverlight through at least Silverlight 3. WMA 10 Pro 32 Kbps is a good alternative for Silverlight + WMP11+ compatibility. For Silverlight + WM9+ compatibility, the not-that-great low bitrate WMA 9.2 modes are the only option; 48 Kbps mono VBR is about the lowest that voice content sounds great in to my ears.

Voice is a 1-pass CBR only codec, unlike WMA Standard and Pro. WMA Voice replaced the now-deprecated ACELP.net audio codec.

Windows Media Audio 10 Pro (LBR)

While the new video codec features introduced with Format SDK 11 were exciting, the biggest technical leap was Windows Media Audio 10 Professional (WMA Pro 10), which offers up to twice the compression efficiency of WMA 9. The original WMA Pro 9 was launched in 2003 as part of Windows Media 9 Series, and offers great audio quality and efficiency at 128 Kbps and up. WMA Pro 9 supports up to 7.1 channels, up to 24-bit sampling, and frequencies up to 96 kHz. But the high minimum bitrate (128 Kbps) took 9 Pro out of the running for lower-bitrate web video. Thus, most streaming video projects kept using good old WMA 9.

WMP 11 added a new frequency interpolation mode to WMA Pro, and incremented the name to WMA 10 Pro. With the new mode, a "baseband" version of the audio is encoded as normal WMA Pro 9 at half the selected frequency, with additional data that tells how the higher frequencies are synthesized. This gives a stream that's backward-compatible with older decoders, but provides enhanced quality with a newer decoder.

Table18.1 Windows Media Audio 9.2 Bitrate-Constrained Modes.

Argument	Bitrate	Sampling rate	Channel	Modes
320_44_2	320	44	2	CBR
256_44_2	256	44	2	CBR
192_48_2	192	48	2	CBR, VBR
192_44_2	192	44	2	CBR, VBR
191_48_2	192	48	2	Low Delay
160_48_2	160	48	2	CBR
160_44_2	160	44	2	CBR, VBR
128_48_2	128	48	2	CBR, VBR
128_44_2	128	44	2	CBR, VBR
127_48_2	128	48	2	Low Delay
96_48_2	96	48	2	CBR, VBR
96_44_2	96	44	2	CBR, VBR
95_48_2	96	48	2	Low Delay
80_44_2	80	44	2	CBR
64_48_2	64	48	2	CBR, VBR
64_44_2	64	44	2	CBR, VBR
63_48_2	64	48	2	Low Delay
48_44_2	48	44	2	CBR
48_32_2	48	32	2	CBR
48_44_1	48	44	1	CBR, VBR
40_32_2	40	32	2	CBR
32_44_2	32	44	2	CBR
32_32_2	32	32	2	CBR
32_22_2	32	22	2	CBR
32_44_1	32	44	1	CBR
24_32_2	24	32	2	CBR
22_22_2	22	22	2	CBR
20_22_2	20	22	2	CBR
20_16_2	20	16	2	CBR
20_44_1	20	44	1	CBR
20_32_1	20	32	1	CBR
20_22_1	20	22	1	CBR
16_16_2	16	16	2	CBR
16_22_1	16	22	1	CBR
16_16_1	16	16	1	CBR
12_8_2	12	8	2	CBR
12_16_1	12	16	1	CBR
10_16_1	10	16	1	CBR
10_11_1	10	11	1	CBR
8_11_1	8	11	1	CBR
8_8_1	8	8	1	CBR
6_8_1	6	8	1	CBR
5_8_1	5	8	1	CBR
0_8_1	0	8	1	CBR

Table18.2 Windows Media Audio 9.2 Quality VBR modes.

Argument	Typical bitrate	Sampling rate	Channel
Q98_44_2	300–480	44	2
Q98_48_2	300–480	48	2
Q90_44_2	200–310	44	2
Q90_48_2	200–310	48	2
Q75_44_2	110–180	44	2
Q50_44_2	70–115	44	2
Q25_44_2	85–145	44	2
Q10_44_2	60–95	44	2
	45–75		

Table 18.3 Windows Media Audio Voice 9 Modes.

Argument	Bitrate	Sampling rate	Channel
20_22_1	20	22	1
16_16_1	16	16	1
12_16_1	12	16	1
10_11_1	10	11	1
8_8_1	8	8	1
5_8_1	5	8	1
4_8_1	4	8	1

WMA Pro 10 provides up to two times the efficiency of WMA 9.2, so at 64 Kbps it can provide similar quality to WMA 9.2 at up to 128 Kbps (see Table 18.3). However, if only the old decoder is used, you only get the lower-quality baseband audio, which suffers from the lower maximum sampling rate. It'll be intelligible, but not an entertainment-grade experience.

WMA 10 Pro is appropriate to use once the majority of your customers are using WMP 11+, Silverlight 2+, Windows Mobile, Flip4Mac or another player with full support. Because it's an enhancement of the older codec, WMA 10 Pro won't trigger a codec download—users get the new codec only if they install a player that supports it. And if you're using WMA 10 Pro at 128 Kbps or above, including all the multichannel modes, you're really still making a WMA 9 Pro bitstream, and will retain full compatibility and fidelity with WMP 9 and other decoders without LBR.

So even if targeting WMP 9, use WMA 10 Pro if you can dedicate at least 128 Kbps to audio, as it will outperform WMA 9.2 at 128 Kbps and up.

The superior efficiency of WMA 10 Pro has made it the default codec for WMV used with Silverlight, and for VC-1 Smooth Streaming files.

That 64 Kbps Comparison Test

There's an oft-cited listening test comparing advanced audio codecs at 64 Kbps run by audio compression enthusiast and expert Sebastian Mares a few years back. It found that WMA 10 Pro was good, but was outperformed by HE AAC. However, the test compared 1-pass CBR encoded WMA 10 Pro to fixed quantization encodes of the other codecs, allowing up to a 10-percent bitrate variance! Needless to say, bit-for-bit VBR beats CBR significantly, and WMA 10 Pro would have been much more competitive had it been tested in Quality VBR like the others. What's more frustrating for me is that WMA 10 Pro is unusual in that it has a real 2-pass VBR mode—it was the only codec at the time that could give a bitrate-accurate VBR encode.

I only mention it as the test is referenced so often; I mean no disrespect to Mr. Mares, who has done a great service in his well-designed double-blind tests over the years. But not using the same rate control modes keeps the WMA 10 Pro results from being valid.

- The Hydrogen Audio test: http://www.listening-tests.info/mf-64-1/results.htm
- An apples-to-apples 1-pass CBR comparison between WMA 10 Pro and HE AAC: http://www.microsoft.com/windows/windowsmedia/forpros/codecs/comparison.aspx

Table 18.4 Windows Media Audio 10 Pro Modes.

Argument	Bitrate	Sample rate	Channels	BitsPerSample	Modes	LBR?
768_96_8_24	768	96	8	24	CBR	No
768_48_8_24	768	48	8	24	CBR	No
768_96_6_24	768	96	6	24	CBR, VBR	No
768_48_6_24	768	48	6	24	CBR, VBR	No
768_44_6_24	768	44	6	24	CBR, VBR	No
640_96_6_24	640	96	6	24	CBR, VBR	No
640_48_6_24	640	48	6	24	CBR, VBR	No
640_44_6_24	640	44	6	24	CBR, VBR	No
440_96_6_24	440	96	6	24	CBR, VBR	No
440_48_6_24	440	48	6	24	CBR, VBR	No
440_44_6_24	440	44	6	24	CBR	No
440_96_2_24	440	96	2	24	CBR, VBR	No
440_88_2_24	440	88	2	24	CBR, VBR	No
440_48_2_24	440	48	2	24	CBR, VBR	No
440_44_2_24	440	44	2	24	CBR, VBR	No
440_48_6_16	440	48	6	16	CBR, VBR	No
440_44_6_16	440	44	6	16	CBR, VBR	No
384_48_8_24	384	48	8	24	CBR	No
384_96_6_24	384	96	6	24	CBR, VBR	No
384_48_6_24	384	48	6	24	CBR, VBR	No
384_44_6_24	384	44	6	24	CBR	No
384_96_2_24	384	96	2	24	CBR, VBR	No
384_88_2_24	384	88	2	24	CBR, VBR	No
384_48_2_24	384	48	2	24	CBR, VBR	No
384_44_2_24	384	44	2	24	CBR, VBR	No
384_48_6_16	384	48	6	16	CBR	No
384_44_6_16	384	44	6	16	CBR	No
383_48_6_24	383	48	6	24	CBR, VBR	No
256_96_6_24	256	96	6	24	CBR, VBR	No

(*Continued*)

Table 18.4 (Continued)

Argument	Bitrate	Sample rate	Channels	BitsPerSample	Modes	LBR?
256_48_6_24	256	48	6	24	CBR, VBR	No
256_44_6_24	256	44	6	24	CBR	No
256_96_2_24	256	96	2	24	CBR, VBR	No
256_88_2_24	256	88	2	24	CBR, VBR	No
256_48_2_24	256	48	2	24	CBR, VBR	No
256_44_2_24	256	44	2	24	CBR, VBR	No
256_48_6_16	256	48	6	16	CBR	No
256_44_6_16	256	44	6	16	CBR, VBR	No
256_48_2_16	256	48	2	16	CBR, VBR	No
256_44_2_16	256	44	2	16	CBR, VBR	No
255_48_6_24	255	48	6	24	CBR	No
192_96_6_24	192	96	6	24	CBR, VBR	No
192_48_6_24	192	48	6	24	CBR, VBR	No
192_44_6_24	192	44	6	24	CBR	No
192_96_2_24	192	96	2	24	CBR, VBR	No
192_88_2_24	192	88	2	24	CBR, VBR	No
192_48_2_24	192	48	2	24	CBR, VBR	No
192_44_2_24	192	44	2	24	CBR, VBR	No
192_48_6_16	192	48	6	16	CBR	No
192_44_6_16	192	44	6	16	CBR, VBR	No
192_48_2_16	192	48	2	16	CBR	No
192_44_2_16	192	44	2	16	CBR	No
191_48_6_24	191	48	6	24	Low Delay	No
191_48_2_24	191	48	2	24	Low Delay	No
160_48_8_16	160	48	8	16	CBR	No
160_48_6_16	160	48	6	16	CBR	No
160_48_2_16	160	48	2	16	CBR	No
160_44_2_16	160	44	2	16	CBR	No
128_96_6_24	128	96	6	24	CBR, VBR	No
128_48_6_24	128	48	6	24	CBR, VBR	No
128_44_6_24	128	44	6	24	CBR	No
128_96_2_24	128	96	2	24	CBR, VBR	No
128_88_2_24	128	88	2	24	CBR, VBR	No
128_48_2_24	128	48	2	24	CBR, VBR	No
128_44_2_24	128	44	2	24	CBR, VBR	No
128_48_8_16	128	48	8	16	CBR	No
128_48_6_16	128	48	6	16	CBR	No
128_44_6_16	128	44	6	16	CBR, VBR	No
128_48_2_16	128	48	2	16	CBR	No
128_44_2_16	128	44	2	16	CBR	No
127_48_2_24	127	48	2	24	Low Delay	No
96_48_2_16	96	48	2	16	CBR	Yes
96_44_2_16	96	44	2	16	CBR	Yes
80_48_2_16	80	48	2	16	CBR	Yes
80_44_2_16	80	44	2	16	CBR	Yes
64_48_2_16	64	48	2	16	CBR	Yes
64_44_2_16	64	44	2	16	CBR, VBR	Yes
48_48_2_16	48	48	2	16	CBR	Yes
48_44_2_16	48	44	2	16	CBR, VBR	Yes
32_32_2_16	32	32	2	16	CBR	Yes

Windows Media Audio 9.2 Lossless

The WMA 9.2 Lossless codec is, as the name implies, a lossless audio codec. A lossless audio codec's output is bit-for-bit identical to its input. Essentially, it's a more efficient alternative to PCM (uncompressed) encoding, and functionally equivalent to zipping up a .wav file. The flip side of lossless encoding is that there's no bitrate control possible—each second of audio takes as many or as few bits as it needs. Hence the codec is only available in Quality VBR mode. Perfect silence takes up very little bandwidth, while white noise takes up as much as uncompressed. Typical savings are around 2:1 for music and 4:1 for multichannel TV/movie soundtracks.

In general, WMA Lossless shouldn't be used for WMV files for distribution, given the high and unpredictable data rate. WMA 9 and 10 Pro can provide incredible-sounding audio at much lower bitrates, and transparent compression at still lower bitrates than WMA Lossless. Honestly, the only difference a user would get between Lossless and a high bitrate WMA Pro is the placebo effect of *knowing* it is lossless.

Legacy Windows Media Audio Codecs

ACELP.net

The WMA codecs have been standard since the 9 series launch, but there are some older codecs you many find in legacy content.

ACELP.net (Algebraic Code-Excited Linear Prediction), like the other low-bitrate voice codecs, provides intelligible speech at low bitrates. Microsoft licensed it from the now-defunct Spiro Labs. There's a lot of old content inside corporate media libraries that use ACELP.net, but the encoder components haven't been available for the last few version of the Format SDK, nor does Windows ship with a decoder anymore.

VoxWare MetaSound and MetaVoice

These were the pre-WMA/ACELP audio codecs for Windows Media in the mid-1990s. As the name implies, they were general-purpose, low-bitrate speech codecs, respectively. While decent for their era, they were not competitive with WMA and ACELP in quality, and took substantially more CPU power to decode. It has been many years since either shipped with Windows.

CHAPTER 19

Ogg

The Xiph.Org Foundation's Ogg format and codecs are an attempt to make competitive public-domain media technologies that are unencumbered by patents and therefore broadly usable in open source products. The project started with audio and has added video in recent years.

The Ogg codecs haven't become popular enough to be included in that many mainstream products so far, but the lack of any licensing requirements means free software updates are broadly available to add Ogg support. That said, Xiph's assertions as to the patent-free nature of their codecs haven't been proved, nor is that kind of proof really possible. They've had no trouble so far.

Why Ogg?

Avoid Licensing Costs

The essential point to the Ogg formats is to be free to implement and play by everyone.

Preference for a "Free" Format

There are markets with a preference for public-domain formats like Ogg. The most notable is Wikimedia, owner of Wikipedia, which has supported Ogg formats for some time and has announced that they will be dramatically expanding that use.

Native Embedding in Firefox and Chrome

Firefox, Opera, and Google's Chrome have added Ogg video and audio support, with native support of the proposed <video> tag in current HTML5 drafts.

Why Not Ogg?

Lower Compression Efficiency

The Ogg codecs currently require at least twice as many bits/second as the best video and audio codecs available. This can increase bandwidth costs and reduce reach when higher resolutions, quality, and durations are used.

Not Broadly Supported

While Ogg is designed to be broadly supported, so far the MPEG-LA licensed codecs dominate video playback, with similarly royalty bearing audio codecs also broadly used. The Ogg formats haven't crossed the chicken-and-egg threshold of broad use driving broad support, particularly in devices. While some do support Ogg Vorbis, none do Ogg Theora yet. It will be hard to get Theora support in devices unless support is added to ASICs.

This may or may not change on the desktop; despite Firefox and Google's forthcoming support, neither Apple nor Microsoft have indicated any plants for support, and their browsers dominate the Mac and Windows systems, respectively. Of course, it's easy to download software players for PCs.

HTML5 itself doesn't currently mandate any particular codecs, and is likely some years away from being finalized.

Ogg File Format

The Ogg format was created along with the Ogg Vorbis format, competing with the MP3 file format. It's rather basic, but it works. One notable lack is a formal metadata mechanism; metadata instead needs to be included in the codec bitstreams themselves.

OGV

Ogg was originally designed as an audio format, so required some tweaking to the specification to add video support. These are .ogv files.

OGM

Early efforts to put video in Ogg by non-Xiph programmers resulted in another format called OGM. This has been deprecated, although some of those older files are still rolling around.

MKV

The Matroska format (named after the Russian nesting dolls) is also able to contain the Ogg codecs. It offers much deeper metadata support than native Ogg, particularly for multiple language audio and captioning.

Ogg Vorbis

Ogg Vorbis was created as a patent-free, royalty-free, open-source audio codec as an alternative to MP3. The project was initiated in 1993, the bitstream frozen in 200, and Version

1.0 was released in 2002. Overall, Vorbis is more efficient than MP3 and in the general ballpark of WMA 9.2 and AAC-LC in compression efficiency, and thus competitive for music recording. One factor complicating comparisons is that Vorbis is normally used in VBR mode, while other codecs default to CBR. While Vorbis is capable of CBR and ABR modes, their use is not recommended. So a VBR encode at the same data rate is probably a better comparison point.

The mainline Vorbis version has improved in recent years, but for most of the last few years, the highest-quality encoder implementation has been the current version of the AoTuV version of the codec (short for Aoyumi's Tuned Vorbis). Ogg has merged AoTuV improvements back into Vorbis in that past, and are expected to do so again.

Ogg Speex

Ogg Speex is a low-bitrate speech codec using the same Code-Excited Linear Prediction (CELP) model seen elsewhere. Speex is more flexible than many other voice codecs, supporting ABR and VBR encoding, bitrates up to 44 Kbps, and sample rates up to 32 KHz. It's most commonly used in videoconferencing, and is less seen in files. It is included in Flash 10, including in flv files.

Ogg FLAC

FLAC is the Fast Lossless Audio Codec—certainly the most straightforward name for anything in Ogg. It was released in 2001, and officially joined the Ogg family in 2003.

It is what it sounds like—a fast, lossless audio codec akin to Apple Lossless from QuickTime or WMA Lossless from Windows Media.

Lossless codecs are pretty much all the same, being lossless and all. The only real difference is in speed and compression efficiency, and compression efficiency doesn't vary by more than 10 percent or so between implementations for typical content; about 50 percent compression is typical for music, and up to 75 percent for soundtracks. This is quite a bit better than just zipping up a .wav file.

FLAC is extremely flexible, and can handle nearly any integer bit depth and up to 7.1 channels. And true to its name, it is quite fast to encode or decode.

Being lossless, the only real thing to configure with FLAC for compression is how much CPU time to increase efficiency. The difference in size is small and the difference in encode time is big, but if it's something that's likely to be broadly distributed, you can use the –best flag for slowest, highest quality for realistic use, or even the -e mode, which is a really slow exhaustive mode that could improve compression possibly another half percent past –best.

I'm not a fan of lossless compression for distribution; once the content is perceptually lossless, more bits are wasted bits. But lossless is great as a production format to reduce transmission time. I definitely will use a lossless codec like FLAC for storing anything I may want to transcode later.

That said, there are certainly those who very much enjoy the psychological effect of having lossless audio content, primarily for music content.

Ogg Theora

Ogg's first effort at a video codec was called Tarkin (announced in 2000). It was a wavelet-based distribution codec, adding to the list of promising-sounding wavelet video codecs that have yet to become useful.

Tarkin didn't make much progress before, On2 decided to open-source their VP3 video codec (a much earlier version of the VP6 codec from FLV files), waiving all patent claims for it. Ogg decided to put Tarkin on hold and work from the VP3 source. On2 and the Xiph foundation Ogg's effort, announced Theora in 2002, with a bitstream format locked down in 2004, but the 1.0 release didn't happen until 2008. Since then, lots of progress has been made in improving compression efficiency of the codec. There were big limitations in the On2 encoder source code which kept Theora from delivering competitive quality, particularly in detail preservation. Xiph's Monty has had a very interesting series of blog posts detailing how they've evolved their implementation to improve quality, which is a three-scoop sundae of codec nerditry. And there's plenty of basic optimizations that have yet to be implemented in Theora. As of 1.1, it still lacks real rate-distortion optimization, lookahead rate control, modern motion search capabilities, and multithreading.

However, it's clear that the core architecture of the Theora bitstream is going to keep it from ever being competitive to H.264 or VC-1. These things are hard to predict, but based on the tools Theora supports (it doesn't even have B-frames!), at best it might wind up as a codec on par with MPEG-4 part 2. While there are many ways in which Theora is different from normal codecs, like encoding from bottom to top instead of top to bottom, there are presumably more about avoiding existing patents than any thing that could provide a big improvement in compression quality.

That may be more than enough for simple embedding of short videos in a web page, and there's been interest in Wikipedia and other organizations about using Theora for that. But the bandwidth savings of more advanced codecs are going to be worth a whole lot more than the actual costs of MPEG-LA license fees for companies doing high volumes or high bitrate content.

Ogg Dirac

The lure of wavelets is strong. They're so good for still images, we think, that it doesn't matter if motion estimation isn't quite as good as with DCT block-based codecs. How hard can it be? we think.

Pretty hard, it always turns out.

Dirac is a BBC-led effort to create a license-free wavelet-based video codec. They've had a lot of buzz, and are now working with Xiph to have Dirac under the Ogg umbrella as well.

There have been some interesting demonstrations of Dirac as a production format. I'd expect it to work well there, since intra-only coding works fine with wavelets—SMPTE is working on I-frame-only Dirac as VC-2. But for a distribution format, we need much better compression efficiency, and that's where Dirac has yet to deliver. There is a research project called Schroedinger working on building a Dirac encoder that's competitive with H.264/VC-1 at typical web bitrates. But it hasn't come close to producing anything competitive yet, nor is there a clear strategy for how that could be possible.

At this point, Dirac implementations aren't any better than Theora, and are a lot slower on encode and decode and less broadly supported. Dirac won't be of interest unless it can be significantly better than Theora for something.

Encoding OGV

At this point, the primary tool for OGV encoding is ffmpeg2theora, a command-line tool based on ffmpeg and available as source code and compiled for major platforms. A variety of GUI front-ends are available for it, but so far all just expose the modes listed next.

It only has a few options that are specific for the video and audio codecs themselves. Key options as of Theora 1.1 include:

- -v/--videoquality sets a Quality VBR level from 0 to 10
- -V/--videobitrate specifies bitrate in Kbps
- --speedlevel specifies quality versus speed 0–2 with 0 the slowest and highest quality
- --soft-target relaxes rate control requirements, making it more like an audio ABR encode. Fine for progressive download, but not for streaming.
- --optimize is the same as -speedlevel 0, giving the slowest, highest-quality encode
- --two-pass Enables two-pass encoding. Really required for decent encoding, particularly with CBR buffers
- -F/--framerate specifies frame rate in frames per second
- --keyint specifies keyframe interval in frames
- --buf-delay Specifies buffer in frames to make a CBR encode. Defaults to the GOP length for 1-pass encodes and the whole file for VBR encodes (making them unconstrained VBR). Unusually, there's no way to specify peak bitrate.

- -a/--audioquality specifies VBR audio quality from –2 to 10; the default of 1 is generally decent
- -A/--audiobitrate specifies an audio bitrate in Kbps, ranging from 32–500 Kbps
- -C/--channels specifies the number of output channels
- -H/--samplerate specifies sample rate in Hz

Ogg Tutorial

This tutorial shows creating an embedded tutorial graphic for player embedding.

Scenario

We're making an OGV with Theora video and Vorbis audio for embedding in a web page collecting public domain video content. We're limited to the following maximum specs:

- 640 × 480 frame size
- 30 fps frame rate
- 30 MB per file

Three Questions

What Is My Content?

The source is a U.S. government video of Bill Clinton giving a speech, with a montage of Air Force planes. It is 1280 × 720p60, 3:45 in duration, and varies a lot in detail and motion.

Who Is My Audience?

The audience is composed of viewers of this public resource web site. We can't make any predictions as to decoder power. It's up to the web site itself to make sure that users have the proper decoders.

What Are My Communication Goals?

We want as good quality as we can get within the specs of the web site, and reasonable buffering time.

Tech Specs

The web site has a maximum frame size of 640 × 480 and a maximum frame rate of 30 fps. With 3:45 and 25 MB, we can have a total bitrate of 1066 Kbps.

(*Continued*)

Since the source is 16:9, the output frame size should be 640 × 360. And we'll divide the source frame rate of 59.94 to get to 29.97.

Since we're not streaming we don't need a tight buffer. But the first part of the clip is a lot more complex than the second, which can make for a poor progressive download experience. So we'll use 2-pass mode with a relatively long buffer window. In other codecs, we might have just set a peak bitrate of 1500 Kbps, but Theora 1.1 doesn't support that mode.

Settings

ffmpeg2theora.exe -o ArmFrcs_2p_950-100.ogv -V 950 -A 100 --optimize --two-pass -x 640 -y 360 -F 29.97 -K 150 -H 44100 ArmFrcs_720p60.avi

Specifying:

- -V 950: Video bitrate 950 Kbps
- -A 100: Audio bitrate 100 Kbps (1050 total; a little headroom on our 1066)
- --optimize: Slowest, highest-quality encode (still real-time on my workstation)
- -x 640 -y 360: 640 3 360 frame size
- --two-pass: Use two pass encoding, making it a bitrate limited with unconstrained peaks
- --buf-delay 600: A 20 second buffer at 30 fps; balance of allocation flexibility and flatter peaks.
- -F 30: 30 fps frame rate
- -K 150: Keyframe every 150 frames (5 seconds)
- -H 44100: Resample audio to 44.1 KHz

And there we are! The output file is within 1% of our target size, and looks more than good enough for our use; although there is some edge ringing and a lot of texture detail loss, we don't have much obvious blocking. Theora has come a long way in the last few years.

CHAPTER 20

RealMedia

RealNetworks (then Progressive Networks) founded the streaming industry as we know it. RealPlayer v1 was launched in 1995 and was the first real-time streaming media player. Video was introduced in RealPlayer 4 in 1997, and the company quickly dominated streaming video as it had done for streaming audio. While there were other formats for distributing audio and video on the Internet, RealNetworks was the innovator and the market share leader for streaming delivery until Windows Media displaced it in the early 2000s.

Historically, the main sources of Real's revenue were server software and the Plus version of the player software. But competition from full-featured free players like WMP forced Real to put more functionality into the free version of the player, reducing the truly useful features in the upsell version. And both Windows Media Services and QuickTime Streaming Server were free, eroding the rather massive per-user pricing RealServer used to demand.

Thus, RealNetworks has long-since shifted to content as a revenue driver, particulary the Rhapsody music service, SuperPass subscription service, and RealArcade casual game service. In parallel, they offer their Helix multiformat severs and hosting services, although they open-sourced much of Helix in 2003. The Helix brand extends to open-source implementations of the RealMedia SDK and RealMedia Producer compression tool.

So, while RealNetworks remains a vital company, their proprietary media format has been quickly fading from the scene. It's still seen in some enterprise environments and in European markets, but has only a sliver of its former dominance. The single largest place it's still used appears to be in Chinese Internet cafés, where RMVB (RealMedia Variable Bitrate) has long been the de facto standard.

Why RealMedia?

Since it's not really a mainstream format anymore, if you need to use RealMedia, you proabably already do, and know why—typically an environment where RealMedia is the best format deployed on all target devices. RealPlayer was the first mainstream media player widely support on UNIX operating systems like Linux and Solaris, and the open source Helix Player has meant that RealVideo support is more common on Linux and even more obscure platforms.

RealMedia Format

The RealMedia format, often called a RealVideo file even if it only contains audio, has the extension .rm. RealAudio, or .ra, an older version of the format, is no longer in use. VBR-coded RealMedia files use the extension .rmvb, although they're not really a new file format. That's just an indication that the files shouldn't be streamed, as they aren't CBR.

RealPlayer

The current version of RealPlayer as of this writing is RealPlayer SP beta (replacing RealPlayer 11; see Figure 20.1) for Mac, Windows, and Linux. RealPlayer has long been widely ported, and it is currently also available for Linux, Palm, Windows Mobile, and Symbian. The Mac/Windows versions have broad support for formats beyond RealMedia, tapping into QuickTime and DirectShow APIs to handle anything the default media players can handle. Performance for complex H.264 files seems notably poor in SP compared to WMP and QuickTime Player, however.

While recent RealPlayer versions have been generally clean and smooth to use, there's lingering animosity over past versions. Historically, the two biggest objections were to the hard upsell to the paid version (currently $39.99), and the cluttered interface. A common user complaint was that people couldn't find the free player's link on Real's site, making it potentially problematic to send a RealMedia link or file to a less tech-savvy friend or relative. The current player is quite clean and the upsell quite appropriate now.

The paid RealPlayer SP Plus version has reasonable value for its price, mainly focused on DVD/CD burning and device transcoding and sync. There are some gimmicky features I'm

Figure 20.1 The new RealPlayer SP has a very clean interface; a big change from RealPlayers past.

surprised that they're still hawking, like "video EQ" to change brightness and saturation during playback. If the video was shot and compressed correctly to begin with, those filters could only be used to make the playback worse.

RealPlayer Mobile

There's also an older RealPlayer Mobile available for Nokia's 9200 Series and 7650, and Windows Mobile devices. RealVideo 9 support is only in a "Preview Release" despite having shipped with RealPlayer in 2002.

Helix DNA Client

The Helix DNA Client isn't a product, but a technology that RealNetworks licenses to other companies, particulary in the mobile space, to build players on top of. This is a commercial product for Real, and is updated regularly unlike RealPlayer Mobile.

RealVideo for Streaming

RealVideo and RealAudio were both designed for real-time streaming, and were the best solution for streaming to modem users. That modem user has largely been abandoned by web video, but the underlying technology is still very capable of low bitrate streaming over lossy and high-latency connections.

SureStream

RealVideo's SureStream supports dynamic switching between multiple data rates for both audio and video, up to eight each. The video bands can vary by data rate and frame rate, but not frame size. The audio bands can vary completely. Each band is given a target connection type, and when the RealPlayer connects, it starts with the stream that matches the specified connection type in the user's player. Once the stream starts, the server will dynamically adjust it up or down to the largest streams that fit within the target data rate. Although audio and video are defined in pairs, SureStream may end up playing the audio track from a higher-quality band than the video, if it can fit within the target bandwidth.

RealMedia streaming uses an implementation of the RTSP protocol, and requires Helix Server (formerly RealServer).

The biggest drawback to the Helix Server historically was its price. It's gotten a lot cheapter over the years; for quite a while server pricing was a major motivation for content companies to switch to QuickTime or Windows Media. Helix Server pricing is now "call to talk to a representative"—it's not clear what deployments cost.

Beyond RealMedia, there's also the Helix Mobile Server targeting 3GPP video. They've done some innovative work around bitrate switching, and Mobile appears to be the larger focus for development and business.

RealVideo for Progressive Download

Earlier versions of RealVideo weren't well suited to progressive download. That changed with the introduction of .rmvb and the 2-pass VBR mode of RealVideo 8. However, current versions do a fine job with it. RealVideo files encoded for progressive download shouldn't use SureStream, of course. A file played locally will only play the highest data rate audio and video stream, so including more than one just makes the file larger.

RMVB remains in popular use in some Asian countries, primarily for pirated or user-generated content.

RealMedia Codecs

RealVideo v10

RealVideo 10 is the current name for the RealVideo encoder; the bitstream is the same as RealVideo 9. RV 9/10 is somewhat like H.264 with in-loop deblocking and B-frames, but with a different quantization method. Following RealVideo tradition, it does a very nice job of getting soft before it gets blocky. Compared to RV9, RV10 has better rate control and motion search, and has been particularly improved for maintaining quality in high-motion scenes.

The RV9/10 decoder has been included in RealPlayer since 2003, so it's a safe default codec for all desktop players now, although some older mobile devices are still stuck on RealVideo 8.

One unique control in the RealVideo codecs is four levels of image quality versus frame rate tradeoff. In order of higher frame rate, these are: Smoothest Motion, Normal Quality, Image Quality, and Slide Show. For most content, I find Smoothest Motion offers the best experience, and always use it by default.

RealVideo 10 is pretty fast to encode and decode. All-around, it was a very promising low-bitrate codec, but it arrived just as RealMedia was seeing a huge market share decline after the release of Windows Media 9 Series. This resulted in the latest RealVideo 10 encoding implementation not being broadly supported in commercial compression tools.

RealVideo NGV

RealNetworks had been working on a new codec called NGV (for Next Generation Video) for a few years, possibly to be called RealVideo 11. The project lead has moved on to another position inside of Real and there haven't been any details about it since early 2008.

RealAudio Codecs

RealPlayer was an audio-only solution until v4, and it has always had a strong focus on audio. Instead of offering a few codecs with different settings, Real has always presented audio codecs as a list of content type, data rate, and channel pairs.

Some are marked "High Frequency." These options support higher frequency response, but may increase artifacts when handling complex audio content. You may need to experiment to determine which version is better suited for your particular content.

SureStream works for audio-only content, so a Helix Server can automatically provide the optimum data rate version to a given user.

There are five basic audio codec families still in use.

RealAudio 10

RealAudio 10 is Real's name for AAC (mostly –LC), and thus generally the best audio codec in RealMedia. Its only drawback is that it requires RealPlayer 10 for playback, which could exclude some machines that have gone a long time without upgrades, or old mobile players. It's probably safe to use for most projects today. There are four modes:

- RealAudio 10 Stereo Music: standard stereo audio.
- RealAudio 10 Stereo Surround: stereo encoded to be ProLogic-safe (as described in RealAudio Surround later in this chapter).
- RealAudio 10 5.1 Surround: a full 5.1 codec.
- RealAudio 10 Plus: HE AAC v1, by far the best low-bitrate RealAudio codec, if supported by the target player. It's not clear which RealPlayer versions support the full HE AAC decoder.

RealAudio Voice

This is a speech codec. It's a fine example of the genre, and is unusual in that data rates go up to 64 Kbps. That data rate is overkill for actual voice content, but 32 Kbps is an excellent choice for audio books and the like.

Stereo Music: RealAudio 8

The Stereo Music–RA8 codecs were introduced in RealPlayer v8.5. They were a substantial enhancement in providing cutting-edge compression at low data rates. There are actually two different sets of algorithms under the RA8 name: a version of Sony's ATRAC-3 codec from MiniDisc players (enhanced for RTSP streaming) is used for 96 Kbps and higher data rates.

This provides quality from decent-for-streaming at 96 Kbps to archival at 384 Kbps. Below 96 Kbps, a RealNetworks propritary bitstream is used. In most apps that use the RealSystem SDK, the in-house codecs are referred to as "RA8" and the ATRAC-3—derived codecs are designated "RA 8" (with a space between the "A" and the "8"). Overall, RA8 Stereo is somewhere around MP3 and AAC-LC efficiency, although its low-bitrate artifacts are arguably less annoying than those of MP3.

Although I usually recommend using mono over stereo for lower data rates, the superior quality of the RA8 Stereo codecs over the mono Music codecs means that RA8 should be used instead.

RealAudio Surround

RealAudio Surround is a tweaked version of the normal RealAudio 8 codec that maintains the high frequencies and phase relationships required to maintain Dolby ProLogic matrixed surround sound for playback through a ProLogic reliever. Note this isn't a true 5.1 solution; it just preserves existing surround information encoded in two channels.

RealAudio Surround requires higher data rates and delivers lower compression efficiency than the normal RA8 codecs, and should therefore be used only with ProLogic content that you need to retain for some reason.

RealAudio Music

Music was Real's mono audio codec. It's ancient and bad; RealAudio 10 or at least 8 should be used instead.

Stereo Music

The pre-RA8 Stereo Music codec is also no longer useful. RA8 offers superior efficiency and a much broader range of data rates.

RealVideo Encoding Tools

The Real compression tools and SDKs haven't been updated for ages, but still work. Beyond the first-party products, many long-lived compression tools continue to include RealMedia support, although it's not supported in newer products.

RealProducer Basic

RealProducer Basic is the free version of the RealVideo encoder, available for Windows and Linux. It's easy for novices to use, but it doesn't support fine tweaking of encoding settings; even SureStreaming is limited to just two streams. Professional results require Helix Producer Plus.

Real Producer Plus

RealProducer Plus is the paid version of RealProducer. Beyond all the necessary quality tuning controls, it adds a very useful bandwidth simulator feature to RealPlayer (Figure 20.2). This allows you to simulate what the stream would look like at different actual data and packet loss rates, and see what the lower band streams look like.

As a compression tool, it's pretty basic, and hasn't seen a significant upgrade in a long time.

Carbon

Of my daily tools, Carbon has the fullest RealMedia implementation (Figure 20.3), with the exception of Producer Plus's Fast/Normal/Slow speed/quality control.

Easy RealMedia Producer

The most full-featured RealMedia encoder that I know of is the freeware Easy RealMedia Producer (Figure 20.4). It's a simple GUI on top of the RealMedia Helix SDK. It exposes all the knobs of that, and so has many controls even beyond those in ProducerPlus. This includes fine-grained control over complexity (it can go to 100, where the "High" setting in Producer Plus maps to 85), turn on B-frames, force all frames to be encoded, and other neat tricks. It's really impressive; I wish I'd had it back when RealMedia was the bread and butter of my business.

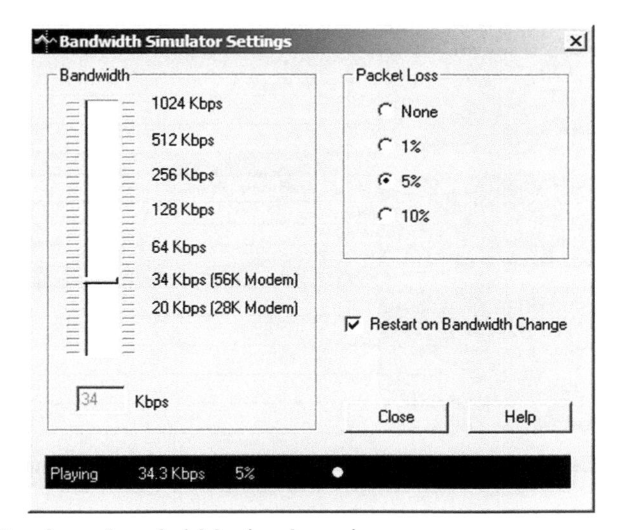

Figure 20.2 The RealProducer bandwidth simulator is an easy way to test streaming performance under adverse network conditions, including packet loss.

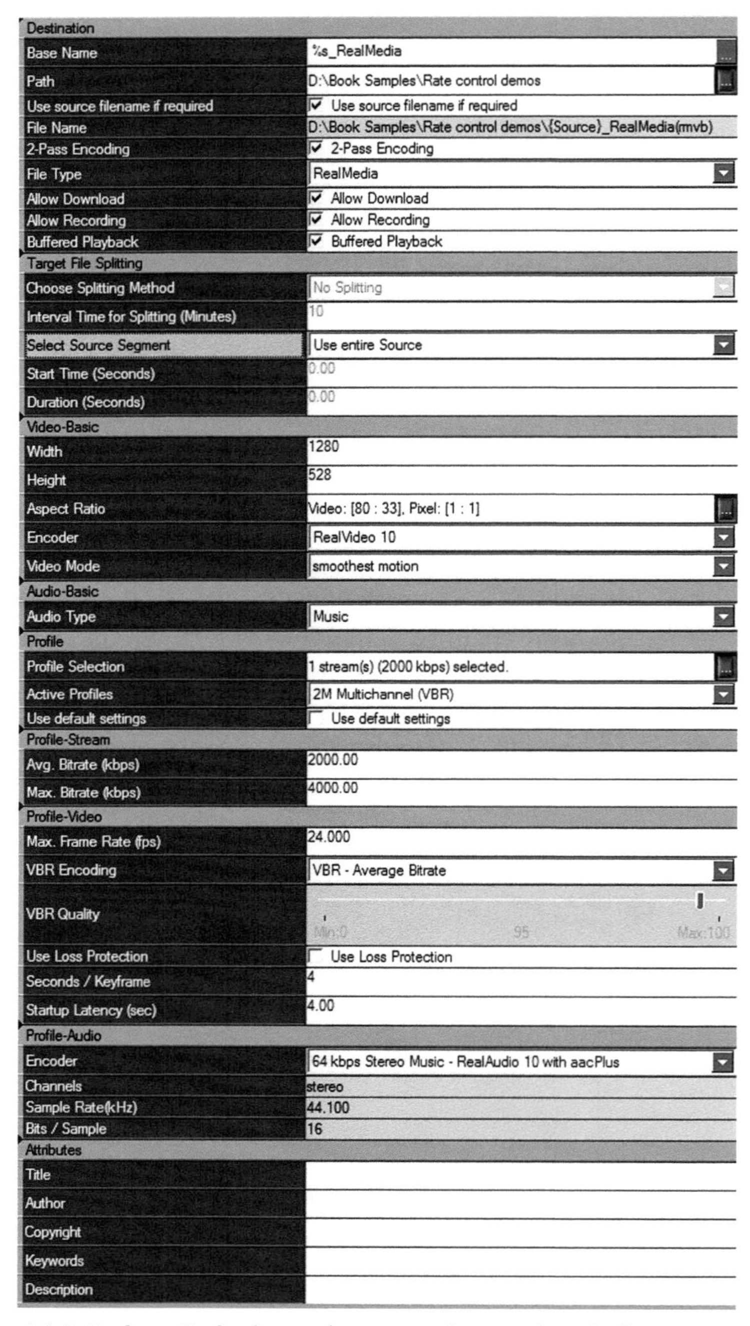

Figure 20.3 Carbon Coder has quite a complete RealMedia implementation.

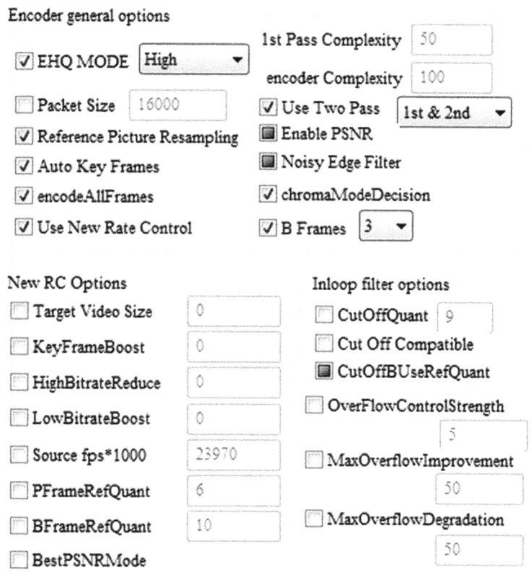

Encoder general options

EHQ MODE High
Packet Size 16000
Reference Picture Resampling
Auto Key Frames
encodeAllFrames
Use New Rate Control

1st Pass Complexity 50
encoder Complexity 100
Use Two Pass 1st & 2nd
Enable PSNR
Noisy Edge Filter
chromaModeDecision
B Frames 3

New RC Options

Target Video Size 0
KeyFrameBoost 0
HighBitrateReduce 0
LowBitrateBoost 0
Source fps*1000 23970
PFrameRefQuant 6
BFrameRefQuant 10
BestPSNRMode

Inloop filter options

CutOffQuant 9
Cut Off Compatible
CutOffBUseRefQuant
OverFlowControlStrength 5
MaxOverflowImprovement 50
MaxOverflowDegradation 50

Figure 20.4 Easy RealMedia Producer exposes advanced codec settings added for RealVideo after the last official SDK release.

CHAPTER 21

Bink

Bink, from RAD Game Tools, is the most popular video format used in PC and console games. And RAD has been in that business for years; Bink launched back in 1999, itself a follow-up to the company's successful Smacker codec, which was already version 2.0 in 1995. Thus the codec, tools, and SDK are tightly focused on satisfying the professional game market. Bink does a fine, if unspectacular, job playing cut scenes, but its real strength is its extremely flexible SDK for incorporating video and audio into games as sprites, textures, or anything else for which a series of bitmaps can be used. The decoder itself is quite lightweight as well, leaving CPU power and memory to prepare for the next level while the video is playing. The encoder is Windows-only, but decoders are available for Win32, Mac OS, Linux, Xbox 360, Sony's PlayStation 2 and 3 and PSP, Nintendo's Wii and DS, plus older consoles.

If all this has left you drooling, you might be the right target audience for Bink. If not (and that will be most of you), Bink isn't something you'll need to worry about.

Why Bink?

You're making a computer or console game, and you need an API that offers fine control over the playback experience. Bink gets bonus points if you need a similar experience over a wide range of hardware.

Why Not Bink?

You're Not Making a Game

If you're not making a game, Bink doesn't have much to offer.

You Need High-Compression Efficiency

Bink is a MPEG-4 part 2 Simple Profile caliber codec in terms of efficiency. VC-1 or H.264 have much better efficiency, and hence would allow more video or better looking video on a disc.

File Format and Codecs

The Bink file format includes both a video and an audio codec, wrapped in a format with the .bik extension.

The Bink video codec is tuned for lightweight decode as a primary goal. It can use a variety of techniques to encode, selected adaptively, including wavelet and DCT. It can get quite blocky under bandwidth stress.

The audio codec is a fixed-quality VBR of roughly AAC-LC efficiency. The default of 4 sounds quite good; lower (less compressed) may not offer much of an improvement, but may not take that many bits either. You can get away with higher when every bit counts for sprite animation and other cases where there might be dozens of hours of content on the same disc remember to encode mono sources as mono; lots of games add all stereo effects at runtime.

Encoder

Bink files can only be encoded with the free Windows-only Bink encoder. The Bink encoder has lots of options, but in most cases there are only a few you'll use. Most controls relate to preprocessing, but the tool's not a great preprocessor, so most of the time you'll to bring in preprocessed source ready for encoding.

In setting data rate, note that the units used are Bps—bytes per second, not Kbps. This specifies the total rate of audio and video data; audio uses as much as it uses, and video gets what's left. Since Bink is largely played off optical media, sustained throughput requirements are an important constraint.

Bink doesn't have 2-pass encoding, but can do lookahead rate control up to 64 frames. This does an analysis X-number of frames ahead before allocating bandwidth, yielding better rate control. Just use the full 64. Older versions of Bink could have lower quality with longer lookaheads, which the first edition of this book recommended against, but that issue has long since been addressed.

Bink's encoder has good multithreading, and encoding times are perhaps 10–15:1 for 720p24 on a Core 2 8-core system.

Keyframe insertion can use either or both a "keyframe at least every" value and a sensitivity control. Using the Hint window, frame type, size, and so on can be manually controlled per frame.

Once a file has been encoded, the Analyze command displays a data rate graph of your file, to look for data rate spikes and compare to peak bitrates for various devices. Since the spikes are where the codec tried the hardest, they're also the most likely to still any quality issues.

Playback

Bink files are played back either via the Bink Player application or from a game incorporating the SDK. Either way, there are a wide variety of tweaks available. These include full-screen playback, a wealth of different hardware display modes.

Business Model

The Bink business model is unique, and a big part of its success. First, all the tools are free. You can incorporate video playback in your app for free, but the Bink logo will be displayed. To include video without the logo requires a per-title fee or a site license. There's also a SDK license that allows for much deeper integration into encoding tools. A full multiplatform, multi-title site license is also available.

Bink Tutorial

This tutorial covers making a Bink file for a multiplatform game title.

Scenario

We're working on a PC/Xbox 360/PS3 game, and are ready to compress the CGI opening title sequence. The game is HD, so we'll be encoding to 720p.

We have 110 MB of space left for the opening scene after all the assets (down from the 150 MB promised in "final" allocation budget; a bunch of extra textures for multiplayer got added at the last minute). We're going to be preloading other content as it plays, and thus need to keep the peak bitrate to 6 Mbps.

Three Questions

What Is My Content?

The source is a 1920 × 1080p24 Lagarith compressed AVI file straight from the animators. It's 4:05; the cinematic team went long again. And they also left some extra black at the end we need to trim off.

Who Is My Audience?

My audience is both the players of the game and the developers. The players want it to look and sound awesome, but I also need to stick within the technical constraints so the code guys can work their magic at the same time.

(*Continued*)

What Are My Communication Goals?

We want to deliver a great audio/video experience with minimal artifacts. And we don't want to get yelled at for going over our file size or peak bandwidth.

Tech Specs

With 100 MB and 4:05, that leaves us 448 KBytes/sec (same as 3591 Kbps, but Bink measures in Kbytes). Our peak of 6000 Kbps means a peak of 750 Kbytes/sec.

It's not a lot of bits, but audio's going to be a pretty small part of it. We'll bump up audio quality to 1, and use the rest for video.

Settings

There's not much to do from there. We set peak and average where we want them. There's no random access during the scene, so we'll use percentage changed as the only way to add new key frames. And we specify an End frame to trim off the excess black at the end of the source. And we specify an End frame to trim off the excess black at the end of the source. Final settings are shown in Figure 21.1.

And once we've done our encode, we need to go back and make sure we hit our target bitrate. Bink includes this very handy analysis graphing tool (Figure 21.2). We can see we're well under the 750 KBytes/sec target.

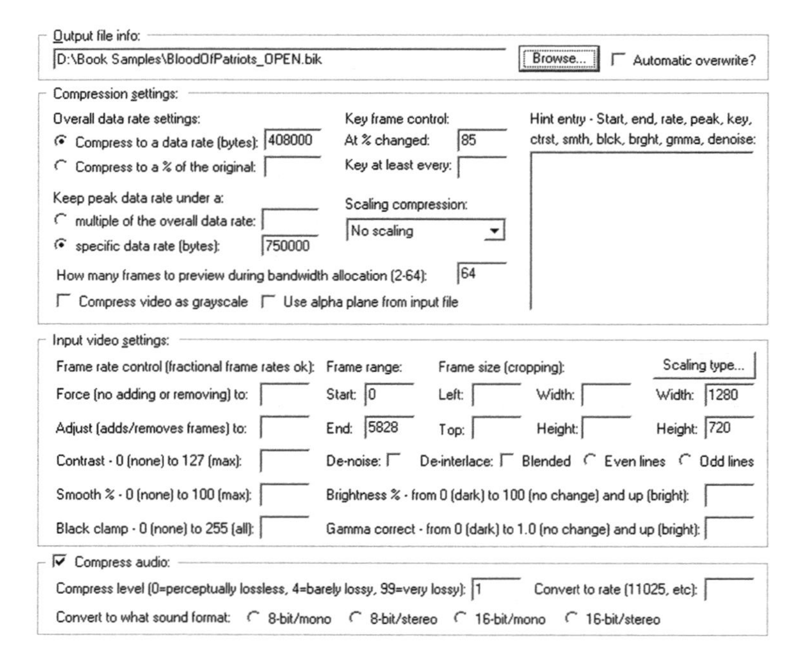

Figure 21.1 Final settings for our tutorial. The layout can be pretty confusing; the height/width on the far right set final frame size, and the middle height/width are for cropping.

(Continued)

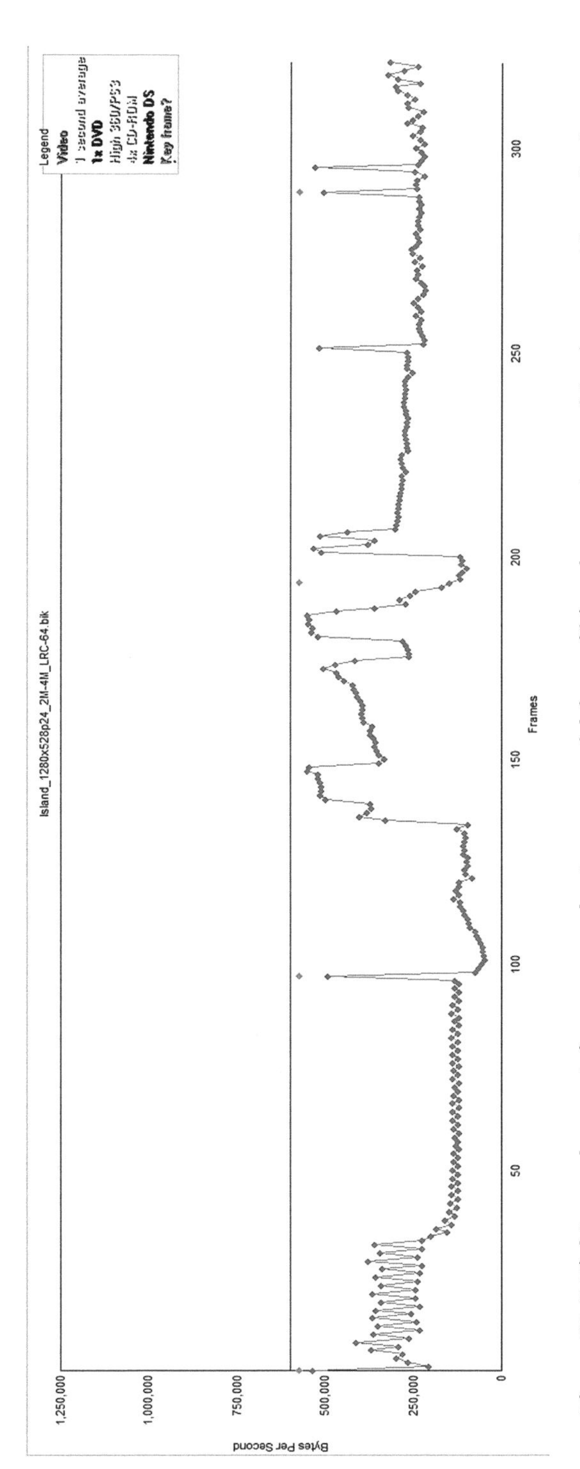

Figure 21.2 Bink's graphs aren't just pretty—they're very useful for verifying that you're not blowing your bitrate budget.

CHAPTER 22

Web Video

For many (most?) people, video compression equates with putting video on the web. Back during the CD-ROM epoch, worthwhile web video was "a year away" for quite a few years. But in 1998, my video compression business went from being 80 percent CD-ROM in January to 80 percent web in December. This was frustrating in some ways—we'd finally gotten CD-ROM to approach broadcast quality, so optimizing video for tiny windows and narrow bandwidths was like starting over with QuickTime 1.0 (thus starting the third iteration of the Postage Stamp Cycle – every five or so years, compression starts over at a postage stamp size. HD, CD-ROM, web, and most recently mobile. Video watches should hit around 2012). Today, web video is at least as good on average as CD-ROM video of a decade ago, and at its best, the web today delivers an HD experience impossible with the discs of old.

That people were willing to put up with postage stamp–sized video meant that web video had to offer something compelling. And it did. The web promised a searchable, immediately available universe of content. No hunting around the video store for the disc you want. No limit to how much video could be put in a given project.

We're seeing a lot of that promise come true today, and with real business models behind them in services like Hulu and Netflix. The story of the next few years is going to be how quickly the web catches up and (perhaps hard to imagine, but we'll be there in a few years) eventually surpasses Blu-ray in quality.

Connection Speeds on the Web

The critical factor in putting video on the web, and the limiting factor for the experience, is connection speed. Fortunately, broadband has gone from the ideal to the baseline; it's hard to remember that we actually streamed video to modems. This is good, because modems really stink for video. Fortunately, anyone with a modem has long since given up being able to watch video on the web, and the web video industry has given up on them.

But what broadband means varies a lot by country, and within a country. Akamai released a "State of the Internet" report at the end of 2008; the portion of the Internet audience that could receive particular bitrates in a few select countries is shown in Table 22.1.

The United States is pretty close to the global average, with most people able to receive at least 256 Kbps, quite a few still under 2 Mbps, and a quarter above 5 Mbps. Global

Table 22.1 Average Internet Connection Speed by Country.

	Average Speed	Below 256 Kbps	Above 2 Mbps	Above 5 Mbps
United States	3.9 Mbps	4.8%	63%	25%
South Korea	15.2 Mbps	0.2%	94%	69%
UK	3.5 Mbps	1.6%	81%	8%
India	7 Mbps	26%	3.7%	0.6%
Global	1.5 Mbps	4.9%	57%	19%

broadband leader South Korea blows everyone else away with an average of 15 Mbps and 69 percent able to do 5 Mbps. The UK is heavily clustered in the 2–5 Mbps band, with few above or below that range. And India, the big country with the slowest connection, has a quarter of Internet users below 256 Kbps, and very few above 2 Mbps.

Compressionists need to understand and adapt to the global and local markets for Internet content. In South Korea, sweating 800 Kbps video would be as atavistic as tweaking your logo for the Netscape web-safe palette. But 800 Kbps may be the high-end stream in India.

And just because someone is accessing web video through a T3 line at work doesn't mean they're able to access video at T3 speeds (45 Mbps). Office bandwidth may be shared among dozens or hundreds of users, not to mention other functions like web and email servers that are given higher network priority than external video.

And as Internet use grows in the home, multiple people can be using the same pipe at the same time, and each can be doing multiple things with that. Even if a household is provisioned at a healthy 10 Mbps, if a couple of kids are downloading music, mom is installing system updates, and dad is installing a new *World of Warcraft* patch while trying to watch some video, maybe only 1 Mbps of that 10 is reliably available over the duration of the stream.

Wireless is another world as well. Some devices can get 2 Mbps on the go already, and are getting faster. But that same device might only get 40 Kbps out in the boonies or on the subway, and might get nothing at all in an elevator.

So, a decade into the web video era, we have incredible bandwidth, but it's unevenly distributed, not just by place, but by time. Much of the challenge and craft of web video is in how we can take advantage of bandwidth when it's there, but still provide a best-effort experience when it's not.

Kinds of Web Video

Web video is a single phrase that can mean a lot of different things. First, we'll break down web video into its major categories, and then talk in more detail about how it works.

Downloadable File

A downloadable file is the simplest form of video on the web. The audience uses a file transfer mechanism such as FTP or BitTorrent to download the file. No attempt is made to play it in real time.

This method is used for commercial download movie services and game trailers, but probably accounts more for pirated content than anything else. The advantage of downloadable files is that there is absolutely no expectation of real-time performance. The limit to data rate is how big a file a user is willing to download; if they're willing to wait a few days for the best possible HD experience, that's up to them. In an extreme example, I once spent two *months* downloading 110 GB of (legally provided) Nine Inch Nails concert footage via BitTorrent.

The downside is that potentially long wait, which runs rather counter to the web expectation of "Awesome! Now!" Even the services that formerly offered downloads only are adding immediate playback modes. For example, Xbox Live Marketplace was download-only originally, then added progressive download to allow the downloaded part of a video to be watched, and is adding 1080p adaptive streaming that should be available by the time you're reading this.

Progressive Download

Progressive download delivers an experience between downloadable files and classic real-time streaming. Like downloadable flies, progressive download content is served from standard web and FTP servers. The most important characteristic of progressive download is that transmission may not be real-time, and it uses lossless file transfer protocols based on the Internet standard TCP (Transmission Control Protocol), typically the web standard of HTTP (Hypertext Transport Procotcol) or (more rarely) the older FTP (File Transfer Protocol).

All content transferred over the web is broken up into many small packets of data. Each individual packet can take a different path from the server to the client. TCP is designed to always deliver every packet of the file (which is why you don't ever find the middle of an email missing, even with network problems). If a packet is dropped, it is automatically retransmitted until it arrives, or until the whole file transfer process is canceled. Because of this, it is impossible to know in advance when a given packet is going to arrive, though you know it will arrive unless transfer is aborted entirely. This means immediate playback can't be guaranteed, but video and audio quality can be.

A progressive download file can start playing while it's partially transmitted. This means less waiting to see some video, and gives the user the ability to get a taste of the content, and the option to terminate the download, should they decide they don't want to see the whole thing. This ability to play the first part of the video while the rest is being transmitted is the core advantage of progressive download over downloadable files.

Figure 22.1 A progressive download player, showing the playhead position and how much content has been buffered so far.

At the start of the progressive download, the user initially sees the first video frame, movie transport controls, and a progress bar indicating how much of the file has been downloaded. As the file is downloading the user can hit play at any time, but only that portion of the file that's been transmitted can be viewed. YouTube is probably the best-known example of a service that uses progressive download.

Most progressive download players are set to start playing automatically when enough of the file has been received that downloading will be complete before the file finishes playing. In essence, the file starts playing assuming the transmission will catch up by the end (Figure 22.1). This works well in most cases, but when a file has a radically changing data rate, a data rate spike can cause the playhead to catch up with the download, so the next frame to be played hasn't downloaded yet. Thus playback pauses for a time to catch up. The major web video players indicate how much of the video has been downloaded by tinting the movie controller differently for the downloaded portion of the movie. As long as the playhead doesn't catch up with the end of the gray bar, you won't have playback problems.

Most systems start playing a file when the remaining amount of data to be transmitted, at the current transmission rate, is less than the total duration of the file. From this, we can derive the following formula to determine the buffering time, that is, how long the user will have to stare at the first frame before the video starts playing. Note connection speed is the actual download speed of the media; so could be 1 Mbps off the 10 Mbps family connection example.

This formula has some interesting implications:

$$\text{delay} = \left(\frac{\text{encodedrate}}{\text{transferrate}} - 1\right) \times \text{duration}$$

1. If the transmission speed is greater than the bitrate, then the buffer time is nil—a progressive download clip can start playing almost immediately in that case.
2. As the ratio between the clip's data rate and the connection speed changes, start delay can change radically. A 5-minute 3000 Kbps clip will:
   - Play immediately at 3000 Kbps or higher
   - Buffer 2.5 minutes at 2000 Kbps
   - Buffer 5 minutes at 1500 Kbps
   - Buffer 45 minutes at 300 Kbps

Longer duration has a linear impact on start delay—a two-minute movie trailer with a six-minute wait is *much* more palatable than a two-hour movie with a six-hour wait. Which is exactly what a 4 Mbps movie would get with a 1 Mbps transfer.

Classically, you could do rapid random access within the downloaded section, but couldn't move the playhead beyond it. In the last several years, many web servers have enabled HTTP byterange access, and a compatible player can then start playback from any point in the file, buffering on from there. Implementations vary; some players flush the buffer of what's been downloaded when doing that kind of random access, so the first part of the file would need to be downloaded again. But when byterange access is available it addresses what was the biggest single limitation of progressive.

In both modes, if the whole clip has been watched, the whole clip should have been buffered, so a second playback shouldn't require any network access. This is a win for the user (patience can be traded for a high-quality experience) and the content provider (if someone wants to watch that dancing monkeys clip 49 times in one sitting, you only had to pay to send them the bits once).

The flip side of this historically was that the movie would be sitting right there in the browser's cache, making it very easy for users to save it. Most of the time that's probably a feature, not a bug; by default Windows Media Player and QuickTime Player Pro have offered a "Save As" option for progressive download content. But for content owners who want to maintain control over who watches their video and where, it's been a concern, although different platforms have offered different ways to keep content out of the browser cache. In the end, meaningful content protection has always required real DRM, be it progressive download or streaming.

Real-Time Streaming

The other side of classic web video is real-time streaming. Its defining characteristic is, of course, that it plays in real time. No matter the duration of the file, it'll start playing in a few seconds, and assuming stable and sufficient bandwidth, it will play all the way through whatever its duration. Each time the playhead is moved, the video will rebuffer, but once playback starts, if all goes well it shouldn't stop until the stream is over.

My years in this industry leave me with a hollow feeling of terror at the words "if all goes well...."

Real-time streaming requires specific streaming video server software, either vendor- and protocol-specific—like with Windows Media, Flash, and RealVideo—or with interoperable standards-based options like RTSP for MPEG-4 and QuickTime. Such servers are required to support the protocols used; RMPT/RMPTe for Flash, with the others based on RTSP.

The classic Real Time Streaming Protocol (RTSP) can use UDP (User Datagram Protocol) packets, not just the TCP used in progressive download and web content. The salient difference between UDP and TCP is that UDP doesn't automatically retransmit a dropped packet. Instead, if a packet of data goes missing, it's up to the player to notice and potentially request it again from the server. While that may sound like a bug, it can actually be a feature.

When TCP automatically retransmits dropped packets, it keeps trying until either the packets arrive or TCP gives up on the whole transfer. And due to the structure of TCP, packets from later in the file than the missing packet aren't available to the receiving application until the missing packet arrives.

So, when using TCP over a slow and lossy network, you can get into cases where some packets get dropped, and their retransmission uses additional bandwidth, crowding out new packets and further increasing the number of dropped packets. And nothing after the dropped packets is playable. And when the playhead gets to the point of the first dropped packet? Wham, no more video, and perhaps the dreaded "...buffering...buffering..." messages of yore.

With the classic UDP streaming model, it's up to the player to figure out when and if to request dropped packets be retransmitted. And when a packet is associated with a frame that needs to play in 100 ms and the average transmission time is 200 ms, the player knows to just not re-request, and trust the decoder will be able to display *something* even without what was in that packet (for an example of what that looks like, see Color Figure 22.2). Coupled with buffers in the server and player to give time for retransmission requests to be made, and to let bitrates average out a bit, UDP promised to tame the wild packets of the consumer Internet into a decent video experience.

That was the vision. And it worked pretty well. But lossy high-latency connections like 56 Kbps modems had are the exception, not the rule these days. And that UDP could actually make it from server to player could never be counted on outside of internal networks inside organizations.

Real-time streaming solutions needed to fall back to TCP or even HTTP transmission when passing through firewalls or routers that don't support UDP. And once a streaming technology has to work over HTTP (and it will for a big chunk of users), the practical difference between real-time streaming and progressive download starts to blur. Originally, progressive download always transmitted the entire file from start to finish, while streaming servers offered random access and bandwidth negotiation, even over HTTP. But web servers that support byte serving can provide random access as well now.

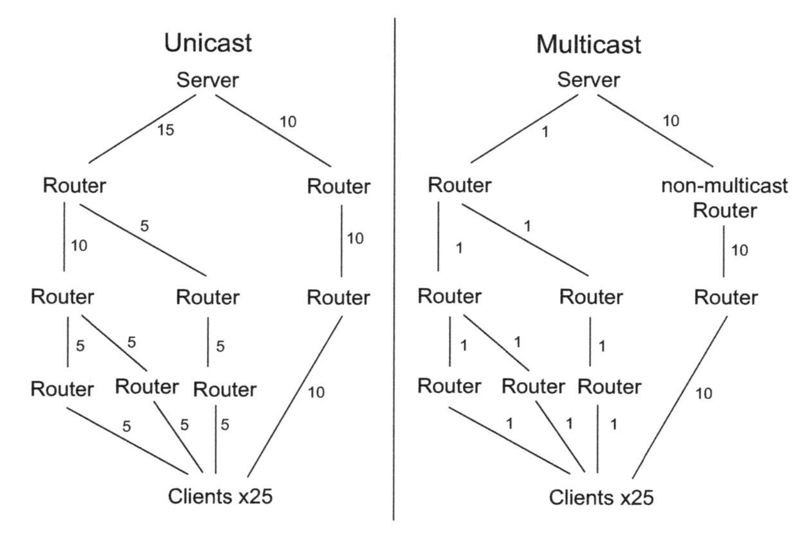

Figure 22.2 A very simple comparison of unicast to multicast. Note that it just takes one nonmulticast router to eliminate much of the value of multicast.

The key differentiated features between progressive download and streaming today are MBR, live, and multicast.

Multiple bitrates for real-time streaming

One streaming feature of Windows Media and RealMedia is Multiple BitRate encoding (MBR) . The content is encoded at multiple bitrates, and the player dynamically switches to the highest bitrate playable at the available transmission speed. For example, an MBR file encoded with 200, 400, and 800 Kbps streams would have the 400 Kbps played at 700 Kbps, and the 200 Kbps played at 350 Kbps.

There are two big challenges with the classic MBR model.

First off, a real-time protocol had no way to know how much extra bandwidth was available in order to increase bitrate. Knowing when to throttle bitrate down is easy; when only 300 Kbps of data is arriving out of a 400 Kbps stream, the actual bandwidth is 300 Kbps. But if data is sent continuously, there's no way to measure how much extra bandwidth could be used when the stream is going well. In practice, MBR tends to switch down if necessary but may never switch back up again. That might be okay for short content, but it's not a good experience if a two-hour movie keeps looking worse the longer it plays.

Second, there was always a pause at stream switching, since the new bitstream acted just like a random access seek. For on-demand, this meant a delay in playback, but for live this meant several seconds or video was simply never seen by the user.

That said, MBR was still a lot better than being stuck in the lowest common denominator with every user is limited to the minimum supported bitrate of. It was a great step forward, but never really paid off in its classic implementation.

The most successful streaming MBR implementation today is Flash Media Server's Dynamic Streaming, discussed later in this chapter and in Chapter 26.

Webcasting

Because progressive download can't control time, just quality, it's not capable of live broadcasting (also called webcasting). Webcasting started as a difficult, finicky, and low-quality experience. But, in the inevitable demonstration of Moore's Law, improved hardware and software brought webcasting to maturity in rapid fashion. Faster processors capable of running more efficient codecs, plus the general increase in available bandwidth, have made a huge difference. Today live streaming is capable of matching the quality of good SD broadcast TV around 1500–2500 Kbps.

Multicasting

With progressive download and non-UDP streaming, each user is sent their own copy (a "unicast") of the content. If 1000 viewers are watching a stream, 1000 streams need to be served. And with 1,000,000 users, 1,000,000 need to be streamed, with 1000 times more bandwidth to provision and pay for. Thus unicast can prevent streaming from developing the economies of scale of broadcast. And if more users try to access the content than there is server and network capacity, the experience can get bad quickly. Being *too* popular wasn't ever a problem in television, but it can and has been with streaming.

Enter multicast stage right, coming to save the day (Figure 22.2). With the IP multicast protocol, only one copy of any given packet needs to be sent to any given router, no matter how many users on the other side of the router are requesting it. That router then sends a single copy of each packet to each additional router that is subscribed to the IP multicast. This should dramatically reduce the bandwidth requirements of live streaming as the number of users scales up.

However, for multicasting to work as advertised, every router between server and client needs to be multicast-enabled. Even though modern routers support multicasting, many older routers don't, or if they do, their operators don't have multicasting turned on. In particular, many ISPs and corporate networks don't allow multicast traffic into their networks. And it only takes one nonmulticast router between server and player to block multicasting for every client connecting through that router.

In most countries, multicast isn't reliably available for delivering to consumers (although the UK is making great progress to universal multicast). It's really only used in enterprise LANs and WANs. But it really shines there. Without multicast, it's impossible to deliver corporate events to tens of thousands of desktops; the unicast bitrates overwhelm the internal routers. Even if a building has 2 Gbps of bandwidth, 1,000 people trying to watch 2500 Kbps live video isn't possible (2.5 Gbps!), let alone all the other data that might be flowing on that pipe. But with multicast, only a single instance of the stream hits the building, and 5,000 people could watch it with little impact.

The biggest drawback of multicasting is that it doesn't support on-demand; multicasting requires users to be receiving the same content at the same time.

Even simple things like pausing don't work in most multicast implementations. Other features might not be available, like retransmission of dropped packets.

Peer-to-Peer

Peer-to-peer (P2P) is sometimes presented as the savior of streaming. The core idea is obvious enough: since users all have upstream and downstream bandwidth, why not have each user send bits on to another user instead of requiring everyone to pull unicast from the source?

Nice idea in theory. But the simple version of this doesn't work well in practice; lots of people are behind firewalls or NAT (network address translation) routers that prevent computers from easily talking to each other. Fully peer-to-peer systems like BitTorrent may work well for files, since users without good access just get slow speeds. But once you've got a real-time playback requirement, a greater percentage of bits have to come from dedicated servers instead of from peers.

So, P2P can pay off in improved costs and quality by adding some new places for content to come from. But it's not a revolutionary improvement for real-time media delivery; in real-world applications, perhaps only 50 percent of bits will come from peers, which may not even offset the costs coming from the greater complexity of P2P systems, and getting users to install P2P clients.

Adaptive Streaming

The new game in the internet video town is adaptive streaming. From a high level it can sound like a complex mishmash of progressive download, streaming, and general oddness. But as strange as it may sound in theory, it's proving incredibly powerful in practice.

My Microsoft colleague Alex Zambelli describes adaptive streaming as "streaming adaptive for the web instead of trying to adapt the web to work with streaming."

Adaptive streaming can be defined in a number of ways, but for me it has three salient characteristics:

- Content is delivered via HTTP as a series of small files.
- Seamless bitrate switching is enabled via switching between different series of those small files at different bitrates.
- The files are small enough to be cached by the existing proxy cache ecosystem for scalability.

The first company to put these features together was the pioneering Move Networks. In their original model, they encoded the same content in multiple bitrates (over a dozen for HD) in three-second Closed GOP chunks. So there's a new file every three seconds for every bitrate.

On the client side, they have a browser plug-in that decides which bitrate to pull the next chunk from. And since these are being read as small files, not as a continuous stream, the problem of measuring available bandwidth gets solved by just measuring the download time per chunk. If a 300 Kbps three-second chunk downloads in one second, then there's about 900 Kbps of bandwidth that moment. By always keeping the next few chunks in the buffer, a drop in bandwidth can be detected soon enough to start requesting at a lower bitrate.

The short GOPs of fixed duration is the next innovation. Since they're the same duration in each data rate, every bitrate starts with a new Closed GOP at the same time. And as they're Closed GOPs, each is independently decodable without reference to any other chunks.

Thus, switching between bitrates is just a matter of feeding the video bitstreams continuously to the decoder, appending those demuxed from each chunk. As long as the decoder can handle resolution switches on the fly, there's not a moment's hesitation when switching between bandwidths.

This can also be used to make startup and random access time a lot faster as well, since a lower bitrate can be requested to start quickly, and then ramped up to the full bitrate.

So, the net gain is that adaptive streaming offers users a welcome hybrid of the best of progressive download and real-time streaming:

- HTTP means it doesn't get stopped by firewalls.
- MBR encoding and stream-switching with no pauses.
- Accurate measurement of available bandwidth for optimum bitrate selection.
- Nearly immediate startup and random access time.
- The "just works" of progressive download with the "no wait" of streaming.

There's one last facet to adaptive streaming that neither progressive nor real-time streaming had: proxy caching. It's not something non-sysadmins may think much about, but tons of web content isn't delivered from the origin web server each time. Instead, popular content gets cached at various intermediate places throughout the network. The goal is to save bandwidth; if an ISP or office has 1000 people all reading CNN at the same time, it's wasteful to keep pulling in that same PNG of the CNN logo over and over again. So a special kind of server called a proxy server or proxy cache is placed between the users and inbound traffic. It tracks and temporarily stores files requested by every computer on the network, If there's a second request for the same file, it gets served from the proxy cache. And everyone wins:

- The user gets the file faster from the cache than if it had come from the original server.
- The ISP doesn't have to pay for the inbound bandwidth.
- The original web site doesn't have to pay for as much outbound bandwidth.

There's a maximum size for files to be cached, however, and it's almost always smaller than video files. But with adaptive streaming, files are only a few seconds each, and so easily fit inside the caches.

Thus, lots of people watching the same video at the same time, or even within a few hours of each other, can be watching from the proxy cache. This provides some of the scalability of multicast, but without having to be live, and without needing special router configuration. It's just part of the web infrastructure already. Pretty much every ISP, business, school, government office, and other organization is going to have a proxy cache already installed and running.

And those caches aren't just at the edge. The content delivery networks (CDNs) all use proxy caches inside their own network to speed delivery. So adaptive streaming works very well for them, particularly compared with having to install and maintain specific servers for each format at potentially thousands of locations. (For more about CDNs, see the "Hosting" section later in this chapter.)

The general experience with proxy caching is that the more popular content gets, the fewer bits per viewer are pulled from the origin server or even the CDN. This restores some of the broadcast model of unexpectedly popular content being a cause for celebration rather than crisis. Since fewer bytes per user need to leave the CDN in the first place, bigger audiences should lower per-user bandwidth costs.

So, is adaptive streaming the holy grail? Not quite. It's a leap forward for many scenarios, but it has some limitations:

- Companies don't typically have internal proxy caches, so WAN broadcast still requires multicast.

- An intelligent client is required. There's a lot of heuristics involved in figuring out the right chunk to grab next. Adaptive streaming hasn't been built into off-the shelf devices and operating systems. It either requires a custom app like on the iPhone or a platform that allows custom apps, like Silverlight.
- There can be a *huge* number of files involved. A decent-sized movie library in multiple bitrates can get to a quarter-billion files pretty quickly. This becomes extremely challenging to maintain, validate, and transport.

The future of adaptive streaming

There's a lot of innovation in adaptive streaming these days, with new implementations coming from Microsoft in Silverlight and Apple for the iPhone and QuickTime X. One trend we're seeing is single files per bitrate on the server, which are divided into chunks only as they're requested. So, that hour-long video could only be eight files instead of 24,000. Since most content in a large library isn't being viewed at any given time, this avoids the management cost of incredible numbers of small files, while still offering all the scalability.

Is Flash Dynamic Streaming Adaptive Streaming?

Flash's Dynamic Streaming is a bit of an odd duck in our taxonomy here. It doesn't support UDP, just TCP and HTTP, so it can't do multicast. It has multiple bitrates, but they're continous streams, so it doesn't leverage proxy caches. It is capable of gapless or nearly gapless bitrate switching.

Overally, I'd say RTMP may be best thought of as the culmination of the classic proprietary protocol MBR stream switching arcitecture, working better than anything had before—but without the increased scalabilty.

What About Videoconferencing?

Streaming formats—even when doing live broadcasting—introduce latency in the broadcast. This means that web broadcasters are simply not viable for videoconferencing because there will typically be a delay of several seconds. Even the lowest latency web protocols like Adobe's RTMP still have an uncomfortable amount of end-to-end delay for easy conversation; it's more like a satellite phone conversation on CNN. The web formats use this latency to optimize quality, which is one reason why they look better than videoconferencing at the same data rate.

Videoconferencing requires videoconferencing products, which are generally based around the H.323 standard.

Hosting

Web video, by definition, needs to live on a server somewhere. What that server is and where it should live vary a lot depending on the task at hand.

Server type is simple to define. Progressive download lives on a normal web server, and real-time streaming uses an appropriate server that supports the format, such as Windows Media Services running on Windows Server for Windows Media and Flash Media Server for Flash. Adaptive streaming that uses discreet chunks may just need a web server, or a specific server may be required to handle chunk requests if large single files are used on the server. For QuickTime and MPEG-4, there are a variety of servers to choose among, although for both formats many of them are based on Apple's open-source Darwin Streaming Server.

The biggest question is where to put your video: on an in-house server, a hosting service, or on your own server in a co-location service? In general, you want your files to be as close to the viewer as possible.

In-House Hosting

Having your media server inside your facility makes most sense for content that is mainly going to be viewed on an internal network. In-house hosting only makes sense for providing content outside your network on extremely small or large scales. Because most businesses are provisioned with fixed bandwidth, your simultaneous number of users is limited to how much bandwidth you have. If you want to handle a high peak number of users, you'd need to buy far more bandwidth than you'd use at nonpeak times. If you're Google or Microsoft, this might make sense.

But otherwise, external-facing web content should never be on an organization's normal Internet connection.

Hosting Services

With a hosting service, you rent space on their servers instead of having to provide or configure servers yourself. Hosting services are the easiest and cheapest way to start providing media to the public Internet. Hosting services are also much easier to manage, and can provide scalability as your bandwidth usage goes up or down. You are typically billed for how much bandwidth you use in a month, so huge peaks and valleys are much less of a problem.

High-end hosting services like Akamai, Level 3, and Limelight describe themselves as Content Delivery Networks – CDNs. A CDN isn't just a single server, but a network of servers. The connections between the servers are high-speed, high-quality, and multicast-enabled. The idea is to reduce the distance packets need to travel on the public Internet by putting edge servers as near to where customers are (a.k.a. all around the "edge" of the Internet) as possible. For multicasting, this means that multiple streams only need to be distributed from each local server. For static content, files can be cached at the local server, so they would only need to be transmitted once from the center. CDNs will deliver both the best total cost and the best user experience for the vast majority of projects and businesses publishing outside their own networks.

Co-location

A co-location facility is a hybrid between in-house and hosting services. You provide your own server or servers, but put them in a dedicated facility with enormous bandwidth.

Co-location makes sense when you have enough volume that it's worth the cost and complexity of managing your own servers. Because you provide more management, co-location is typically somewhat cheaper per bit transmitted than a hosting service. The biggest drawback to co-location is that you need to physically add more machines to gain scalability.

Tutorial: Flexible MPEG-4 For Progressive Download

Scenario

Our company has an large library of training videos for our extensive line of home gardening products. We'd like to get these published to the web. These are sold globally by a variety of affiliates and independent distributors, so we want to be able to provide files to our partners that they can host easily and use in a variety of media players.

The Three Questions

What Is My Content?

The content is mainly marketing and training materials for a variety. There's a lot of older content produced on BetaSP and DV 480i 4:3, with a smattering of content produced by a European division in 576i25. Newer content is HD 16:9, produced in a mix of 24p, 30p, and 30i (whatever the camera operator felt like that day, apparently). The clips are pretty short, ranging from 2–10 minutes.

Who Is My Audience?

We really have two audiences. The initial audience are the many local companies who are going to use these files. Some of them have standardized on different media players, including Flash, Silverlight, and QuickTime. The final audience is the people around the world who will be watching the clips.

What Are My Communication Goals?

We're making these clips available to help our affiliates sell more products, and to improve customer satisfaction and reduce the cost of telephone support by providing

(*Continued*)

visual demonstration of product use. We'll also save some money by having to buy and ship less physical media.

We want these files to "just work" in a wide variety of scenarios and players, over a reasonable variety of bandwidths. The video and audio quality needs to make the content clearly understandable, and to generally be attractive and clean, reflecting our brand.

We want random access to be spritely so users can easily scrub around the tutorial clips while reviewing a how-to procedure.

These files could be used for years to come, so we want something that should be compatible with commonly available players in the future. We also want to define a relatively bulletproof process that won't require a lot of per-clip fiddling. We've got a whole lot of files to publish!.

Tech Specs

Given that these are short clips and we don't have much knowledge or control over affiliate's hosting, progressive download is an obvious choice. That way we can guarantee quality of playback, if not buffering time.

MPEG-4 files with H.264 video and AAC-LC audio are compatible with Flash, Silverlight, QuickTime, and Windows Media Player in Windows 7, and should be compatible going forward.

Our content can be pretty complex. By definition, it's full of foliage, and a handheld shot of an electric pruner going crazy on a bush can take a lot of bits to look decent. We're going to need more bits/pixel than average. For easy progressive download playback, we're going to target 1000 Kbps total, allocating 96 Kbps to audio at 44.1 KHz stereo. While we might have used 128 Kbps for content with more music, our soundtracks aren't central to the experience; we mainly need the voiceover to be clear and any short stings of music to not be distractingly bad. That leaves 904 Kbps for video. We'll use a total peak bitrate of 1500 Kbps to keep progressive playback smooth and decode complexity constrained, yielding a 1404 peak video bitrate, assuming CBR audio.

For easy embedding, we'll stick with a maximum 360 line height, so 16:9 will come out as 640 × 360 and 4:3 as 480 × 360. It always bugs me when 4:3 gets 640 × 480 and 16:9 gets only 640 × 360; the 16:9 source is generally of higher quality, and so can handle more pixels per bit, not less. And yes, we could have done Mod16 (640 × 352 and 464 × 352), but these values are at least Mod8, and let us have perfect 16:9 and 4:3 aspect ratios.

For preprocessing, we'll crop eight pixels left/right with 4:3 setting; they generally include at least some content originally from BetaSP with horizontal blanking we need to get rid of. Eight left/right is still well outside of the motion-safe area, so we won't lose any critical information. The HD sources are all digital and full-raster, so no deinterlacing is needed there.

We'll have frame rate match the source up to 30p; 24p, 25i, 30i, and 30p stay the same.

For good random access, we're going to use four-second GOPs with two B-frames. With 30p source, that'll give us 120 frames per GOP, of which 80 are B-frames, so the worst-case number of reference frames would be 40 in any GOP.

(*Continued*)

For H.264, our choices are Baseline or High Profile. High Profile is compatible with all the web players, but not with most devices. With careful encoding, we could try to make iPod-spec .mp4 files so they could be played there as well. However, there's a significant efficiency hit in that. In order to keep quality up at our bitrate we'll use High Profile. If a user wants to watch on a device, these files are short and will transcode quickly as-is.

We've got quite a lot of content to encode, so we're going to tune for a good balance between quality and speed.

Because our source varies in aspect ratio, frame rate, and interlaced/progressive, we want settings that adapt to the source without making us burn a bunch of time on per-file tweaking. Episode, Expression Encoder, and Compressor all have adequate adaptive preprocessing to let us to have one preset for 4:3 480i and another for all the 16:9 sources.

Settings in Episode

We'll start with the H.264_480_360 and H.264_Widescreen_640 × 360 presets, with the following changes (Figure 22.3):

- Output format to .mp4 from .mov (.mov would have worked in our software players, but .mp4 is a little more broadly supported)
- H.264 General
  - VBR using Peak Rate
  - Peak Rate: 1404
  - Average rate: 904
  - Natural and Force Keyframes with 120 keyframe distance (4 sec @ 30p)
  - Number of reference frames: 4 (a little faster, and just as good with video source)
  - Number of B-frames: 2
- H.264 Profile and Quality
  - Encoding Profile: High
  - Entropy Coding: CABAC (with 1.5 Mbps peak, decode complexity should be okay)
  - Display aspect ratio: 1:1 (square-pixel)
  - 2-pass interval: 500 frames (best variability)
  - Encoding speed versus quality: 50 (much faster and nearly as good)

(*Continued*)

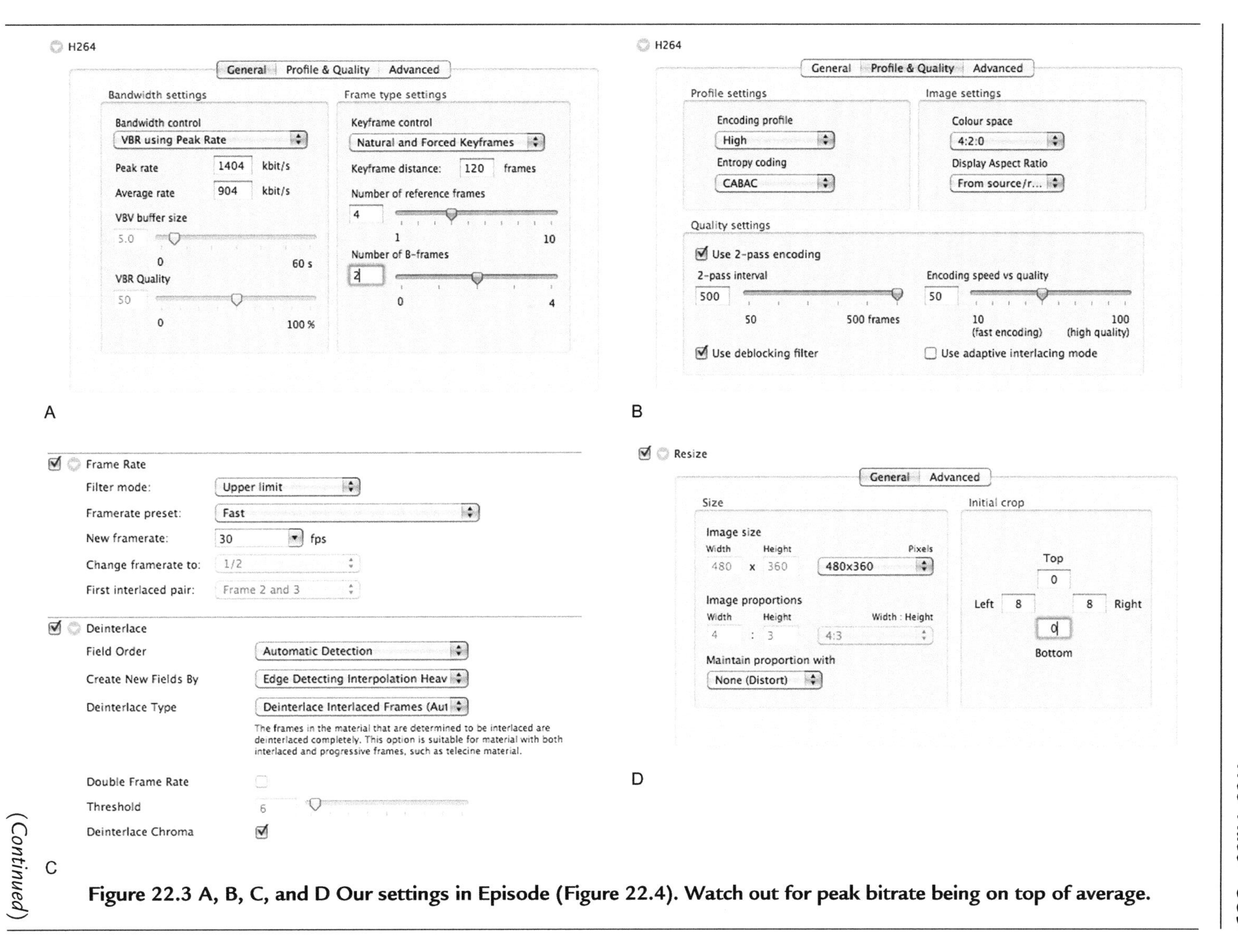

Figure 22.3 A, B, C, and D Our settings in Episode (Figure 22.4). Watch out for peak bitrate being on top of average.

(Continued)

- Frame Rate
  - Upper limit of 30
- Deinterlace
  - Create New Fields By: Edge Detecting Interpolation Heavy (best quality outside of the very slow Motion Compensation mode)
  - Automatic mode for the 16:9 preset as well, since it'll turn off automatically with progressive source
- Resize Advanced
  - (16:9) Preprocessing: Lowpass for large downscales (the 4:3 are all SD, so doesn't matter there)
- Audio
  - Bitrate to 96 Kbps
  - Volume: Normalize

Settings in Expression Encoder

We can start with the H.264 Broadband VBR preset. EEv3 is Main Profile only, but supports CABAC and multiple reference frames at least.

- Key Frame Interval to 4
- Average bitrate to 904
- Video
  - Peak bitrate to 1404
  - 4:3
    - Width 480, Height 360
    - Video Aspect: 4:3
  - 16:9
    - Width 640, Height 480
    - Video Aspect 16:9
- H.264 Settings

(*Continued*)

Figure 22.4 A, B, and C Our settings in Expression Encoder. Make sure that you have height and width still at 0 and 480 after setting the 4:3 crop.

- Audio
  - Bitrate: 96 Kbps
- Enhance
  - (4:3) Crop Left 8, Width 704, top 0, Height 480
  - Volume Leveling: On (Normalization in Expression Encoder)

(Continued)

Settings in Compressor

Unfortunately, Compressor doesn't provide full access to H.264 features when going to a MP4 file. So we'll have to create a .mov and use a batch utility to remux to .mp4 after. It's a relatively simple AppleScript for those so inclined.

Compressor can't do High Profile, so this will be Main Profile only. Also, we can't set multilple reference frames or the B-frames value. However, QuickTime doesn't apply CABAC, either, so the decoding will be easier, hopefully compensating for more reference frames.

We can start with the H.264 LAN preset, with these modifications (Figure 22.5):

- Video
- Key Frames every 120 frames
  - Data Rate: Restrict to 904
  - We don't have any explicit peak buffer control; we'll leave it on Download and hope it comes out with a reasonable peak.
    - We need multipass on to get this option
  - By using Frame Reordering we'll get 1 B-frame
- Audio
  - Rate: 44.1 KHz
  - Quality: Best (it's really not meaningfully slower)
  - Target Bit Rate: 96 Kbps
    - Stick with average bitrate mode as this is progressive download
- Filters
  - Audio: Dynamic Range On (This is Compressor's Normalize filter. The defaults are nice, and allow greater tuning. The noise floor is particularly useful with noisy source, but needs to be calibrated per-source.)
- Geometry 4:3
  - Crop Left and Right of 8
  - Frame Size 480 × 360 (Custom 4:3)
- Geometry 16:9
  - Frame Size 640 × 360 (Custom 16:9)

(*Continued*)

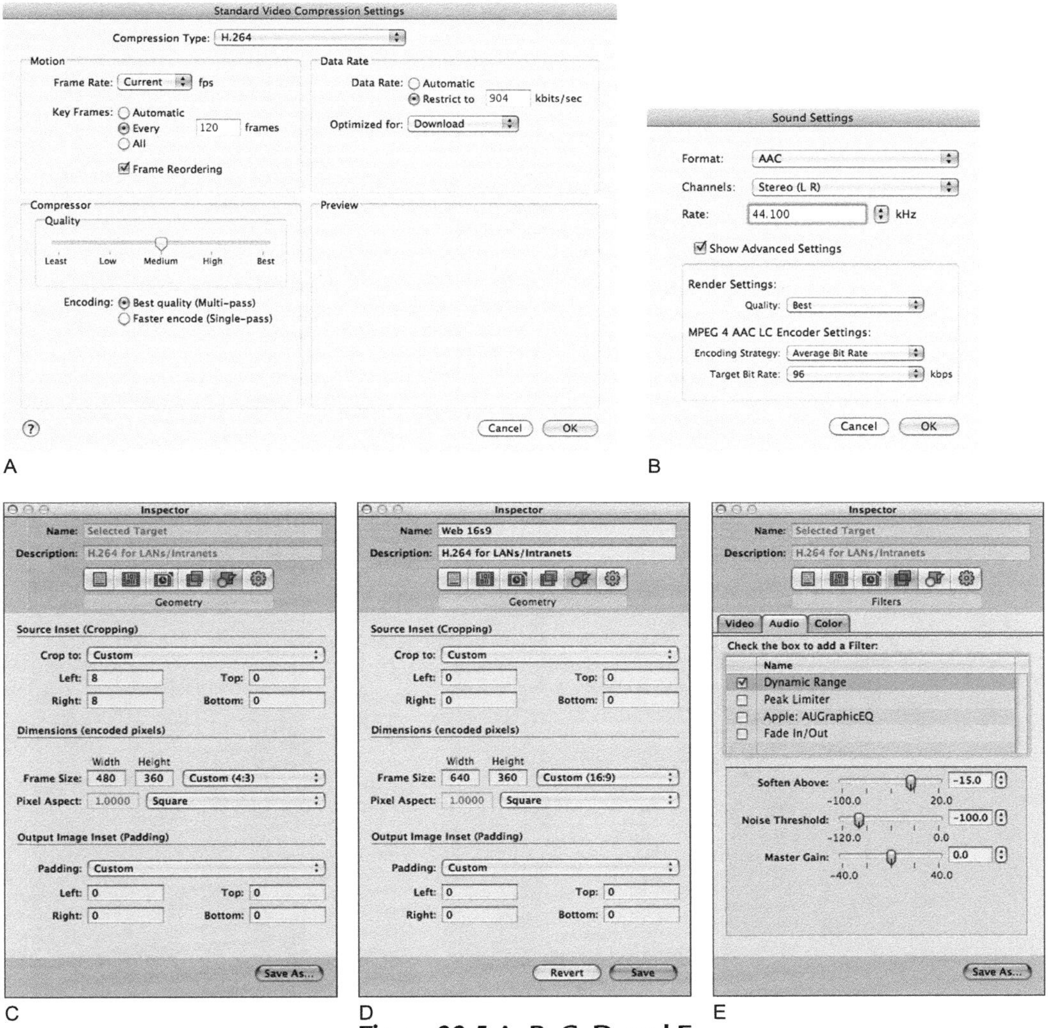

Figure 22.5 A, B, C, D, and E

Our settings in Compressor. I sure wish I could get to the advanced features targeting .mp4.

CHAPTER 23

Optical Disc: DVD, Blu-Ray, and ROM

Introduction

This chapter is about video delivered on optical discs, particularly DVD and Blu-ray disc (BD), and also general principles that apply to Video CD, CD- and DVD-ROM, and other optical media–based usage.

While discs may seem somewhat atavistic in the Internet era, they're still a huge portion of content delivery. And Hollywood studios sequence their releases, so there's a particular "DVD window" (which includes Blu-ray) that comes directly after theatrical release but before content is made available for video-on-demand (VOD)/pay-per-view (PPV), then premium cable, and lastly broadcast.

Since DVD makes up such a huge portion of Hollywood revenue (although it has shrunk some in recent years), it gets a lot of focus from the studios, who make a lot more per disc than they do per viewing in the VOD/PPV or later windows.

The combination of big markets, and hence big budgets, plus the interoperable interactivity on DVD and Blu-ray means that disc get lion's share of complex authoring. So discs are where we see scene-by-scene compression tuning, director's commentaries, multiple languages for audio and captions, nicely designed menus, and a panoply of extras.

From a compression perspective, DVD and Blu-ray have very precise specs, which makes things a lot more straightfoward than web encoding. However, they also have high expectations, which can make those limitations chafe in some cases.

I hope we'll see the "DVD window" evolve into more of a "premium window" so interactivity (and high budgets!) can have their bits liberated from their prison of spinning polycarbonate. This is a project I've been personally working on for a while now, and 2010 should see a number of downloadable movies with all the features of the DVD version.

But for the time being, DVD and Blu-ray are where we see premium interactive design and innovations, and the budgets that can afford those efforts.

Characteristics of Disc Playback

The physical differences between Internet delivery and discs have an impact on how we compress. With discs, we typically get a lot of (but not limitless) capacity with quite high peak bitrates. However, due to the relatively slow rotational speed of optical discs, GOPS and buffers are smaller than we use for network-distributed media in order to deliver decent random access. And with discs, we have a precise maximum number of bits we can use, without the "maybe we can get away with 1500 Kbps" gambles of web video.

The combination of fixed maximum capacity and fixed maximum bitrate means that there are two basic strategies for rate control on discs, depending on capacity of the disc and duration of the content:

- When the duration of the content is such that the disc won't be filled when encoded at maximum bitrate, CBR can be used.
- When the duration of the content means that the average bitrate needs to be lower than the peak, 2-pass VBR yields the optimum quality.

However, particularly with DVD, there's a big difference between discs that are stamped in a factory (called "replicated") and those burned one-at-a-time in a drive (called "duplicated"), forcing a tradeoff between breadth of compatibility and peak bitrate. I'll talk later in this chapter about how that works in practice.

The interactive layers of DVD and Blu-ray are well specified, which make them more feasible to use than typical PC media players, although still behind Flash and Silverlight. Today, DVD's interactive technology may seem nearly as primitive as an abacus, but a decade of refinement of authoring technologies targeting that allow for quite rich experiences.

Blu-ray has very deep interactivity support via Java (BD-J), although we still haven't seen easy-to-use interactive authoring tools for BD-J like we have for DVD.

A Cunning Trick for Multiple-File 2-pass VBR

2-pass VBR works great when there's one big clip that dominates most of the disc. But a compilation disc that has multiple shorter subjects would yield a per-clip 2-pass VBR. This can waste bits on the easy clips and starve harder clips, yielding inconsistent quality.

One simple solution is to concatenate all the files together into one big source, and encode them at once, so rate control is applied over the whole clip. Make sure there's a chapter mark at the start of each clip, and chapter navigation can be used to present the content as individual clips.

DVD

DVDs were probably the first place most people saw digital video without palpable artifacts. It was definitely a quality jump over broadcast, cable, and particularly VHS of its era. And despite the predictions of some film buffs, it could look a lot better than the Laserdisc. Laserdisc was the best ever composite consumer format, but DVD was component, could do higher bitrate 5.1 sound, and had much better interactivity.

DVD Tech Specs

DVDs can be either single- or dual-sided, and each side can be either one or two layers. The second layer has somewhat lower capacity than the first.

Most DVDs are the classic 12 cm diameter, like CD and Blu-ray. A smaller 8 cm size is also available. However, these small discs don't work reliably in disc-load drives like those in laptops. While cute, they're no cheaper to make than 12 cm discs, and I recommend against them for compatibility reasons. Table 23.1 shows the various DVD types and their capacities.

DVD introduced the UDF (Universal Disc Format) file system, an updated version of which is used with Blu-ray. UDF is very handy for ROM discs as well, as it is well-supported by all operating systems.

One nice physical property of DVD is that each side is symmetrical, with the reflective surface in the center of the disc. This contrasts with CD, where the reflective surface was at the bottom of the plastic, next to the label, and Blu-ray, where the reflective surface is under just 0.1 mm of protective coating. This makes DVDs easier to manufacture and quicker to cool, allowing more discs an hour through a machine and a cheaper per-disc price than either CD or BD.

MPEG-2 for DVD

DVD video uses a tight subset of MPEG-2, with each disc being either NTSC or PAL. DVD is very capable of producing the highest quality of any SD consumer video format. The spec maximum bitrate is 9.8 Mbps for video and audio and 10 Mbps total for audio, video, and subtitles. In practice, experts recommend using a peak of at most 9.6 Mbps for replicated discs. DVD supports VBR, and high-end discs almost always use 2-pass VBR to produce optimal quality within the disc capacity.

While the encoded video on nearly all DVDs is 720x width, other modes are supported for 4:3 content (720x is required for 16:9). MPEG-1 bitstreams encoded to Video CD spec are

Table 23.1 DVD Types and Capacity.

Name	Sides	Layers	Diameter	Capacity
DVD-1	1	1	8 cm	1.46 GB
DVD-2	1	2	8 cm	2.66 GB
DVD-3	2	1/1	8 cm	2.92 GB
DVD-4	2	2/2	8 cm	5.32 GB
DVD-5	1	1	12 cm	4.70 GB
DVD-9	1	2	12 cm	8.54 GB
DVD-10	2	1/1	12 cm	9.40 GB
DVD-14	2	2/1	12 cm	13.24 GB
DVD-18	2	2/2	12 cm	17.08 GB

Table 23.2 Legal DVD Encoding Modes.

Format	Standard	Width	Height	Aspect Ratio
MPEG-1	NTSC	352	240	4:3
MPEG-2	NTSC	720	480	4:3 or 16:9
MPEG-2	NTSC	704	480	4:3
MPEG-2	NTSC	352	480	4:3
MPEG-2	NTSC	352	240	4:3
MPEG-1	PAL	352	288	4:3
MPEG-2	PAL	720	576	4:3 or 16:9
MPEG-2	PAL	704	576	4:3
MPEG-2	PAL	352	576	4:3
MPEG-2	PAL	352	288	4:3

also supported for compatibility with existing content. But even if 320 × 240/288 encoding is appropriate (like trying to cram six hours of talking-head video on a disc), MPEG-2's greater efficiency and support for VBR make that a better choice. Legal DVD encoding modes are shown in Table 23.2.

Many encoders for DVD only support the 720x modes.

Quality expectations for DVD are quite high, especially compared to other MPEG-2 applications. Typically, any obvious blocking or ringing artifacts are considered unacceptable. "Good enough" data rates with professional encoders for 24p/25p sources are around 4–5 Mbps with peaks up around 8 Mbps. MPEG-2 VBR encoders may provide a minimum bitrate, used to make sure that the VBR algorithm doesn't strip easy scenes of too many bits. Interlaced DVDs can be either top or bottom field first. I recommend setting the output field order to the same as the source to avoid unnecessary conversions. Interlaced content requires roughly 20 percent higher bitrate than progressive.

The maximum GOP length of DVD is 18 frames for NTSC and 15 for PAL, with the defaults being 15 and 12 respectively. This makes for quick random access, but a risk of rapid keyframe strobing at lower bitrates.

Aspect Ratio

MPEG-2 for DVD supports both 4:3 and 16:9, and the player will convert as needed for the attached display (assuming correct configuration).

16:9 is still a taller ratio than most films. Because of this, even 16:9-encoded MPEG-2 for DVDs will have some letterboxing, although much less than the same content in 4:3. 16:9 can use 33 percent more active pixels than 4:3 with wide-screen content—especially important when upscaled on computers or HD displays.

You can gain a slight increase in compression efficiency by aligning the letterbox edges with macroblock or block boundaries. The very sharp edge between the letterbox and the film content can cause noticeable ringing; MPEG-2 is more prone to this than more recent codecs. See Table 6.1 for some typical combinations.

Progressive DVD

It is a lingering tragedy of the age that MPEG-2 on DVD doesn't support MPEG-2's progressive sequence mode; only interlaced sequences are used, so there's no way to do "real" 24p encoding. However, interlaced sequences do support both progressive and interlaced frames. For NTSC discs, this can be used to encode the 3:2 telecine pattern in a way that's easy for players to reverse (see page 115 for a more thorough discussion of 3:2 pulldown and inverse telecine).

24p DVD uses a MPEG-2 sequence with repeat_field tags. These let the encoder not encode the duplicate field in the "3" phase of "3:2 pulldown." This doesn't need to be handled manually—the encoder itself should correctly tag the output if the source is 24p.

Then, on playback a "progressive" DVD player will see the pattern of repeat_field tags and reassemble the original 24p content.

PAL compressionists have it easier; 24p source is sped up to 25p and encoded as a series of progressive frames. Furthermore, they get 20 percent more pixels a frame (576 instead of 480).

NTSC 30p content can also be encoded as 30p easily, by just encoding it as a series of progressive frames. This is often a better choice for 60p sources without a lot of high motion (25p can be better for 50i for the same reason). For sports, going from 60p or 50p to 30i or 25 can deliver much smoother motion, but it always feels shameful and cruel to rend innocent progressive frames into interlaced fields.

With interlaced displays are on their way out, progressive discs look a *lot* better than interlaced on a progressive screen, particularly for PC or HD playback.

Multi-Angle DVD

One much-hyped feature of DVD is its ability to present multiple camera angles that can be selected and viewed on the fly. It sounds cooler in theory than it is in practice. Multi-angle is complex to author, and even when done perfectly, the speed of angle transitions can vary quite a bit depending on the player. To have seamless multi-angle switching, the GOPs in each stream must align perfectly. So you'll normally need to compress with:

- Max peak of 8 Mbps per angle (up to 5)
- Closed GOP

- Auto keyframe/I-frame insertion/Scene detection off (to keep GOP cadence intact)
- Any chapter marks need to be the same frame in all bitrates

Unlike multiple audio streams, which come right out of video top bandwidth, each angle gets its own bandwidth. In theory, each can be VBR, but that makes quite complicated; many products require multi-angle tracks to be CBR. Needless to say, five 8 Mbps CBR angles eat up disc capacity very quickly.

Multi-angle is an interesting technology, but not much used outside of the adult industry (I'll leave that proof as an exercise for the reader). Thus it's not supported by many tools outside of the high end.

Multi-angle is one of those finicky things you want to have successfully authored *before* you commit to a project using it.

DVD Audio

DVD's practical mux limit is 9.6 Mbps is for video plus all audio tracks. All DVD audio codecs are CBR, so every bit allocated to audio comes out of your video peak bitrate. You therefore need to be judicious in how many audio tracks you include and how many bits to give them, particularly with harder-to-encode interlaced video.

The PCM, Dolby Digital, and DTS codecs are covered in more detail along with MPEG-2 on page 173.

PCM (mandatory)

PCM audio is mandatory on DVD players, but I strongly recommend against it. Even something as simple as stereo audio eats up 1.5 Mbps of your precious bandwidth. Dolby Digital encodes much more efficiently.

Historically, PCM audio was mainly seen in consumer-authored discs created with tools that didn't license AC-3 encoders from Dolby. However, any DVD authoring tool worth using today includes at least stereo AC-3.

PCM support does offer some impressive numbers, but they're largely specsmanship for anything outside of the rare audio-only "Audio DVD" format, particularly 96 KHz. Supported options are:

- 48 or 96 KHz
- 1-6 channels
- 16 or 24-bit

Dolby Digital (mandatory)

Dolby Digital (AC-3) is really the default codec for DVD; most of the discs in the world use AC-3 and nothing else.

DVD has a pretty typical implementation, supporting up to 448 Kbps (less than the codec's 640 Kbps max as supported on Blu-ray, but an upgrade from Laserdisc's 384 Kbps). Good defaults are 224 Kbps for stereo and 384 Kbps or 448 Kbps for 5.1 mixes. If there's only a single track and the project is audiocentric, you can max out stereo at 448 Kbps; that's still more than 1 Mbps saved versus PCM.

But even with AC-3's better compression, be careful when providing a lot of audio tracks. One 448 Kbps track isn't a problem, but add English, Spanish, and French at 448 Kbps plus a commentary track at 224 Kbps and your peak bitrate is 1568 Kbps lower.

I normally only encode the original language at 448 Kbps. Since dubbed audio generally sounds lousy anyway, I'll reduce bitrate on those if needed to keep video quality up; audiophiles will be listening to the original language track anyway. And a commentary track can go down to 128–196 Kbps as long as it's mainly speech and doesn't mix in too much of the soundtrack.

Note that DVD only plays back one audio track at a time. If you want to present commentary with the soundtrack in the background, you'll need to mix all of that together before compression.

DTS (optional)

The DTS audio codec is optional on DVD. This means any DVD using DTS also needs to include PCM or AC-3.

While DTS, particularly in its higher 1536 Kbps mode, can be more transparent than AC-3 at 448 Kbps, that's a big price in bandwidth budget, particularly with an AC-3 fallback stream still required. Most DTS on DVD uses the lower 768 Kbps mode, which isn't consistently better than 448 Kbps AC-3.

Due to DTS being an optional codec, for most content the reduction in video bitrate and capacity isn't worth the theoretical advantages. One obvious exception would be music-centric content, like concert videos. In those cases, a premium audio experience at the top DTS data rate may be well worth it.

MPEG-2 Layer II (mythical)

MPEG-2 Layer II is often listed at a DVD audio codec, but it's almost never seen in the wild, and isn't mandatory in NTSC players. AC-3 is more efficient than Layer II, and Layer II isn't universally supported by receivers, so there's no reason to use it.

DVD Interactivity

That DVDs are as interactive as they are is a testament to more than a decade of hard work on the part of authoring tool vendors much more so than to the underlying technology, of which there really isn't much. I doubt you can buy a wristwatch that doesn't have more processing power than the first DVD players.

In the end, DVD can play a single video and a single audio track at a time. Subtitles are just 4-bit indexed color TIFF files, as are most other overlay graphics. Total available memory for programming is 256 bits. Not Mbits. Not Kbits. Just bits: a set of sixteen 16-bit numbers. And programming them can feel like using an abacus wearing mittens.

But wow, some impressive experiences have been built on top of that simple base. A key innovation was the use of video for menus, so the very simple 4-bit TIFF overlays aren't needed.

DVD Mastering

VOB files

DVD video doesn't actually use MPEG-2 files, but incorporates the MPEG-2 video and audio information into Video Object (VOB) files, multiplexing (muxing) the video, audio, and subtitles together. These individually are limited to 1 GB in size, so if you look at the disc's file structure, you'll see a series of these for movie-length content.

Replication

A replicated disc is mass-produced from a "glass master," as opposed to the one-at-a-time discs burned in duplication. The replicator creates the glass master from a provided disc image. Traditionally, this was provided on a DLT tape, but most replicators now also support DVD-R masters.

DVD replication has become dirt-cheap and broadly available with the maturity of the technology. Short runs of 300–500 discs can be purchased for under $1,000, with the cost per disc dropping as volume goes up (that's just for the disc, without printing or packaging).

The replicator is responsible for collecting a $0.04 MPEG-2 license fee per disc, paid to MPEG-LA.

While dual-layer replication was initially finicky and expensive, it's now only a slight premium over single-layer, and actually cheaper per GB.

Beyond being cheaper as discs runs get into the thousands, replicated discs are also more reliable, particularly in older players, and more particularly with higher peak mux rates and dual layers.

Duplication

Duplication ranges from a laptop burn of a single copy of a disc to professional-grade automated burning of hundreds of discs by a replication company.

For the usual political–technical reasons, there's two camps of writable DVD discs: the "dash" family originated by Pioneer and the official writable format of the DVD Association (call it a "minus" at your peril!) and the "plus" family later introduced by Philips and Sony.

There are those in the industry who still take this split very personally, but there are pragmatic reasons to pick a particular format for a particular application. The good news is that PC DVD-ROM drives of today support reading and writing all the formats well. It's dedicated DVD video players, particularly older ones, where compatibility concerns creep in.

DVD-R

DVD-R was the first writable disc format. The original version was called "DVD-R for Authoring" and clocked in at only 3.95 GB. I worked with Intel on the first DVD encoded on a PC instead of the refrigerator-sized real-time encoders of the day. We burned test discs on one of the first three DVD-R drives in the United States. The blanks were $40 each, took several hours to burn and verify, and failed about half the time. When burning a disc we'd have everyone clear off that floor of the building to minimize any vibrations that might increase the risk of a bad burn.

Fortunately, this was soon replaced with today's "DVD-R for General," which everyone calls just DVD-R today. With further fortune, it has the same 4.7 GB capacity of a DVD-5.

DVD-R is the most reliable burnable disc for DVD players, particularly older ones that predate the later writable formats. Even so, some early players have trouble with higher bitrates and Open GOP; my first-generation Sony 700 DVD player only liked Mitsui Gold DVD-R discs, and would spit out anything else. If content can handle it, I generally recommend a maximum mux rate of 6.5 Mbps with DVD-R discs, as this improves reliability.

There is a dual-layer variant called DVD-R DL. It is much less compatible with older players. Quite a bit more expensive, and burns much slower than single-layer. I stay away from it for DVD unless it's really the only way to get the movie on the disc, and if I think the audience is mainly going to have newer players or be using PCs.

DVD-RW

DVD-RW is the rewritable version of DVD-R. Thus, you can add to or rewrite the disc after an initial write. This is handy for PC use, but doesn't offer much for DVD authoring, since the disc is mastered in one pass anyway. Given how cheap DVD-R discs are, and that DVD-R burns at a higher speed, DVD-RW doesn't have a clear niche for DVD video.

DVD+R

DVD+R was only introduced in 2002, and was more focused on replacing CD-R discs for PC use than competing with DVD-R in movie recording. Thus the "+" formats are better suited to PC file storage, but have lower compatibility with older players. Since they don't have any advantages when it comes to burning movie discs, I don't recommend DVD+R for DVD.

DVD+R has a dual-layer variant called DVD+RW DL. Even less compatible for DVD video.

DVD+RW

DVD+RW, as you figured out from the title, is the "+" rewritable variant. +RW predates +R, and was the impetus for the format. It was originally launched in 1997 as a 2.8 GB disc (big for its time), but quickly abandoned until 2001 when a 4.7 GB version emerged. Its biggest difference from DVD-RW is support for random access writes instead of DVD-RW's track-based writes, making it much faster and easier to make a lot of little changes to a disc.

Thus, it's the best "floppy replacement" of the disc types. As USB thumb drives are quickly taking over that market, however, the DVD+RW is seeing rapidly declining use.

DVD-RAM

DVD-RAM is an early cartridge-based rewritable disc format. While the oldest of the bunch, it's the least compatible (not even including the cartridge thing) of the writable discs. I used it for backups and client delivery many moons ago, but its low capacity makes it ill-suited for anything today.

Tip: No Sticky Labels!

So, disc labels. It's great to have a labeled disc, with a nice professional logo, not just a scrawl with a Sharpie. It looks great for the client.

But whatever you do, don't use a sticker label! They have some big downsides:

- They make the disc a little heavier than the spec, making the drive motor work harder.
- They're hard to get perfectly centered, and if they're not, they're unbalanced and can cause wear and noise while spinning.
- If it's not perfectly adhered all the way around the edge, a little corner could peel up and turn into 9200 RPM flypaper inside of the DVD mechanism of your customer.

To get a nice label on a burned disc, you've got two good options:

- Using discs with a white matte finish, an inkjet printer with a 12 cm disc tray can produce a very high-quality image; often better than a silkscreened replicated disc.

- Using LightScribe or LabelFlash compatible discs and burners, a monochromatic image can be burned on the label side by putting the disc in the drive upside down. However, these images are less attractive and less durable than inkjet ones (particularly when exposed to bright light), and more suitable for consumer and one-off projects.

Blu-ray

Introduction

Blu-ray was originally conceived of as a HD version of the DVD recorders then very popular in Japan, used for off-air recording like VHS once was. Thus it was designed as a writable format from the first instance, avoiding DVD's painful experience trying to retrofit a writable format into a replicated disc technology. Blu-ray eventually evolved into a packaged media format once they had replication and DRM figured out.

Blu-ray launched into a fierce format war with the rival HD DVD format. Each format delivered similar experiences but with their own strengths and weaknesses. There was a lot of sound and fury from corporations and enthusiasts, since both sides felt HD discs couldn't take off until the format war was resolved, and pushed hard for a swift victory. In the end, HD DVD was defeated in the boardroom, not the lab.

It remains to be seen whether Blu-ray is the last great optical format or the first casualty of the HD Internet.

For the readers of this book, the biggest challenges in Blu-ray may come from the expense and complexity of authoring—in particular, the requirement that all replicated discs require AACS encryption.

Blu-ray certainly isn't going to provide the "optical floppy" and backup roles that CD-ROM and to a lesser degree DVD-ROM did. Hard drives are already much cheaper than Blu-ray per GB, flash memory is approaching that point quickly, and adoption of BD-ROM drives in PCs is much slower than it was for CD- and DVD-ROM. The success of Blu-ray largely hinges on its success for video content.

Blu-Ray Tech Specs

Similar to DVD, Blu-ray discs can be either single- or dual-layered, and single or dual-sided, in either 12 cm or 8 cm size. Unlike DVD, both layers have the same data density of 25 GB (so dual-layer is 50 GB). So far all movie discs have been 12 cm single-sided, with a mix of

BD-25 and BD-50 capacity. The 8 cm discs have been used mainly in camcorders, but haven't really caught on and appear to be fading out in favor of flash memory storage.

Red-laser media can also be used for Blu-ray content, with those discs being called either BD-5 or BD-9; these use the DVD physical formats with the BD logical format. Some of the first generation Blu-ray players weren't able to play those with their original firmware; it's unclear if a meaningful number of those are still in use.

Player profiles

While Blu-ray is lauded for its interactive features, not all players support those, and none of the first generation. There have been three profiles of video players so far. In the rush to compete with HD DVD, Blu-ray launched with a simplified "Grace Period" Profile 1.0 with limited features. This has since been replaced by Profile 1.1 (a.k.a. "Final Standard" or "Bonus View") and Profile 2.0 "BD-Live" players (see Table 23.3).

There is also a BD-Audio profile (3.0, although it has fewer features than 1.0) for audio-only disc. There has been little interest in BD-Audio so far, unsurprising given the staggering lack of customer interest in the previous DVD-Audio and SuperAudio CD formats.

The different levels of functionality can make authoring complex, as the disc has to be able to hide or expose features based on what the player can handle.

Additionally, Profile 1.0 players without a network connection have no easy way to get firmware updates, and so may have issues with current discs. Increasingly, publishers aren't bothering to test compatibility with older firmware in older players, instead offering instructions for how to find current firmware. This requires a user to download a disc image on a PC and burn a DVD-R.

The most popular and capable Blu-ray player so far has been the PS3, sold at a loss by Sony, hoping to make it back on game sales. The PS3's Cell processor, which hasn't proved anything special for games per se, is an excellent media processor able to handle full BD playback, including dual-channel decoders, in software. The PS3 has had continual firmware updates keeping even first-generation models competitive with the latest high-end dedicated Blu-ray players.

Blu-Ray Video Codecs

Blu-ray has three mandatory video codecs: MPEG-2, VC-1, and H.264. The initial wave of Blu-ray discs used MPEG-2, and didn't deliver great quality, although this was as much due

Table 23.3 Blu-Ray Profiles.

	Profile 1.0	Profile 1.1	Profile 2.0
Internet connection	None	None	Required
Local storage	Optional	256 MiB min	1024 MiB min
Secondary video decoder	Optional	Required	Required
Secondary audio decoder	Optional	Required	Required

to poor mastering and QA as to poor compression. It turned out that Sony's compression lab was using expensive, high-end Sony HD CRT displays with professional calibration. Properly calibrated with uniform gamma, CRT produces a softer image than a 1080p LCD, and its much lower black level hides blocking-in-black issues that are painfully revealed on LCDs. After the poor reviews of their first round of titles, Sony bought a bunch of consumer displays to use in their QA process.

Today, most commercial Blu-ray discs use either H.264 or VC-1, providing much more efficient encoding. While an early goal for Blu-ray was to use MPEG-2 to record ATSC or DVB broadcasts for later playback, hard drive storage has become much cheaper than Blu-ray discs per GB. And so PVRs or Media Center–style recording dominates most markets.

At this point in the market, there are a number of compression tools with "Blu-ray" settings that aren't fully compliant with the Blu-ray spec; the output files won't mux correctly or pass validation. Generally the high-end tools are fine, but make sure others have been able to use lower-end tools for real Blu-ray production before you spend a long time authoring.

Blu-ray supports a big array of video formats, spanning SD to HD (Table 23.4). In a welcome innovation, it supports real 24.000 fps as a frame rate, as well as the more traditional 23.976.

You can get a *lot* of SD video onto a Blu-ray with H.264. At 1 Mbps VBR (an aggressive H.264 High Profile bitrate for SD interlaced, but not insanely so), about 100 hours including low-bitrate audio can fit on a BD-50.

Table 23.4 Supported Video Formats and Specs for Blu-Ray.

Resolution	Frame rates	Aspect ratio	Codec
1920 × 1080	50i 59.94i	16:9	Any
1920 × 1080	23.976p 24.000p	16:9	Any
1440 × 1080	50i 59.94i	16:9	H.264 and VC-1 only
1440 × 1080	23.976p 24.000p	16:9	H.264 and VC-1 only
1280 × 720	50p 59.94p	16:9	Any
1280 × 720	23.976p 24p	16:9	Any
720 × 480	59.94i	4:3/16:9	Any
720 × 576	50i	4:3/16:9	Any

For typical use, even BD-25 is ample for long-form content. BD supports a maximum mux rate of 54 Mbps and a maximum video bitrate of 40 Mbps—nice to have for MPEG-2 but overkill for VC-1 and H.264.

I think DVD-9 could be a fine medium for quite a lot of content. That's about 8 Mbps for two hours of video, which is within reach of 1080p24 with good encoders.

Tip: 2-Sec GOP at Low Bitrates

Normally Blu-ray requires a GOP length of 1 second. But if the peak bitrate of the encode is 15 Mbps or less, a 2-second GOP is permitted.

Lower peak bitrates are never a bad thing when targeting DVD-R playback, and when trying to cram a lot of content on a disc, 2-sec GOPs with H.264 or VC-1 can improve low bitrate efficiency and reduce keyframe flashing.

Blu-Ray Audio

Blu-ray has a confusing selection of audio codecs, many of them optional. With optional codecs, there's no requirement for players to include a decoder, and so the user may need a receiver to decode that bitstream. And since the player can't do any mixing in that case, using an outboard decoder means that any sound effects or audio commentary stored in a separate audio track can't be heard.

So, Blu-ray audio really should be decoded and mixed in-player, and then passed out along with the decoded and composited video frames. The best and easiest option is to simply have HDMI carry out synced uncompressed video and audio. Having six discreet analog audio outputs also works, albeit with a much greater tripping hazard. In cases where HDMI audio isn't an option, a receiver without HDMI 1.1, a fallback to TOSLink output may be required. In that case, 5.1 audio would be re-encoded from the mixed version to AC-3 or DTS.

PCM (mandatory)

Straight-up uncompressed audio is as simple as it gets. While the bitrates are high, they're generally fine for simple titles; even 7.1 24-bit 48 KHz can fit into the 14 Mbps between the 40 Mbps max video bitrate and the 54 Mbps max mux rate. Many early BD titles included both 5.1 PCM and 640 Kbps AC-3.

Blu-ray's PCM goes up to 7.1 channels, 24-bit per channel, and 96 KHz. I remain of the belief that anything beyond a well mastered 48 KHz 20-bit is mainly useful for annoying bats, but don't discount the placebo power of big numbers, and the marketing value thereof.

The downside to PCM comes when trying to pack a whole lot of stuff onto a single disc, particularly a red-laser BD.

Dolby Digital (mandatory)

Dolby Digital is a mandatory codec, and is included on most discs. BD's AC-3 goes up to 640 Kbps maximum, compared to the DVD and ATSC maximum of 448 Kbps. Unless you're targeting DVD-5/9, the full 640 Kbps should be easy to fit.

DTS-HD High-Resolution Audio (mandatory)

One big change from DVD is that DTS has become a mandatory codec, after being optional on DVD. This means a disc can use only DTS, without an AC-3 or PCM fallback.

While the higher bitrate 1.5 Mbps flavor of DTS was rarely used on DVD due to the high bitrate, it's much more feasible on Blu-ray media. I can't see using the half-rate mode of DTS over 640 Kbps AC-3; AC-3's greater efficiency should give it better quality if there's any apparent difference.

The HD extension to DTS allows up to 96 KHz 24-bit (again, mainly useful for annoying bats) at up to 6 Mbps. DTS-HD is a good intermediate for content where 640 Kbps AC-3 might not be enough, but lossless is too expensive in bits.

Dolby Digital Plus (optional)

Dolby Digital Plus was a great codec on HD DVD, but is much less applicable in its Blu-ray implementation. BD DD+ is merely an enhancement layer to add two extra channels to take a 5.1 AC-3 stream to 7.1 audio. Given the dearth of 7.1 audio sources in general, DD+ on BD is unlikely to see wide use.

DD+ can use up to 1 Mbps extra, but since it'll only contain two channels compared to the 640 Kbps used for 5.1, in practice it'll be much less.

Dolby TrueHD (optional)

Dolby TrueHD is Dolby's lossless codec. In its Blu-ray implementation, it's really two different bitstreams muxed together—a normal AC-3 track and the lossless track, using the Meridian Lossless Prediction algorithm from the old DVD-Audio format. Thus, TrueHD is always backward-compatible to old players, although no bits are saved through this mechanism.

TrueHD on Blu-ray supports up 24-bit, with up to 7.1 with 48 KHz, and up to 96 KHz with stereo only.

DTS Master Audio (optional)

DTS Master Audio (DTS-MA) is the DTS lossless codec. Like TrueHD, it includes a backward-compatible core. Unlike TrueHD, the lossless layer is encoded as the difference between source and the core encode, making for a more efficient encode.

DTS-MA goes up to 24-bit and 96 KHz for stereo and 48 KHz up to 7.1. The bitrate can go up to peaks of 24.5 Mbps.

Blu-Ray Interactivity

HDMV

Initial Blu-ray titles used the HDMV interactivity layer, which provided a DVD-like authoring environment. It's largely being displaced by the much more capable BD-J.

HDMV may become memorable as the last major media technology to use indexed color graphics; HDMV image overlays were 256-indexed color palette PNG.

BD-J

BD-J is the "real" Blu-ray interactivity layer, and increasingly the only one supported by authoring tools.

BD-J is an implementation of Java that runs on BD players, and working with it directly is software engineering, not menu design. Vendors are building design layers that sit on top of BD-J to provide richer design environments that don't require coding, like they did with DVD.

Blu-Ray Mastering

Replication

Blu-ray struggled for a while with BD50 replication, which frustrated many proponents, since that offered a hypeable capacity advantage over HD DVD's 30 GB max.

AACS

Perhaps the most frustrating aspect of Blu-ray is the mandatory requirement for AACS encryption on replicated discs (red laser too), which includes some pretty stiff fees. And even these fees are a big reduction from what they were until the summer of 2009 (disclosure: thanks in part to lobbying from the Interactive Digital Media Association, of which I'm a board member).

This adds to the cost and complexity of using replicated BD discs; coupled with the high costs of BD-R media, it makes business models based around short-run titles (think wedding videos) a lot less economically feasible than with DVD (Table 23.5). DVD-R is the only feasible choice today for short-run Blu-ray content in many cases.

Duplication

Duplication on Blu-ray has a couple of big advantages over DVD.

Table 23.5 Mandatory AACS Fees.

Fee Category	Original	Current	Note
AACS Content Provider License	$3,000 one-time	$500/year	$3000 eliminates yearly fee
AACS Title Key Certificate	$1,300/master	$500/master	Per master, not per title
AACS per-disc Royalty	$0.04/disc	$0.04/disc	Uncapped

First, since Blu-ray was designed as a writable format, all players can handle all BD-R discs, both 25 GB and 50 GB.

And more importantly, AACS isn't required, which makes low runs a lot more cost-effective.

But there's a big downside; BD-R blanks are a lot more expensive than BD replication. We'll hopefully see prices fall over time as we did with CD-R and DVD-R.

DVD-R

Blu-ray works fine on red-laser burnable media, including DVD-9. And with VC-1 and H.264, you can actually get a couple of hours of pretty good-looking video on a single disc.

In a compressionist showoff move, I once did a two-hour 1080p movie on DVD-9 with MPEG-2, using 8.5 Mbps ABR. It wasn't "Blu-ray quality" but held its own against a lot of broadcast ATSC MPEG-2. Modern VC-1 and H.264 implementations can do very nicely at these bitrates at 1080p24.

If you stick to peak bitrates of 15 Mbps or less, you can use a two-second GOP, which can help efficiency on a small disc. Also, while Level 4.1 H.264 requires four slices in the Blu-ray spec, lower bitrates are Level 4.0 and so can use single-slice encoding for a little better efficiency.

BD-R

BD-R is the write-once flavor of BD. It was originally very expensive, but BD25 blanks have dropped to about $4 in volume as this book was written (still more expensive per GB than DVD-R). Dual-layer are also available and have good compatibility, but are about four to five times as expensive.

BD-RE

BD-RE is the rewritable flavor of BD. It's less than twice the cost of BD-R, so it can be useful for test discs. It's available in BD25 and (quite expensive) BD50.

There had been high hopes for BD-RE to be the next generation floppy/CD-RW/DVD + RW sneakernet/home backup format, but this seems unlikely unless prices drop dramatically and Blu-ray drives become much more common in PCs.

HD DVD – A Lament By Way of Full Disclosure

While I've spent most of my time at Microsoft on the codec and then Silverlight teams, I actually joined the company as part of the HD DVD team, before the format launched. Although we got steamrollered in the end, it was an interesting time, and a great format.

Some of the features of HD DVD I miss in the Blu-ray era include:

- DVD-based manufacturing process, so a single production line could produce either kind of disc with similar costs.
- Higher bar of mandatory audio codecs, including the real Dolby Digital Plus codec, which was stellar at 1.5 Mbps. Audio was always decoded and mixed in-player.
- No low-end profiles: All players included an Ethernet port, local storage, and secondary video decoder.
- A web-like authoring workflow using markup and JavaScript.
- AACS was optional for both replicated and duplicated discs.

HD DVD was a much better format for smaller video shops doing short run content, which in aggragate makes up a huge part of the video industry. We'll see if Blu-ray finds a home there, or whether they'll stick with DVD until they can fully transition to the web.

Tutorial: 24p DVD

Scenario

You're a compressionist by day, but a proud parent all the time.

Your daughter is in the school musical, along with many other kids. The theater department has had its budget cut again, risking the chance of next year's play being cancelled, which your younger daughter is eager to be in. The parents have agreed to make a DVD of the play to sell as a fundraiser.

Other parents will shoot and edit the event, but you're on tap to make the DVD.

The Three Questions

What Is My Content?

The content is being shot with several AVCHD 24p cameras.

The editor uses Final Cut Pro and will provide content back as 720p24 ProRes .mov files, with a stereo mix.

(*Continued*)

Who Is My Audience?

The addressable market is largely grandparents and other relatives of cast members. Fortunately, there's a large chorus and many extras, and with preorders, 500 discs looks like a reasonable sales target. By selling the discs at $20/each and using a cheap ("cost-effective") duplication service, we should net enough to save the theater program. The show must go on!

What Are My Communication Goals?

This is your kid that you're compressing here! This doesn't just have to look good to the other grandparents—it's got to look good to you, too.

Tech Specs

Since we're doing DVD-R for cost reasons and as these are general consumers, we should stick to DVD-5. To be really safe, we should keep the peak bitrate to 6.5 Mbps.

While we just have a stereo mix, as a musical the audio is central to the experience. We'll bump up the stereo bitrate a bit to 256 Kbps.

The source is 16:9 720p, so we'll encode 16:9, scaled down to DVD's 720 × 480. The video editor (overinspired above and beyond the call—like you) carefully color-corrected in 709 color, so you'll want to convert to 601. The source is 24p, which we'll need to properly 3:2 encode on DVD.

The whole show with titles and credits runs a not-unreasonable 127 minutes.

With our total disc capacity of 4.7 GB, that would give us a total bitrate of 4.727 Mbps. Less the audio, we're left with 4.461 Mbps for video. With our target 6.5 Mbps for maximum compatibility with DVD-R media, less audio, our video peak will be 6.244 Mbps: kind of on the low side, although this is 24p. If it yields artifacts (particularly in scenes with *our* kid), we can bump it up a bit until we hit a good quality/compatibility combo.

Get the Gnome Digital Spreadsheet

My friend Bruce Nazarian, when he's not busy being the tireless President and CEO of the DVDA/IDMA (the organization that help get AACS rates down a lot) and running Gnome Digital, also is the founder of invaluable disc authoring site Recipe4DVD.com. You can download this great Excel spreadsheet he made for planning on DVD bit budgets there (see Figure 23.1). And while you're at it Bruce and I would both welcome you to the IDMA/DVDA web site: **http://www.idmadvda.org**

It's a great not-for-profit organization by and for content creators trying to do make video be more interesting than just a rectangle with stereo audio. And we always welcome new members and sponsors.

(*Continued*)

Destination	
Base Name	%s_MPEG2_DVD_NTSC(Mastering)
Path	D:\Book Samples\DVD Tutorial
Use source filename if required	☑ Use source filename if required
File Name	D:\Book Samples\DVD Tutorial\{Source}_MPEG2_DVD_NTSC(Mastering)(m2v)
Target File Splitting	
Choose Splitting Method	No Splitting
Interval Time for Splitting (Minutes)	10
Select Source Segment	Use entire Source
Start Time (Seconds)	0.00
Duration (Seconds)	0.00
Stream-Basic	
Stream Format	DVD(MPEG Program/Elementary Stream)
Stream Type	MPEG-2 Elementary Stream
File Extension (Video)	m2v
Video-VOB Output	
Use Marker Point if exist	☑ Use Marker Point if exist
Video-Basic	
Video Standard	NTSC 720 x 480
Width	720
Height	480
Frame Rate(fps)	23.976p -> 29.97i (2-3 pull-down)
Interlacing	Upper/Top Field First
Aspect Ratio Code	16 x 9
Quality/Speed	Mastering Quality
Use GRID Encoder	☐ Use GRID Encoder
Time Code Type	Automatic (Same as source time code type)
Use Closed GOP	☐ Use Closed GOP
Video-Bitrate Control	
Bitrate Type	VBR (Variable Bitrate)
Number of passes	2 pass
Video Bitrate(kbps)	4400
Max Bitrate(kbps)	6200
Min Bitrate(kbps)	500
Video-Advanced	
Closed Caption	Don't use
Profile/Level	MP @ ML
Put Sequence Header on each GOP	☑ Put Sequence Header on each GOP
VBV Buffer Size (KB)	224
Max GOP size	15
GOP Structure	Automatic
Enable Scene Detection	☑ Enable Scene Detection
Picture Structure	Always Frame
Chroma Format	4:2:0
Intra DC Precision	10
Use Strict GOP bitrate control	☑ Use Strict GOP bitrate control
Create DVD Compatible Stream	☑ Create DVD Compatible Stream
Use Sequence Display Extension	☐ Use Sequence Display Extension
Audio-Basic	
Use Audio	Don't Use

Select whether to optimize encoding for highest speed, highest quality, or a balanced between the two. In the ultimate quality mode, the encoder performs all kinds of methods to improve the picture quality, but it might take up to 10-20 times longer than other modes.

Figure 23.1 Bruce Nazarian's DVD bit budget spreadsheet makes it easy to plan out complex projects with subtitles and multiple audio tracks.

(*Continued*)

Encoding in ProCoder/Carbon

First, we need to make sure we've got the ProRes decoder installed, which is a free download from Apple. We'll then apply the 709 to 601 video filter to the source.

Since we're going to be importing into an authoring tool later, we want our output to be a MPEG-2 elementary stream (.m2v) and a Dolby Digital file (.ac3).

This being our kid, we'll use the NTSC DVD Mastering Quality preset as a starting point. Mastering Quality takes a little longer, but this is a fine if it takes overnight. We need to make some modifications:

- Frame rate to "23.976 -> 29.97i (2–3 pull-down)"
- Aspect Ratio Code to 16 × 9
- Used Closed GOP to On (more compatible with DVD-R)
- Video Bitrate to 4400 Kbps
- Max Bitrate to 6200 Kbps
- Min Bitrate to 500 Kbps (don't want blocking on easy scenes)
- Intra DC Precision to 10; a little slower, but can help reduce banding in those back-of-the-theater shadows
- Use Audio to Don't Use; we're doing AC-3 separately

And then we add a AC-3 target. The editor has told us that dialnorm was –17 dB in the source, and we nod politely and pick that preset. See Figures 23.2 and 23.3.

The defaults here are dialed in for exactly what we want; a rare pleasure. It's even 256 Kbps.

Encoding in Compressor

Coming into Compressor from a Final Cut source is really easy. We also could have imported the .mov straight into DVD Studio Pro and set the encoding settings there. Either approach gives the same quality.

We'll start with the "DVD: Best Quality 150 minutes" preset, which will set up both .m2v and .ac3 outputs.

For AC-3, we just need to do two things:

- Up bitrate to 256 Kbps
- Change dialnorm to –17 dB

For MPEG-2, Compressor correctly figures out that we're doing 23.98p 16:9. For Quality, we're already set to 2-pass VBR Best and Best motion estimation, the slowest, highest-quality options. We just change a couple settings:

- Average Bitrate to 4.4
- Maximum Bitrate to 6.2

(*Continued*)

Destination	
Base Name	%s_AC3
Path	D:\Book Samples\DVD Tutorial
Use source filename if required	☑ Use source filename if required
File Name	D:\Book Samples\DVD Tutorial\{Source}_AC3(ac3)
Audio-Basic	
Encoder Mode	Dolby Digital (AC-3)
Channels	2
Sample Rate(kHz)	48.000
Data Rate	256kbps
Dialogue normalization	-17dB
Bitstream Mode	Main Audio Service
Audio - Preprocessing	
Bandwidth-limiting lowpass filter	☑ Bandwidth-limiting lowpass filter
LFE lowpass filter	☑ LFE lowpass filter
DC highpass filter	☑ DC highpass filter
Digital de-emphasis	☐ Digital de-emphasis
90-degree phase shift	☑ 90-degree phase shift
3dB attenuation	☐ 3dB attenuation
Audio - Bitstream Information	
Center downmix level	-3dB
Surround Mix	-3dB
Dolby Surround Mode	Disabled
Copyright Bit	☑ Copyright Bit
Original Bitstream	☑ Original Bitstream
Audio - Extended Bitstream Information	
Alternate bitstream syntax	☐ Alternate bitstream syntax
Stereo downmix preference	Lt/Rt downmix
Lt/Rt Center mix level	-3dB
Lt/Rt Surround mix level	-3dB
Lo/Ro Center mix level	-3dB
Lo/Ro Surround mix level	-3dB
Dolby Digital Surround EX™ Mode	Disabled
Audio - Dynamic Range Compression meta...	
Global Dynamic Range Compression profile	None
Use Advanced Dynamic Range Compressi...	☐ Use Advanced Dynamic Range Compression mode
Dynamic range control Line mode profile	Film Standard
Dynamic range control RF mode profile	Speech
Audio Production Information	
Audio Production Information exists	☑ Audio Production Information exists
Peak Mixing Level	105 dB SPL
Room Type	Small Room, Flat Monitor
Target File Splitting	
Select Source Segment	Use entire Source
Start Time (Seconds)	0.00
Duration (Seconds)	0.00

Figure 23.2 Our MPEG-2 settings in Carbon Coder. The help text rather exaggerated how slow Mastering Mode is; it's been perhaps only 2x slower in my use.

(Continued)

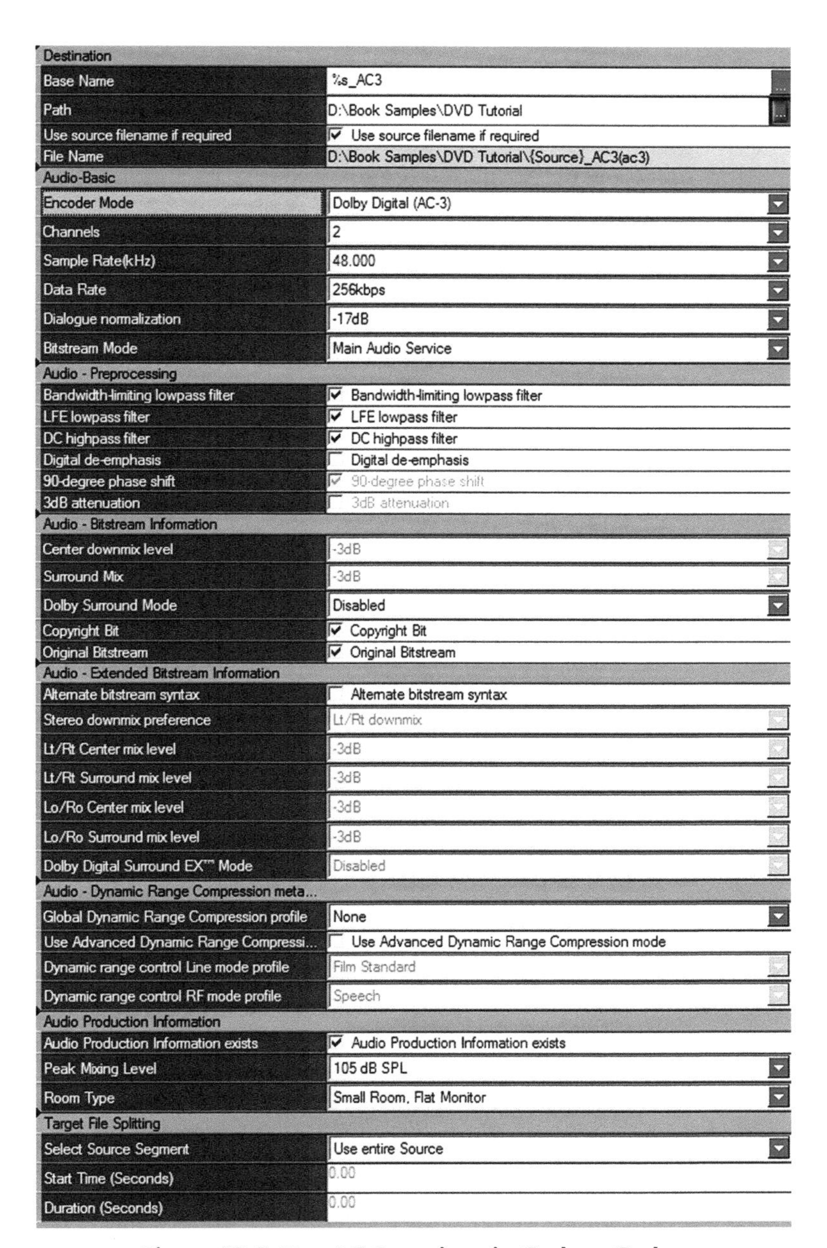

Figure 23.3 Our AC-3 settings in Carbon Coder.

(Continued)

And that's it if we're going to be authoring in DVD Studio Pro. But if we were using Encore or another product, we'd need to uncheck "Add DVD Studio Pro metadata." That option makes the files quicker to import into DVDSP, but makes the file incompatible with other authoring tools.

Compressor automatically does the right thing with chapter metadata, and should do the 709 to 601 correction as well. See Figures 23.4 through 23.7.

Encoding in Adobe Media Encoder

In Adobe Media Encoder, we start with the NTSC 23.976 Widescreen High Quality preset (Figure 23.8).

Like Compressor, AME correct sets us up as 16:9 23.976p. Metadata is set correctly. We do have some initial changes:

- Raise Quality to 5 (slowest, highest quality)
- Change Bitrate Encoding mode to 2-pass VBR (much better than 1-pass for longform content)
- Minimum Bitrate at 0.5 Kbps
- Target Bitrate at 4.4
- Maximum Bitrate at 6.2

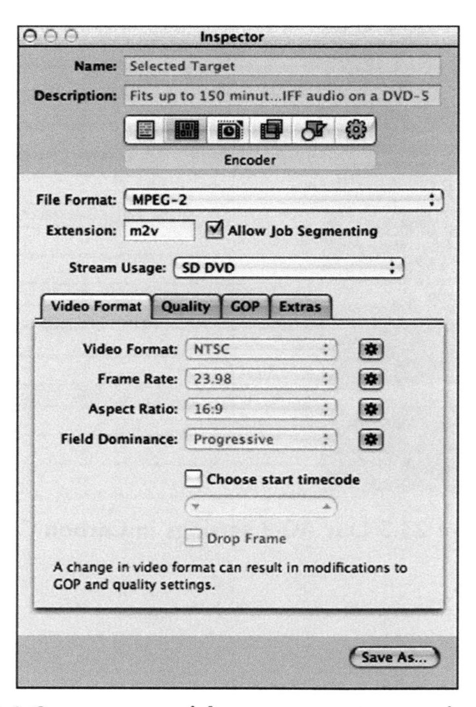

Figure 23.4 Compressor picks up our source settings exactly.

(Continued)

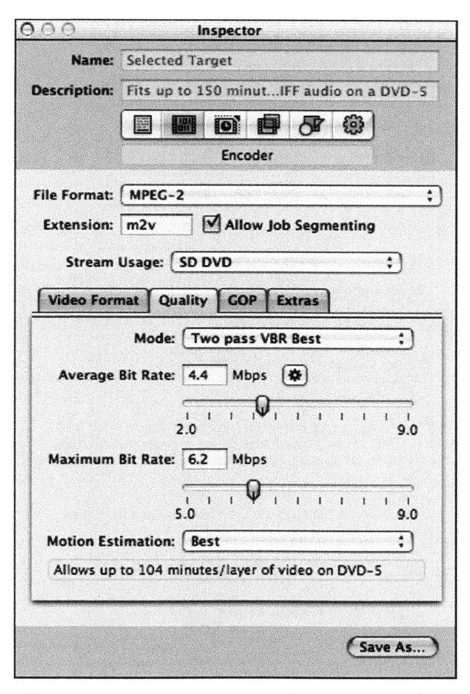

Figure 23.5 Even with quality turned up to max, Compressor's MPEG-2 is still pretty fast, and quality is still not as good as the leading tools.

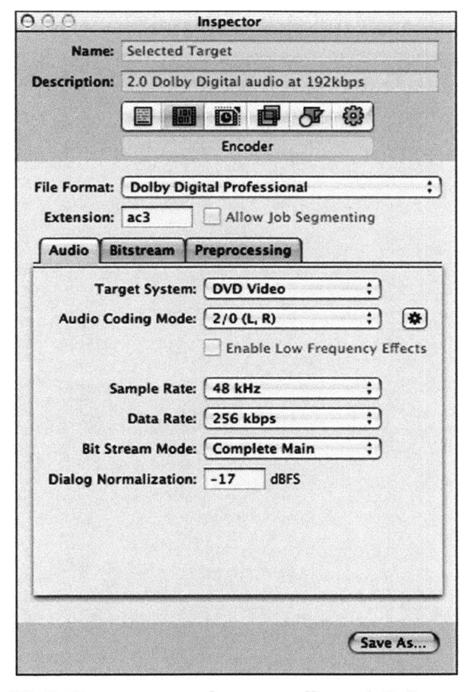

Figure 23.6 Compressor has excellent AC-3 controls.

(Continued)

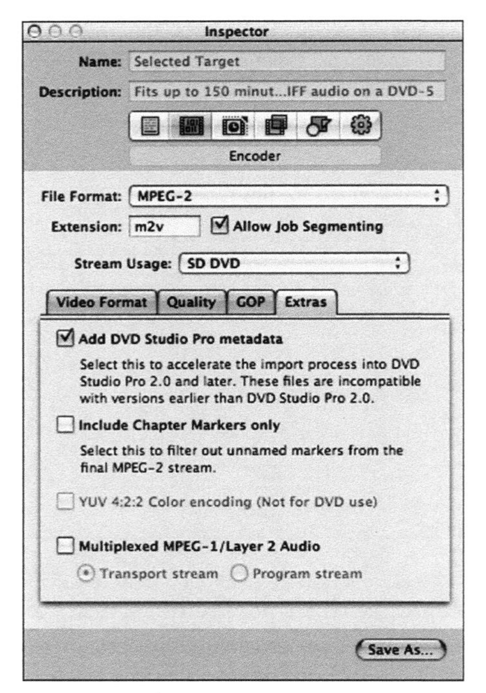

Figure 23.7 The default metadata setting here makes .m2v files incompatible with authoring products other than DVDSP. Beware!

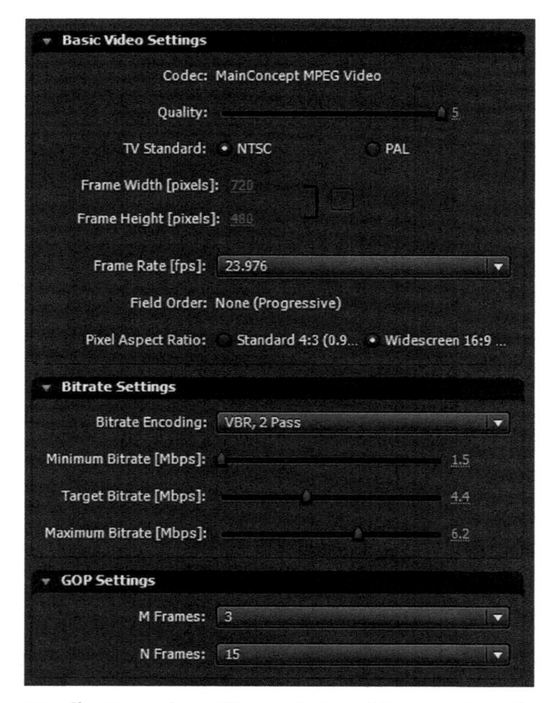

Figure 23.8 Adobe Media Encoder offers all the video settings in a nice single pane.

(Continued)

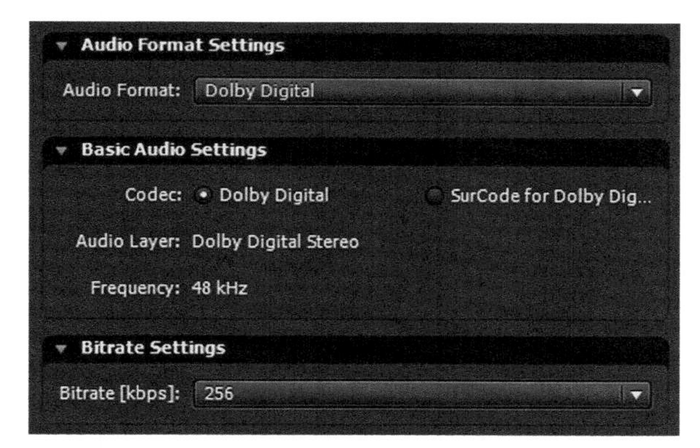

Figure 23.9 Very few Dolby Digital options are exposed by default. We don't have dialog normalization, for example.

For audio, we'll do two things:

- Switch from PCM (yuck!) to Dolby Digital (Figure 23.9)
- Raise bitrate to 256 Kbps

CHAPTER 24

Phones and Devices

Introduction

This chapter is about making content that plays back on non-PC, non–optical disc devices. That may sound more defined by what it's not than by what it is, but the categories are pretty well-defined in practice.

While there are a lot of terms of art, the world of devices breaks down into basically two families: stuff that mainly runs off batteries, and stuff that mainly runs off AC power.

Phones and Portable Media Players

Your portable battery-powered devices include media-capable phones, iPods, Zunes, and similar small-screen devices suitable for use on the go. They're designed for long battery life and hence low-watt operation. They're generally going to have smaller screens than SD, although we're seeing a few 640 × 480 and beyond displays creep into high-end phones. But given normal viewing distances, most viewers aren't going to get value out of incredible pixel counts on the display.

The main codec used in portable devices is H.264 Baseline profile, with VC-1 Main Profile and MPEG-4 part 2 also somewhat common. We're starting to see some devices support the more efficient H.264 Main and High profiles. There's also a de facto "Baseline + B-frame" profile equivalent supported by some devices and chipsets; B-frames don't have much of a decoder complexity impact and can be very helpful for compression. Unfortunately, documentation is often poor on what devices support what kinds of files, and some devices explicitly block playback of files not encoded as part of their ecosystems.

We're on the cusp of a revolution in portable devices, as GPU-derived System-on-a-Chip (SOC) devices like NVvidia's Tegra and AMD's Imageon make HD decode and HDMI output possible with small devices. The Zune HD is the first entrant of this new generation, with high-end phones with similar features to follow.

There are of course some audio-only devices left, typically very small, low-priced units like the iPod Shuffle.

The average mobile phone isn't as capable as most portable media players (PMPs) for media playback, but the market is shifting in that direction, led by Apple's iPhone. Some of this is simply a logical outgrowth of the development of smartphones; once you have a good-sized color screen and fast processor for handling the web and email, media support is a relatively trivial addition. And PMP chipsets are starting to migrate into phones, providing the same functionality.

Optimal Viewing Distances for Device Screens

We're seeing the resolution of portable media devices go up faster than the screen size. This can pay off in crisper text, but beyond a certain point it won't provide a better video experience given the distance at which most folks view the players.

Table 24.1 shows the screen sizes and optimal viewing distances for the most popular PMPs. The industry standard guideline of an optimal pixel size relative to the viewer is 1/60 of a degree, which is the point where individual pixels blend together for someone of average vision. Get any closer, and the pixel edges themselves can be seen. I also include the maximum distance the screen can be away and still fill as much of the visual field as a movie screen would from the back row of a THX-certified theater (which simply isn't feasible for PMPs). Note that there's no downside to having a higher resolution on the device, it's just that beyond the optimal distance, dpi matters less than screen size.

As you can see, our portable devices, whatever their resolution, aren't going to provide the same kind of detail as a normal HD display.

Table 24.1 Typical PMP Screen Sizes and Optimal Viewing Distances, Compared to 60" Flat Panel.

Device	Size (in)	Width	Height	Aspect ratio	Optimum distance	THX maximum
iPod nano	2	320	240	4:3	17"	3.5"
iPod classic	2.5	320	240	4:3	22"	4.3"
iPod touch/iPhone	3.5	480	320	3:2	21"	6.4"
Zune 4/8/16	1.8	320	240	4:3	16"	3.1"
Zune 80/120	3.2	320	240	4:3	28"	5.5"
Zune HD	3.3	480	272	16:9	21"	6.2"
PlayStation Portable	4.3	480	272	16:9	27"	8.1"
Archos 5	4.8	800	480	5:3	18.2"	8.8"
37" SD CRT display	35	720	480	4:3	150" (13')	61" (5')
60" 1080p LCD	60	1920	1080	16:9	225" (19')	113" (9')

Consumer Electronics

There's not a great term of art for "video playback devices you plug in that aren't full-fledged computers." They're sometimes called set-top boxes (STBs), appliances, and extenders. For this chapter I'm going with CE (consumer electronics) devices.

Modern CE device have much more capable ASIC decoders than portable devices, typically including H.264 High Profile, VC-1 Advanced Profile, and MPEG-2. CE devices are also much more likely to support interlaced encoding than a portable device; all STBs do, but a few devices are progressive only, or support interlaced SD but only progressive HD.

For outputs, everything new supports HD now. For backward compatibility, most have composite, S-video, and component, and current devices have HDMI as well.

5.1 audio support is also common, and is an area in flux, reflecting the changing state of consumer interconnects. Classic analog stereo is a given. Beyond stereo, there are three approaches to multichannel audio output; any given device may do one or multiple.

The simplest is bitstream pass-through, where the device doesn't have its own encoder, but just passes PCM stereo, Dolby Digital, or DTS audio bitstreams out to a receiver. This is cheap and simple, but prevents the device from doing any kind of audio mixing, like button feedback. Bitstream pass-through is most commonly provided via TOSLink, but some devices support it via digital RCA or HDMI.

The rarest in practice is multichannel analog—six analog outputs (often combined into three stereo pairs for space reasons). This is how most Windows PCs deliver multichannel output. It's not seen too often in audio gear outside of the high end, though it was the only multichannel output for DVD-A and SuperAudio CD. Given all the cabling required, receivers tend not to have more than one set of multichannel analog inputs, if even that. And since that's generally the best way to hook up PC surround audio, any user that uses the same setup for gaming and media isn't likely to have anywhere to plug in another device.

Why Portable Devices?

Portable media devices have been a hallmark of consumer media technology since the transistor made the pocket-sized radio possible in 1957. From that through Sony's Walkman, the battery-powered CD player, MP3 players, and on to today's iPods, Zunes, and media-capable phones, we've had a good half-century of being able to carry our music with us in some form.

In the mid-2000s, some early video-on-the-go devices started emerging; I still have a Creative Zen Portable Media Center device that worked surprisingly well in 2004, but definitely wasn't pocket-sized. But there's a whole lot more bits to create, store, and decode with video than audio, so it took a few iterations of Moore's Law to get the cost of authoring, storage, and watts for decoding down to the point where video could be a low-incremental-cost bonus feature to a music player. Originally players that supported video all used tiny hard drives, but the capacity of flash memory keeps going up, and it's possible to fit many hours of video on an all-solid-state device (although with a higher price per GB for HD-based models).

The bigger problem for portable video playback is screen size. With good headphones, even an iPod nano can make a sound as big as the world. But a small device can't have a big screen, so there's an unavoidable tension between portability and visual experience. And historically, portable devices have had pretty weak LCD displays using 15- or 16-bit color (so 5–6 bits per channel), and thus a lot of banding or dithering. This is poised to change with the OLED displays used in a few phones and the new Zune HD.

But for those who fly, commute on mass transit, or treadmill at the gym, the risk of eyestrain, neck cramps, and tired arms can be worth watching what you want when you want it. And heck, maybe someday video projection inside sunglasses will finally go mass-market, like they've been promising for years.

The dedicated portable media players typically include analog video output. Originally those were all composite and all terrible, although some have analog component now. However, even that output is generally pretty terrible with lots of noise. Also, none do 60p yet, nor interlaced; 24p content gets output with very ugly judder where every sixth frame gets repeated, instead of the smoother 3:2 pulldown.

HDMI output should be much better; digital is a lot easier to deliver than analog in a small device.

Why CE Devices?

The outboard CE device is nearly as venerable as the portable music player, with its roots deep in the analog era. The first cable set-top boxes were invented in the 1960s.

Soon followed an explosion of ways to watch media at home, starting with some amazing failures like DiscoVision through the revolutionary VHS and high-end Laserdisc to the industry-changing DVD. But the set-top box didn't vanish in the face of home playback; it kept getting enhanced to support dozens, then hundreds of channels; went from analog to digital to HD; and was adapted for satellite and ITPV use as well, competing against the cable industry where the technology was born.

And for all we talk about DVD as a driver of the entertainment industry, the classic instant-on, watch-what's-on experience has long commanded many more eyeball-hours. But those lines are blurring. As we see more home recording and pay-per-view functionality in the set-top boxes, it's not clear if there's anything sustainably unique about the traditional, linear viewing experience offered by cable and satellite over what streaming to a CE device could deliver.

As TVs and computer monitors increasingly become the same thing, many of the same experiences can be delivered from a PC running Windows Media Center, XBMC, or similar software. Until recently content providers' DRM rules haven't allowed live cable/satellite video broadcasts to play directly on PCs (although many enterprising viewers found ways around that). But even that line in the sand is being erased with native CableCard support for Windows 7 Media Center finally announced.

But we've been seeing more and more PC-centric experiences working on CE devices, like Netflix's on-demand streaming via the Roku digital video player, and a broad array of Media Center Extender and Digital Living Network Alliance (DLNA) devices that allow a user's Windows Media Player or (non-DRMed) iTunes library to be browsed from a PlayStation 3, among many other clients.

How Device Video Is Unique

The biggest difference between device video and general PC video is that the format specifications are much tighter, with files generally working perfectly or not playing at all (better) automatically scode to a spec file. However, this still burns could introduce artifacts, and often results in a bigger file than one optimally encoded for the device.

Unfortunately, these specifications aren't always clearly documented, and there are lots of things that should work in theory but don't in practice, or shouldn't work in theory but do (sometimes, in some places) in practice.

CE devices and their much more powerful ASICs can handle a much broader range of content than portable devices, and often can play a wide array of existing files.

Getting Content to Devices

Devices have a panoply of ways to load, stream, or connect to content. Because they're not meant to be a primary device, they assume users have a different primary device they use to manage and store content.

Attached Storage via USB

A number of devices, particularly CE, support playback of content via USB storage (like hard drives and flash drives). There are some caveats, though.

The first is that only FAT32-formatted drives are broadly compatible. Most CE devices won't support NTFS- (Windows) or HFS+ (Mac)-formatted drives. However, files on FAT32 are limited to a maximum size of 4 GiB. That's normally enough for a two-hour movie in 720p (around 4.7 Mbps average bitrate), but generally not enough for 1080p.

Older versions of Windows can only format FAT32 partitions with a maximum size of 32 MB; laughably small for today's hard drives, but this is fixed in Windows 7.

The second issue is speed of flash drives. While they're getting faster, models even a couple of years old can have quite slow read speeds. If your file is stuttering while playing back off a USB flash drive, the drive may not support the peak bitrate, and might not be a problem with the decoder or content.

Sideloaded Content

Sideloaded content is content on a mobile device transferred from a computer. This is the classic sync model used with most devices to date, as they're able to take advantage of the faster downloads and larger storage capacity of a PC.

All Zunes and the iPod Touch/iPhone now support finding and downloading music straight to the device via WiFi, but most users still do their music browsing and downloading on a PC.

Progressive Download to Devices

Progressive download can be challenging on phones that don't have much internal storage or RAM, as the entire file can't be cached. This removes the big "rapid random access" advantage of progressive download. A PC may be able to do progressive download of a 1 GB file, but that same file could offer terrible performance on a phone.

Standard Streaming to Devices

Many phones support streaming protocols, most commonly MPEG-4 RTSP for 3GPP files and the Windows Media streaming protocol. The classic streaming protocols were designed for the high-latency, lossy Internet of the past, which isn't that far from the mobile world today.

Traditional streaming protocols aren't used in non-phone devices; Wi-Fi sideloading is as far as they get today. The biggest technical barrier is the power requirements of keeping the radio on for sustained periods of time while also running the screen and decoder chips; this can run down a battery quickly.

Adaptive Streaming to Devices

Adaptive streaming for devices is in its infancy, but clearly of growing interest. Smooth Streaming is coming to the Xbox 360, and is being actively implemented by several STB vendors. Apple's adaptive streaming technology launched in June 2009 on the iPhone and QuickTime X, and an Apple TV implementation seems inevitable, although still unannounced.

It's not clear if an all-HTTP delivery mechanism is ideal for wireless networks, however. I'd love to see what would be possible with MBR switching of an adaptive streaming chunked format coupled with UDP and multicast.

There are interesting possibilities for adaptive streaming in aggressively prefetching data as quickly as bandwidth allows, and then turning off the power-hungry radio receiver hardware for several minutes until it's time for the next bunch of chunks to download.

Sharing to Devices

Devices are often on a LAN with a PC or other device storing content. So it's possible to use that reliable, low-latency, high-speed local bandwidth to "stream" almost like a local file. This allows a big content library to exist on the PC's cheap storage, with multiple client devices able to consume that as needed.

This works the best with wired Ethernet connections, of which the slowest old home router these days provides at least 100 Mbps, easily enough for multiple HD streams. Wi-Fi can be more of a challenge, as obstructions and interference inside a house can reduce bitrates (it's not fun when microwaving the popcorn kills movie playback). And older protocols like 802.11b are a lot slower than newer variants like 802.11g and 802.11n.

Universal plug and play

Universal Plug and Play (UPnP) is set of networking protocols from the eponymous industry group, pushing easy interoperability between different devices and platforms. One of their many efforts is the UPnP AV standards.

Many devices can connect to an UPnP AV "Media Server" and its library of content. The most common of these is Windows Media Connect, built into WMP11 and later, making it very easy to share content libraries from Windows to compatible devices and players. There's no default UPnP Media Server on Macs, but a number of third-party ones are that can share non-DRM content from the iTunes library.

On the client side, many CE devices support UPnP AV, including Xbox 360 and PS3, and some recent Sony and Samsung displays. On the software side, WMP 11+ (so XP with an optional install, and Vista+ out of the box) is also a client, as are many common players like VLC.

DLNA

The Digital Living Network Alliance is an industry consortium defining interoperable media sharing across different platforms and devices. It's notable for supporting Windows Media DRM sharing to non-Microsoft devices like the PlayStation 3.

DLNA builds on top of the UPnP and other specifications.

(Disclosure: I created much of the DLNA test library of WMV files.)

Windows Media Player/Center and Zune

Windows Media Player/Center and the Zune desktop software offer somewhat deeper interoperability, including DRM support (for content with proper rights) when using other Windows computers or the Xbox 360 as a client. This can include richer navigation of libraries.

Media Center Extenders are specific devices that interoperate with Windows Media Center, and can provide the full GUI of Media Center as well as media playback. This enables library management, recording setup, et cetera to be done from the couch, not just the computer room.

Windows 7 adds live transcoding and streaming to devices. You can browse the media library on your home PC from anywhere in the world, and then it'll transcode and broadcast it to your device in real-time in a format the device can handle over the available bandwidth. If you've got a TV tuner in your PC, this would allow watching a live local channel on your phone anywhere in the world.

iTunes

Apple's iTunes is capable of media sharing to the Apple ecosystem of products, like other copies of iTunes, iPhone/iPod Touch, and AppleTV.

DRM-protected iTunes content is playable only on Apple devices or via iTunes, however. Only Apple is able to publish iTunes content. The Apple walled garden can be a seamless experience if Apple provides want you want as a publisher and device vendor, but with a significant sacrifice in flexibility.

The Walled Garden

There are also lots of devices, particularly classic STBs, that only play content from the specific service that provided them. This is called a "walled garden"—it may be beautiful, but nothing comes in or gets out.

As a compressionist, you may be asked to provide content to a walled garden spec. The most common of these are the CableLabs specs, but many operators have their own particular ones. Be wary of these projects if you don't have a tool that explicitly supports the precise format in question; even if your MPEG-2 encoder *looks* like it supports all the needed parameters, there's often little things that cause validation errors.

Devices of Note

There's a ton of devices out there, but there's a few major brands that are most targeted with encoded content, because they're available in high enough volume.

iPod Classic/Nano/Touch and iPhone

Apple's iPod line has three main tiers: the screenless iPod shuffle, the 320 × 240 iPod classic and iPod nano, the iPhone-derived iPod touch, and the iPhone itself (Figure 24.1).

While screen sizes vary (the iPod nano and iPod classic are 320 × 240 while the iPhone and iPod touch are 480 × 320), the same content will play back in any current iPod with a screen.

Figure 24.1 The Apple iPhone 3GS (which looks exactly like the Apple iPhone 3G, and features the same screen size as the iPod touch).

All the iPhone/iPod devices released since the iPod 5G in late 2005 support the same basic parameters:

- .mp4 or .mov wrapper (self-contained only)
- Up to 640 × 480p30
- H.264 Baseline
  - 1.5 Mbps max with Level 3.0
  - 2.5 Mbps max with 1 reference frame
    - Apple calls this "Baseline Low-Complexity"; not a term from H.264 spec
- MPEG-4 part 2 Simple Profile up to 2.5 Mbps
- AAC-LC audio up to 48 KHz stereo

Apple didn't make authoring that easy, though. Originally any iPod video file more than 320 × 240 requires a special string be added to the file indicating it was compatible. Things are more flexible with devices since the late 2007 launches of the iPod Classic/Touch and iPhone. They increase H.264 support to:

- 720 × 480p30 and 720 × 576p25
- 5 Mbps max Level 3
- 10 Mbps peak
- 10 Mb VBV

The iPhone 3GS has hardware capable of decoding 720p in both H.264 High Profile and VC-1 Advanced Profile, but Apple hasn't provided any way to sync HD content.

Apple TV

Apple TV is a relatively minor product compared to the massive success of the iPod and iPhone. It's a simple playback device with limited HD support, mainly compatible with iTunes Store–style content.

It's quite limited in what it can play back, particularly compared to other CE devices:

- .mp4 or .mov wrapper (self-contained only)
- H.264 Main Profile (constrained to Baseline + B-frames)
  - Up to 5 Mbps peak
  - Progressive only
  - CAVLC only—no CABAC
  - Max of 1280 × 720p24 or 960 × 540p30
- MPEG-4 part 2 Simple Profile
  - Up to 3 Mbps peak
  - Max 720 × 432p30
- AAC-LC audio up to 160 Kbps
- AC-3 audio for 5.1 audio

Given these constraints, most content HD content targeting the Apple TV needs to be explicitly encoded for it. Main Profile without CABAC is quite a bit less efficient than the standard High Profile, so most existing HD files will use features that are incompatible with the Apple TV.

Zune

The original Zune (the boxy Zune 30) used software decode, and was limited to 320 × 240p30 WMV (VC-1 Main Profile) with a peak bitrate of 1.5 Mbps. The decoder can drop frames with very high motion at the full 320 × 240p30, but generally 320 × 176p30 and 320 × 240p24 always play back reliably.

The second-generation Zune 2 line introduced in 2007 (the Zune 4, 8, 40, 80, and 120) also have a 320 × 240 screen, but a much more powerful decoder. It handles the full range of NTSC and PAL SD sizes, including non-square-pixel support. Supported are:

- Windows Media
  - VC-1 Main Profile Main Level

    - Up to 720 × 480p30 or 720 × 576p25 3 Mbps peak
    - WMA Standard or Pro up to 192 Kbps
- MPEG-4 (.mp4)
    - H.264 Baseline
        - Up to 720 × 480p30 or 720 × 576p25 2.5 Mbps peak
    - MPEG-4 part 2
        - Simple Profile
        - Up to 720 × 480p30 or 720 × 576p25 4 Mbps peak
    - AAC-LC audio up to 192 Kbps 48 KHz

Zune HD

The Zune HD (Figure 24.2) is a quite different device than the first generations of Zune, so I'll give it its own entry.

It incorporates the NVidia Tegra chip, the first in a new generation of video SOC chipsets with big improvements in HD file playback for battery-powered devices. It also has a nice OLED 24-bit display, eliminating the banding of older Zunes and most other portable media players. And its dock provides a HDMI port, making it a good portable HD source. It supports:

- Windows Media
    - VC-1 up to Advanced Profile Level 2

Figure 24.2 The Zune HD is a very different device than earlier Zune models.

- Up to 720 × 480p30 or 720 × 576p25 10 Mbps peak
- Up to 1280 × 720p30 14 Mbps peak
- WMA Standard up to 192 Kbps
- WMA Pro up to 384 Kbps 48 KHz 5.1

- MPEG-4 (.mp4)
  - Max 4 GB file size
  - H.264 Baseline + B-frames (superset of AppleTV spec)
    - Up to 720 × 480p30/720 × 576p25 10 Mbps peak
    - Up to 1280 × 720p30 up to 14 Mbps peak
  - MPEG-4 Part 2
    - Advanced Simple Profile Level 5
    - Up to 720 × 480p30 or 720 × 576p25 4 Mbps peak
  - AAC-LC audio up to 256 Kbps 48 KHz

- AVI/Divx (.avi/.divx)
  - Max 4 GB file size
  - Divx 4/5 Home Theater profile
  - Up to 720 × 480p30 or 720 × 576p25 4 Mbps peak
  - MP3 audio up to 192 Kbps 44.1 KHz

Xbox 360

The Xbox 360 has a three-core CPU, and uses software decoding, not an ASIC. Like many software players can often play content well beyond the listed specs. These are what's guaranteed to work, and it will attempt to play back higher rates, potentially with dropped frames. It can often handle real-world H.264 content with 15 Mbps peaks seems to work fine.

Around this time this book hits the shelves, Microsoft will be launching a new adaptive streaming service for the Xbox 360, delivering up to 1080p.

The Xbox 360 can play back content from a FAT32-formatted drive connected via a USB port, shared content (including DRM) via the Window Media Player and Zune media sharing features, and via other products using the same protocol. For example, Connect360 is a third-party app that allows access to (non-DRMed) content shared from iTunes. Note FAT32 playback is limited to 4GB, but file sharing can use files of any size. Official specs are:

- Windows Media
  - WMV 7, 8, 9, and VC-1 Simple, Main and Advanced profiles

  - Up to 1280 × 720p60 or 1920 × 1080p30 30 Mbps peak
  - WMA Standard, Pro, and Lossless up to 48 KHz 5.1

- MPEG-4 (.mp4 and .mov wrappers)
  - H.264
    - Baseline, Main, and High up to Level 4.1
    - Up to 1280 × 720p60 or 1920 × 1080p30 10 Mbps peak
    - AAC-LC stereo
  - MPEG-4 Part 2
    - Simple and Advanced Simple profiles
    - Up to 1280 × 720p30 5 Mbps
  - AAC-LC stereo

- AVI/Divx (.avi wrapper)
  - Simple and Advanced Simple profiles
  - Up to 1280 × 720p30 5 Mbps
  - AAC-LC stereo
  - Dolby Digital up to 5.1

PlayStation Portable

The PlayStation Portable (PSP) is mainly a handheld game machine, but it's also quite popular for media playback. The disc-based models have larger screen than the other PMPs we're discussing here—a full 4.3 inches. The new PSP Go is a still large 3.8 inches.

Sony being, well, Sony, launched the PSP with a proprietary disc format called UMD—Universal Media Disc—both for game playback and for playback of a proprietary movie format. UMD is a red-laser 60 mm disc in a cartridge and can store up to 1.8 GB. UMD enabled much bigger games than competing handhelds using cartridges had at the time.

However, after an initial burst of sales, the UMD disc format quickly lost consumer interest. And powering the motor and laser for an optical disc is a significant power drain in a battery-powered handheld. In 2009, Sony introduced the PSP Go, which replaces the UMD with 16 GB of internal flash memory to download games. They continue to sell the current UMD-based PSP 3000, but clearly the future of media playback on the PSP is via downloads, network sharing, or flash storage.

The PSP uses Sony's proprietary Memory Stick flash memory format, which can be quite a bit more expensive than more common flash storage types. Still, given the PSP's high-quality screen, users were quickly drawn to the PSP for playback. Original firmware blocked playback of video files more than 320 × 240 (presumably to avoid competition with the UMD experience), but that "feature" was quickly hacked and eventually formally dropped.

The PSP is still a very fussy player, and pretty much plays only content specifically designed for it. Uniquely, it requires CABAC entropy coding, and so isn't compatible with CAVLC or Baseline at all. Thus iPod and PSP files are wholly incompatible; the Zune HD is the only other portable media player capable of playing PSP H.264 files. Supported are:

- MPEG-4
  - H.264 Main Profile (CABAC only!)
    - But not using pyramid B-frames, which are incompatible
  - 720 × 480, 352 × 480, and 480 × 272 frame sizes
  - MPEG-4 part 2 Simple Profile
  - AAC-LC
- AVI (really just for still camera video)
  - Motion JPEG
  - PCM or μ-Law

PlayStation 3

Sony's PS3 aimed to build on the PS2's dominance of the living room gaming to living room media consumption, with Blu-ray support a key feature. This hasn't paid off so far; the cost and complexity involved in being a best-of-breed game machine and Blu-ray player has left the PS3 as the third place console, a far cry from the overwhelming first place the PS2 held. While the innovative Cell processor in the PS3 hasn't really lived up to hopes as a uniquely powerful CPU for console games, it has proven to be a video decoding monster. Thus the PS3 isn't really limited by decoding horsepower, but by the formats and codecs that have been implemented.

It can even do H.264 at 1080p60, probably the first CE device to do so, although as of Firmware 2.80, that forces 1080i output due to a bug of some sort. PS3-supported media types include:

- MPEG-4
  - H.264 up to 1080p60 4:2:2 (!)
  - AAC-LC
- MPEG-1
- MPEG-2 Program and Transport stream
  - H.264
  - MPEG-2

- AAC-LC, AC-3, or PCM
- PCM
- Camera AVI
  - Motion JPEG
  - PCM or μ-Law
- DivX/Xvid/AVI
  - MPEG-4 part 2
  - MP3 audio
- WMV
  - VC-1/WMV 9
  - WMA Standard
  - DRM-free only

Formats for Devices

MPEG-4

The MPEG-4 file format is the most broadly supported for devices, although that's split up in H.264, Part 2, and 3GPP variants.

H.264

H.264 has become the most common codec for devices, given its adoption in broadcast industry for CE devices and the Apple iPod/iTunes ecosystem. All current PMPs support it, as do new phones with media playback.

H.264 is normally supported in the .mp4 wrapper.

Note that content purchased for the iTunes Store and protected with Apple's FairPlay DRM are only playable via Apple devices and the iTunes software, plus a few specific Motorola phones. These will have the .m4v extension, although other files with that extension may not include DRM.

MPEG-4 part 2

MPEG-4 Part 2 Simple Profile was the original video codec used for the iPod. And since the H.264 decoder ASICs generally include part 2, devices often contain support for part 2.

However, it's rarely the best choice. Most devices that do it also do H.264, and there are more H.264-only than part 2-only devices out there. And the Apple ecosystem is Simple Profile only, setting a quite low bar for compression efficiency.

It's common for devices that support part 2 to support lower maximum bitrates and frame sizes than for H.264 and VC-1—for example, Apple TV and the Zune HD only do Part 2 in standard definition. What's worse is that many devices that claim part 2 support don't publish detailed specs, so it can be trial and error to figure out what works. For example, SP + B-frames is pretty common in non-Apple decoders (B-frames are by far the most useful and simplest to implement feature from ASP).

3GPP

The 3GPP formats are a variant of MPEG-4, and use H.264 or MPEG-4 part 2 video.

The biggest difference is on the audio side, where 3GPP often includes HE AAC and voice-only codecs like AMR and CELP. 3GPP and its myriad codec choices are covered in Chapter 12.

Unfortunately, device support varies widely, although we're seeing convergence around H.264 Baseline and HE AAC v1 for new phones as chipsets get better. Hopefully, in a few years a single file and publishing service can address all mobile devices, but we're not there yet. Too often, video delivery to phones must be handled by the carrier, specifically encoding for the devices they support.

Windows Media and VC-1

Windows Media remains a popular choice for device media. One big reason is that it's the only widely supported licensable DRM. Beyond Windows itself, Windows Media DRM protected files can play on Zunes, Windows Mobile and CE devices, many other devices using the licensable Windows Media and Windows Media DRM porting kits, and Silverlight. That's led to Windows Media becoming the underlying technology for services the offer device compatibility like Netflix's streaming service and Blockbuster and Amazon's download services.

For historical chipset reasons, portable device VC-1 is mainly Main Profile. This will presumably change as Silverlight and Smooth Streaming comes to devices (although H.264 certainly works with Smooth Streaming and PlayReady DRM). CE devices generally support Advanced Profile and hence interlaced.

Most devices support WMA Standard and Pro, although some non–Windows Mobile/CE devices don't support decoding the Pro LBR modes yet.

AVI/DivX/Xvid

DivX Networks was an early leader in standardizing device media playback with their DivX-certified licensing program, which was particularly popular in off-brand DVD players.

These generally follow the profiles specified by DivX Networks, although many devices (particularly portable ones) don't carry the explicit DivX certification.

Normally these AVI files use MP3, but some also contain AC-3 for multichannel audio.

The lower compression efficiency of MPEG-4 part 2 + MP3 generally means this should only be used when specific devices are targeted for which this is the best option. It's quickly becoming a legacy technology. The new Divx profiles are based on H.264 in a MKV wrapper.

Audio-Only Files for Devices

For all this talk of video, music listening accounts for a lot more hours of use than video watching for a typical device. Audio "just works" for the most part, so there just isn't as much to cover.

Any device is going to support MP3, of course, and most support .mp4a (AAC-LC only MPEG-4 files). WMA support is quite common, with Standard (WMA 9.2) most common, with WMA Pro and Lossless in many devices, particularly those based on Windows Mobile. Apple's devices (but not many others) also support the Apple Lossless codec.

While open source advocates have been singing the praises of Ogg Vorbis for many years—and it is a more efficient codec than MP3—it's not supported by many mainstream devices. The FLAC lossless codec is sporadically supported as well.

All that said, most portable audio devices are used to play MP3 files, even though it's the least efficient of the broadly used lossy codecs. But guaranteed interoperability with every device make MP3 the safe choice, and storage capacity is cheap enough today to make the compression efficiency difference immaterial for most music collections.

Generally portable players only play back stereo audio, although some can fold down multichannel to stereo for playback.

CE devices rarely store audio, but many can play back the same formats via streaming and sharing.

Encoding for Devices

For portable devices, the core tension is between encoding for the screen or for the output. If encoding for personal use on a single device, or a class of devices that only does 320 × 240, encoding more pixels than that is a waste of time and bits. If encoding for a broad class of devices that vary in screen size, encoding for the size of the biggest screen can be a good compromise. So, a 16:9 file targeting iPod and Zune playback could use H.264 Baseline 1.5 Mbps at 480 × 272. That will play 1:1 on the Zune HD and iPod touch/iPhone, and would get scaled down to 320 × 180 on the iPod classic/nano and Zune 4/8/80/120.

However, not all devices do a good job of scaling the video, either in performance or quality. In particular, phones not of the current cutting-edge generation often don't use good hardware scalers, so ideally encode a size where no scaling on playback is required (like 320 × 176 on a 320 × 240 display). Sometimes a 2x scale mode is faster than a fractional ratio.

It's also possible, of course, to make a generally compatible file that'll play on most devices and most media players (as in the H.264 tutorial). These generally follow the iPod 5G specs, which are the most restrictive in use today. Still, they can do a decent job at standard definition, and so offer some future proofing. 2.5 Mbps is generally sufficient for most content with H.264 Baseline with one ref, the maximum the iPod 5G supports. For when faster downloads or more efficient use of precious space is appropriate, 3 references and 1.5 Mbps peak can be better.

Unfortunately there's no H.264 setting that'll play in both the iPod and PSP; the iPod can't use CABAC, and PSP requires it. So, if you want to make a single file that can play on every PMP, it'll have to be MPEG-4 part 2 Simple Profile.

For CE devices, things are easier. The major ones all support H.264, MPEG-4 part 2, and VC-1 Windows Media in their ASIC or software decoders, although support for particular file formats and audio codecs can vary. In general, a High Profile H.264 file with stereo AAC-LC audio will work everywhere, but some devices like the Xbox 360 don't handle 5.1 AAC-LC. Conversely, the PS3 doesn't do WMA Pro, which the Xbox 360 can use for 5.1.

Devices Tutorial

We covered making generic iPod-compatible H.264 files back in the H.264 chapter. So for this chapter, we'll look at how to encode screen-optimized clips for the standard-sized Zune and iPod models.

Scenario

We work for a skateboarding company's marketing department. We've got a great new line of decks coming out in the fall, and a viral video campaign to back it up. To that end, we've got a good 10-minute highlights video that we want our customers to install on their mobile devices to show off to their friends.

Three Questions

What Is My Content?

The source was shot on HDV at 1080i30, edited in Premiere Pro, and saved out as a 16:9 1440 × 1080i30 anamorphic Cineform Prospect HD file. It veers from static close-ups over ethereal flutes to crazy motion, wild effects, and a pounding beat.

Who Is My Audience?

The audience is boys (and some girls) and their toys. They've got allegiance to their devices, so we want to offer content tuned to offer a perfect experience on all iPod,

(*Continued*)

Zune, and PSP models. Since that audience is always looking for eye candy to justify their purchases, we think being a great demo clip will up viewership and our brand.

However, these guys go through a lot of content; we don't want to waste so many bits that they delete the file. So, we want to be as small as we can be while still retaining excellent quality.

What Are My Communication Goals?

Our brand is all about technical brilliance while being extreme *and* awesome. So the stuff has to look and sound just great, while keeping the file size reasonable for quick downloads and long life on the device.

Tech Specs

Preprocessing is pretty easy. We're taking that big 1080i source down to 320 × 176 or 480 × 272 for our Mod16 output resolutions. Deinterlacing quality doesn't matter so much.

We want to make an optimal file for the primary mobile devices in our audience, and each flavor. So, we're targeting the following devices:

- iPod 320 × 240
- iPhone/Touch 480 × 320
- Zune 320 × 240
- Zune HD 480 × 272
- PSP 480 × 272

While in theory we could get away with doing 480 × 272 and 320 × 176 MPEG-4 Simple Profile part 2 to cover all of these, the quality of that codec may not be up to our goals on all our devices. So, we're going to suck it up and make a variants customized to each device.

- iPod classic/nano: These have smaller screens, and in the case of the nano, less storage than the iPhone/touch. So we're going to use three ref frames.
  - 320 × 176 (16:9 is 320 × 180, so Mod16 rounded)
  - H.264 Baseline 2-pass VBR 1 Mbps ABR 1.5 Mbps PBR (750 KB buffer)
  - Three reference frames
  - AAC-LC 192 Kbps
- iPhone/iPod touch: The bigger screen is going to want more bits, and hopefully most people have one with decent storage. We're going down to one ref frame in order to bump up peak bitrates and still have a file that'll work on the 5G iPod.

(*Continued*)

Since Apple doesn't put version numbers on the iPods, users often don't know which model they have.

- 480 × 272 (16:9)
- H.264 Main-pass VBR 1.5 Mbps ABR 2.5 Mbps PBR (1250 KB buffer)
- One reference frame
- AAC-LC 192 Kbps

- PSP: Ah, H.264 Main Profile. We can do quite a bit more:
  - 480 × 272 (16:9)
  - H.264 Main 2-pass VBR 1.5 Mbps ABR 2.5 Mbps PBR (1250 KB buffer)
  - *CABAC on, pyramid B off (as required by PSP) Three reference frames, max three B-frames
  - AAC-LC 192 Kbps
- Zune 30 and 4/8/16/80/120: This is going to need to work on the original Zune 30 as well as the more recent models, so we can just stick with those constraints.
  - 320 × 176
  - VC-1 Main Profile 1.0 Mbps ABR 1.5 PBR (750 KB buffer)
  - WMA 9.2 192 Kbps
- Zune HD: This could play the iPhone file with no problem, but we can push the video quality up higher, since we have access to higher peak bitrates. Either H.264 or VC-1 would be fine, but we'll go with WMV as Zune owners are more likely to be WMV partisans.
  - 480 × 272
  - VC-1 Main Profile 1.5 Mbps ABR 4 Mbps PBR (2000 KB buffer)
  - WMA Pro 192 Kbps

Simple enough. Beyond those, we're going to go for awesome video quality and extreme slowness on our encodes as well. Low resolutions always encode pretty quickly anyway.

Expression Encoder

Expression Encoder 3 has become a pretty darn capable H.264 encoder for devices; the lack of High Profile or HE AAC hasn't been a limitation.

(*Continued*)

We've got pretty good presets to start with for most of these:

- H.264 iPod/iPhone touch
- A modified version of this for the PSP
- H.264 iPod classic/nano
- VC-1 Zune 1
- VC-1 Zune HD

From there, we can make some extreme performance choices:

- Complexity 5
- Threads Used: 1
- RD Optimization and 4 × 4 ME Partition Level on for H.264
- True Full Chroma and Adaptive Motion Match for VC-1

Those will increase encoding time at least 20x over the defaults, with perhaps a 10 percent improvement in efficiency. But hey, we're extreme. It's what we do. Don't try those settings with HD video, of course!

Here's a more realistic "slower better better" set of choices:

- Complexity 4
- Threads used: 2 for 320 × 176, 4 for 480 × 272 (following our "At least 64 lines per slice" rule)
- Adaptive True Chroma and Adaptive Motion Match for VC-1

Which would look nearly, and probably indistinguishably, extreme.

Those would be five to ten times faster than the insane settings, but within 1–2 percent of their quality. Awesome, certainly but arguably not extreme. See Figures 24.3 through 24.7.

Carbon

Carbon has full-featured implementations of both the VC-1 Encoder SDK and Main Concept H.264 SDK, and so has a lot more knobs and dials for mobile encoding than other tools.

(*Continued*)

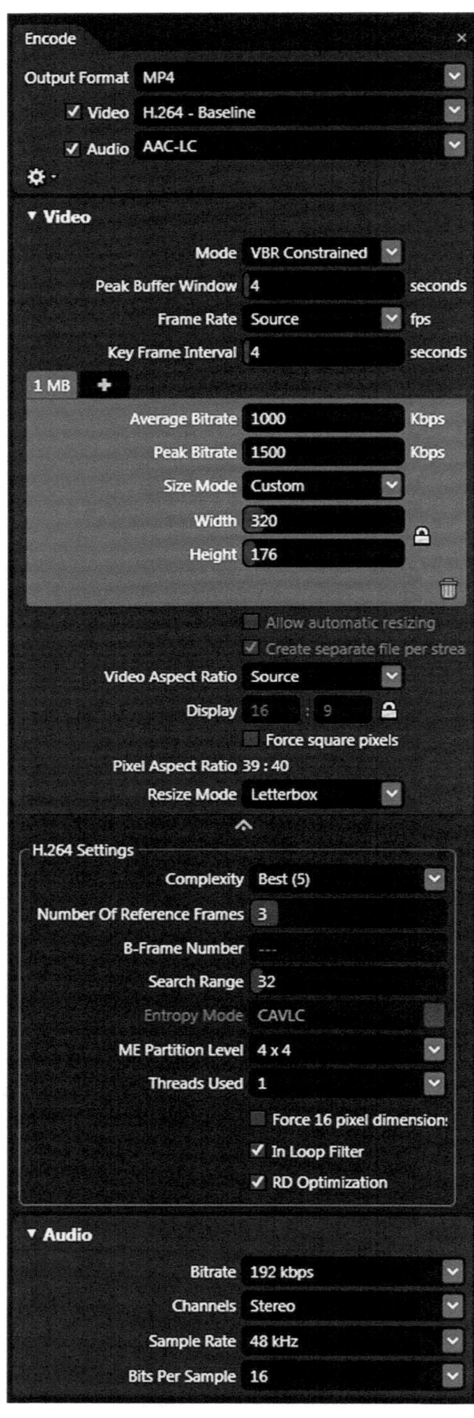

Figure 24.3 H.264 settings in Expression Encoder for the iPhone and iPod touch.

(Continued)

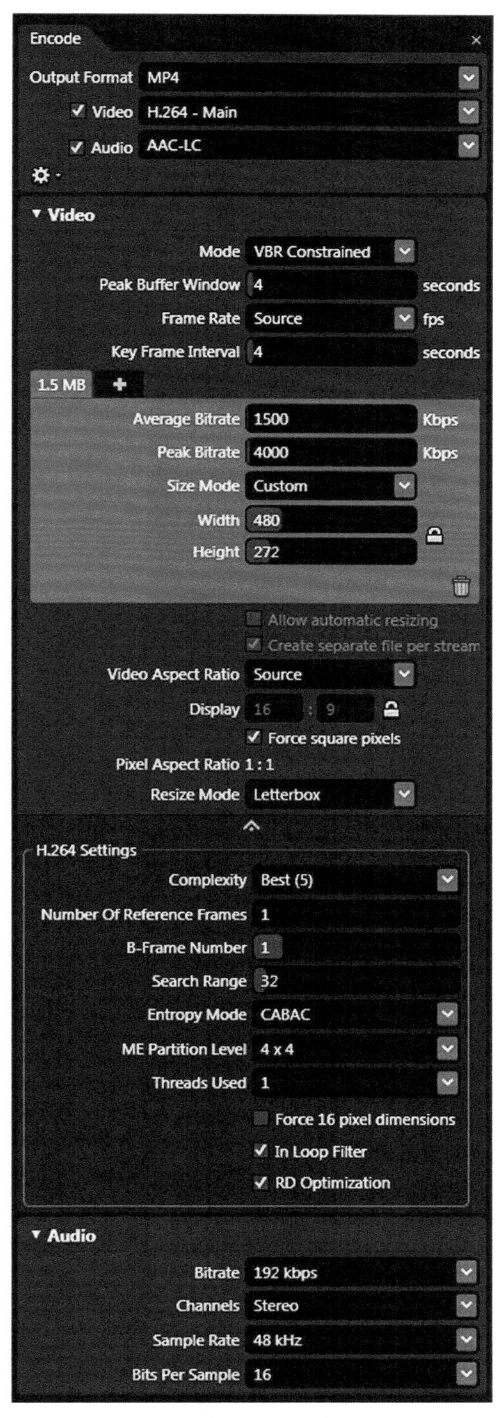

Figure 24.4 H.264 settings in Expression Encoder for the PSP.

(Continued)

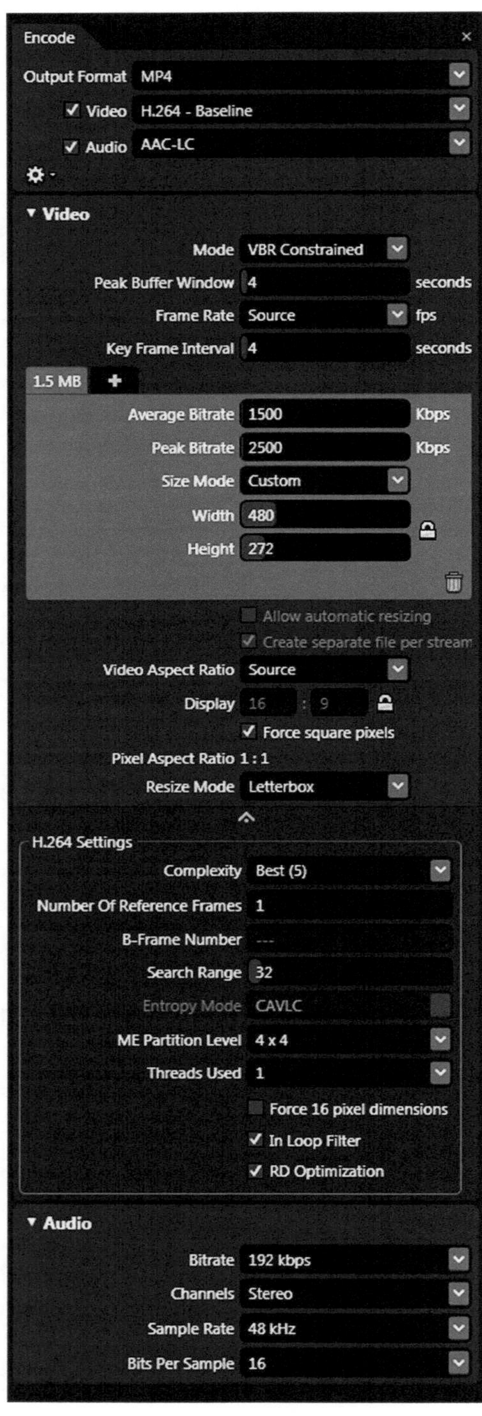

Figure 24.5 H.264 settings in Expression Encoder for the iPod classic/nano.

(Continued)

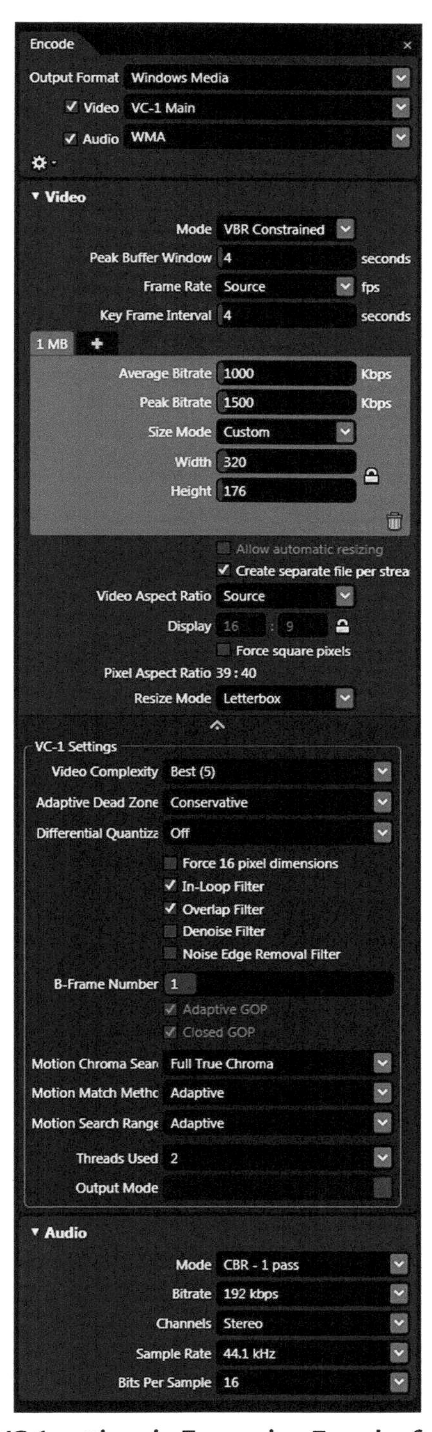

Figure 24.6 VC-1 settings in Expression Encoder for the Zune 1.

(Continued)

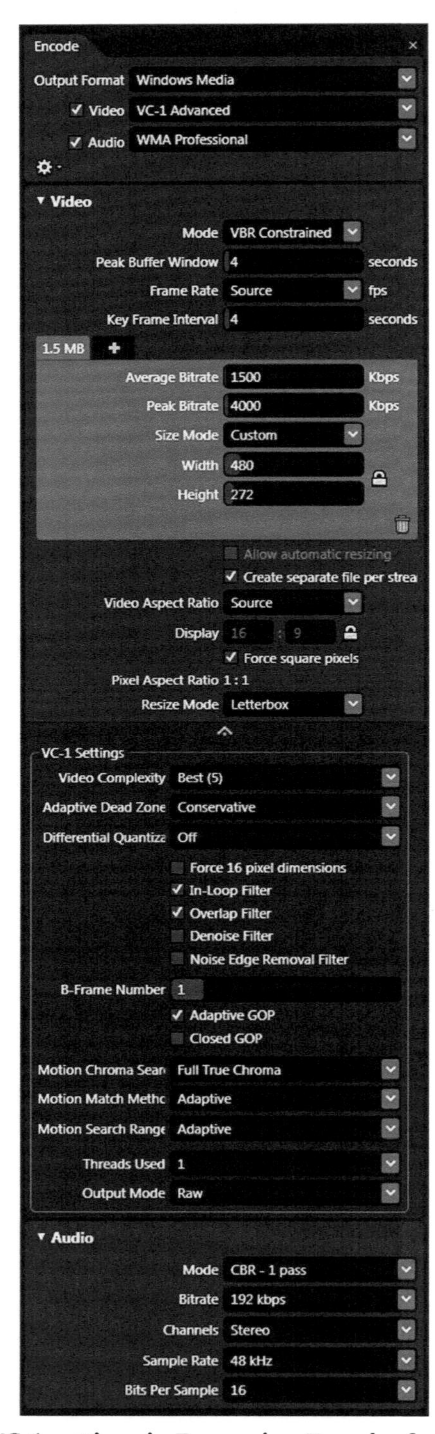

Figure 24.7 VC-1 settings in Expression Encoder for the Zune HD.

(Continued)

Unfortunately, the Carbon presets aren't very well suited to what we want. We'll just start with the System H.264 and VC-1 presets and modify from there. Note that Main Concept can do adaptive B-frame placement, so we enter 3 here:

H.264

- Stream Type: MPEG-4 System
- Device Compatibility: Sony PlayStation Portable for PSP
- Video Bitrate Mode: VBR 2-pass
- VBV Buffer Size: 0 (determined by Level)
- Profile: Baseline for iPod/iPhone, Main for PSP
- Level: 3.0
- Size of Coded Video Sequence: 120 (4-sec GOP @ 29.97 fps)
- B-frames (PSP only)
  - Number of B-pictures: 3
  - Use Adaptive B-frame placement: On
  - Reference B-pictures: On
  - Allow pyramid B-frame coding: Off (not supported by PSP)
- Slices: 1
- All "Use fast" options: Off
- Use rate distortion optimization: On
- Adpative Quantization Mode: Complexity (Keep the backgrounds smooth)
- Adaptive Quantization Strength: –50
- Audio Bitrate: 192

VC-1

- Complexity: 5
- Profile: Main
- Number of threads: 1

(Continued)

- GOP size: 120
- VBV Buffer as above
- Closed GOP: Off
- Overlap Filter: On
- Audio Code: WMA 9.2 for Zune 320 × 240 and WMA Pro for Zune HD
- Channels: 2
- Sample Rate: 48 KHz
- Bits/Sample: 16
- Audio Bitrate: 192 Kbps

Carbon's able to encode this quite a bit faster with its Insane settings than EEv3's, both because it can encode all four output streams in parallel (big time saver when doing single-threaded encoding) and because Main Concept doesn't have any modes as insanely slow as EEv3's "RD Optimization." See Figures 24.8 through 24.12.

(*Continued*)

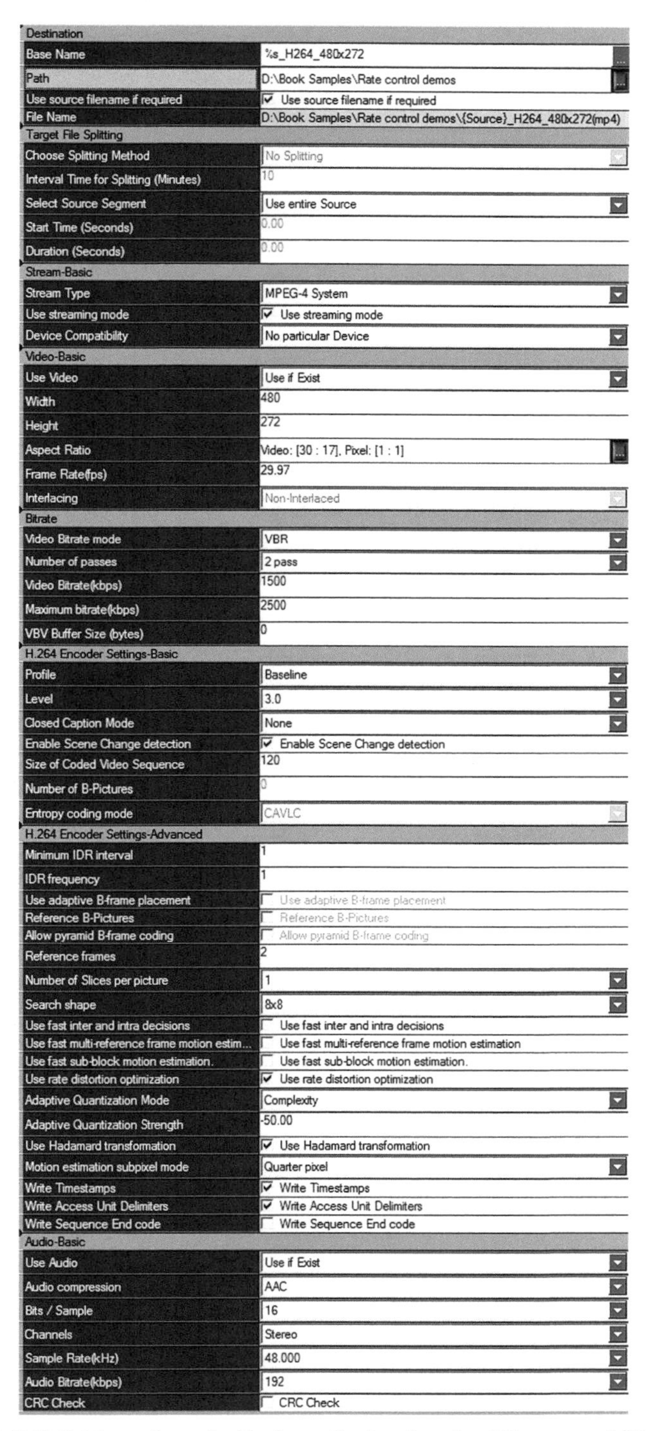

Destination	
Base Name	%s_H264_480x272
Path	D:\Book Samples\Rate control demos
Use source filename if required	☑ Use source filename if required
File Name	D:\Book Samples\Rate control demos\{Source}_H264_480x272(mp4)
Target File Splitting	
Choose Splitting Method	No Splitting
Interval Time for Splitting (Minutes)	10
Select Source Segment	Use entire Source
Start Time (Seconds)	0.00
Duration (Seconds)	0.00
Stream-Basic	
Stream Type	MPEG-4 System
Use streaming mode	☑ Use streaming mode
Device Compatibility	No particular Device
Video-Basic	
Use Video	Use if Exist
Width	480
Height	272
Aspect Ratio	Video: [30 : 17], Pixel: [1 : 1]
Frame Rate(fps)	29.97
Interlacing	Non-Interlaced
Bitrate	
Video Bitrate mode	VBR
Number of passes	2 pass
Video Bitrate(kbps)	1500
Maximum bitrate(kbps)	2500
VBV Buffer Size (bytes)	0
H.264 Encoder Settings-Basic	
Profile	Baseline
Level	3.0
Closed Caption Mode	None
Enable Scene Change detection	☑ Enable Scene Change detection
Size of Coded Video Sequence	120
Number of B-Pictures	0
Entropy coding mode	CAVLC
H.264 Encoder Settings-Advanced	
Minimum IDR interval	1
IDR frequency	1
Use adaptive B-frame placement	☐ Use adaptive B-frame placement
Reference B-Pictures	☐ Reference B-Pictures
Allow pyramid B-frame coding	☐ Allow pyramid B-frame coding
Reference frames	2
Number of Slices per picture	1
Search shape	8x8
Use fast inter and intra decisions	☐ Use fast inter and intra decisions
Use fast multi-reference frame motion estim...	☐ Use fast multi-reference frame motion estimation
Use fast sub-block motion estimation.	☐ Use fast sub-block motion estimation.
Use rate distortion optimization	☑ Use rate distortion optimization
Adaptive Quantization Mode	Complexity
Adaptive Quantization Strength	-50.00
Use Hadamard transformation	☑ Use Hadamard transformation
Motion estimation subpixel mode	Quarter pixel
Write Timestamps	☑ Write Timestamps
Write Access Unit Delimiters	☑ Write Access Unit Delimiters
Write Sequence End code	☐ Write Sequence End code
Audio-Basic	
Use Audio	Use if Exist
Audio compression	AAC
Bits / Sample	16
Channels	Stereo
Sample Rate(kHz)	48.000
Audio Bitrate(kbps)	192
CRC Check	☐ CRC Check

Figure 24.8 H.264 settings in Carbon Coder for the iPhone and iPod touch.

(*Continued*)

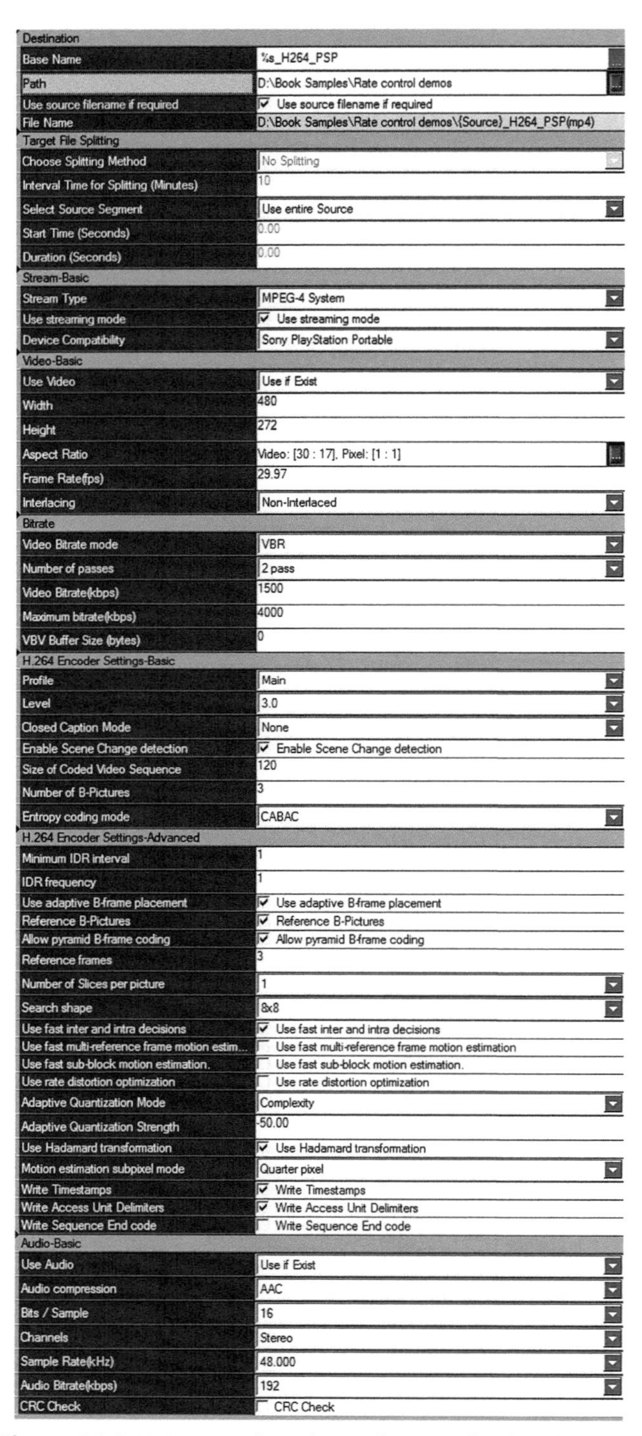

Destination	
Base Name	%s_H264_PSP
Path	D:\Book Samples\Rate control demos
Use source filename if required	☑ Use source filename if required
File Name	D:\Book Samples\Rate control demos\{Source}_H264_PSP(mp4)
Target File Splitting	
Choose Splitting Method	No Splitting
Interval Time for Splitting (Minutes)	10
Select Source Segment	Use entire Source
Start Time (Seconds)	0.00
Duration (Seconds)	0.00
Stream-Basic	
Stream Type	MPEG-4 System
Use streaming mode	☑ Use streaming mode
Device Compatibility	Sony PlayStation Portable
Video-Basic	
Use Video	Use if Exist
Width	480
Height	272
Aspect Ratio	Video: [30 : 17], Pixel: [1 : 1]
Frame Rate(fps)	29.97
Interlacing	Non-Interlaced
Bitrate	
Video Bitrate mode	VBR
Number of passes	2 pass
Video Bitrate(kbps)	1500
Maximum bitrate(kbps)	4000
VBV Buffer Size (bytes)	0
H.264 Encoder Settings-Basic	
Profile	Main
Level	3.0
Closed Caption Mode	None
Enable Scene Change detection	☑ Enable Scene Change detection
Size of Coded Video Sequence	120
Number of B-Pictures	3
Entropy coding mode	CABAC
H.264 Encoder Settings-Advanced	
Minimum IDR interval	1
IDR frequency	1
Use adaptive B-frame placement	☑ Use adaptive B-frame placement
Reference B-Pictures	☑ Reference B-Pictures
Allow pyramid B-frame coding	☑ Allow pyramid B-frame coding
Reference frames	3
Number of Slices per picture	1
Search shape	8x8
Use fast inter and intra decisions	☑ Use fast inter and intra decisions
Use fast multi-reference frame motion estim...	☐ Use fast multi-reference frame motion estimation
Use fast sub-block motion estimation.	☐ Use fast sub-block motion estimation.
Use rate distortion optimization	☐ Use rate distortion optimization
Adaptive Quantization Mode	Complexity
Adaptive Quantization Strength	-50.00
Use Hadamard transformation	☑ Use Hadamard transformation
Motion estimation subpixel mode	Quarter pixel
Write Timestamps	☑ Write Timestamps
Write Access Unit Delimiters	☑ Write Access Unit Delimiters
Write Sequence End code	☐ Write Sequence End code
Audio-Basic	
Use Audio	Use if Exist
Audio compression	AAC
Bits / Sample	16
Channels	Stereo
Sample Rate(kHz)	48.000
Audio Bitrate(kbps)	192
CRC Check	☐ CRC Check

Figure 24.9 H.264 settings in Carbon Coder for the PSP.

(Continued)

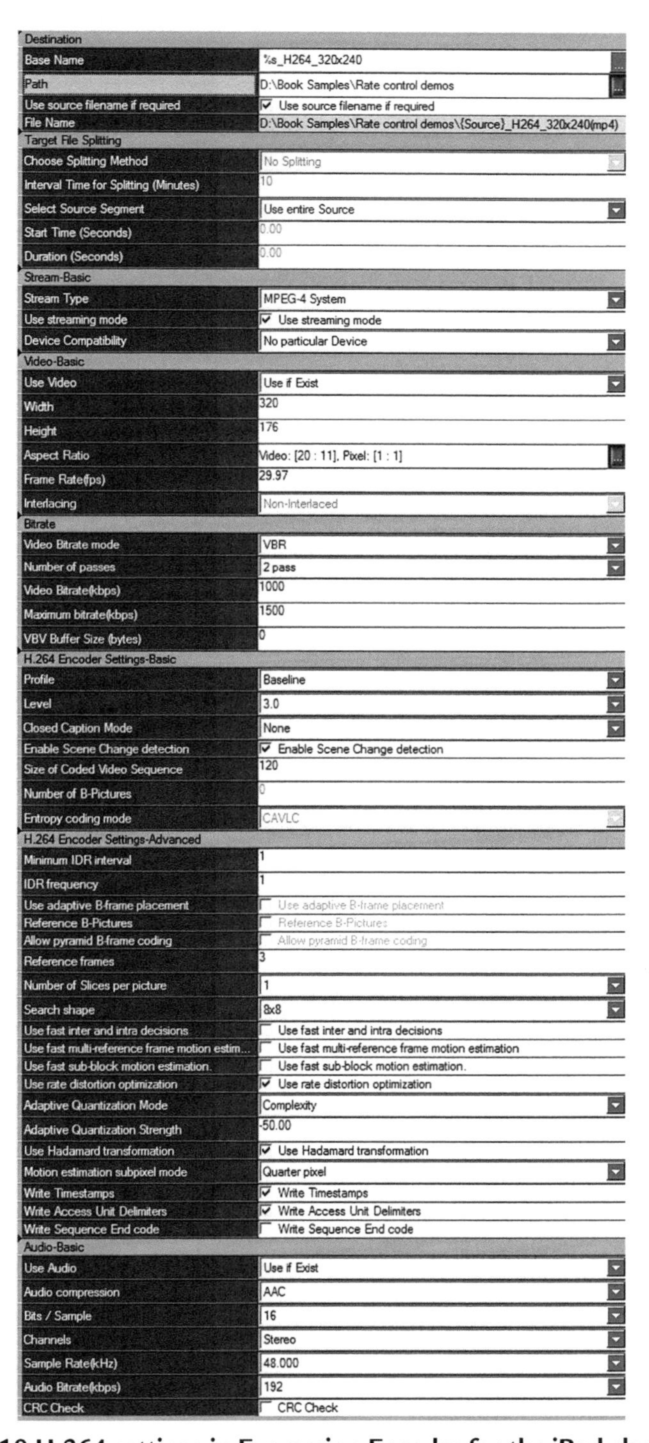

Destination	
Base Name	%s_H264_320x240
Path	D:\Book Samples\Rate control demos
Use source filename if required	☑ Use source filename if required
File Name	D:\Book Samples\Rate control demos\{Source}_H264_320x240(mp4)
Target File Splitting	
Choose Splitting Method	No Splitting
Interval Time for Splitting (Minutes)	10
Select Source Segment	Use entire Source
Start Time (Seconds)	0.00
Duration (Seconds)	0.00
Stream-Basic	
Stream Type	MPEG-4 System
Use streaming mode	☑ Use streaming mode
Device Compatibility	No particular Device
Video-Basic	
Use Video	Use if Exist
Width	320
Height	176
Aspect Ratio	Video: [20 : 11], Pixel: [1 : 1]
Frame Rate(fps)	29.97
Interlacing	Non-Interlaced
Bitrate	
Video Bitrate mode	VBR
Number of passes	2 pass
Video Bitrate(kbps)	1000
Maximum bitrate(kbps)	1500
VBV Buffer Size (bytes)	0
H.264 Encoder Settings-Basic	
Profile	Baseline
Level	3.0
Closed Caption Mode	None
Enable Scene Change detection	☑ Enable Scene Change detection
Size of Coded Video Sequence	120
Number of B-Pictures	0
Entropy coding mode	CAVLC
H.264 Encoder Settings-Advanced	
Minimum IDR interval	1
IDR frequency	1
Use adaptive B-frame placement	☐ Use adaptive B-frame placement
Reference B-Pictures	☐ Reference B-Pictures
Allow pyramid B-frame coding	☐ Allow pyramid B-frame coding
Reference frames	3
Number of Slices per picture	1
Search shape	8x8
Use fast inter and intra decisions	☐ Use fast inter and intra decisions
Use fast multi-reference frame motion estim...	☐ Use fast multi-reference frame motion estimation
Use fast sub-block motion estimation.	☐ Use fast sub-block motion estimation.
Use rate distortion optimization	☑ Use rate distortion optimization
Adaptive Quantization Mode	Complexity
Adaptive Quantization Strength	-50.00
Use Hadamard transformation	☑ Use Hadamard transformation
Motion estimation subpixel mode	Quarter pixel
Write Timestamps	☑ Write Timestamps
Write Access Unit Delimiters	☑ Write Access Unit Delimiters
Write Sequence End code	☐ Write Sequence End code
Audio-Basic	
Use Audio	Use if Exist
Audio compression	AAC
Bits / Sample	16
Channels	Stereo
Sample Rate(kHz)	48.000
Audio Bitrate(kbps)	192
CRC Check	☐ CRC Check

Figure 24.10 H.264 settings in Expression Encoder for the iPod classic/nano.

(Continued)

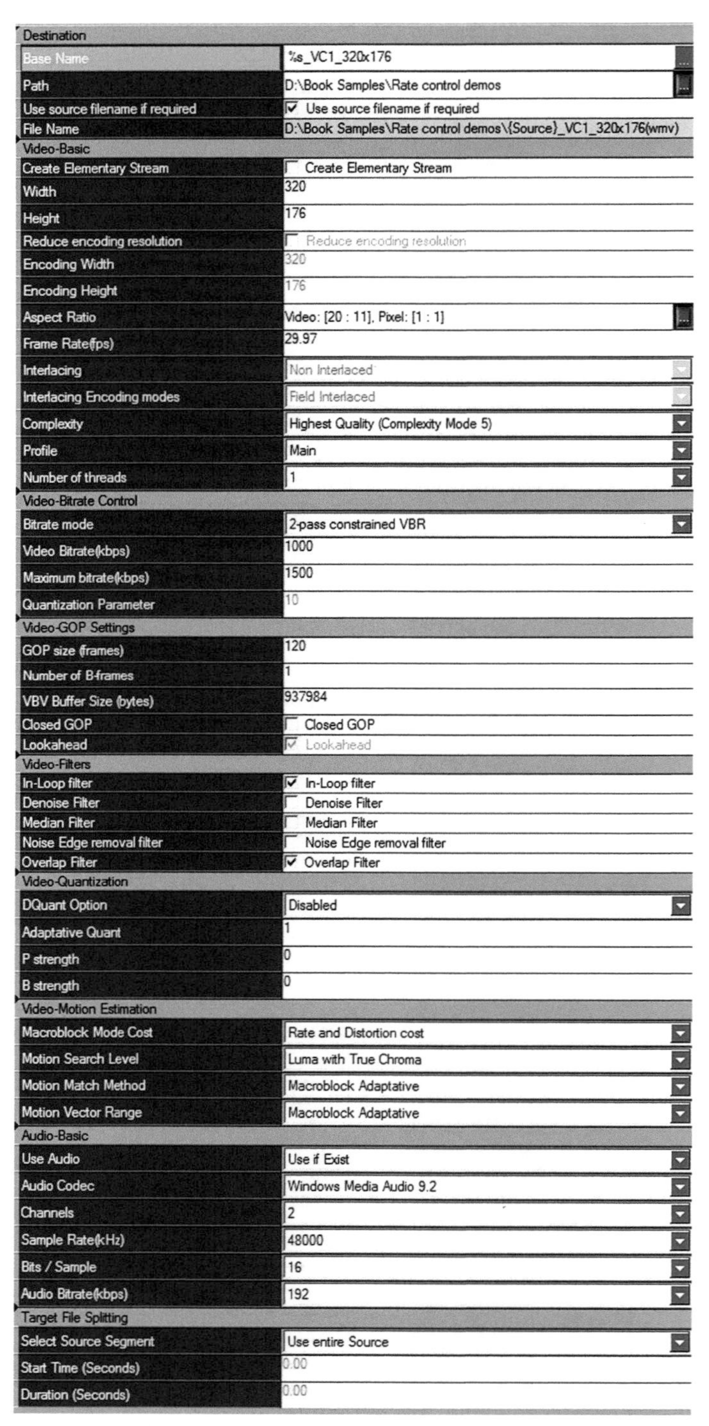

Destination	
Base Name	%s_VC1_320x176
Path	D:\Book Samples\Rate control demos
Use source filename if required	☑ Use source filename if required
File Name	D:\Book Samples\Rate control demos\{Source}_VC1_320x176(wmv)
Video-Basic	
Create Elementary Stream	☐ Create Elementary Stream
Width	320
Height	176
Reduce encoding resolution	☐ Reduce encoding resolution
Encoding Width	320
Encoding Height	176
Aspect Ratio	Video: [20 : 11], Pixel: [1 : 1]
Frame Rate(fps)	29.97
Interlacing	Non Interlaced
Interlacing Encoding modes	Field Interlaced
Complexity	Highest Quality (Complexity Mode 5)
Profile	Main
Number of threads	1
Video-Bitrate Control	
Bitrate mode	2-pass constrained VBR
Video Bitrate(kbps)	1000
Maximum bitrate(kbps)	1500
Quantization Parameter	10
Video-GOP Settings	
GOP size (frames)	120
Number of B-frames	1
VBV Buffer Size (bytes)	937984
Closed GOP	☐ Closed GOP
Lookahead	☑ Lookahead
Video-Filters	
In-Loop filter	☑ In-Loop filter
Denoise Filter	☐ Denoise Filter
Median Filter	☐ Median Filter
Noise Edge removal filter	☐ Noise Edge removal filter
Overlap Filter	☑ Overlap Filter
Video-Quantization	
DQuant Option	Disabled
Adaptative Quant	1
P strength	0
B strength	0
Video-Motion Estimation	
Macroblock Mode Cost	Rate and Distortion cost
Motion Search Level	Luma with True Chroma
Motion Match Method	Macroblock Adaptative
Motion Vector Range	Macroblock Adaptative
Audio-Basic	
Use Audio	Use if Exist
Audio Codec	Windows Media Audio 9.2
Channels	2
Sample Rate(kHz)	48000
Bits / Sample	16
Audio Bitrate(kbps)	192
Target File Splitting	
Select Source Segment	Use entire Source
Start Time (Seconds)	0.00
Duration (Seconds)	0.00

Figure 24.11 VC-1 settings in Expression Encoder for the Zune 1.

(Continued)

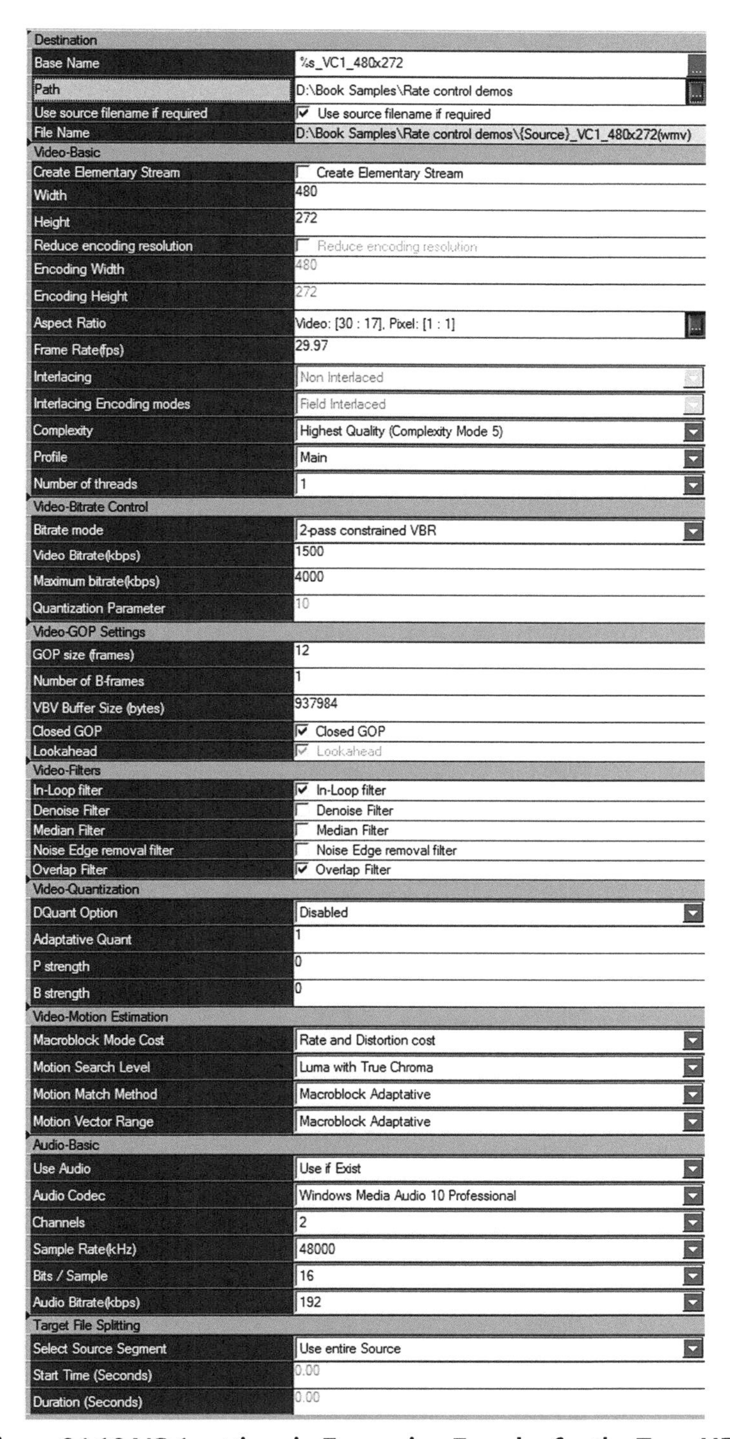

Figure 24.12 VC-1 settings in Expression Encoder for the Zune HD.

CHAPTER 25

Flash

Introduction

Flash is often called an RIA (Rich Internet/Interactive Application) platform. There's plenty of argument about what exactly that means, but I think of it as enabling rich applications for the web capable of doing much more than HTML. As a platform, a vast industry writes applications for Flash that run in a secure "sandbox" without access to the underlying computer or its data. Flash apps can't do anything damaging to the underlying computers (or at least they're not supposed to; there've been a number of exploited security faults in Flash over the years) so users don't need to install anything or type in an admin password. Flash apps should "just work" like anything else in a web page.

In this Flash competes with the other mainstream RIA platforms of JavaFX and Silverlight, and potentially with HTML5 as it is standardized and supported in browsers.

Early Years: Flash 1–5

It's sometimes hard to remember that Adobe's Flash has only been a video platform for half of its existence. Flash started out as FutureSplash Animator, a simple vector graphics animation layer. I first saw it at the RealNetworks 1996 conference, pitched as a way to deliver animated content with compressed audio over the slow modem connections of the era.

Macromedia acquired FutureSplash and launched Flash 1.0 later in 1996. Following a similar course to Macromedia's Director, Flash slowly gained broader functionality, including programmability via the ActionScript scripting language.

Good bundling deals, a consistent cross-platform/browser experience, and a small runtime (quick to download even over modems) led Flash into becoming a ubiquitous plug-in, and the SWF format became a standard for internet advertising chafing at the limitations of Animated GIF.

Unsurprisingly, many got excited about the possibility of getting video playback to work in Flash (similar efforts existed for Java). QuickTime even supported Flash as a track type to add richer navigation on top of QuickTime media playback, and Windows Media integration was also possible. But those required both Flash and the other media player be installed.

Video Is Introduced: Flash 6–7

Macromedia listened to its audience, and launched basic media support in 2002's Flash MX (6). This introduced FLV as an ingest format for the Flash application; video assets needed to be bundled into the SWF. The initial codec was the simple but fast and small Sorenson Spark H.263 implementation, with MP3 audio. The first version of Flash Communication Server MX, the predecessor of today's Flash Media Server, was released later that year.

Initially, Flash Video was more interesting as a way to get video inside Flash-centric experiences than it was interesting for video-centric experiences. Things got a lot better with Flash MX 2004 (7), which enabled progressive download of FLV files and performance improvements.

VP6 and the Video Breakout: Flash 8–9

Video in Flash came of age in 2005 with Flash 8 (thankfully reverting to version numbers), which introduced On2's much more efficient VP6 codec, and was coupled with the rebranded Flash Media Server 2.0. However, the two most important events for Flash in 2005 were arguably in the corporate, rather than technological, world: Adobe acquired Macromedia, and YouTube was founded.

Flash utilization for video grew rapidly in this period, eating into the market share of Windows Media only a few years after it had largely replaced RealMedia.

The RealMedia/QuickTime/Windows Media competition had focused on lower-level technologies like codecs, playback performance, and hosting costs. While Flash was weaker on those metrics, Flash won on critical new ones. Flash changed the game with a focus on seamless cross-platform, cross-browser support and rich interactivity and design in the player.

The adoption of VP6 did have some downsides, however; as a proprietary codec only On2 provided professional-grade encoder implementations. And On2 wasn't beholden to Adobe or Flash; it was there to make money for its shareholders. As the popularity of Flash grew, so did the cost of VP6, with many companies chafing at the cost of using it, and some choosing to deliver lower-quality video with the consumer-grade implementation provided by Adobe. VP6's encoder was slow as well, and remained single-threaded long after competing codecs could utilize all of a 4-core CPU. YouTube skipped VP6 entirely.

The H.264 Era: Flash 9–10

Adobe's corporate culture had been more standards-focused than Macromedia's, and so as it looked at next generation media technologies, MPEG-4, H.264, and AAC were obvious choices. There was a big, existing ecosystem of companies making interoperable encoders, so Adobe wouldn't be caught in a single-vendor bind like it had with VP6. While Adobe had a big business in video authoring tools, they didn't have any compression-specific products to undercut. And clearly these were great codecs. Adobe had the mobile world in its sights as well, and hardware support for H.264 and AAC was rapidly being added to those devices.

MPEG-4 and H.264 support was added in the big Flash 9 Update 3 (version 9.0r115) late in 2007.

One part of MPEG-4 that Adobe didn't add support for was RTSP-based streaming. The company decided to leave its own RTMP as the only supported streaming protocol in Flash.

The Future: Mobile and CE Devices

Adobe's "Open Screen Project" is an ambitious effort to make Flash as ubiquitous in phones and other consumer electronics devices as it is on the PC. This builds on the significant traction of Flash Lite in various devices. Adobe has also announced GPU-accelerated hardware decode is coming in Flash, promising it will address long-stranding complaints about poor video decode performance.

Why Flash?

Ubiquitous Player

The biggest single reason Flash has been adopted is its ubiquity on Mac and Windows PCs; most users have a reasonably current version installed, and it's an easy upgrade for most who don't. Adobe's surveys suggest that 98 percent of consumers have some version Flash installed on at least one computer they use although many may be a version or two behind. Enterprise penetration is similar, although tends to update more slowly.

Adobe's working hard to get Flash broadly deployed on mobile devices as well, although that's at a much earlier stage, and with much bigger differences in media support per device.

Uniform Rich Cross-Platform/Browser Experience

Beyond being everywhere, Flash can do a lot with the nonvideo features of Flash. This includes animations, overlays, synchronized data displays, and branded controls.

And perhaps most importantly, advertising. SWF has long been popular in web advertising, so being able to integrate Flash advertising with video playback was an important factor for ad-funded video sites.

Excellent Codec Support

Flash has great codec support with H.264 High Profile and AAC (up to HE v2). This also makes it compatible with lots of existing content in the QuickTime and MPEG-4 worlds.

Why Not Flash?

Higher Total Cost of Ownership for Streaming

Flash Media Server is certainly capable, but it's also expensive to scale up for large audiences due to its unicast-only approach to content delivery. Adaptive Streaming technologies can deliver significantly lower cost per user for large audiences. FMS also has a higher per-server license fee than competing technologies; both Mac OS X Server and Windows Server 2008 include full streaming support as part of the OS.

Playback Performance

Flash has long had a reputation for sluggish and choppy playback, and even after GPU acceleration for compositing was added in Flash 10, things still get pretty dodgy as frame sizes go up.

Scaling to full-screen is particularly problematic, especially for higher-resolution screens (my main ones are 1920 × 1200 and 2560 × 1600, so I really feel this pain). It gets really slow, with lots of dropped frames, and typically with nearest-neighbor scaling artifacts.

We can hope further refinements of GPU acceleration will address these issues; the equivalent feature in Silverlight works fine on the same systems. Adobe has announced they'll be adding GPU video decoding in a future version, which should further help.

Flash 10's Mac and Linux versions perform somewhat poorly compared to the Windows version on the same hardware, and particularly with video. It's not uncommon to see 3x higher CPU utilization booted into Mac OS X versus Windows in Boot Camp.

Flash for Progressive Download

Flash handles progressive download quite nicely with both FLV and MP4 content. ActionScript players can enable byte-rate request based random access, providing a more streaming-like experience à la YouTube.

Flash for Real-Time Streaming

Flash streaming uses Adobe's RTMP protocol, which the company has now published. The primary RTMP server is Adobe's own Flash Media Server, but Wowza Media Server Pro and the open-source Red5 also support Flash-compatible RTMP delivery.

RTMP uses TCP with HTTP fallback, but unlike other streaming protocols doesn't support UDP or multicast. Designed and introduced later than RTSP and its many implementations implementations, it's tuned for the much more reliable and lower-latency Internet of today. FMS is lauded for providing very low broadcast delays, generally less than two seconds and sometimes even subsecond. The relatively small buffers that implies suggests that RMTP might not perform as reliably on slow or lossy connections, but those are rare and becoming rarer.

Flash takes a somewhat different model for content protection compared to other platforms. Instead of encrypting the content before serving, FMS's RTMPe protocol applies unique encryption in real-time for each user's unicast-delivered stream. This increases server-side CPU load per user by perhaps 25 percent. Also, as the files or streams are unencrypted before they arrive at FMS, there's some additional vulnerability compared to systems where encryption is applied with encoding.

RTMPe requires FMS, and remains proprietary to Adobe.

Note that neither the Flash client nor FMS supports standard MPEG-4 RTSP streaming; RTMP and RTMPe are the only supported real-time protocols.

Dynamic Streaming

FMS 3.5 introduced a new MBR switching technology called Dynamic Streaming. It's a multiple bitrate stream switching technology implemented on RTMP.

Compared to older MBR implementations like Intelligent Streaming (Windows Media) and SureStream (RealMedia), Dynamic Streaming offers much smoother bitrate transitions and better measurement of available bandwidth (the Achilles' heel of past MBR techniques). The maximum bitrate is determined by "bursting" data to the client periodically instead of keeping a continuous stream of bytes. Thus 5 seconds of content could be sent to the Flash client as quickly as possible, and if it only took one second to arrive, then available bandwidth may be five times higher than the media bitrate. Bitrate switches are done automatically to match network conditions and CPU decoding performance (if at least 20 percent of frames are dropped at the current stream). They can triggered programmatically as well—for example, to switch between camera angles.

In terms of the customer experience, Dynamic Streaming seems to be in the ballpark of Adaptive Streaming solutions like Move Networks and Smooth Streaming. The bigger differences are going to be in the proxy caching inherent in adaptive streaming, which the unicast-only RTMP can't provide.

However, as I write this about nine months after FMS 3.5's release, there simply aren't any high-profile Dynamic Streaming sites to judge the experience on. Everything seems promising in theory, and I'm very curious to see what it's like in practice. The Adobe/Akamai demo player yields a lot of dropped frames and has trouble making it to HD bitrates even on my 8-core monster workstation, but that could be due to player tuning or other issues.

Authoring Dynamic Streaming

Dynamic Streaming is simpler than adaptive streaming solutions on the authoring side, since it has no requirement for aligned GOPs or fragmented files; it works fine with existing FLV and MPEG-4 content. This is achieved with a couple of mechanisms:

- Waiting until the next I-frame in the stream is switched to.
- If the content doesn't have matching timelines, or the GOP length is longer than the client buffer, generating a new I-frame server-side to switch to.

So no particular care should be required to enable video stream switching. Audio stream switching is another matter; there's typically a "pop" if there's any audio ongoing as audio bitrates are switched. The only way to avoid this is to use the exact audio input and settings for all streams. The efficiency of HE AAC comes in handy here; even 48–64 Kbps can sound quite good, and can be paired with both low- and high-bitrate video.

Flash for Interactive Media

Flash is an interactive and rich media format that includes video. Flash itself is a topic for many, many books, and most people have seen plenty of Flash movies, so I won't belabor the topic here. Simply stated, Flash can do pretty much anything you can image on the interactive side, with sufficient code and design effort.

Flash for Conferencing

While I'm generally not talking about videoconferencing in this book, Adobe's Real Time Media Flow Protocol provides some interesting functionality for Flash. RTMFP, unlike RTMP is UDP based and can directly connect two users to improve latency. It applies P2P techniques to enable two clients to communicate directly. However, it doesn't support multicast.

I've imagined that Adobe would merge some of this functionality with RTMP, but so far it's strictly been for videoconferencing between small numbers of users.

Flash for Phones

Macromedia was working on Flash Lite for phones and other mobile devices well before it was acquired by Adobe.

Even though Flash is a relatively small piece of code by PC standards, it's a lot for a phone, so Flash Lite has always supported only a subset of the full Flash. And the different screen and control types of a phone are generally a poor fit for a Flash app that assumes keyboard, mouse, and 1024 × 768 display.

Thus, Flash Lite is more about reusing the skills and tools of Flash authoring than having current Flash applications "just work" on devices.

That said, Adobe's Open Media Project (OMP) is a big effort to bring a much fuller implementation of Flash to forthcoming generations of powerful phones and other CE devices. It'll be a while before we see what exactly OMP is going to provide for content authors.

The current version of Flash Lite is 3.1 (based on Flash 8), which offers a good base of media features. It's available for both phones and "digital home" devices like set-top boxes, and it offers several key features:

- F4V and FLV support, with H.264 Baseline, VP6, and H.263 decode
- Reference software decoders for H.263 and VP6 (much slower and power-hungry than an ASIC decoder); H.264 support requires hardware
- Up to 720p30 decode with compatible hardware (not common yet)
- ActionScript 1 and 2 support (ActionScript 3 is a notable lack compared to full Flash)

Note that, unlike the desktop, phones don't get automatic Flash updates. Thus older devices have older versions of Flash Lite; there are many more phones running Flash Lite 2.0 than 3.1 as I write this.

While Flash Lite supports all of the listed codecs, it's up to the handset vendors to integrate playback, and they may not include them all (particularly VP6). And phones vary widely in screen size and decoding horsepower; it's nearly impossible to make video files that "just work" on all Flash Lite devices.

Adobe has done a good job of integrating mobile authoring and design of Flash Lite with Flash. Their Device Central app (bundled with Flash and Creative Suite) emulates a wide range of devices running Flash Lite, including media playback.

Going forward, Adobe promises great things for Flash on phones and CE devices as part of the Open Screen Project, with much more of Flash 10 included. It has yet to yield shipping products, so it's too early to say what it'll mean for content delivery.

Formats and Codecs for Flash

In general, H.264 and AAC are the best codecs available in Flash, so you should use F4V unless you have a compelling reason to use FLV. It's a three-way choice based on video codec.

FLV with H.263

Pros:

- Lowest decoder complexity
- Supports alpha channels

- Most compatible with older devices
- Required for Flash 7 or earlier

Cons:

- Lowest compression efficiency by far

FLV with VP6

Pros:

- Alpha channel support
- Much better compression efficiency than H.263
- Adaptive postprocessing does good job of hiding artifacts
- Support back to Flash 8

Cons:

- Best results require expensive Pro version of VP6
- Slower encoder
- VP6 + MP3 less efficient than H.264 + AAC, particularly at low bitrates
- Least likely to be supported on devices

F4V

Pros:

- Best compression efficiency
- Allows AAC audio; HE AAC v2 is about 3x as efficient as MP3 at low rates
- Broad device support

Cons:

- No alpha channel support
- Highest decode complexity (but can be tuned for perf/quality tradeoff)
- Some H.264 uses may require licensing payments to MPEG-LA (see H.264 page 227)
- Require Flash 9 Release 3 or later

FLV

FLV and its encoding options are fully covered in Chapter 15.

MP3

Flash can play a standalone .mp3 file quite nicely, all the way back to Flash 4.

AAC offers better efficiency, particularly at lower bitrates, but there's no reason not to continue using existing MP3 assets if they provide adequate quality and size. New audio-only content should be .m4a.

F4V

MPEG-4, H.264, and AAC are covered in detail in their respective chapters. This section just documents the specifics of their Flash implementations.

A .f4v file is just a .mp4 that contains Flash-compatible video and audio. There's no requirement to use the .f4v extension; .mp4 can be used interchangeably. This is generally simpler for content meant to play in other media players as well as Flash.

Flash can play AAC-only MPEG-4 (.m4a) files just fine. Flash calls them ".f4a" by default, but .m4a works fine as well.

H.264 in Flash

Flash supports a broader range of H.264 profiles and modes than many PC players, although some aren't of much practical use. Flash's 8-bit rendering pipeline keeps 10-bit High Profile from offering any image quality improvement, although there's no real downside to it either. And although Flash can decode interlaced H.264, it lacks a deinterlacer, leaving those ugly horizontal lines painfully obvious.

- Baseline: Mainly for devices
- Main: Only need in unusual case of Main-only device compatibility
- High: The best choice for PC playback
- High 10-bit 4:2:0 and 4:2:2

In general, your choice is going to be Baseline when targeting devices and High 4:2:0 otherwise. Flash video playback can be pretty hardware demanding, so make sure to test playback on your minimum spec hardware. You may want to turn off CABAC at higher bitrates, and definitely want to keep peaks as low as you can with sufficient quality.

AAC in Flash

Flash's MPEG-4 support came along with extensive AAC support. I recommend that the following bitrate ranges use the particular profiles. AAC-LC can be the most transparent at high bitrates and HE v2 the least, but the lower the bitrate, the better the more complex codecs perform.

- < 128 Kbps: AAC-LC
- 64–128 Kbps: HE AAC v1 (spectral band replication)
- > 64 Kbps: HE AAC v2 (parametric stereo)

Note that Adobe Media Encoder labels HE AAC as AAC+.

Flash's that audio pipeline is limited to stereo playback only. While it can decode 5.1 AAC audio just fine, it'll be mixed down to two channels of output.

ActionScript Audio Codecs

Flash 10 introduced new audio APIs with much finer control over audio generation and playback. This can support audio decoders in ActionScript 3, like an audio-only version of Silverlight's Raw AV MediaStreamSource.

So far, there have been just some demonstration releases of Ogg Vorbis decoding using this API.

Encoding Tools for Flash

This section covers tools for Flash-compatible MPEG-4 creation with Flash-specific functionality. FLV tools are covered in Chapter 25, and general H.264 compression tools are covered there.

Adobe Media Encoder

Adobe Media Encoder is bundled with the Creative Suite products. It's a simple tool designed for nonprofessional encoders, and does a good job of helping those new to compression make reasonable choices.

AME uses the quite capable Main Concept H.264 implementation, and includes all three AAC flavors. However, it offers little configuration beyond bitrate, Profile, and Level. It doesn't even expose 2-pass CBR.

Sorenson Squeeze

Squeeze was the first professional-grade compression tool with Flash support, from when Flash 6 added Sorenson Media's Spark codec.

Squeeze continues to have good Flash support for FLV (including perhaps the best H.263 implementation) and F4V. Squeeze's H.264 also uses Main Concept underneath, but exposes more parameters than AME for quality tuning.

Squeeze has a very handy ability to publish a SWF player along with the encode, using its own or custom templates.

Rhozet Carbon/Adobe Flash Media Encoding Server

Carbon is one of the few tools to actually offer a .f4v extension, albeit as part of its standard H.264 target. F4V otherwise is the same as Carbon's MPEG-4 Program Stream.

Like AME, Carbon uses Main Concept's H.264 implementation, but exposes many more parameters for finer-quality tuning. Its simultaneous rendering makes for faster throughput of Dynamic Streaming files, particularly on highly multicore systems.

Adobe's Flash Media Encoding Server is a version of Carbon with all the non-Flash-compatible formats ripped out. It's otherwise identical in functionality.

Adobe Flash Media Live Encoder

Adobe's Flash Media Live Encoder (AFMLE) is a free (beyond needing FMS for it to be useful) live encoder for Flash, as the name suggests. It's available for Windows and Linux, and supports standard capture cards and devices for each.

AFMLE, not just a mouthful of an acronym, also supports FLV (H.263 and VP6) and H.264, with multiple bitrate encoding. However, it's a pretty basic webcasting tool. It's easy to learn and use, but isn't meant for carrier grade use.

For some reason, its H.264 is limited to Main Profile, and its AAC is LC and HE v1 only, no v2.

Most high-end live encoding products—like those from Inlet, ViewCast, and Digital Rapids—also support live Flash encoding. Since FMS doesn't support RTSP, FMS support needs to be added to each encoder.

Flash Dynamic Streaming Tutorial

As Flash supports FLV and H.264 playback, the tutorials in those chapters are relevant. This tutorial covers authoring multiple bitrate content for a Dynamic Streaming website.

(*Continued*)

Scenario

We're working for a web media company publishing TV and movie for consumers, using Flash.

While there's still a big library of 480i content we deal with, increasingly we're getting 1080p24 masters. We're moving from FLV to F4V as our encoding format, and introducing HD video delivery at the same time.

We want to find good encoding settings for Dynamic Streaming, scaling from our current minimum bitrate of 300 Kbps to a new maximum of 3 Mbps 720p (3 Mbps max determined by our aggregate bandwidth costs).

The Three Questions

What Is My Content?

The content is Hollywood produced feature film and TV content. We're starting with 1080p24 content for this project.

Who Is My Audience?

The audience is general consumers of web video content. Thus they'll have a broad range of hardware and connection speeds.

What Are My Communication Goals?

We want to make sure that we provide a good basic experience on older machines with slower bandwidth, ramping up to a higher-end experience on faster machines.

Tech Specs

Since this is H.264, Dynamic Streaming, and HD, we'll require Flash 10 for best playback performance.

Our bitrate range is 300–3000. As a rule of thumb, I like to have around a 1.5x ratio between bitrates when doing MBR, which we get with 6 bitrate bands.

We'll then need to figure out appropriate frame sizes for each, scaling from an easy to decode 320 × 176 at the low end to the full 1280 × 720 at the high end.

We'll use the same audio bitrate and other settings for each band, in order to avoid popping sounds with bitrate switching. Audio bitrate makes up a bigger proportion of total bitrate as bitrate drops. Since this is entertainment content, we'll go for 64 Kbps HE AAC v1. That will preserve stereo separation better than HE AAC v2 while still leaving 236 Kbps for video out of the 300 Kbps band, and still provide a "HD" quality sound at 3000 Kbps.

Since we're looking at volume production, we're not going to go crazy on encoding settings; we'll pick a good balance between quality and encoding time. Since we're targeting an 8-core machine, we can tweak the number of slices on each bitrate in order to keep the machine going full blast, while maximizing the efficiency of CABAC for low bitrates.

(*Continued*)

To enable easy stream switching, we're going to use the following settings:

- CBR
- Closed GOP
- 3-second adaptive GOP
- 3-second buffer

With a Closed GOP, the first frame of a GOP won't reference the last frame of the previous GOP, and so we won't need to run simultaneous decoders at switching points.

By sticking to three seconds for GOP length and buffer, switching and rebuffering time shouldn't be more than three seconds, even for a stream right at available bandwidth. Of course, FMS will try to buffer enough to make it more seamless than that, but three seconds should offer similar switching performance to Move Networks and Smooth Streaming.

If we wanted to be really seamless on switching, we could have turned off Scene Detection, and would have gotten exactly 72 frames a GOP. However, the quality hit from that can be pretty big. Normally the scene detection result is the same for all streams, so the cadence should reset at every edit. FMS does a pretty good job of switching between unaligned streams, so the quality gain is worth the chance of greater server load if the streams get out of cadence here and there.

We'll use High Profile everywhere, as the most efficient choice, and set the appropriate level based on frame size and bitrate. For slices, we want to have something over eight total, so that with the simultaneous encoding we're going to use all our CPU, but the lower-resolution bands take less relative time. Plus there's overhead from preprocessing and audio. The following settings should be pretty close. Having two slices at higher bitrates reduces decoder latency by enabling multithreading, enabling us to cold-start from a stream switch a little faster on slower machines.

Overall, this works out as shown in Table 25.1.

Carbon/AFMES Settings

In Carbon, we can start with the H.264 448 K 24p preset, and modify from there. We can make some basic modifications:

- Height 176
- Frame rate 23.976

Table 25.1 Bitrates, Frame Size, Levels, and Slices for our Tutorial Project.

Total bitrate	Audio	Video	Width	Height	VBV (KB)	Level	Slices
300	64	236	320	176	112	2.0	1
500	64	436	432	240	187	2.1	1
800	64	736	512	384	300	2.2	2
1200	64	1136	720	400	450	3	2
2000	64	1936	960	544	750	3.1	2
3000	64	2936	1280	720	1125	3.1	2

(*Continued*)

- Number of passes: 2-pass (Main Concept gets keyframe strobing in 1-pass CBR)
- VBV Buffer size
- Video Bitrate: 236
- VBV Buffer: 112000 (Carbon takes as bytes, not KBytes or Kbits as in other tools)
- Size of Coded Video Sequence: 72
- Use fastmulti-reference motion estimation on 1200 Kbps and above (speeds up encodes with a good number of reference frames, with a little quality hit; speed hit is proportional to frame size, so we'll leave off for the lower bands where every last bit is needed)
- Use fast inter/intra-mode decisions: Off for 500 and 300 Kbps (also frame size proportional; let's squeeze out as much low bitrate quality as possible)
- Adaptive Quantization Mode: Complexity (don't want blocky backgrounds)
- Adaptive Quantization Strength: –50 (a good midrange value)
- Number of B-pictures: 3
- Entropy Coding Mode: CABAC
- Use adaptive B-frame placement: On
- Reference B-pictures: On
- Allow Pyramid B-frame coding: On
- Reference Frames: 4 (the most as are useful for nonanimation content)
- Slices: 1

The other settings are good defaults for balanced quality/performance.

For audio, we'll simply do these three settings:

- Audio compression: HE-AAC version 1
- Sample Rate: 44.1 KHz
- Audio Bitrate: 64 Kbps

And with that, we can fill out the rest of our matrix. See Figure 25.1. We could have used CAVLC instead of CABAC at higher bitrates to reduce decode complexity, but will count on Dynamic Streaming to just switch to a more playable bitrate for slower systems.

And there we go. Total encoding time is 3–4x real time, so each workstation can produce six to eight hours of programming a day.

(*Continued*)

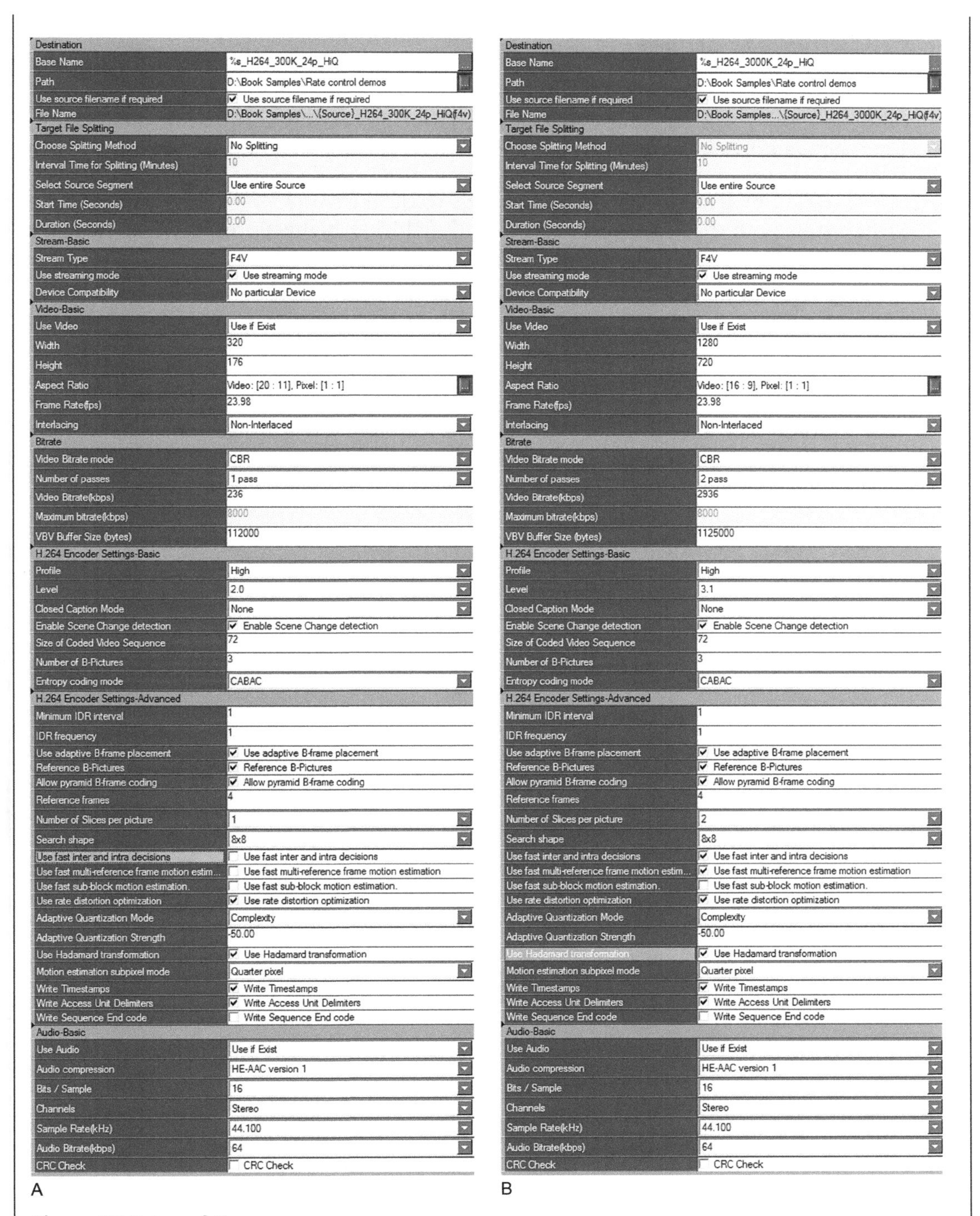

Setting	A	B
Destination		
Base Name	%s_H264_300K_24p_HiQ	%s_H264_3000K_24p_HiQ
Path	D:\Book Samples\Rate control demos	D:\Book Samples\Rate control demos
Use source filename if required	✓ Use source filename if required	✓ Use source filename if required
File Name	D:\Book Samples\...\{Source}_H264_300K_24p_HiQ.f4v	D:\Book Samples...\{Source}_H264_3000K_24p_HiQ.f4v
Target File Splitting		
Choose Splitting Method	No Splitting	No Splitting
Interval Time for Splitting (Minutes)	10	10
Select Source Segment	Use entire Source	Use entire Source
Start Time (Seconds)	0.00	0.00
Duration (Seconds)	0.00	0.00
Stream-Basic		
Stream Type	F4V	F4V
Use streaming mode	✓ Use streaming mode	✓ Use streaming mode
Device Compatibility	No particular Device	No particular Device
Video-Basic		
Use Video	Use if Exist	Use if Exist
Width	320	1280
Height	176	720
Aspect Ratio	Video: [20 : 11], Pixel: [1 : 1]	Video: [16 : 9], Pixel: [1 : 1]
Frame Rate(fps)	23.98	23.98
Interlacing	Non-Interlaced	Non-Interlaced
Bitrate		
Video Bitrate mode	CBR	CBR
Number of passes	1 pass	2 pass
Video Bitrate(kbps)	236	2936
Maximum bitrate(kbps)	8000	8000
VBV Buffer Size (bytes)	112000	1125000
H.264 Encoder Settings-Basic		
Profile	High	High
Level	2.0	3.1
Closed Caption Mode	None	None
Enable Scene Change detection	✓ Enable Scene Change detection	✓ Enable Scene Change detection
Size of Coded Video Sequence	72	72
Number of B-Pictures	3	3
Entropy coding mode	CABAC	CABAC
H.264 Encoder Settings-Advanced		
Minimum IDR interval	1	1
IDR frequency	1	1
Use adaptive B-frame placement	✓ Use adaptive B-frame placement	✓ Use adaptive B-frame placement
Reference B-Pictures	✓ Reference B-Pictures	✓ Reference B-Pictures
Allow pyramid B-frame coding	✓ Allow pyramid B-frame coding	✓ Allow pyramid B-frame coding
Reference frames	4	4
Number of Slices per picture	1	2
Search shape	8x8	8x8
Use fast inter and intra decisions	Use fast inter and intra decisions	✓ Use fast inter and intra decisions
Use fast multi-reference frame motion estim...	Use fast multi-reference frame motion estimation	✓ Use fast multi-reference frame motion estimation
Use fast sub-block motion estimation.	Use fast sub-block motion estimation.	Use fast sub-block motion estimation.
Use rate distortion optimization	✓ Use rate distortion optimization	✓ Use rate distortion optimization
Adaptive Quantization Mode	Complexity	Complexity
Adaptive Quantization Strength	-50.00	-50.00
Use Hadamard transformation	✓ Use Hadamard transformation	✓ Use Hadamard transformation
Motion estimation subpixel mode	Quarter pixel	Quarter pixel
Write Timestamps	✓ Write Timestamps	✓ Write Timestamps
Write Access Unit Delimiters	✓ Write Access Unit Delimiters	✓ Write Access Unit Delimiters
Write Sequence End code	Write Sequence End code	Write Sequence End code
Audio-Basic		
Use Audio	Use if Exist	Use if Exist
Audio compression	HE-AAC version 1	HE-AAC version 1
Bits / Sample	16	16
Channels	Stereo	Stereo
Sample Rate(kHz)	44.100	44.100
Audio Bitrate(kbps)	64	64
CRC Check	CRC Check	CRC Check

A B

Figure 25.1 A and B
The 300 Kbps (25.1A) and 3000 Kbps (25.1B) settings in Carbon. Other than bitrate and frame sizes, we're using somewhat slower-but-better settings at the lower bitrates.

(Continued)

Adobe Media Encoder Settings

I'd think more people would use AFMES than AME for this kind of industrial-strength encoding job, but AME certainly can make Dynamic Streaming–compatible content.

However, we don't have nearly the same access to settings in AME, so we'll probably get somehow lower coding efficiency, and less control over decode complexity. It'll also encode faster, but we could have tuned AFMES to do the same.

We can start with the F4V–HD 720p (Flash 9.0r115 or higher) preset. We'd then modify to the following settings for video:

- Frame rate: Same as Source
- Level: 3.1
- Bitrate Encoding: CBR
- Bitrate: 2.936 (it'll only show it as 2.94)
- Set Key Frame Distance: 72

And for audio:

- Codec: AAC+ Version 1
- Bitrate: 64

See Figure 25.2.

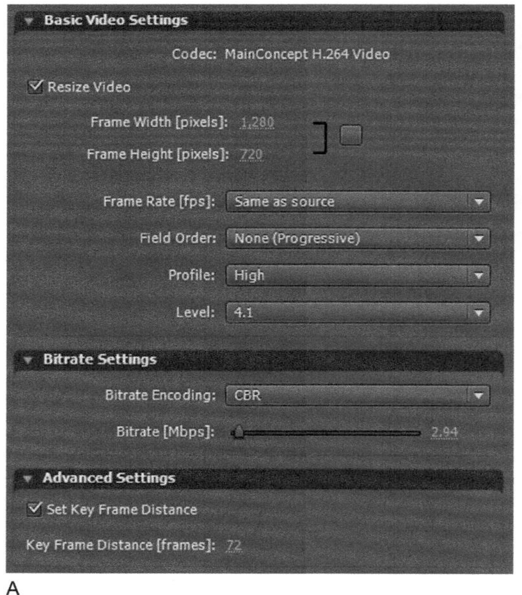

A

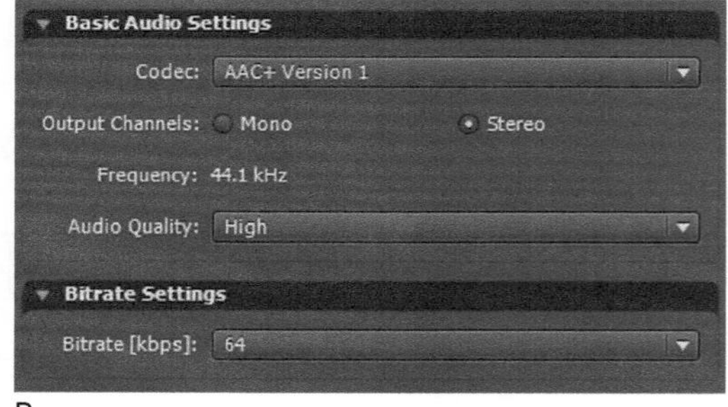

B

Figure 25.2 A and B
AME video (25.2A) and audio (25.2B) settings. They're fully compatible, but with much less room for tweaking. It bugs me that I can't see the third decimal in bitrate here; sometimes the difference between 50 Kbps and 59 Kbps matters!

CHAPTER 26

Silverlight

There are a few different ways people think of Silverlight. (Disclosure: I'm the Video Strategist on the Silverlight team at Microsoft.)

Folks coming from the Windows Media world may think of it as a replacement for the Windows Media Player OCX used for embedding WMP in web pages. For that audience, Silverlight provides several key improvements:

- Good cross-platform and cross-browser support, a notable weakness in the WMP OCX
- Windows Media DRM on the Mac for the first time in years
- RIA functionality for much richer interactivity and design

Those coming from the RIA industry may think of Silverlight as another competitor to Flash and JavaFX. To that market, Silverlight offers several unique offerings:

- Support for the huge .NET and Visual Studio ecosystems
- Windows Media support
- Adaptive Streaming via Smooth Streaming

And existing .NET developers consider Silverlight the portable version of their platform, bringing them:

- Cross-platform/browser support
- A much smaller installer than .NET

Silverlight is available for Windows and Intel Mac computers today. Microsoft is collaborating with Novell on its open-source implementation for Linux called Moonlight. Mobile versions have been announced, and Silverlight on Xbox has been demonstrated.

History of Silverlight

NET

While Silverlight itself is only a few years old, it's fundamentally an implementation of the Microsoft .NET Framework, first released in beta in 2000.

.NET was designed to provide a more flexible, secure, and easy-to-develop-for base for Windows applications. Core to this was the adoption of managed code. Like Java bytecode, .NET managed code is an intermediate language that doesn't get executed directly on the processor, and so is much less capable of accidentally or maliciously harming the underlying system. Unlike the Java or Flash runtime interpreters, .NET's Common Language Infrastructure (CLI) supports a wide variety of languages. The most common language used in .NET is C#, a variant of C, but many other languages can be compiled to CLI-compatible code.

.NET also contains a large set of built-in functions, the "Class Library," that a programmer knows will be available for use in particular versions of .NET.

While .NET started as a Microsoft technology, substantial portions of it have been released as standards, open source, or under an Open Specification Promise. The C# language and CLI are a license-free ECMA spec, like JavaScript.

.NET 3.0, released in 2006, introduced a new presentation layer called the Windows Presentation Foundation (WPF), that defines user interfaces via a XML language called XAML (eXtensible Application Markup Language), making it dramatically easier to design and maintain rich interfaces. This came along with the Expression Studio products for authoring XAML for .NET and Silverlight (and, of course, compression via Expression Encoder). The Expression apps are themselves WPF apps; they were used to develop themselves.

Looking at Silverlight from a .NET perspective, it's essentially the CLI + XAML + Media Foundation (described in the Media on Windows chapter). Portable WPF support was seen as Silverlight's defining feature; the codename for Silverlight was WPF/E, for WPF/Everywhere.

Silverlight 1.0

Silverlight 1.0 was released in 2007. It had XAML and media, but lacked the CLI; all programming had to be done via the browser's own JavaScript engine. Thus it was mainly used for media playback.

Silverlight 1.0 was the last version to support PowerPC Macs (Apple shipped the last PowerPC Mac in 2006).

Silverlight 2

Silverlight 2 was released in the fall of 2008, after Beta 2 had been used for NBC's successful Beijing Olympics streaming coverage. Some call Silverlight 2 the "real" version of Silverlight as it added the CLI, programmable with .NET. It included some notable additions for media:

- The MediaStreamSource API, enabling new protocol and file format support to be written in managed code

- Smooth Streaming was launched later in 2008, based on MediaStreamSource
- WMA 10 Pro audio decoding
- Windows Media and PlayReady DRM support

Silverlight 3

Silverlight 3 was released summer 2009. It didn't include anything as disruptive as the CLI, but added many new media features:

- MPEG-4 file format with H.264 High Profile and AAC-LC codec support
- video and audio decoder decoders via MediaStreamSource
- GPU compositing and scaling for better media performance
- General media playback improvements—Core 2 Duo 2.5 GHz can play 1080p24 6 Mbps VC-1
- Out-of-browser support, allowing Silverlight applications to install and run as local apps on Mac and Windows

The Future

A few Silverlight 4 features have been announced:

- Native Windows Media multicast support
- Offline DRM (no net connection required to install or play DRM'ed content)

There are some other great media features in Silverlight's future, but I can't include more details here. Check out Silverlight.net and the <URL for book updates> for announcements.

Microsoft has long been in the device world, and has versions of .NET for devices in the .NET Compact and Micro frameworks. Silverlight Mobile has been announced for Windows Mobile and Symbian handsets (specifically Nokia's Series 60 and 80). Intel has also announced and demonstrated Silverlight for Moblin, their Linux-based device platform Silverlight for the Xbox has also been demonstrated, as has Silverlight on Intel's Moblin platform.

Why Silverlight?

Uniform Cross-Platform/Browser Experience

For those with lots of Windows Media content or services, Silverlight is the easiest way to get those delivered in the browser for Macs and non-IE Windows browsers.

Moonlight on Linux is rapidly catching up with the released versions of Silverlight for Linux support, and had Smooth Streaming support working in early 2009.

Broad and Extensible Media Format Support

Silverlight 3 supports a wide array of media files directly:

- Windows Media
- MPEG-4 with H.264 and AAC
- MP3

And it's also very extensible, using a feature called MediaStreamSource (MSS). MSS can be used to implement new protocols or file formats in native code, added seamlessly to a Silverlight player. And with the new Raw AV feature in MSS, video and audio decoders can also be implemented in managed code. Some early demonstrations have included Ogg Vorbis playback and a Commodore 64 emulator (which I'd never thought of as a codec, but it works).

Smooth Streaming

IIS Smooth Streaming is an adaptive streaming technology, using IIS as the server, with Silverlight as the first platform supporting it. Smooth Streaming offers seamless bitrate switching and proxy cachability for live and on-demand content. It's the first adaptive streaming technology for a RIA architecture.

.NET Tooling

There were about four million registered .NET developers as of 2009; that's a big pool of talent familiar with development with C#, XAML, and their associated tools. Finding skilled .NET developers and teams is a lot easier than for Flash and JavaFX.

.NET is particularly popular in "Line of Business" applications that corporations build and use internally.

Silverlight Enhanced Movies

Silverlight 3 provides a great platform for interactive video experiences like those provided for DVD and Blu-ray. This is a particular passion of mine, and a project I've been working on for a while. Microsoft is providing a reference player to movie studios that supports menus, extras, and multi-langauge audio tracks and subtitles, enabling the DVD and Blu-ray experience and beyond via disc, download, or Smooth Streaming.

Leading UK retailer Tesco is the launch partner, and will include Silverlight enhanced versions of specific titles starting fall 2009.

Why Not Silverlight?

Ubiquity

Silverlight's had over 400 million installs in its first year and a half, but it'll take a little while to catch up with Flash and Java's ubiquity. For the time being, a greater portion of users will need to install or update Silverlight than Flash.

Silverlight 2 and 3 are also not available for PowerPC Macs. They've dropped below 1 percent of total web use, but can be a vocal minority. Apple stopped selling PowerPC computers in 2006, and the newest of them are now outside of three-year extended warranties; that minority will continue to shrink.

Performance

Like its RIA cousin Flash, the overhead of the rich rendering model increases system requirements for playback compared to simpler players like Windows Media Player.

For systems with compatible GPUs, Silverlight 3 makes this much better by offloading scaling and compositing operations, leaving the CPU largely free for the video decoder. However, there is a significant minority of PCs without sufficiently capable GPUs to take advantage of this. And due to browser limitations on the Mac, Silverlight 3 only uses the GPU on Mac OS X in full-screen mode.

Playback overhead can be substantially reduced by not scaling the video rectangle, called a "Fast Path" player. This is the recommended behavior for non-GPU-enabled systems.

Silverlight's .NET runtime is also much faster than ActionScript in Flash. CPU-heavy applications can be many times faster in Silverlight. This is a big part of what makes Raw AV codecs possible.

Silverlight for Progressive Download

Silverlight supports progressive download for WMV, MP4, and MP3 files. If the files are on a web server that enables byte-range access, Silverlight will use that for random access.

Silverlight for Real-Time Streaming

Silverlight works as a player with Windows Media Services. Silverlight 3 always uses the HTTP fallback mode of WMS, which needs to be turned on server-side.

Silverlight can be more responsive than WMP, in fact. By always using HTTP fallback, it can skip several seconds of protocol negotiation that WMP sometimes uses. Silverlight also has a programmable buffer size, as described in Chapter 16, enabling much lower latency playback than out-of-the-box WMP.

Silverlight doesn't have native support for MBR stream switching, but it does support the URL nomenclature for selecting a particular bitrate. The following code would request the 100 Kbps band from content.wmv:

```
mms://server/content.wmv? WMContentBitrate=100000
```

Thus, stream-switching logic can be implemented in managed code for cases where the same stream needs to work with both WMP and Silverlight; we used this with the 2009 NCAA March Madness. For Silverlight-only projects, it's easier to just use Smooth Streaming.

While Silverlight 3 doesn't have UDP support, Silverlight can receive Windows Media multicast streams when the Project Starlight plug-in is also installed. It is available free as compiled binaries and source code at Microsoft's Codeplex web site. Silverlight 4 will add native multicast support.

Silverlight doesn't have native RTSP support for MPEG-4 files. However, it could be implemented using MediaStreamSource, similar to Starlight.

IIS Smooth Streaming

While Silverlight has good compatibility with existing content libraries, most media content created specifically for Silverlight is now using the IIS Smooth Streaming adaptive streaming technology.

Smooth Streaming isn't so much a feature of Silverlight as a feature that runs on Silverlight. The hard work in the client is all managed code using MediaStreamSource, like Project Starlight. Beyond that, it uses specially structured MPEG-4 files and a module for the IIS web server. Smooth Streaming support is coming to other platforms as well, including Xbox. Full specs for the protocol and file format have been published under Microsoft's Open Specification Promise, allowing third parties to build interoperable implementations without charge or restriction from Microsoft.

The Smooth Streaming File Format

Smooth Streaming uses the MPEG-4 format using the fragmented subtype (fMP4). It's described in more detail in Chapter 11, but the core reason fMP4 was used is that it allows a header per fragment instead of one big header for an entire file (see Figures 26.1a and 26.1b). Since most MPEG-4 players don't support fMP4 yet, Smooth Streaming uses a .ismv extension.

Each set of media files also has two XML files that describe the content, and tell the Silverlight client how to request the right chunk and IIS where to find it in the ISMV files.

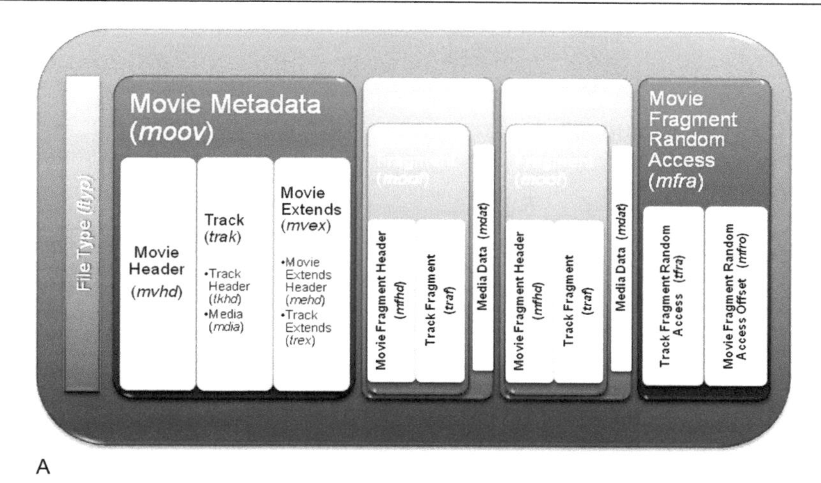

A

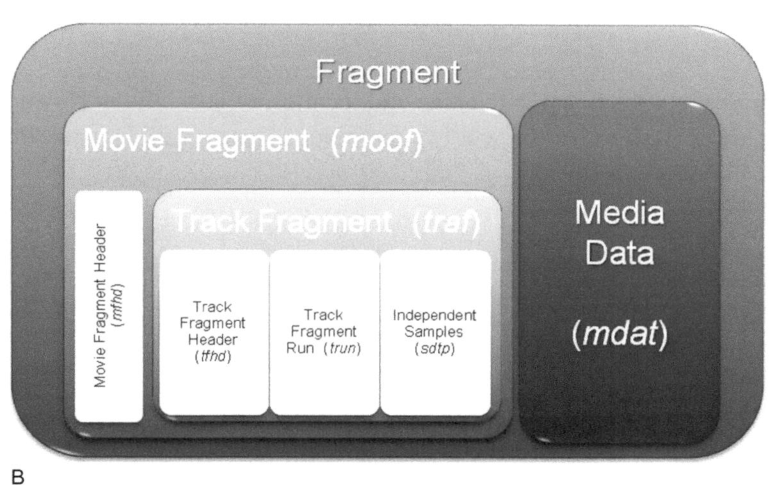

B

Figure 26.1 (A) The structure of a fragmented MPEG-4 file (courtesy of Alex Zambelli). (B) The structure of a fragment of fragmented MPEG-4; there is one fragment per chunk delivered with Smooth Streaming.

The smaller file is the .ism, which is a SMIL 2.0 file just listing the available bitrates, like this:

```
<?xml version=“1.0” encoding=“utf-16”?>
</--Created with Expression Encoder version 3.0.1308.0-->
<smil xmlns=“http://www.w3.org/2001/SMIL20/Language”>
  <head>
    <meta
    name=“clientManifestRelativePath”
      content=“Foo.ismc”/>
  </head>
  <body>
```

```
<switch>
  <video
  src="Foo4000.ismv"
  systemBitrate="4000000">
  <param
  name="trackID"
  value="2"
  valuetype="data"/>
  </video>
  <video
  src="Foo_2800.ismv"
  systemBitrate="2800000">
  <param
  name="trackID"
  value="2"
  valuetype="data"/>
  </video>
```

The larger is the .ismc, which stores the size and location of each chunk for each bitrate, as well as the display properties for the whole experience. The overall data and the first couple of streams look like this:

```
<SmoothStreamingMedia
  MajorVersion="2"
  MinorVersion="0"
  Duration="1448948980">
  <StreamIndex
    Type="video"
    Chunks="109"
    QualityLevels="10"
    MaxWidth="1280"
    MaxHeight="692"
    DisplayWidth="1280"
    DisplayHeight="692"
    Url="QualityLevels({bitrate})/Fragments(video={start time})">
    <QualityLevel
    Index="0"
      Bitrate="4000000"
      FourCC="WVC1"
      MaxWidth="1280"
      MaxHeight="692"

      CodecPrivateData="250000010FE37E27F1598A27F856718048800000010E5A2040"/>
    <QualityLevel
      Index="1"
      Bitrate="2800000"
      FourCC="WVC1"
```

```
MaxWidth="1280"
MaxHeight="692"

CodecPrivateData="250000010FE37E27F1598A27F85671804880000001 0E5A2040"/>
<QualityLevel
 Index="2"
 Bitrate="2000000"
 FourCC="WVC1"
 MaxWidth="1280"
 MaxHeight="692"
```

And a few chunks for a given bitrate look like this:

```
<c
  n="0"
  d="22522482">
  <f
  i="0"
  s="67"
  q="4170"/>
  <f
  i="1"
  s="72"
  q="4486"/>
  <f
  i="2"
  s="67"
q="4170"/>
```

An actual playable directory of Smooth Streaming content as exported from EEv3 looks like Figure 26.2.

Beyond the .ism, ismc, and .ismv files, there are also the following files:

- Default.html is a web page that embeds and launches the player. Lots of player parameters like display size, default caption state, GPU acceleration, and offline playback support are controlled from HTML to make it easy to change player behavior after deployment.
- The Thumb.jpg is a thumbnail to represent the file in a media browser.
- MediaPlayerTemplate.xap is the Silverlight media player itself, including all the assets and managed code. A default template can be used as is, modified in Expression Blend, or a new one created from scratch. Source code is provided with Expression Encoder.
- Preview.png is a preview image of the media player's GUI.

Name	Type	Size
Default.html	HTML Document	13 KB
Fame_Tapered_Trailer.ism	ISM File	6 KB
Fame_Tapered_Trailer.ismc	ISMC File	154 KB
Fame_Tapered_Trailer_230.ismv	Expression Encoder Smooth Streaming File	4,901 KB
Fame_Tapered_Trailer_330.ismv	Expression Encoder Smooth Streaming File	6,789 KB
Fame_Tapered_Trailer_475.ismv	Expression Encoder Smooth Streaming File	9,294 KB
Fame_Tapered_Trailer_682.ismv	Expression Encoder Smooth Streaming File	12,775 KB
Fame_Tapered_Trailer_980.ismv	Expression Encoder Smooth Streaming File	17,897 KB
Fame_Tapered_Trailer_1408.ismv	Expression Encoder Smooth Streaming File	24,841 KB
Fame_Tapered_Trailer_2023.ismv	Expression Encoder Smooth Streaming File	35,646 KB
Fame_Tapered_Trailer_2907.ismv	Expression Encoder Smooth Streaming File	49,631 KB
Fame_Tapered_Trailer_4176.ismv	Expression Encoder Smooth Streaming File	71,796 KB
Fame_Tapered_Trailer_6000.ismv	Expression Encoder Smooth Streaming File	92,623 KB
Fame_Tapered_Trailer_Thumb.jpg	JPEG Image	16 KB
MediaPlayerTemplate.xap	XAP File	102 KB
Preview.png	PNG File	16 KB
Settings.dat	DAT File	43 KB
SmoothStreaming.xap	XAP File	58 KB

Figure 26.2 The standard directory structure of a Smooth Streaming encode plus player from Expression Encoder 3.

- Settings.dat is a log file recording the compression settings used making the media. This can be useful to troubleshoot or recreate past jobs.
- SmoothStreaming.xap is the Smooth Streaming heuristics. They're in a separate file, making it possible to update without changing the media player and vice versa.

Tip: Expression Encoder 3 Installs ISMV Playback

Installing EEv3 (including the free version) also installs a DirectShow filter for ISMV files so that WMP and other DirectShow-based players can be used for testing and local playback. If you're not on Windows 7, you'll need to have a third-party H.264 and AAC decoder installed to play .ismv files using those codecs.

CBR Smooth Streaming: v1

The initial implementation of Smooth Streaming encoder shipped with Expression Encoder 2 SP1, using the VC-1 Encoder SDK. The only encoder mode that assured all streams would start new GOPs at the same time was 1-pass CBR, which had been tuned for broadcast applications with that requirement.

Encoding in this mode uses 2-second GOPs, so with 24p content every 48th frame starts a new GOP. Additional I-frame may also be inserted mid-GOP to helps quality, but the cadence will be preserved.

Since Smooth Streaming tries to maintain a buffer of several chunks, some bitrate variability per chunk is allowed, so a 5-second buffer is used to improve quality.

Since 1-pass CBR is used for live, this model was easy to adapt to live encoding, as used in the live encoders from Inlet, Envivio, Digital Rapids, and announced by ViewCast, Anystream, and Telestream.

1-pass CBR is also used in the initial H.264 Smooth Streaming implementation in Expression Encoder 3.

VBR Smooth Streaming: v2

The next generation of on-demand Smooth Streaming uses 2-pass VBR encoding. And it's not just variable bitrate, but also can vary GOP length and frame size to deliver a more efficiently encoded and consistent experience.

Microsoft's Expression Encoder 3 was the first product with the new Smooth Streaming Encoder SDK (SSE SDK). This initial version is VC-1 only, but the techniques are quite applicable to H.264. The SSE SDK introduces some core innovations for Smooth Streaming encoding that improve encoding time and quality:

- A single analysis first pass suffices for any number of bitrates, speeding encoding time.
- Variable chunk duration, adjusting to align with scene changes, reducing keyframe flashing. Typical GOP length remains 2 seconds, but can go up to 4.
- Variable frame size; the coded resolution of a given frame is reduced to the largest frame that wouldn't show bad artifacts. As hard-to-encode scenes typically have a lot of motion, and hence motion blur, this reduced resolution is generally not detectable. The resolution reduction can be horizontal, vertical, or both depending on the axis of motion. Reducing the pixels processed on the most challenging scenes speeds encoding and decoding.

A

B

Figure 26.3 (A) A frame encoded at 1.2 Mbps 848 × 480 using the CBR codec. (B) The same encoded at 848 × 480 max with the SSE SDK VBR, which encoded this chunk at 640 × 366.

- Variable bitrate encoding, so each bitrate bad has a peak and average. The heuristics client can the best chunk based on individual file size, giving hardest content more bits while keeping the maximum bandwidth utilization predictable.
- An improved core VC-1 implementation the Windows 7 with 8-way threading, faster per-thread performance, and improved quality with noisy backgrounds.

See Figure 26.4 for how this works in practice. VBR audio can also be used with VBR video, taking advantage of WMA 2-pass VBR modes.

Compared to the VC-1 Encoder SDK, the SSE SDK offers far fewer parameters to tweak. It baked in the settings that testing showed provided optimum quality in the lowest encoding time.

Parallelism is handled by encoding multiple simultaneous chunks. So the SSE SDK can easily saturate a 16-core machine, and probably beyond that.

Watch out for memory use, however; a 1080p encode can run the risk of exceeding the default 2 GiB process limit in 32-bit Windows. Running under 64-bit Windows lets 32-bit apps like EEv3 access 4 GiB of memory, which should be ample even for 2 K encoding.

Authoring Smooth Streaming

The initial round of Smooth Streaming encoding tools use the SSE SDK. But now the Smooth Streaming file format and manifests have been published, other vendors are building their own compatible implementations.

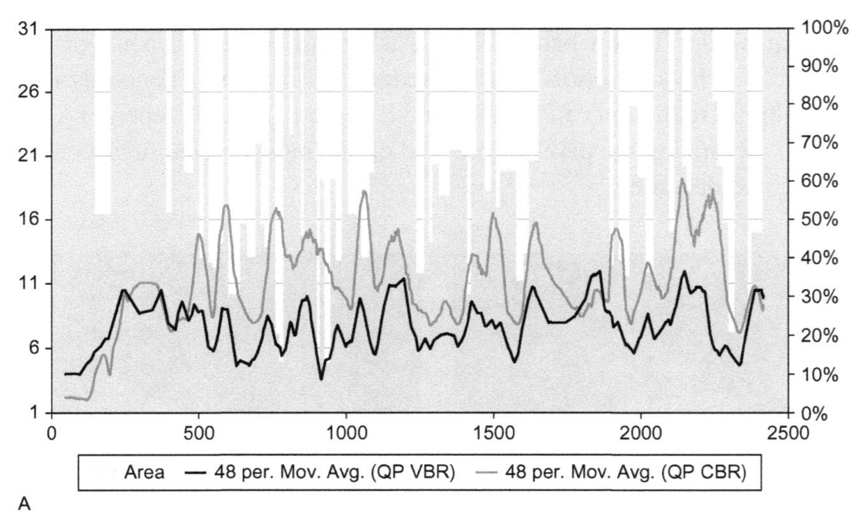

Figure 26.4 (A) Comparing the QP of a CBR and a VBR encode with a particularly challenging piece of content, we see that the VBR encode reduces frame size to keep QP from getting too high. For easier sections, CBR and VBR have similar QP and VBR has full frame size. Harder sections drive down frame size dramatically; places it drops to below 5 percent of the CBR.

Row Labels	Mean Bytes	Mean QP	Max QP
0	4689	2.2	3.5
1	4334	2.1	3.5
2	4920	3.6	10
3	19648	4.4	9
4	7806	9.1	17
5	8853	10.8	18
6	10212	11.1	18
7	15640	8.9	18
8	11807	8.2	12
9	8625	11.1	18
10	13372	11.7	26
11	10978	15.2	31
12	7227	11.4	31
13	7466	8.2	15
14	15442	10.2	31
15	10121	16.3	31
16	8172	12.6	31
17	11227	15.0	31
18	10946	13.6	31
19	8963	10.6	18
20	10098	9.2	18
21	11217	17.7	31
22	8581	10.8	27
23	11545	14.3	27
24	8766	12.5	29
25	7277	8.9	14
26	12174	8.0	13
27	9689	9.3	15
28	12297	8.5	13
29	9845	13.1	24
30	10759	14.8	31
31	8702	11.3	29
32	9934	7.8	13
33	10621	15.1	31
34	6745	12.1	21
35	5093	10.5	18
36	6107	9.0	15
37	14336	9.3	16
38	17506	9.8	14
39	9406	15.2	31
40	10260	7.8	11
41	9388	12.1	16
42	12787	10.4	22
43	10170	13.6	31
44	9723	16.6	31
45	10612	15.8	31
46	8538	17.2	31
47	10391	8.8	16
48	7441	8.2	13
49	10475	10.3	18
50	16172	7.9	12
Grand Total	**10039**	**10.9**	**31**

B

Chunk	Frames	Width VBR	Height VBR	Aspect	Mean Bytes	Mean QP	Max QP
0	48	1280	528	2.42	1124	4.0	5
1	48	1280	528	2.42	1124	4.0	4.5
2	24	1280	528	2.42	1481	5.7	6.5
3	14	1280	528	2.42	1279	4.9	5.5
4	58	864	400	2.16	7067	6.7	7.5
5	58	1280	528	2.42	7971	10.4	12
6	44	1280	528	2.42	11673	8.7	10
7	48	1280	528	2.42	12593	9.0	10
8	30	1280	528	2.42	14699	11.4	13
9	22	864	400	2.16	4631	4.0	4.5
10	36	1280	528	2.42	10562	10.4	12
11	34	1008	416	2.42	5256	7.2	8
12	20	1280	528	2.42	16106	12.4	14
13	14	576	464	1.24	5728	5.3	6.5
14	15	848	528	1.61	11040	7.6	9
15	13	560	464	1.21	15056	6.0	7
16	22	704	352	2.00	8305	4.5	5
17	16	592	496	1.19	15449	10.9	13
18	15	1280	528	2.42	16139	12.4	14
19	38	704	288	2.44	10574	4.3	5
20	23	896	368	2.43	5443	5.7	6.5
21	30	800	336	2.38	5618	4.0	4.5
22	14	1280	368	3.48	7324	7.5	9
23	13	928	352	2.64	11166	6.2	7
24	17	640	448	1.43	15152	7.2	8
25	18	1280	528	2.42	14931	11.3	13
26	14	496	272	1.82	16005	3.9	4.5
27	15	432	192	2.25	9682	2.2	2.5
28	14	1280	528	2.42	15447	10.3	12
29	18	1216	400	3.04	6001	5.7	6.5
30	21	1280	528	2.42	14761	8.7	10
31	13	1120	272	4.12	14879	5.7	6.5
32	23	1280	528	2.42	16515	12.4	14
33	16	1280	528	2.42	15529	8.9	10
34	14	384	208	1.85	16098	1.7	2
35	15	976	416	2.35	13747	6.2	7
36	18	368	224	1.64	10208	2.7	3
37	17	1280	320	4.00	10234	6.7	7.5
38	23	816	320	2.55	5888	6.7	7.5
39	17	1280	528	2.42	15139	9.7	11
40	28	768	448	1.71	5824	3.9	4.5
41	15	880	368	2.39	15957	10.4	12
42	32	928	288	3.22	15654	9.7	11
43	14	1088	384	2.83	11779	6.7	8
44	21	1280	528	2.42	5685	4.4	5
45	18	1280	528	2.42	11641	6.2	7
46	21	1280	528	2.42	16539	11.4	13
47	28	1280	528	2.42	14616	10.4	12
48	18	1280	528	2.42	12710	11.3	13
49	30	1280	528	2.42	7671	11.4	13
50	24	752	320	2.35	6947	4.0	4.5
51	17	1280	528	2.42	5044	8.9	10
52	13	832	352	2.36	7082	4.6	5.5
53	22	1008	432	2.33	8660	6.2	7
54	35	944	400	2.36	9314	7.0	8
55	37	1152	400	2.88	10372	7.1	8
56	21	1040	432	2.41	9230	6.7	7.5
57	13	704	288	2.44	8552	4.0	4.5
58	20	1088	416	2.62	10771	8.9	10
59	26	1280	528	2.42	14639	10.4	12
60	22	960	400	2.40	8275	6.7	7.5
61	13	1280	528	2.42	15982	10.4	12
62	17	1008	352	2.86	9405	6.7	7.5
63	48	944	448	2.11	15093	8.0	9
64	25	736	304	2.42	6734	4.4	5
65	20	688	400	1.72	5736	5.2	6.5
66	14	832	528	1.58	7391	6.9	8
67	16	832	528	1.58	13891	11.3	13
68	40	1280	528	2.42	14897	10.4	12
69	56	1280	528	2.42	8928	8.0	9
70	48	1280	528	2.42	5720	8.0	9
71	30	1280	528	2.42	5835	8.7	10
72	16	1280	528	2.42	14069	9.6	11
73	23	1088	528	2.06	9007	10.4	12
74	20	1280	528	2.42	16100	13.4	15
75	13	992	256	3.88	15557	10.4	12
76	16	1280	528	2.42	14206	11.3	13
77	14	912	288	3.17	6415	4.4	5
78	27	672	352	1.91	15003	9.8	12
79	22	1200	448	2.68	7343	5.7	6.5
80	29	1024	400	2.56	5955	6.7	7.5
81	24	832	288	2.89	8785	4.5	5
82	20	1280	528	2.42	13038	9.6	11
83	28	1280	528	2.42	13655	7.9	9
84	22	896	352	2.55	15008	5.2	6.5
85	25	1280	528	2.42	11315	9.5	11
86	14	1280	336	3.81	8558	7.1	8
87	15	960	400	2.40	12645	6.2	7
88	16	1280	528	2.42	15695	13.3	15
89	22	1280	528	2.42	16340	12.4	14
90	30	1280	528	2.42	13473	8.9	10
91	19	1136	480	2.37	15636	13.4	15
92	16	960	448	2.14	8518	8.9	10
93	15	896	368	2.43	12380	8.0	10
94	25	704	192	3.67	13333	4.5	5
95	42	1280	528	2.42	7861	6.4	7.5
96	14	656	272	2.41	9208	3.9	4.5
97	39	784	400	1.96	3775	5.0	5.5
98	39	1280	528	2.42	12608	10.5	12
99	31	1280	528	2.42	9257	10.5	12
Grand Total	**2408**	**1071.980066**	**449.1495017**	**2.41**	**10172**	**7.7**	**15**

C

Figure 26.4 (Continued) (B) A table showing average size and minimum and max QP per chunk of the same challenging encode. We see the worst-case QP 31 in a number of chunks. (C) The same source encoded with VBR. Due to very fast editing, the average GOP and chunk duration is much shorter. We can see how challenging chunks get scaled down, with 16 frames down to 368 × 224. The worst case is QP 15. Note that the aspect ratio of the encoded rectangle varies depending on direction of motion, T from 1120 × 272 (4.12:1) to 832 × 528 (1.58:1).

While the stock Smooth Streaming defaults are several bands of video and one of audio, there are a lot of other configurations and features that can be used.

Bitrates

Smooth VBR is quite a bit different than a traditional MBR approach. Since there's no cost to switching between bitrates, there's no need or even attempt to stick to a single bitrate. If watching a clip with 3000 and 2000 Kbps bands with a 2500 Kbps connection, the client can switch between 2000 and 3000 chunks in whatever pattern provides the optimal viewing experience.

,It's a good idea to keep a relatively consistent ratio between average and peak bitrates when targeting Silverlight. The EEv3 default is 1.5x, which allows some variability but still reduces decode complexity of the hardest chunks by reducing frame size.

The places where you'd diverge from the standard ratio are at the lowest, and potentially highest bitrates.

Once the user is getting the lowest bitrate, there's no way to go lower, so if you have a 300 Kbps minimum, you'll want peak and average to be equal and with video total 300 Kbps. Otherwise there may be a sustained 450 Kbps period, and the user will see buffering. The SSE SDK is happy to have average and peak be the same value. I then normally use a 1.25x ratio for the second to lowest band to taper up to the full 1.5x.

If the highest bitrate is HD, decode complexity can be the limiting factor. If going up to, say 1080p24 6 Mbps, I'd want have peak = average there as well; a 9 Mbps peak can be asking for trouble. Also, with a flatter bitrate, hard chunks will get more aggressively downscaled, balancing out all the extra vectors in complex scenes that are harder to encode. If doing a flat top rate, I'll use a 1.25x for the second bitrate to taper to the full VBR.

For 720p24 or below, bitrates and frame sizes are enough lower that it's generally fine to stick with the 1.5x ratio; 720p has only 40 percent of the pixels as 1080p.

Frame size

Because of automatic frame resizing, we can be a lot more aggressive about frame rate; if the frame's too big for a given chunk, it'll be reduced as needed. So we can avoid the lowest-common-denominator frame sizes where it's almost too big a frame for the hardest bits, but quite a bit smaller than would work for the easy majority of typical content.

If we weren't worried about decode complexity or encoding time, we could just leave all bitrates at the max frame size and see what comes out. And it's often good practice to have more than one top bitrate share the same frame size. However, we do care about decode complexity, and as a frame's maximum size sets a maximum cap on decode complexity, it's a good idea to specify a lower frame size for bitrates that are going to get scaled down a bunch anyway. My rule of thumb is to reduce frame size at least to the point where 75 percent of chunks are left at full size.

Chunk/GOP duration

The default chunk length of two seconds is going to be fine for most content. By default, the SSE SDK will extend up to twice the default length to align with a natural keyframe, and use shorter chunks with rapid cuts.

Captions

Smooth Streaming supports synchronized captions as part of the manifest. These can be in a variety of languages, including Unicode languages like Chinese. It's up to the Silverlight player to determine when and how they're displayed. Since captions are delivered in advance as part of the index, a Silverlight player can make the entire script's captions available for navigation or search.

Multiple language audio

Smooth Streaming can support multiple audio tracks, normally used for alternative languages. The same functionality could be used for audio commentary, with commentary and main audio in different tracks and mixed together at the player.

Audio-only

Smooth Streaming also works for audio-only applications, live and on-demand; it's just an .ismv file with no video.

Live smooth streaming

Smooth Streaming was launched as an on-demand technology, with a live beta announced soon after. The live beta was used in a number of successful live events through the summer of 2009, including Wimbledon and the Tour de France, and shipped summer 2009.

Functionally, the live encode appends chunks as created to an .ismv and manifest. Thus a live event can be a "PVR in the cloud" with users able to play, pause, and scrub anywhere between the start of the event and the current time.

Silverlight for Interactive Media

Silverlight can deliver much more complex media presentations than WMP ever could. This can include multiple camera views, with thumbnails using the lowest Smooth Streaming bitrate. Silverlight can also use synchronized triggers to control client-side behavior.

> **Tip: Simulated Live Encoder**
>
> You don't need a real live encoder to experiment with hosting and playback of Live Smooth Streaming. The Live IIS module comes with an emulator that loops pre-encoded ISMV files pushing them server as if it were a live encoder.

Given the depth that XAML and .NET are capable of, a Silverlight media player doesn't have to be just a media player, and can be a real application that happens to involve media. There are video editors and multicamera video monitoring products using Silverlight already.

Silverlight for Devices

Silverlight has been announced for Windows Mobile, Symbian, Moblin, and Xbox 360, but few details have been released yet.

The Xbox 360 demo was specifically of the use of Silverlight for advertising as part of Xbox Live.

Formats and Codecs for Silverlight

In general, format choice in Silverlight is driven by what compatibility is required, if any. Windows Media offers easy interoperability with WMP clients, and MPEG-4 with QuickTime, Flash, and other MPEG-4 players. Smooth Streaming is Silverlight-only, for the moment and can offer a compelling experience for Silverlight-only content.

Windows Media

Silverlight's Windows Media support includes the mainstream video and audio codecs:

- Windows Media Video 9 Advanced Profile, no interlaced (VC-1 AP)
- Windows Media Video 9 (VC-1 SP, MP)
- Windows Media Video 8
- Windows Media Video 7

The most widely used codec not listed is WMV 9 Screen. However, VC-1 does better than WMV 9 Screen with modern Aero Glass and Mac OS X user interfaces.

WMV9 would be used for any new content, as it's much more efficient than the very old WMV 7/8 codecs. For most Silverlight tasks, Main and Advanced profiles offer equivalent quality, and Main is more compatible with older versions of WMP.

On the audio side, Silverlight supports the following codecs:

- Windows Media Audio 7-9.2
- Windows Media Audio 9-10 Professional (including the 10 Pro LBR modes)

Silverlight 3's audio pipeline is stereo-only. It can decode 5.1 and 7.1 WMA Pro audio, but converts it to stereo for playback.

For content targeting Silverlight 2+ and/or WMP 11+, WMA 10 Pro's 32–96 Kbps range is very high quality; 64 Kbps is the default for Smooth Streaming, and sound great for most content. Those modes provide reduced fidelity when played on older versions of WMP. If broader compatibility is needed, WMA Pro at 128 Kbps and higher is excellent and highly compatible. Below 128 Kbps, WMA 9.2 is the compatible choice, but audio fidelity is worse than WMA 10 Pro at lower bitrates. If WMA 9 must be used, remember that 2-pass VBR can offer a big quality boost over CBR at the same bitrate.

Silverlight 2+ can decode WMV files using Windows Media DRM. However, the Silverlight client needs to contact a PlayReady license server for authentication. Don't sweat it if you're not using DRM. If you are, a DRM service provider should already have implemented the required infrastructure to issue proper licenses to both Windows Media Player and Silverlight. Silverlight 4 will include a secure license store enabling Offline DRM.

MPEG-4 and H.264

MPEG-4 support was added in Silverlight 3. The initial implementation supports H.264 (High, Main, and Baseline profiles, progressive scan only) and AAC-LC (mono or stereo only).

H.264 High Profile can outperform VC-1 in efficiency, particularly as bits per pixel goes down. Decoder requirements are higher with H.264, particularly when expensive features like CABAC are used.

Silverlight 3 doesn't support DRM or RTSP streaming of MPEG-4 content; both require WMV today. The built-in .mp4 playback is progressive download (with byte-range request support) only. Extensions to other protocols are possible using MediaStreamSource.

Smooth Streaming

Smooth Streaming currently requires codecs go in pairs; H.264 with AAC, or VC-1 with a WMA codec.

The differences in encoder implementations are a lot bigger today than in the bitstreams themselves, but that area is evolving quickly.

VC-1 and WMA

Silverlight 2 introduced Smooth Streaming with VC-1 Advanced Profile and WMA Standard or Pro. VC-1 Simple and Main profiles are not supported, as they aren't MPEG-4 file-format-compatible.

At the moment, the SSE SDK for VC-1 offers dramatic quality improvements over both the old CBR VC-1 mode and EEv3's H.264 mode, which is CBR Main Profile only.

VC-1's lower decoder complexity means that CPU-bound computers will be able to decoder higher resolutions than H.264. This is further helped by the SSE SDK's variable frame sizing, since the most complex, high-motion scenes get encoded with fewer pixels making decoder complexity "flatter" than with traditional encodes.

H.264 and AAC

H.264 High Profile is a more efficient codec than VC-1, and the EEv3 Main Profile CBR Smooth encodes look better than its VC-1 CBR encodes at the same bitrate.

But the combination of adaptive GOP placement, VBR, dynamic frame size, and lower decode complexity makes the SSE VC-1 implementation a better choice in most cases when targeting Silverlight on a PC. This will change as H.264 implementations become available that integrate SSE SDK techniques.

While H.264 decode complexity will be somewhat higher than VC-1, this can be mitigated by adjusting encoding settings per band. For example, using CABAC at low bitrates (where it pays off most in quality) and CAVLC at higher bitrates (where CABAC has the biggest impact on decode complexity) can help 1080p playback.

H.264 Smooth will be required for Silverlight on devices with only H.264 decode, of course. H.264 Baseline works perfectly in Smooth, unlike VC-1 Simple/Main.

Silverlight 3's AAC decoder only handles AAC-LC, so audio bitrates need to be a little higher to match WMA 10 Pro quality. I recommend using 96 Kbps minimum, and ideally 128 Kbps.

MP3

Silverlight can also do progressive download of MP3 files. WMA 10 Pro can match that quality at less than half the bitrate, but there's no reason not to use existing MP3 files if they're of adequate quality and bitrate.

Raw AV

Silverlight 3's Raw AV pipeline makes it possible to add new codecs, formats, and protocols to Silverlight using managed code. They are compiled into a .dll like any other Silverlight code, and built into a player. There's no install or user action required; the user experience is identical to playback with the built-in formats.

The big drawbacks to Raw AV decoders are that they're roughly half the speed of a native decoder, and someone has to make them. There's been initial work by third parties on MPEG-2 and Ogg Theora/Vorbis players using Raw AV, and some companies are offering Raw AV decoders for license.

Silverlight 3 doesn't support UDP, so any protocol would need to support HTTP or TCP. One example is Edgeware, which has made a MPEG-2 transport stream Silverlight client that, with their appliance converts an IPTV stream to a Silverlight-compatible HTTP stream.

Encoding Tools for Silverlight

In general, any compression tool that makes MPEG-4 and/or WMV content is already Silverlight-compatible. What's new are tools that take advantage of specific Silverlight features, and which make Smooth Streaming-compatible content (sets of .ismv, ism, and ismc).

Expression Encoder

A core goal of Expression Encoder is making great Silverlight content. It's also very focused on usability, particularly for the designer audience that Expression Studio targets.

Expression Encoder exists more to foster the Silverlight and broader Microsoft ecosystems than to be a big cash cow itself. Hence EEv3 with just WMV encoding is free, and adding Smooth Streaming and H.264 encoding, plus MPEG-2 and MPEG-4 decode, is only $49.

The underlying technology used in Expression Encoder is shared with other compression tool vendors, so similar quality and features should soon be available in other products as well.

Inlet

Inlet is incorporating Smooth Streaming authoring across their line of products. They had Live Smooth in Spinnaker and SSE SDK in Armada before Expression Encoder 3 shipped.

Envivio

Envivio's 4Caster was the first Live Smooth Streaming encoder incorporating PlayReady DRM.

Carbon

Rhozet's Carbon will include SSE SDK by the time this book hits the market.

Digital Rapids

Digital Rapids has added live and on-demand Smooth Streaming to StreamZ 3.1.

ViewCast

ViewCast, creators of the Osprey card and Niagra streaming appliances, will have Live Smooth Streaming soon.

Grab Networks

Agility, from the former Anystream, will have integrated Smooth Streaming by now.

Silverlight Tutorial: Smooth Streaming

Smooth Streaming is the obvious tutorial here. Since the scenarios are simple, I'll be using the same multiple bitrate scenario as we used last chapter for Flash, tuned for Smooth Streaming. I'll includes captioning and chapter marks to keep things from being too repetitive.

Scenario

As previous, we're a web media company looking to publish 1080p24 sources to HD multiple bitrates. Due to new accessibility requirements, we're adding captions to our video, and chapter-based navigation.

Three Questions

What Is My Content?

Still 1080p24 film/video content. This time we have caption files, too.

Who Is My Audience?

We're still targeting a broad swath of consumers on a variety of hardware and software. We're adding captions to increase accessibility, further growing our potential audience.

What Are My Communication Goals?

Again, we want everyone to have a good, high-quality playback experience. With Smooth Streaming, we also increase the reliability and scalability of delivery, and reduce our costs, with proxy caching.

Tech Specs

We'll stick with the same bitrate ranges we used for the Flash encoding. However, we'll encode with Smooth Streaming VBR, so those will only be the averages. As the 3 Mbps cap was to constrain bandwidth costs, we can use higher peaks as appropriate. We'll

(*Continued*)

use the standard 1.5x for the higher rates, tapering down to 1.0x to make sure that 300 Kbps is sustainable with 300 Kbps of bandwidth.

And we can raise our target frame size a bit since we'll get automatic downconversion where needed, and generally lighter-weight decoding with VC-1 (see Table 26.1). That lets us use 720p for the top two bitrates; it'll reduce frame size more often at 720p, but we'll have more detail most of the time.

Audio is can be the straightforward default of WMA 10 Pro 64 Kbps average 96 Kbps peak at 44.1 stereo. 64 Kbps CBR WMA 10 Pro is normally quite transparent, but VBR will let the most challenging parts get a boost for optimum fidelity.

Expression Encoder 3 Settings

That's all pretty straightforward to implement in EEv3. We can start with the "VC-1 IIS Smooth Streaming-HD 720p VBR" preset, delete a couple surplus bitrates, and make the changes described previously (see Figure 26.5).

Which was boring enough that it's good we have metadata to keep us on our toes.

EEv3 can import DXFP, SAMI, and XML formatted captions. Smooth Streaming captions follow the Windows Media/SAMI style captions where there's a script event for text to display, and a second blank one that erases it.

Most captions will be available in one of these formats, or in a format that's easily converted to one of these. But if a customer just dumps a text file on you, it's actually pretty easy to convert that into XML with Excel as long as you can get the right fields into the right columns. Here's a simple example:

```
<ScriptCommands>
<ScriptCommand Type="caption" Time="00:00:15.000" Command="at the. at the
left we can see"/>
<ScriptCommand Type="caption" Time="00:00:17.500" Command=" "/>
<ScriptCommand Type="caption" Time="00:00:18.000" Command="at the right
we can see the"/>
 </ScriptCommands>
```

It's also possible to manually type in the captions directly into EE, but it's not recommended unless you've got a very short project. Dedicated captioning tools are a

Table 26.1 Bitrates and Frame Sizes for Our Silverlight Tutorial.

Total bitrate	Audio average	Audio peak	Video average	Video peak	Max width	Max height
300	64	96	236	236	432	240
500	64	96	436	545	512	288
800	64	96	736	1104	848	480
1200	64	96	1136	1704	960	544
2000	64	96	1936	2904	1280	720
3000	64	96	2936	4404	1280	720

(*Continued*)

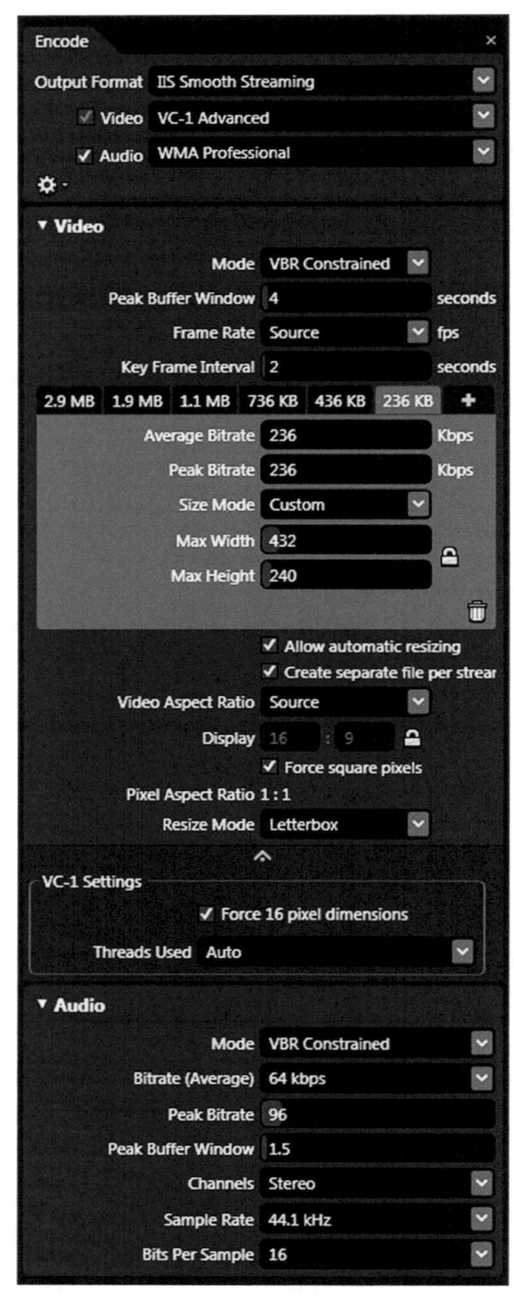

Figure 26.5 Our settings in Expression Encoder. It's nice to have all the settings that can vary as streams in tabs, with the global settings global, so you don't have to keep double checking to make sure audio settings and frame rates are set correctly in six different places.

(Continued)

much better way to go. Public broadcaster WGBH's MAGpie has long been a popular free choice.

Chapter marks in EEv3 can be imported and exported with a similar XML format, which is handy for automating placement. They're also easy to place manually. For a TV episode like this, we'd definitely want a chapter marker after each commercial break. See Figure 26.6.

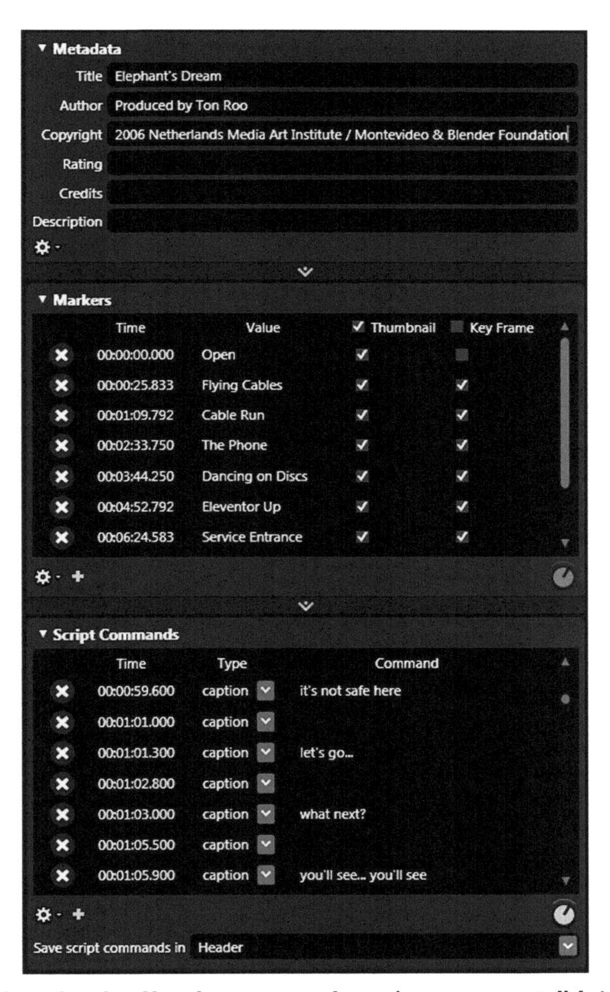

Figure 26.6 Metadata for the file, chapters, and captions set up. I didn't bother to set the first frame as a keyframe, since the first frame automatically is one.

(Continued)

Here's a simple example of chapter markers:

```
<Markers>
<Marker
Time="00:00:25.8333334"
Value="Flying Cables"
GenerateKeyFrame="True"
GenerateThumbnail="True" />
<Marker
Time="00:01:09.7916667"
Value="Cable Run"
GenerateKeyFrame="True"
GenerateThumbnail="True" />
</Markers>
And that's that!
```

CHAPTER 27

Media on Windows

Introduction

This chapter is about media playback on Windows, inclusive of but not limited to Windows Media.

I'm sure renaming it "Windows Media" from NetShow made sense back in the day, but it does make for some challenging textual gymnastics. For example, I can't say "Windows Media APIs"—I have to say "Windows APIs for Media." I offer my sincere apologies for any unpoetic cadences deriving thereof.

A History of Media Features in Windows

DOS

It all started with Microsoft DOS for the IBM PC in 1981. Command-line, character-based display, and no audio that deserves the name (more of a muffled beep at best).

But man, did it run business apps, and a lot of them, and rapidly found a place on desks around the world.

Windows 1–2

Windows 1.0 launched back in 1985, without much market impact. The early versions of Windows ran on top of DOS, and were quite basic and not widely used due to high system requirements and not much software that took good advantage of the GUI. Things perked up a bit with Windows 2 as the first versions of the familiar Microsoft Office apps began to appear, but most applications were still DOS-only.

This didn't slow the runaway success of DOS during this period other companies figured out how to make compatible hardware, driving rapid competition on features and price in the PC market, and rapid growth.

While Macs dominated publishing and other media-related tasks, DOS dominated business applications.

Windows 3.0/3.1

The year 1990 was when Windows 3.0 finally went mainstream, as a critical mass of native Windows apps started to appear, and as faster machines made Windows a great way to work with multiple DOS apps at once.

The first real media functionality arrived in 1991 in the Windows 3.0 MultiMedia Extensions version, preinstalled on "multimedia" PCs—they were required to have sound cards, CD-ROM drives, and at least 640 KB of RAM! This included the first version of Video for Windows, the AVI file format, and Media Player. The extensions were incorporated into Windows 3.1 in 1992, which was the real breakout version.

Windows 95/98/Me

The Windows 95, 98, and Me (Millennium Edition) versions straddled the DOS-based Windows 3.1 and the 32-bit NT-based worlds (more on NT in a moment), offering backward compatibility with old DOS applications and lower system requirements.

Windows 95 (from 1995, natch) is probably the oldest Windows version that would still seem reasonably like a computer to people born since its release. Some core features it introduced:

- Filenames longer than eight characters
- The Start menu
- 32-bit applications, raising the memory a program could easily use from 640 KB to 2 GB

OEM Service Release 1 at the end of 1996 added MPEG-1 playback, making it the highest quality format supported on both Mac and Windows out of the box, and thus a popular CD-ROM codec. Win95 also came with Internet Explorer 2, the first web browser included with an operating system. Windows 95 was the first version really targeted as a multimedia playback platform.

Windows 98 was fundamentally a refined version of the big changes in 95. It introduced DirectShow, which remains the primary API for media authoring and playback for most Windows apps. DirectShow's improvements made media playback a lot smoother.

Windows 98 is probably the oldest version still seen on running machines. And it's the oldest capable of anything like a modern media experience, being the oldest version of Windows that can run WMP 9. Windows machines of this age should all have decent sound, full-color video, and at least a CD-ROM drive. DVD-ROM support was added in Win98 Second Edition a year later, including a DVD player app (which required a separate MPEG-2 decoder). Windows Me (Millennium Edition) was the last non-NT-based version of Windows. While never as popular as 98 or XP, it introduced important media technologies, including WMP 7 and Movie Maker. It suffered from stability and other issues, and most users updated to Windows XP.

And thus ended the era of DOS.

NetShow

NetShow was the start of Microsoft's effort to go beyond playback of local AVI files to streaming. Version 1.0 launched in 1996 as an audio-only technology, with video coming in 2.0 the next year. Things got really interesting after the acquisition of VXtreme, with core personnel and technology driving NetShow 3.0, which included the ASF file format and early parts of what become the MPEG-4 part 2 video codec.

While technically interesting, RealNetworks still had the dominant market share for web streaming.

Two big differences in Microsoft's approach were making media server technology much less expensive than RealNetworks', and a big focus on getting software developers to incorporate NetShow into a wide array of products.

Windows NT

Windows NT was started back in 1988 to build a real 32-bit OS from the ground up without the DOS underpinnings of 3.0 through Me, and all recent versions of Windows from XP on are evolutions of NT.

NT introduced core modern features like:

- Protected memory (one bad program couldn't easily crash the whole computer)
- Multiprocessor support
- The NTFS file system and file names beyond eight characters

This made Windows NT a much more stable and powerful for workstations compared to Windows 95/98 and the Mac OS of the era. Windows NT 3.1 launched back in 1991, starting a decade of NT- and DOS-based Windows coexistence. Windows 95 and NT shared the Win32 API, though, so most applications could run fine on either.

It was followed by NT 3.5 in 1994, notable for big speed improvements. Version 3.5.1 in 1995 was a bigger update than its name implies; particularly for media. This was the first version of NT to see much use for content creation, particularly 3D animation and some video editing products.

NT4 in 1996 was the breakout version, and was widely used in corporations as personal desktops. NT4 adopted the Windows 95 interface and bundled apps, and therefore offered the power and stability of NT with the usability and familiarity of the DOS-based Windows. It had some very important new media features.

- Video drivers moved into the kernel for much better performance
- The first version of DirectX (although without 3D)

A number of formerly Mac-only companies like Avid and Media 100 released high-end video editors for NT, leveraging the greater stability of the platform.

The Windows NT brand ended with NT 4, which only went as far as WMP 6.4.

Windows Media Launches

NetShow was renamed Windows Media in 1999, before the Windows 2000 and Me launches. A key feature was unification of the local-file Media Player with the NetShow player into the single "Windows Media Player"—an architecture and name that remains today.

Windows 2000

Windows 2000 (in 1999) was NT5 with a new brand, and is about the oldest NT still seeing significant use (mainly behind corporate firewalls on managed desktops, although my kids still have a few in their elementary school). The summer 2009 Forrester data shows Windows 2000 as only 1.2 percent of enterprise desktops.

Windows 2000 launched with WMP 6.4, but can install WMP 9 and Silverlight; it's the oldest OS with Silverlight support.

Windows XP

XP was based on the business-focused 2000, but incorporated many consumer features from 98/Me to finally unify Microsoft's product lineup and say goodbye to DOS. XP is the oldest version of Windows in broad use today; most projects can target it as a baseline.

Windows XP introduced critical innovations for HD media playback, including DXVA for hardware-accelerated video decode, GPU compositing for video playback, and 5.1 audio mixing. Windows had gone from being years behind the Mac in media technology to establishing a substantial lead: QuickTime wouldn't get 5.1 audio until 2005 or hardware decode until 2009.

Windows XP originally shipped with "Windows Media Player XP" (really WMP 8). However, the essentially universal Service Pack 2 includes WMP9, so that can be considered the baseline today. WMP 10 was only supported on XP, and XP is the oldest version that can support WMP 11. Everything older is limited to WMP 9 or earlier.

Windows XP got a big upgrade to Movie Maker with good DV camcorder support, which was widely used for consumer and corporate video production.

Windows Media Center was originally implemented as a different edition of XP in 2002, offering a "lean-back" remote control-based interface for watching TV. By the 2005 edition, this expanded to include live analog and digital TV tuners and playing content out to CE device extenders so that the PC doesn't need to be near the TV.

Windows XP was the first iteration of Windows to have a 64-bit version, although little media software outside of 3D animation took advantage of it.

Windows Media 9 Series

The original streaming wars largely ended in 2003 with the release of Windows Media 9 Series. It had a broad swath of deep features, low operating costs, great audio/video quality, and was preinstalled on the dominant OS. RealNetworks soon transformed from a media technology platform company to a content licensing and distribution company.

Beyond Windows itself, device support was ramping up quickly, not only on Windows CE and Mobile, but in a variety of media players coming to compete with the iPod.

But while the experience was great on Windows and embedded in Internet Explorer, there wasn't a clear plan for alternate browser or Mac support.

Ben Waggoner Joins Microsoft

Well, *I* like to think it was an important day in the company's media history! More importantly, this is where this history turns from the observation of an outsider to an insider.

I was hired as a program manager on the HD DVD team focusing on the awesome new Windows Media Encoder Professional Edition.

Alas, HD DVD died and Professional Edition never shipped (although much of its technology made it into CineVision PSE, the VC-1 encoder SDK, and Expression Encoder).

Windows Vista

Windows Vista had a challenging development process and launch, but it contained a lot for content creators and consumers. Vista shipped with WMP 11, so all the advanced WMV and WMA codecs are fully supported. The Home Premium and Ultimate versions included DVD playback for both WMP and Media Center.

Vista introduced the Media Foundation API for media playback, although it was mainly used for protected media playback, with DirectShow remaining the default for other content. The Windows Imaging Component added much deeper still image support, and the HD Photo codec (being standardized as JPEG XR).

The Display Window Manager (DWM) in Vista adds system-wide GPU compositing. This saves CPU power, and also makes the user interface much more responsive under heavy load (like compressing video in the background). This is part of broader DirectX enhancements, which include pixel shader support; GPU processing using Direct3D requires Vista or higher.

Vista also added DXVA 2.0, enabling hardware accelerated decode of H.264 and MPEG-2, making software Blu-ray players possible. But it didn't include out-of-the-box decoders for either other than a MPEG-2 decoder only available for DVD playback in WMP.

On the encoding side, Vista introduced tuning for NUMA architectures; Opteron and Nehalem dual-socket systems run quite a bit faster in Vista than XP.

Vista has been inaccurately maligned for poor performance. Particularly for media authoring and playback, on good hardware it's a definite upgrade from XP.

Windows Server 2008 is built on top of Vista, so it shares Vista's media features. WMP and the other media libraries are included in the "Desktop Experience" feature, which must be installed for media playback or authoring. 2008 R2 is based on Windows 7.

Windows 7

Windows 7 was just finalized as this chapter was written, so check the book and Microsoft web sites for any new details that have popped up since release.

Windows 7 has lots of media enhancements, with a big expansion in media format support. It also has important improvements in hardware support to enable low-power machines with great media playback.

Win 7 includes Windows Media Player 12, with much more pervasive Media Foundation support. WMP 12 is only available as part of Win 7; WMP 11 remains for XP and Vista.

The biggest deal in Windows 7 may be a broad expansion in media format support beyond Windows Media and AVI for both playback and transcoding. Out of the box, it'll handle most of the files in media libraries that formerly required third-party components (it's a good sign when MKV—the Matroska Video format—is the most common request left). The new formats and codecs include the following:

- MPEG-4
- MPEG-4, QuickTime, 3GPP, including AVCHD
- MPEG-4 part 2 (SP and ASP)
- H.264 (up to High, all levels)
- AAC, including LC and HE v1 and v2, and multichannel
- "DivX/xvid" AVI
- MPEG-4 part 2 ASP
- MP3 and MS ADPCM
- MPEG-2
- Program and Transport Streams, including HDV
- MPEG-2 up to 1080i/p
- H.264 up to High
- Dolby Digital, PCM, and MP2

These new formats are available in the mainstream Home Premium, Professional, Enterprise, and Ultimate editions. However, WMV remains the baseline in the developing-country targeted Starter and Home Basic editions.

Windows APIs for Media

Given how long Windows has been around, and the strong emphasis that Microsoft places on backward compatibility, it's not surprising that there are a lot of different generations of the technology.

Video for Windows

Video for Windows (VfW) was the first popular API for playback and encoding in Windows, introduced way back in the 16-bit era for Windows 3.1. It was basic, but did the job. Its last significant update was for Windows NT.

Microsoft deprecated it back in 1996 in favor of DirectShow, but it's still supported and used by a few apps, most notably VirtualDub.

DirectShow and Media Foundation codecs need to explicitly have VfW support turned on, which is why some codecs like Microsoft's DV implementation don't show up in VirtualDub or other VfW apps (if any others are left).

DirectShow

DirectShow had been the mainstream playback and non-WMV video encoding API in Windows since the mid-1990s. It was originally named ActiveMovie, but was added to the DirectX family of technologies and thus renamed DirectShow.

DirectShow was a lot more powerful than VfW, with support for new technologies like multichannel audio, and is also much more extensible. Plug-ins are implemented as DirectShow "filters" or as DMOs—DirectShow Media Objects. These can be codecs, demuxers, video and audio effects, etc. These get combined into a "Filter Graph," which is the flow of the content from demuxing to decoding and display or writing back to a file. The free GraphEdit utility (Figure 27.1) included with the Windows SDK visualizes this, and is handy for troubleshooting or designing a custom workflow.

The flexibility of DirectShow enabled to handle pretty much all pre-Vista media playback, including DVD playback and Media Center.

On the encoding side, DirectShow is mainly used for source decode and AVI writing. Creation of other files is normally handled by other libraries. DirectShow is also heavily used by video apps on Windows like Premiere Pro.

There's a very big ecosystem of DirectShow filters to extend playback features in Windows. There are commercial ones for professional formats like MXF, and free decoders like ffdshow.

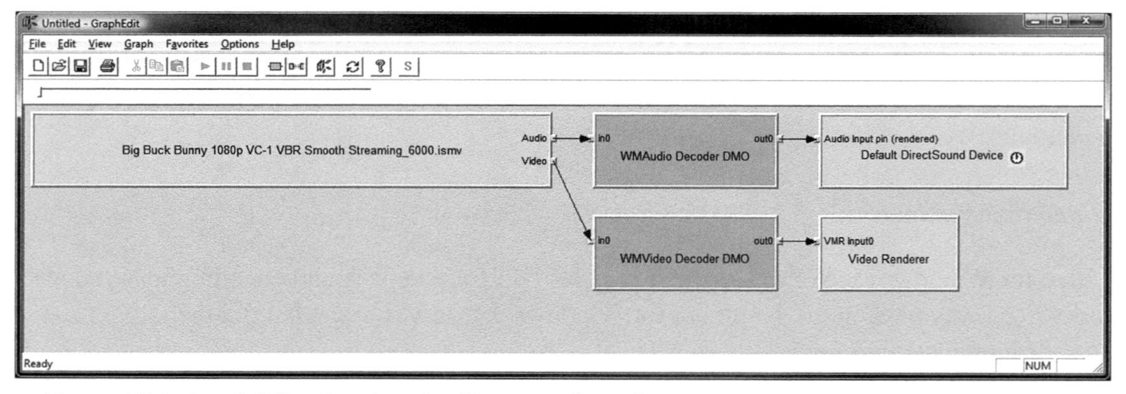

Figure 27.1 GraphEdit, showing the filters used to play a Smooth Streaming .ismv using the demuxer installed by Expression Encoder 3.

DXVA

DXVA is the DirectX Video Acceleration API, introduced with WMP 10 to enable hardware acceleration of video decoding. The initial DXVA implementation was MPEG-2 and WMV only.

DXVA 2.0, in Vista forward, made hardware-accelerated H.264 support possible. However, that was only exposed to specific apps that could pass it the demuxed H.264 bitstream, without WMP or other players getting access to full .mp4 playback.

Whether DXVA is used depends on the driver and the video card. Very old or cheap GPUs may have no or limited DXVA. And the amount of speedup can vary by model of card as well, with newer cards able to offload more decode from the CPU.

Tip: Avoiding .dll Hell

DLL hell is when multiple dynamic link libraries (.dll files) on Windows all claim to do the same thing, so Windows isn't sure which to use.

This can happen with codecs, and it's not uncommon for someone who downloads a lot of content to have multiple H.264 and MPEG-2 decoders installed from difference places. But there's no mechanism in DirectShow to make sure the optimum .dll for the bitstream is picked, so sometimes a buggy or incompatible decoder is used, which leads to some difficult-to-diagnose issues.

The first step to avoiding this is only installing decoders that you know and need. The worst problems are general caused by "codec packs," which are packages of many decoders, often pirated, to enable playback of pirated content. There are some well-tested codec packs like the Combined Community Codec Pack (CCCP), but most others are buggy to the point of malware, and in some cases *are* malware masquerading as a codec pack.

Windows 7 addresses this problem by always defaulting to any Microsoft or hardware-vendor decoder where available. So, if an app wants to use a different software H.264 decoder would need to explicitly add it in its Filter Graph. WMP and MCE on Windows 7 always pick the internal or hardware decoders, barring registry key fiddling.

ffdshow

One popular filter for DirectShow is ffdshow, an implementation of the very flexible ffmpeg open-source player library as a DirectShow filter.

ffdshow enables WMP and DirectShow compatible compression tools to open a wide variety of other formats like Ogg, Snow, Dirac, H.264, other MPEG-4 variants, and others. Versions are installed with Sorenson Squeeze and Rhozet Carbon.

ffdshow is highly configurable through a highly frightening dialog—see Figure 27.2.

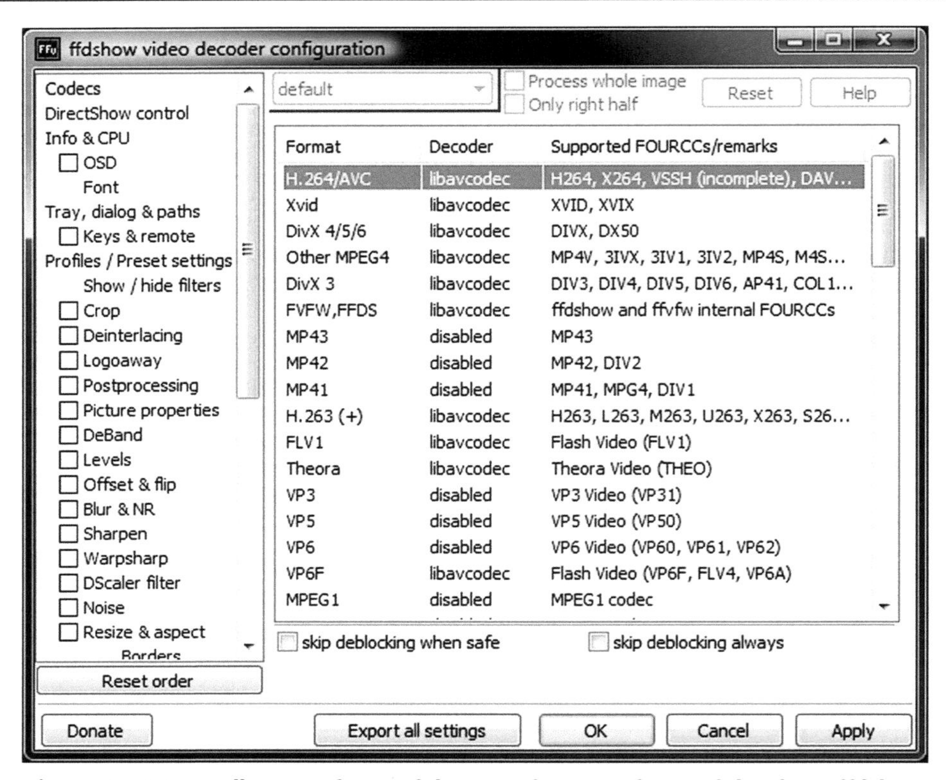

Figure 27.2 A small taste of one of dozens of panes of one of the three ffdshow configuration dialogs.

AVISynth

AVISynth, described in more detail in Chapter 6, is itself a DirectShow filter. So other DirectShow apps see an .avs file as a media file and can scrub, transcode, and so on, even though the file itself is just a text file of instructions for how to synthesize the video.

AVISynth is incredibly powerful, and broadly used in the preproduction and compression industries. That it's Windows only is another reason why complex and high-quality compression workflows are mainly on Windows.

Haali Media Splitter

Haali Media Splitter is a free demuxer that can extract the elementary streams from a wide variety of source formats not supported on older versions of Windows like MPEG-2 transport and program streams, MPEG-4, and Ogg. It also includes an excellent AVI demuxer that sometimes works on very large AVI files that the built-in XP/Vista demuxer has trouble with.

It's also the best-known demuxer for the Matroska format (MKV), often used to deliver rich subtitles. MKV isn't natively supported in any version of Windows. There are other demuxers for MKV, however, including ffdshow.

CoreAVC

CoreAVC is a popular DirectShow H.264 decoder from CoreCodec often used to add H.264 support to older versions of Windows (and is widely available on other platforms). It supports Baseline-High 4:2:0, but not 4:2:2 or 4:4:4.

It was a famously fast software-only decoder in its 1.0 version, and has tons of decoding configuration options. It was one of the first decoders to take advantage of GPU pixel shaders for video playback, using Nvidia's CUDA technology. Since all CUDA-compatible GPUs also include DXVA 2.0 support, on Windows it is probably most useful for XP users.

Media Foundation

Media Foundation (MF) is the new Windows API for media playback and authoring. Just like DirectShow didn't kill VfW, MF doesn't replace DirectShow as much as supplement it. They can share codecs and other technologies, with MF able to use DirectShow DMOs. The native MF equivalent is the Media Foundation Transform (MFT).

MF was introduced in Vista, but was mainly used for protected media playback; DirectShow continued to be used for most WMP playback. MF also contains the functionality of the Windows Media Format SDK for playback.

Windows 7 greatly expands the use of MF, with native implementations of many codecs and formats.

One important addition in Media Foundation is much lower latency and controllable latency for streaming. Traditionally, WMP and other players using the Format SDK had a default five-second buffer with no easy way to configure a lower value. In MF, this is programmatic, all the way down to 0. Silverlight is based on Media Foundation, and this is how it's able to deliver lower latency than WMP.

Another great feature of Media Foundation in Win 7 is an automatic bob deinterlacer for all interlaced codecs. So you get 60p out of 30i files.

Protected Media Path

The Protected Media Path (PMP) is an MF feature added in Vista that protects specifically flagged DRM content from screen captures or unencrypted (non-HDCP) output. This is a contractual requirement from Hollywood for top-tier HD content, with new output protection requirements phasing in every year for the next few years.

There was quite a kerfuffle when Vista was launched about how this feature was going to slow system performance or do other horrible things. However, PMP is only used when playing back DRM-protected media flagged to require it, and has very little overhead (generally any GPU with HDMI output has DXVA 2.0). This urban myth remains oft-quoted but little-tested on sites like Slashdot. PMP will never be applied to content you create unless DRM is applied to that content, which generally means that content won't be available for any platform without output protection.

Hardware-accelerated transcoding

Windows 7's MF implementation adds a new API for hardware acceleration of video decoding and encoding. This is unofficially called SHED, for Secure Hardware Encoding Decoders ("Secure" refers to its ability to transcode and display while preventing capture of the uncompressed intermediate frames). This goes beyond the DXVA model of having decoding circuitry on just the GPU and lets it be implemented as a dedicated chip or part of another system component.

By combining hardware decode, preprocessing, and encode in a single function, transcoding doesn't have to stress the main CPU or memory. This makes converting content for devices much faster. However, the quality of the initial SHED implementations isn't what a hand-crafted encode on an 8-core monster workstation could do. But it's a huge improvement in "good enough, fast enough" particularly on netbooks and other less powerful systems.

SHED products have been announced by Broadcom, ViXs, Toshiba Semiconductor, and Quartics. SHED could also be delivered as an add-on PCI Express card. Exposed as a MFT, this hardware transcoding can be easily called by any Media Foundation application.

Windows Media Format SDK

The Windows Media Format SDK is used to both encode and play back Windows Media content, and is covered in the Windows Media chapter. The most notable recent change is the quality and performance improvements in the Windows 7 encoder over FSDK 11 for XP and Vista.

Most professional WMV encoding products use the VC-1 Encoder SDK, which is a static library added directly to an application, and doesn't have any dependency on the Windows version.

Major Media Players on Windows

Windows Media Player

Windows Media Player is implemented on top of the available media APIs, using them for all presentation. It's more than just a player, of course, with library management features for video and audio content.

WMP can also share content to compatible devices like the Xbox 360 and other copies of WMP.

The most commonly seem versions are 9 (XPSP2 default), 11 (Vista, and recommended update for XP), and 12 (Windows 7).

Generally, anyone who was able to upgrade to 10 has long since updated to 11.

Zune Media Player

The Zune desktop software was originally a slightly modified version of WMP11, but it became a distinct app in its 2.0 version. Like WMP, it offers media library management. But it's got some additional features of note.

Only the Zune client can sync to Zune devices, and it can be used as media player app without owning a Zune. The Zune client provides H.264 and MPEG-4 part 2 decoding for XP and Vista, and can share MPEG-4 files to the Xbox 360, which WMP11 can't. So it can be very useful for XP/Vista users,

Zune is also a good podcast client (one reason for that MPEG-4 playback).

VLC

VLC is the VideoLan Client, an open source media player originally created by French college students. It's available for pretty much every other platform you could imagine as well as Windows; possibly because all the decoders are self-contained.

Thus VLC can't suffer from dll hell. It's a great thing to have on a production machine where you don't want to install a lot of extra DMOs or MFTs.

It does have a nasty habit of taking over a whole lot of file type associations when installed; be careful to uncheck any you want to watch in another player.

Silverlight (Is Not a Media Player)

While Silverlight is certainly used to play media on Windows, it's not a media player itself. You can't double-click or drag a media file to "Silverlight" to play because it's not an application itself, but a runtime that hosts applications.

So, installing Silverlight has no impact on how media gets played outside of Silverlight applications, be they in-browser or out-of-browser. One can certainly build an application that uses Silverlight for media playback, but the desktop icon is of that application, not Silverlight itself.

Windows Media Center

Windows Media Center (generally abbreviated "MCE" from its original name of "Windows XP Media Center Edition") is another media player app, designed for a "10-foot" experience, used with a remote from a couch. It supports the same media playback functions as WMP, but with a UI appropriate for remote control navigation.

MCE also can record live video using available analog or digital capture hardware. This includes MPEG-2 bitstreams from ATSC, DVB, and unencrypted cable ("ClearQAM"—you don't even want to know what QAM stands for). Special CableCard-approved PCs can also view and record (but not transcode, due to contractual restrictions) encrypted premium content via a CableCard.

Media Center Extenders

A Media Center Extender is a remote device that can play content from MCE. This includes the MCE interface; an extender plugged into a TV will offer largely the same experience and navigation as if the MCE system was plugged into the display directly. This enables the MCE machine (hopefully with a lot of storage) to be the hub for recording and storage, with multiple cheaper devices accessing the same library.

The Xbox 360 works as an extender, as do specific devices from other vendors. They're described in more detail in Chapter 25.

Media Formats on Windows

AVI

AVI (Audio Video Interleave) was the original media format from Windows 3.0, and continues to be widely used long after Microsoft tried to deprecate it in favor of ASF and the largely forgotten Advanced Authoring Format—AAF.

The strength and weakness of AVI is its simplicity. Compared to modern formats like MPEG-4, AVI doesn't do a whole lot more than store a video and an audio track. It doesn't even support variable frame speed; each AVI file has an explicit frame rate, and the only way to include content running at a lower frame rate is just to have "dropped frames," leaving the encoded frames with a duration that is a multiple of the file's frame rate.

But that simplicity has meant that third parties could easily develop reasonably complete support, far more easily than a similarly complete parser for MPEG-4 or ASF. This was the genesis of DivX/Xvid, and why AVI files are broadly supported across platforms and players. However, those different implementations can vary in the details; stuff like VBR audio doesn't always work reliably.

But once you want to do interesting things like multiple audio tracks or captioning, AVI can draw up short.

AVI Versions

The original implementation of AVI was limited to 1 GB in size, which seemed laughably high back when it was created: CD-ROM was limited to 650 MB and external hard drives were smaller yet.

That quickly became a big limitation for content authoring and capture applications, so a group of companies called OpenDML, led by Matrox, created extension to allow arbitrary sized AVI files. These are generally called "OpenDML," "AVI 2.0," or "New AVI."

It's been around since 1995, and so there's no reason to use the old AVI type anymore.

In-Box AVI Video Codecs of Note

As AVI has been deprecated for years, there aren't many built-in encoders in Win 7, and they're largely ancient ones.

Generally, AVI should be used for content authoring, not content delivery. "DivX/Xvid" content with MPEG-4 part 2 in AVI is the main remaining use of AVI for content these days.

Uncompressed

There are quite a few different uncompressed video "codecs" in AVI and the media pipelines in Windows. They're obviously not good for delivery, but are quite handy for authoring. Each has its own four character code (4CC), and often there are different 4CCs indicating the same sampling but different arrangements of samples. That's largely transparent, handled under the hood. In general, codecs internally use the "planar" variants, where the image is stored as separate series of Y′, Cb, and Cr samples, instead of the "packed" formats where Y′, Cb, and Cr samples for each pixel are stored together. It rarely matters for compatibility, but you can get a tiny performance improvement by using the planar versions as source for encoding.

But if you're using a tool like VirtualDub, it's good to make sure that you're making the color space conversions you want, for optimum speed and quality, and not forcing an unneeded trip through RGB.

RGB

This is good old 8-bit per channel RGB. There are also 15-bit (5-bit per channel and 16-bit (6-bit G and 5-bit R and B) modes, but those aren't used by any modern codec.

RGBA

This is 8-bit RGB with an 8-bit alpha channel (and so 32-bit per pixel). It's only needed if you actually have an alpha channel.

YV12/NV12/IYUV (4:2:0)

Most delivery codecs are 4:2:0, so that's what you want to deliver to. YV12 is the planar version.

Note that the "Intel IYUV" implementation that ships with Windows is limited to standard-def frame sizes, and fails at HD frame sizes. In the case of VirtualDub, it's better to use "Uncompressed" and specify YV12. Most AVISynth scripts end with either ConvertToYV12 for progressive or ConvertToYUY2 for interlaced.

UYVY/YUY2/YV16/HDYC (4:2:2)

These are the various 4:2:2 codecs. YUY2 is the planar version most used by codecs.

For historical reasons, a lot of video hardware wants to get video in 4:2:2 even though most codecs are 4:2:0. Of course, 4:2:2 is very important when doing interlaced content, so that chroma samples don't get subsampled across fields.

V210 (4:2:2 10-bit)

V210 is a 10-bit 4:2:2 codec. Mainly used for authoring, of course. Authoring codecs like Cineform use V210 with 10-bit 4:2:2 sources.

Microsoft Video 1

Microsoft Video 1 is the AVI equivalent to the QuickTime "Video" codec; an ancient, horrible, 16-bit pre-Cinepak codec without any rate control. I mention it only so you don't use it. I'm kind of amazed that Windows 7 still includes the encoder for it.

Microsoft RLE

This one's an ancient Run-Length Encoded codec, not useful for anything.

Cinepak

Cinepak was the first decent CD-ROM codec, but there's been no reason to use it since the 1990s. Still, it's a lot better than Video 1. See Figure 27.3.

DV25

Windows has included a DV25 decoder since DirectShow was introduced. It supports .dv and .mov wrappers as well as .avi for easy interoperability.

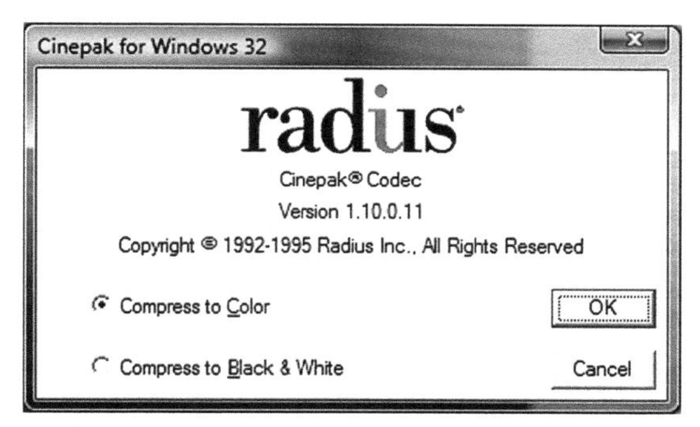

Figure 27.3 The Cinepak configuration dialog has to be the oldest dialog box in Windows 7; Radius hasn't existed for over a decade. I actually used the Black & White mode to good effect on several projects.

In-Box Audio Codecs of Note

PCM

Uncompressed is obviously your go-to codec for authoring purposes.

VfW was stereo only and only went up to 48 KHz 16-bit. DirectShow and Media Foundation apps can author multichannel 96 KHz up to 32-bit float and so on. Just don't try to play them in VirtualDub.

Transporting multichannel audio in an audio-only AVI file can be more compatible than in WAV.

MP3

Windows has included a MP3 decoder for AVI since well before Windows 2000. It's the best in-box audio codec for compression efficiency.

The MP3 encoder in older versions of Windows was quite limited, and only went up to 56 Kbps (not nearly enough for transparency). The decoder has always been full-featured, though, and so third-party encoders like those from DivX used MP3 as the default audio codec.

a-Law/u-Law

These are ancient 8 KHz telephony codecs. Don't use.

IMA/Microsoft ADPCM

These are old 4:1 compressed CD-ROM audio codecs. Revolutionary in the day (we could do 16-bit 44.1 KHz on CD-ROM!), but a lot more lossy than even MP3. Don't use.

Third-party AVI Codecs of Note

Since Microsoft turned its attention to Windows Media, AVI codec development has largely been done by third parties.

Cineform

Cineform is a great, visually lossless wavelet-based authoring codec. I use it heavily, as it offers a master as good as lossless for recompression. It also has great integration with Adobe Premiere Pro, leveraging wavelet subbands for high performance scrubbing and effects previews.

You'll have to pay a fee to use the encoders, but there's free decoders for download. In a bonus, the Mac decoder can read AVI, so a Cineform AVI file is a great mastering format for cross-platform workflows.

The encoders are fast enough for real-time HD capture on a fast computer.

Huffyuv

Huffyuv is an open-source AVI encoder decoder. It's lossless RGB or 4:2:2, using Huffman coding. Unlike many other lossless codecs, it's quite fast to encode and decode, with real-time capture quite possible.

Its biggest limitation is the lack of a proper 4:2:0 mode…

Lagarith

…which was nicely addressed by the Huffyuv-based Lagarith. Lagarith added arithmetic coding for better compression efficiency, and a native 4:2:0 mode.

While arithmetic coding is a lot slower than Huffman encoding, Lagarith is multithreaded for encode and decode, so real-world performance can still be pretty good, although real-time capture can be a stretch.

When I need really mathematically lossless 4:2:0, Lagarith has been my codec of choice for years.

DivX/Xvid

As described in detail in Chapter 12, DivX was born of a hack to enable the MS MPEG-4 ASF codec to work in AVI files. This lead to the whole DivX ecosystem, and the xvid open-source implementation thereof.

MPEG-4 part 2 isn't supported in Windows out of the box before Win 7, but decoders are readily available.

Hardware-specific codecs

Vendors like Black Magic Design and AJA provide their own AVI encoders/decoders for their capture cards. These are generally free downloads for easy interoperability.

WAV

The .WAV file format is the audio-only equivalent to AVI. Audio-only AVI files are certainly possible, and are more reliably supported for multichannel audio.

WAV uses the same set of codecs available for AVI on the system. In practice, that means WAV is mainly PCM, since the only good in-box audio codec for WAV is MP3, in which case you might as well make a .mp3.

Multichannel audio support in WAV can be fragile, with no support in VfW products or pre-XP versions of Windows.

Windows Media

The primary media format for Windows has long been, of course, Windows Media. Because at least WMP 9 is installed on nearly all Windows machines, that makes WMV an easy choice for content that will offer good quality and efficiency with high compatibility across Windows computers. That won't change until Windows 7 is the baseline version, as XP is today. Windows Media is covered in depth in Chapter 16.

Windows 7 finally has the integrated bob deinterlacer on for interlaced VC-1 files, making it possible for viewers to watch interlaced content without having to tell them to set a registry key.

DVR-MS

DVR-MS is the internal recording format of Media Center on XP and Vista. It's mainly used to store the native MPEG-2 and AC-3 bitstreams of an ATSC or other unencrypted digital broadcast. WMP and the Zune client can automatically transcode from DVR-MS to portable devices.

It wasn't easy to use outside of Media Center before Vista, but it's supported in DirectShow and Media Foundation apps on Vista and Windows 7.

MPEG-1

MPEG-1 playback was added way back in a service release for Windows 95. It's a fine implementation, and is the easiest-to-author format still compatible with ancient machines, as the Windows Media encoders from that era aren't available in recent Windows versions. The pre WMP 9 versions didn't handle aspect ratio correction for non-square pixels, so encode as square pixels to make sure the video displays the same on all platforms.

MPEG-2

Windows didn't include its own general purpose MPEG-2 decoder before Windows 7. The Home Premium and Ultimate editions of Vista included a MPEG-2 decoder for DVD playback, it wasn't available to other apps, or even for WMP to play back MPEG-2 files.

PCs sold with DVD drives from Win98 on would include a DirectShow MPEG-2 decoder. There have been many vendors and versions of these, but most allowed the decoder to be used in any app, and supported file-based playback as well.

Windows 7 has built-in MPEG-2 program and transport stream demuxers, including HDV, and a good MPEG-2 decoder. The Win 7 bob deinterlacer is probably most useful here, given all the 30i MPEG-2 in the wild.

MPEG-4

While Windows Media Player 7 included an ISO MPEG-4 Part 2 compliant Simple Profile video decoder, it didn't support the MPEG-4 file format or AAC audio.

Windows 7 is the first version with out-of-box support for .mp4. Making up for lost time, it's extremely complete, including a wide range of features:

- Part 2 Simple and Advanced Profiles
- H.264 through High Profile, including interlaced
- No Level restriction (although it'll be limited by what the decoder can do; most Windows 7 systems will have DXVA decoding)
- AAC inc luding multichannel and HE v1 and v2
- AVCHD
- In-transport stream, .mp4, .mov, 3GP, and AVI files

Tutorial: Preprocessed AVI Intermediate

All of our WMV-related tutorials apply to Windows, and for Windows 7, most of the other formats do as well. So I'm going to show a workflow demo of using VirtualDub to make a lossless preprocessed intermediate.

Scenario

We are writing a book about video compression, and need to make a variety of tutorials for mobile device playback. We have a nice rights-cleared 720 × 480i30 4:3 4:3 anamorphic source file we want to turn into a preprocessed 640 × 480 square-pixel progressive AVI file for use in those tutorials.

The source doesn't have a sound track need to add some audio as well.

Three Questions

What Is My Content?

The source file is an uncompressed 4:2:2 AVI without audio. It's made up of a variety of different source clips, which have different horizontal blanking.

So we're going to insert a rights-cleared audio file as well.

Who Is My Audience?

Me!

What Are My Communication Goals?

I want an interesting-to-encode 640 × 480 square pixel progressive source file that's easy to use in any of my Windows-based compression tools. It should be lossless YV12 so that all encoders will get the exact same source samples, hopefully eliminating any color space conversions that would keep tests from being apples-to-apples.

And it should also have some interesting-to-encode audio.

Tech Specs

Pretty straightforward. I need to take my source, crop out any blanking around the edges, deinterlace it nicely, and scale it to 640 × 480. We need to add that audio source next. We then will save it to the Lagarith codec, in its lossless YV12, with the audio.

VirtualDub

Because we're writing an overdue compression book at 4:43 AM, this will need to be quick and hopefully not dirty. This is a job for VirtualDub.

We have the current version installed, with the old and trusty Smart Deinterlace 2.8 beta 1 deinterlacer.

(*Continued*)

Opening and scrubbing through our source, we see it's quite interlaced. We open that up in Smart Deinterlacer with default settings, and scrub through the clip. There are definitely some places where motion is leaking through without being deinterlaced. This can be a little tricky, as the scenes vary so much, but we tweak knobs until we find something that works reasonably well everywhere. Thank goodness for Smart Deinterlace's "show motion areas" which only shows the parts of the image flagged as motion. Then you can tweak settings until it's not letting interlaced stuff through, but leaving the static parts of the frame alone. See Figure 27.4.

Next up is cropping and scaling. The easiest way to do it in VDub is to set the output frame size first, and then apply the Crop settings to that. Scaling is easy: 640 × 480. We'll go with Lanczos; detail can be a good thing for codec testing.

Figure 27.4 (A) Our final deinterlacing settings are doing a good job of discriminating the moving ping–pong elf–man from the background. (B) Quite a decent job, even with this really high–motion frame.

(*Continued*)

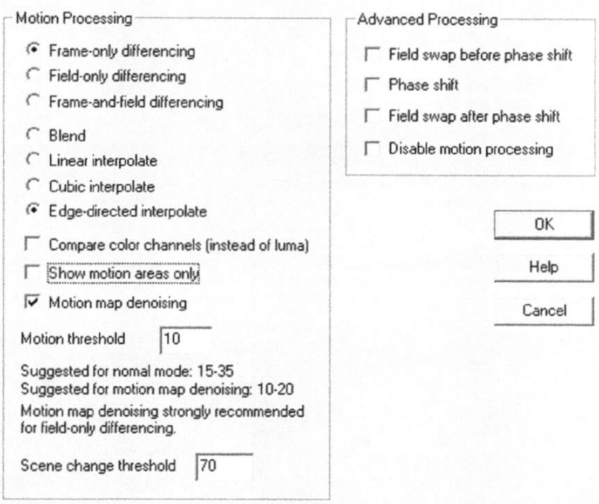

C

Figure 27.4 (Continued) (C) The final settings.

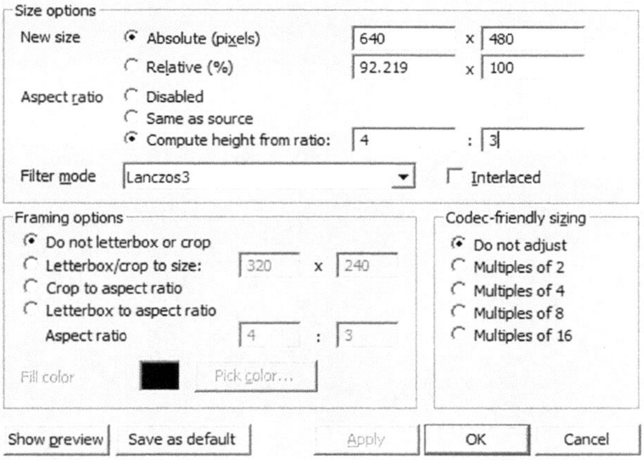

A

Figure 27.5 (A) While we can crop from the Resize dialog, the mode from the Filter dialog is easier.

As it's a mix of sources, we have lots of sections without any horizontal blanking at all, but one sequence with very wide blanking on the left side. Cropping out that much doesn't seem to obviously distort any other scenes, so we'll call that good. See Figure 27.5.

Now encoding settings (Figure 27.6). Lagarith at YV12, easy enough. We're in a hurry, so multithreaded, of course. Null Frames are just frames dropped because they're exact repeats. We'll leave it on, but there are probably no frames like that here.

(*Continued*)

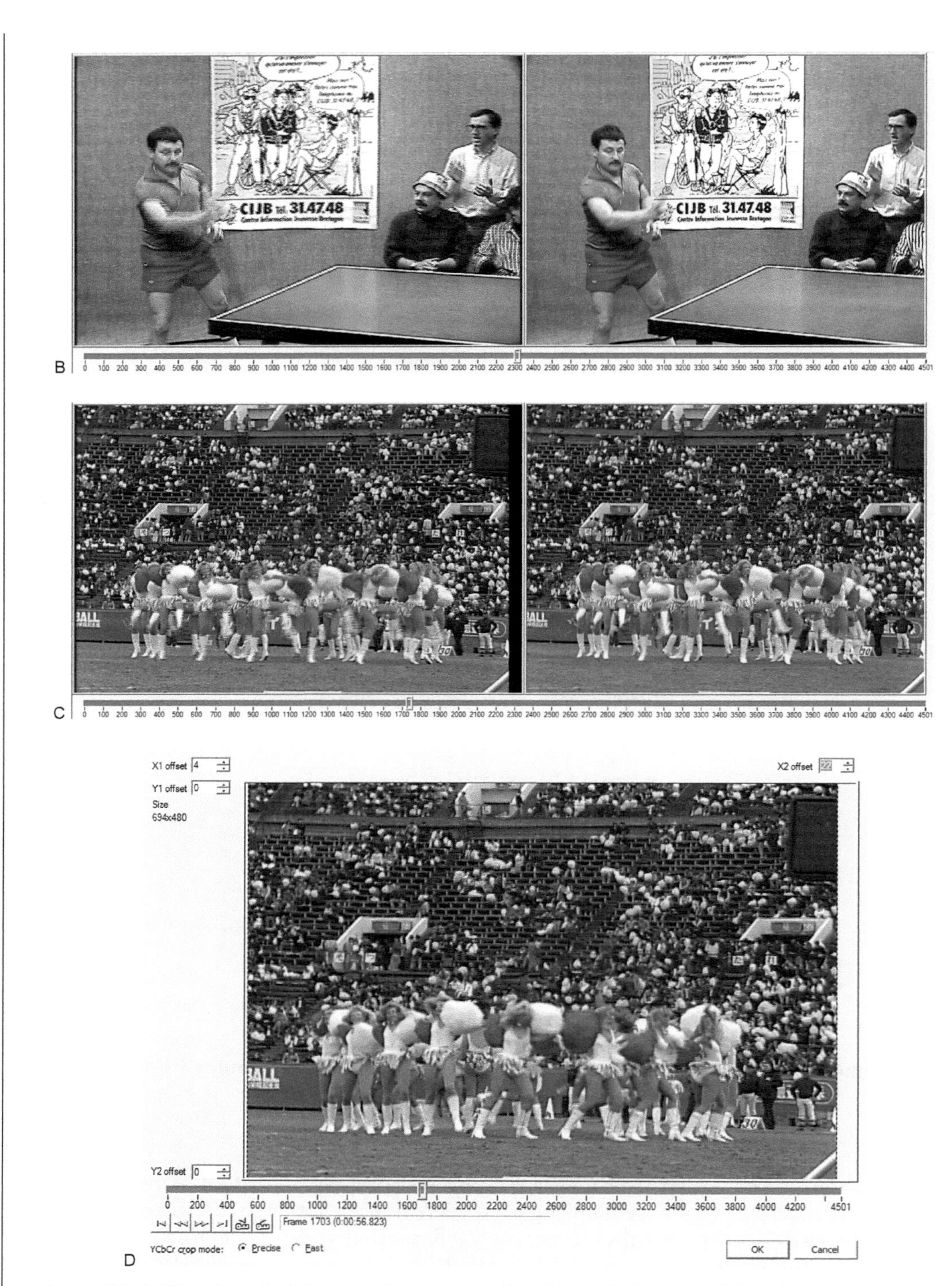

Figure 27.5 (Continued) (B) Our ping–pong shot has a little subtle blanking on the edges, but it'd be easily masked with a standard 8/8 left/right crop. (C) Our cheerleaders are another matter entirely, and require almost 3x the standard crop. (D) I really quite like this cropping dialog. I get my scrub control, and can switch between visual and numeric entry modes.

(Continued)

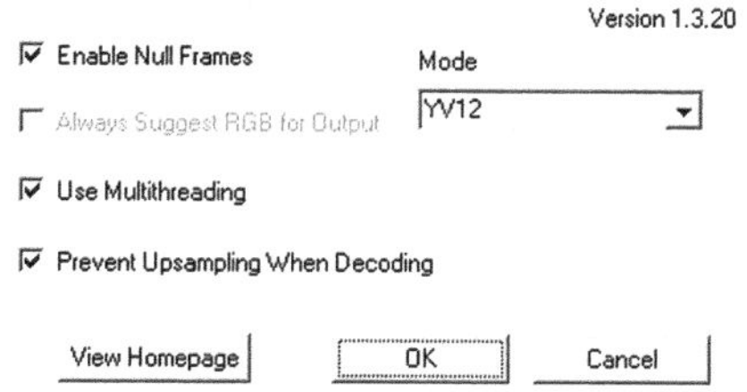

Version 1.3.20
Enable Null Frames
Mode
YV12
Always Suggest RGB for Output
Use Multithreading
Prevent Upsampling When Decoding
View Homepage
OK
Cancel

Figure 27.6 Our Lagarith settings.

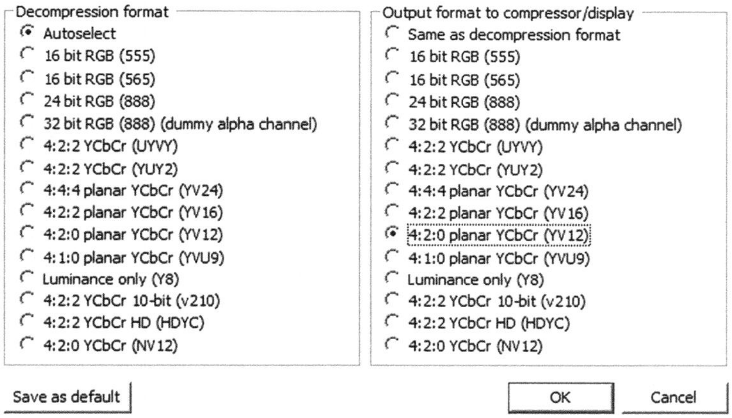

Decompression format
Autoselect
16 bit RGB (555)
16 bit RGB (565)
24 bit RGB (888)
32 bit RGB (888) (dummy alpha channel)
4:2:2 YCbCr (UYVY)
4:2:2 YCbCr (YUY2)
4:4:4 planar YCbCr (YV24)
4:2:2 planar YCbCr (YV16)
4:2:0 planar YCbCr (YV12)
4:1:0 planar YCbCr (YVU9)
Luminance only (Y8)
4:2:2 YCbCr 10-bit (v210)
4:2:2 YCbCr HD (HDYC)
4:2:0 YCbCr (NV12)

Output format to compressor/display
Same as decompression format
16 bit RGB (555)
16 bit RGB (565)
24 bit RGB (888)
32 bit RGB (888) (dummy alpha channel)
4:2:2 YCbCr (UYVY)
4:2:2 YCbCr (YUY2)
4:4:4 planar YCbCr (YV24)
4:2:2 planar YCbCr (YV16)
4:2:0 planar YCbCr (YV12)
4:1:0 planar YCbCr (YVU9)
Luminance only (Y8)
4:2:2 YCbCr 10-bit (v210)
4:2:2 YCbCr HD (HDYC)
4:2:0 YCbCr (NV12)

Save as default
OK
Cancel

Figure 27.7 The #1 VirtualDub mistake is forgetting to set your output color space correctly. It always defaults to RGB.

The next step is easy to forget: make sure that your output color space is set to what you want; it defaults to RGB (Figure 27.7). We want YV12. That gives us a 4:2:2 to 4:2:0 conversion, so later encodes won't need any color space conversion for faster processing and higher fidelity.

Adding audio is trivial: Audio > Audio from a rights-cleared music file. Our audio is longer than video; VDub just cuts it off after the last frame of video. Not an elegant fade-out, but that doesn't matter for codec testing.

And that's it. It's just about a real-time export on a decent machine.

CHAPTER 28

QuickTime and Mac OS

Introduction to Mac

Although I've worked for Microsoft for nearly four years now, I spent the first couple decades of my computing life as primarily a Mac user and fan.

Apple nailed a GUI usable for creative long apps before anyone else, first winning the print market with PageMaker (on top of QuickDraw) and then video with Avid, Media 100, and Premiere (on top of QuickTime). So many of us who grew up in the early days of the industry grew up with the Mac, since it was the first platform to pioneer so many features. Apple and the Mac are in a resurgence today despite, or perhaps because of, the fact that the differences now are much smaller. Macs and Windows machines were long different ecosystems, with different processor families, accessories, display connectors, hard drive formats, media architectures, and so on, and really seemed like different worlds. But now that Macs and Windows run on the same underlying Intel-architecture PC platforms, software has papered over the gaps.

History of the Mac as a Media Platform

Birth of the Mac

The first Mac launched in 1984. There was no video possible with its 512 × 384 1-bit display (each pixel was either black or white). And the specs would make a toaster laugh today: 8 MHz Motorola 68000 processor (first of the 68 K series) with 128 KB of RAM. But its GUI revolutionized graphics with MacPaint and PageMaker, using the very fast and capable QuickDraw API.

Another less-noted Mac innovation appeared in that first model: built-in sound. It was only 8-bit mono at 22.25 KHz (Macs used oddball sample rates for many years). This was nearly a decade before all new PCs shipped with sound at least as good.

Macintosh II

The Macintosh II in 1987 broke the Mac out of the sealed box, switching to a much faster processor, adding NuBus expansion slots, and support for color. And that made it possible to deliver true 24-bit color graphics for around $10,000, a great deal compared to the proprietary UNIX workstations then otherwise required.

The Mac II was the first true PC workstation, established the basic architecture of the Mac for the next decade, and launched desktop content creation.

And as prices dropped, a color display became an assumed feature of a Mac well before that was true of Windows, leading to much early use of Macs as kiosks.

Formation of Avid, Digidesign, and Radius

With the Mac II and its successors, the Mac supported hardware cards for display, capture, and storage. The OS was quite lightweight, so media apps could directly control hardware, while still leveraging the Mac GUI. The add-ons typically cost many times the cost of the Mac itself.

Thus, we saw early digital video and audio companies targeting the Mac chassis in the second half of the 1980s, with products that defined the Non-linear editor and Digital Audio Workstation.

Avid and Digidesign are still well known, but Radius and SuperMac (which Radius acquired) had arguably the most innovative legacy of all. Founded by key veterans of the Mac development team, they pioneered some critical technologies:

- Multiple-monitor displays (Radius Full Page Display)
- CPU add-on cards (Radius Accelerator)
- CD-ROM video (Cinepak)
- DSP-assisted media processing
- Blade server (Radius Rocket—a NuBus card that was a complete Mac. Worked better in theory than in practice)
- First video capture card (SuperMac's Video Spigot)
- First interlaced video capture card (VideoVision Studio)
- First component video capture card (Telecast)

I've got a fond place in my heart for Radius. My first capture card was the VideoVision, and my first good capture card was the Telecast (later VideoVision SP) with component analog video. My future wife was capturing video on a Mac even before that, and remains the only known person to make money with a Video Spigot.

Macromind Director

Adobe's Director started as Macromind VideoWorks in 1985, and became Director 1.0 in 1989. It quickly became a leading authoring environment for kiosk and CD-ROM titles.

Director was Mac-only until version 4 (1994), although 3 (1992) added the ability to make players for Windows.

System 7

Good OS color integration didn't come until System 7 in 1991. This was a big upgrade, adding better multitasking and making media apps possible. Mac OS was based on the System 7 architecture until Mac OS X shipped a decade later.

QuickTime 1.0

QuickTime launched as an upgrade to System 7 in 1991. While primitive today, it was for many of us the first time we saw video on a computer screen (Video for Windows was a year away).

QuickTime 1.6 in 1993 introduced the Apple Compact Video codec (later renamed Cinepak, and actually from Radius). CD-ROM video was finally possible.

Products like Director quickly adopted QuickTime for their video playback.

The Multimedia Mac

By the early 1990s, desktop Macs evolved into being good multimedia machines out of the box, far ahead of the average business-configured Windows PC. They all included at least 8-bit 22 KHz stereo audio and color displays.

The modern multimedia machine was really established with the Quadra/Centris AV line in 1993, which had 16-bit stereo up to 48 KHz standard, enough VRAM to run 1024 × 768 at least to 16-bit color, and a 2x CD-ROM drive.

QuickTime 2

QuickTime 2 came out in 1994 for Mac and Windows (though playback-only on Windows). 2.1 in 1995 added IMA Audio, and made the CD-ROM world safe for 16-bit audio.

PowerPC Switch

In the early 1990s the Motorola 68 K architecture that all Macs had used to date was falling behind Intel's 486 and Pentium processors.

Without a clear direction to enhance 68 K, Apple, IBM, and Motorola formed the AIM alliance with ambitious plans for a next generation processor, platform, and operating system to compete against Windows and Intel. After tremendous effort, only the PowerPC processor came to market, which was adopted for all new Macs.

It did deliver on its performance goals for a time, being much faster than 68 K and often faster than the best x86 processor in any given year.

The Birth and Death of Mac Clones

As part of PowerPC adoption plans from 1995–1997 Apple licensed the Mac OS to other companies to build Mac models, notably including PowerPC manufacturer Motorola, graphics vendor Radius, and Power Computing, which invented the build-to-order model. However, these machines tended to be better than Apple's on the high end or cheaper than Apple's on the low end, so Apple's sales plummeted. My company went from buying all Apple to buying mostly clones during that period.

When Steve Jobs returned to Apple, he cancelled all licensing, and purchased Power Computing.

QuickTime 2.5 and QuickTime Media Layer

QuickTime 2.5 was a Mac-only release in 1996 that introduced the QuickTime Media Layer, letting QuickTime be a runtime for richer interactive experiences. It also included a MPEG-1 decoder for PowerMacs.

QuickTime v3

QuickTime 3 was announced and won the best of show at NAB 1997 (the National Association of Broadcasters show), finally promising full authoring as well as playback on Windows. However, that proved much harder than anticipated due to Mac OS dependencies, and a development reboot involved porting much of the Mac OS API inside of QuickTime for the Windows version.

It finally shipped in 1998 (again winning best of show at NAB). QT3 also introduced the Sorenson Video and QDesign Music codecs, making QuickTime an excellent platform for progressive download.

The streaming wars between Windows Media, RealMedia, and QuickTime started heating up in the late 1990s.

QuickTime Enters the Streaming Wars

CD-ROMs were platform-specific, so the playback architecture wasn't seen as that critical. But web video launched during the explosive growth of the web industry, and so was considered highly strategic by Apple and Microsoft. Both entered a heated three-way competition with RealNetworks to become the dominant technology.

In this period, Apple introduced web video and then streaming support, and the innovative QuickTime Media Layer. In the end, Apple lost its focus on web video amongst its attempts to survive, and the QuickTime team was refocused on building a platform for Apple's consumer and professional media apps to run on.

It's only in the last few years that Flash and Silverlight have been able to deliver on what QuickTime Interactive was doing more than a decade earlier.

Mac OS X Begins and Steve Jobs Returns

The System 7 lineage had obviously ran out of steam by the early 1990s. It didn't have good support for multitasking or multithreading; a bug in any app could freeze the entire machine. Apple had spent the better part of decade trying to find their own Windows NT–like way to a modern OS compatible with existing hardware and software.

When in-house efforts foundered, they went shopping for alternatives, and were enamored with NeXT's OpenStep, a UNIX-based object-oriented OS popular in high-end finance and industrial-grade web sites. It was designed with good hardware abstraction, supporting four different CPU types. And NeXT's founder and CEO was Apple's own co-founder and Mac leader Steve Jobs.

And Apple bought NeXT, in order to use OpenStep. A common joke is that Apple though they were buying NeXT, but NeXT really bought Apple. I don't think any stockholder would complain, though.

The G3 Era and the PC Convergence

The G3 processor was originally designed as a low-power consumer CPU, but it wound up being faster than "high-end" PowerPC and x86 processors, even with dual-processors, so Apple used it across their line.

Macs had long been very different architecturally from Wintel machines, using a different expansion bus (NuBus instead of PCI and AGP), drive interface (SCSI instead of ATA), networking port (AAUI instead of 10Base-T), and even VGA plug. These differences meant that Mac peripherals were often much more expensive than PC equivalents built for a much larger market. And Macs were equally infamous for a user-hostile case design that turned simple RAM or drive upgrades into an hour of cursing and bleeding knuckles.

The canonical "Blue and White" G3 model was the first model fully designed after Jobs's return to Apple, and it adopted PC-standard technologies like PCI, ATA, 100Base-T, and USB. It also had a very easy-to-service case design that has evolved into the Mac Pro of today.

QuickTime 4: Streaming and The Phantom Menace

In 1999, the web release of the *Star Wars: Phantom Menace* trailer got 6.4 million downloads (2.2 million QuickTime 4 beta downloads), an incredible number, given the limitations of the early web. It also showed that web-distributed video that requires no apologies was possible; it truly was better-than-VHS quality.

QuickTime 4 itself added real streaming support with the first RTSP implementation, as well as MP3 audio playback.

In an early Apple foray into open source, the source code for the QuickTime Streaming Server was released as the Darwin Streaming Server.

This probably reflected Apple deciding to abdicate the streaming wars and focus QuickTime engineering on providing a better base for Apple's pro and consumer media apps.

Final Cut Pro

Final Cut began life as a Macromedia project led by Randy Ubillos, creator of Premiere. Its architecture was designed to address the then-limitations of Premiere for extensibility and professional applications. While nominally cross-platform, due to Apple's problems in the era, the focus was on Windows development. After Macromedia went into hard times during the CD-ROM to web transition, they sold it to Apple.

Media 100 was on the verge Final Cut for their Windows editor when Apple swept in; it would be a different content creation world if that had happened!

Final Cut started as mainly a DV editor, but grew over the next few versions into a capable HD professional tool. It was also joined by a variety of other video tools like Motion, Compressor, and DVD Studio into the Final Cut Studio suite.

Final Cut was critical to Apple's sale of high-end, high-profit Mac desktops and laptops to video editors.

QuickTime 5

QuickTime 5 came out in 2001, and was one of the shortest-lived major versions. It introduced codec downloading (later removed from QT7), improved QTVR, and added MPEG-1 for

Windows, and in 5.0.2 the Sorenson Video 3 codec, which remained the best in QuickTime until H.264.

For the content creator, the DV decoder in QT5 was finally good enough in quality and performance that there was no need for third-party replacements anymore.

The G4 Era

Motorola's G4 processor introduced AltiVec, a SIMD architecture à la SSE that offered much better performance for media processing when software was tuned for it; most media apps were, and saw big gains (4–8x faster wasn't uncommon for particular filters).

A "Graphite" G4 in 2001 was the first PC with a built-in DVD-R, boosting the early popularity of DVD Studio Pro.

However, with Windows XP in 2001, the Mac lost any advantage in multimedia playback, as XP introduced multichannel audio and GPU accelerated decode.

The G4 rapidly hit a wall, and was stuck at 500 MHz from 1999–2001, while Intel went from the Pentium 3 600 MHz to the Pentium 4 at 1500 MHz.

From that point on, Windows PCs were always a lot faster than PowerPC Macs for general computing, with an even bigger price/performance lead. AltiVec-optimized apps gave the Mac a good niche until Intel's SSE3 P4s developed an unassailable lead even for media processing in 2004. Long-time Mac apps like Photoshop and After Effects were clearly faster on Windows.

QuickTime 6 and MPEG-4

After many years of primarily using proprietary codecs from Sorenson Media and QDesign, QuickTime 6 in 2002 embraced MPEG-4 in a big way, reading and writing .mp4 files, the MPEG-4 part 2 video codec, and AAC-LC audio.

This became central to both development of QuickTime and Apple's device strategy, as the iPod and now the iPhone use only standard video and audio codecs. 3 GP phone support was also added and substantially updated in the various point releases.

QuickTime 6.3 was the last version available for Mac OS 9. QT6 was also notable for skipping Windows with 6.1 and 6.2, with 6.3 returning as cross platform.

Mac OS X, Finally for Real

Mac OS X took years after the NeXT acquisition to actually ship. And it never truly solved the compatibility problems that had bedeviled a new Mac OS for ages. It delivered hardware compatibility by waiting until it could declare every Mac designed before NeXT unsupported, and handled old software by emulating an entire old Mac in software. The underlying OS was much smoother and more stable, but until apps were recompiled as "native" the experience was quite clunky. OS X nominally shipped as 10.0 in 2001, but it wasn't useful for much. 10.1 in 2001 was usable for office tasks, but didn't even support DVD playback. A native version of Final Cut Pro 3 was released a few months after 10.1, but it took some time for other products to arrive. 10.2 from 2002 was the first with enough native apps to make it worth running for content creation, with DVD playback, and offered new fundamental features, particularly Quartz Extreme. Quartz Extreme offloaded GUI compositing to the GPU, saving CPU power and memory bandwidth, which were growing increasingly scarce on the G4.

The G5 Era

Apple introduced the G5 processor from IBM in 2003, which was a big leap in media performance. Beyond having much higher clock speeds, the G5 fixed the increasingly limited G4 memory bus, which hurt G4 performance at least as much as clock speed. The G5 was initially available as dual-processor, and then went to quad.

The G5 made HD in software on the Mac possible, and introduced the aluminum design Mac Pros use to this day.

People eagerly awaited the PowerBook equivalents.

The Device Revolution

Apple was transformed by the iPod in 2001 and again by the iPhone in 2007. While the Mac has regained market share in recent years, this pales in comparison with (and may be driven by) Apple's incredible success with devices. The iPod was an industry game changer, dominating music players nearly to the degree Windows dominates personal computers. The iPhone is a relatively smaller fish in a much bigger sea, but is also very successful. In 2009, Apple has been selling about six iPods and two iPhones for every Mac.

The strong video features of those devices has played a big role in shaping our industry, and Apple's device ecosystem has driven uniform licensing rules for media sales, DRM-free major-label music, and the very name of podcasting.

QuickTime 7 and H.264

QuickTime 7 introduced H.264 (one of the first technologies using it), and was the biggest architectural upgrade since QT3. It took much more advantage of OS X features, particularly leveraging the Core Graphics API for GPU scaling and compositing of video.

Apple's H.264 implementation, while not the best on the market, was the first competitive encoder that Apple has created in-house since QuickTime 1.0. From Cinepak in QT 1.5 to

Sorenson Video 3 in QT6, the best video codec had been a proprietary technology licensed from a third party.

Intel Switch

After a decade of backing the PowerPC, the final straw was the lack of a reasonable upgrade for the PowerBook, which was lagging further behind every year in performance. And the G5 itself wasn't ramping up as promised. Steve Jobs infamously promised a 3 GHz G5 "within 12 months" after the G5 launch, but in three years it never made it past 2.7 GHz.

Intel had been patiently offering Apple help to switch to their processors for decodes, including a couple relatively complete ports of Mac OS before OS X. And of course Mac OS X was derived from the x86 compatible OpenStep, and Apple had maintained secret x86 versions of Mac OS X. And with Intel's Core 2 in development with impressive performance and performance/watt, Apple pulled the trigger.

It went quite well. The one downside is that Classic didn't make the transition, and so the new Macs were only able to run the OS X apps from the last few years; the oldest Mac apps that ran were now from 2001, not 1984.

In the cross-platform content industry, many of us long struggled with either lugging around two laptops to every industry event, or using slow software emulation like Virtual PC. The advent of Intel Macs with Boot Camp and virtualization meant that any Mac can be as good a Windows machine as any. The biggest hole in the Windows-on-Macs story is that Apple's EULA (end-user license agreement) doesn't allow virtualizing Mac OS from Windows—only the other way around.

This Intel switch culminated in Apple's transition from an innovator in computer hardware to an innovator in computer design. While the MacBook and MacBook Pro are fine professional computers, they're assembled from the same commodity parts as any Dell or HP, and aren't available in nearly the same variety of configurations.

Reduced Focus on the Mac and Professional Content Creation

One downside to Apple's device success is a lessened focus on the Mac, which is no longer the core of the company's business (after decades as Apple Computer, in 2006 the company became just Apple).

This has been keenly felt in professional video and audio, where Final Cut Studio (FCS) got the relatively minor FCS 3 release more than two years after FCS 2, after nearly a decade of substantial annual releases. DVD Studio Pro has been on version 4 since 2005 (still with its then-innovative HD DVD support long after that format's demise). One-time high-end effects tool Shake was cancelled after several years of slow development.

This may be a reflection of success as much as anything, however. With the return of Adobe and the recommitment of Avid to the Mac, Apple has less need for first-party products to drive high-end margin sales. The Mac will be a fixture in editing suites for a long time to come.

The Future: Snow Leopard and QuickTime X

Mac OS 10.6 (Show Leopard) is described by Apple as a smaller update in terms of features than past releases, with the focus on optimization, which isn't a bad thing. As we enter the era of quad-core laptops, 16-core workstations, GPUs with 1000+ stream processors, an OS has dramatically different needs and opportunities for optimizations.

Apple's also pulling the plug on PowerPC with 10.6. It will only be supported on Intel Macs, and so all those PowerBooks and G5s (every Mac before and many from 2006) are stuck at 10.5.

Of course, Apple being Apple, many touted 10.6 features are well-branded implementations of things that Windows and Linux have been doing for years. For example, OpenCL is another pixel-shader language similar to NVidia's CUDA and Microsoft's DirectX Compute APIs.

But from a media perspective, anything that helps make content creation, compression, or playback work better is a big deal for us. And now that we've wrung out much of the potential performance from SSE optimizations it's multicore and GPU processing that are the low-hanging fruit.

There are significant enhancements in the core technologies that could make for improved media performance in Snow Leopard:

- Expanded 64-bit support.
- OpenCL pixel shaders for programmable GPU acceleration for recent NVidia and ATI graphics cards. While created by Apple and initially shipping in 10.6, it is in progress as a multivendor standard.
- Grand Central Dispatch API for easier creation and tuning of multicore applications

QuickTime jumped from 7 to X in 10.6. Hopefully Apple will follow past practice and make it available for 10.5 and Windows, but haven't made any indications. Given the scope of architectural changes, I can imagine that much of the work is 10.6-specific. Even if we see a QuickTime X for 10.5 and Windows, there could be significant differences, particularly for PowerPC Macs. Some highlighted features include:

- General optimizations leading to a 2.8x faster QuickTime Player launch times (I hope we see that for Windows, where QuickTime Player can take 20–30 seconds to load).

- HTTP Live Streaming, as described shortly.
- A new QuickTime Player tuned for consumers, but less useful for content authors (more on that shortly as well).
- GPU decode of H.264 (finally!), but only announced for the lowish-end NVidia 9400M, which are in the default configs of all new-in-2009 consumer Mac models. But they're not in any older Macs, or in Mac Pros or build-to-order MacBook Pros or iMacs with upgraded graphics. Hopefully Apple really requires a PureVideo HD VP3–compatible GPU (G98 core or later). The G98 adds only a few features on top of existing cards for accelerated decode; it seems a shame to have everything else fall back to software.
- GPU scaling and compositing (which QT7 already did; it's not clear what's different/improved here).
- Integrated ColorSync for accurate color space conversions. This is a big deal, since incorrect color space conversions have plagued QuickTime since Y′CbCr decode/encode modes were added to some but not all codecs in QT6. At least the Mac finally uses video-standard 2.2 gamma, abandoning their idiosyncratic 1.8 gamma after 22 years.

A personal history of my Macs

Say what you will, but I clearly remember every Apple computer I've owned personally or used as my primary work machine.

It's interesting to note that my nearly two-year-old phone has a faster CPU (400 MHz), more RAM (128 MB) and storage (2 GB) than the first half of my Macs:

- Apple //c: 1 MHz, 128 KB. I really wanted a Fat Mac, but there weren't good programming tools at the time. I hacked a lot of AppleSoft Basic code at home through high school.
- Mac Plus: 8 MHz, 1 MB RAM. For a summer job doing book layout in PageMaker 1.0.
- Mac SE: 8 MHz, 1 MB, 20 MB HD. I got my first real Mac for college. I had a copy of Tetris, which I was pleased to find girls would come over to my dorm room to play.
- Mac IIci: 25 MHz, 4 MB RAM. It was a launch unit, for use in a computer music class I was taking; it was actually faster than the college's MicroVAX at the time. I ran Premiere and Director 1.0 on it, created my first animation and digital video, and wrote innumerable screenplays. It remained my main machine for five years.

- PowerMac 8100, 80 MHz, 64 MB RAM. 20" CRT monitor, Radius VideoVision capture card, and SledgeHammer 4 GB RAID. Our first actual compression workstation, and cost about $20 K with displays and capture gear. The defective BART4 chip in the first-generation PowerMacs left it incapable of keeping audio sync at reasonable bitrates.
- PowerMac 7100, 66 MHz, 48 MB RAM. My home machine for many years. It didn't do much media work, but it sure wrote a lot of articles and email about media, and played a lot of early Bungie games.
- PowerMac 8500, 150 MHz.A compression workstation much faster than the 8100. I was out of the editing room by that point.
- PowerBook 1400. 33 MHz, 16 MB RAM. My first laptop! Poor video bus couldn't play video worth a darn; anything more than 320 × 240 turned into a slideshow.
- PowerBook 3400. 240 MHz, 32 MB RAM. I talked our COO into trading for my 1400 to do better customer demos. It was much faster for almost everything except video playback.
- PowerMac 8600, 2 × 250 MHz, 192 MB RAM. An 8600 upgraded with dual procoprocessors, my first dual-proc compression workstation. Easily encoded 4x more content a day than the system it replaced, my first lesson that skimping on hardware can be very expensive.
- PowerMac 8500 with G3 upgrade card, 266 MHz, 128 MB. My old compression system was dumped as too slow for Photoshop, and came home to replace the 7100. Terran Interactive bought me a processor upgrade as thanks for writing a white paper.
- PowerBook G3: My first Terran Interactive laptop, and my first one that could actually play full-screen video.
- PowerMac G3: 300 MHz, And my first Bondi Blue compression station. It was single processor, but the G3 was so much better it was still faster than the dual 8600. With a Media 100 xs board.
- PowerBook G3 (Bronze): 400 MHz. First with DVD decoding hardware and USB ports. Terran purchased this so I could do DVD playback demos. I won Unreal playing with just the trackpad. Still, my first PowerBook upgrade that wasn't notably faster.
- PowerMac G4 DP 2 × 450 MHz. A monster machine replacing the G3 as my main capture/edit/compression box. Dual processors, and the G4 added AltiVec SIMD that really helped optimized media apps.

- PowerBook G4: 500 MHz. The first-generation Titanium enclosure. Beautiful, but with poor Wi-Fi reception and fragile; the screen snapped off three different times. The G4 was a good boost for media apps. First widescreen display (1152 × 768). So loud under load I'd have to hold a pillow over my head in the hotel for overnight trade show encodes.
- PowerBook G4 800 MHz: Not that much faster in CPU, and a few more pixels (1280 × 854), but a much better graphics card, and most importantly, much quieter.
- PowerMac G5 DP: The only Mac workstation I bought between Terran and Microsoft. Apple's performance lagged a lot in the years the G5 launch and the Intel switch. The 23" monitor I bought with it remains my secondary monitor.
- PowerBook G4 17" 1.333 GHz, 1 GB RAM. With the aluminum body popular for many years. It took some of the screenshots in this book. The 1440 × 900 display seemed huge at the time.
- iMac G4 1.333 GHz.With the 20" display on an arm. Remains the kitchen computer to the day. I can tell when a kid has left Safari open on a page with a Flash banner ad, as the fan goes crazy until I close it.
- MacBook Pro 17" 2 × 2.6 GHz: Microsoft buys me a Mac. The performance boost from the G4 1.33 Ghz to this was probably as big as from the first G3 PowerBook to the last G4. Has spent most of its life running Vista in Boot CampH. ■

Introduction to QuickTime

Apple's QuickTime is the granddaddy of all the media architectures. The first public release of QuickTime 0.9 was back in 1991, when a top-of-the-line computer ran at 25 MHz.QuickTime's impact and legacy are hard to overstate. It laid the foundation for the entire desktop video authoring industry, as well as the practical playback of video on desktop computers. And it's done this with a remarkable degree of backward compatibility. All the files on the QuickTime v0.9 developer's CD-ROM than ran on Mac OS 7 still play back today.

The legacy of QuickTime is increasingly found in MPEG-4 these days; the MPEG-4 file format is closely based on QuickTime. Apple itself has made MPEG-4 its primary file type, particularly on its devices like the iPhone and iPod. Part of this presumably was to offer a clean break from the past and not have to support the tremendous variety of codecs and features a .mov could contain.

Thus, while we'll talk some about QuickTime as a delivery format, this chapter will mainly focus on it as a content creation and playback architecture, and as the primary media architecture on the Mac. Most content created for the Mac can and should be .mp4; there's little reason to use .mov for content delivery anymore.

The QuickTime Format

The QuickTime file format was the basis of the MPEG-4 file format, and so will seem extremely familiar to anyone who's used MPEG-4 much. The biggest difference is nomenclature; what's called a "box" in MPEG-4 is called an "atom" in QuickTime. But they're both the same concept as a unit of the file that can itself contain other units.

Beyond that, we have the same audio and video tracks, and even hint tracks for streaming, all derived from QuickTime for MPEG-4. However, most MPEG-4 players support only a subset of the theoretical features of the QuickTime or MPEG-4 formats. Since most of those features were defined by their implementation in QuickTime, it's unsurprising that QuickTime provides the fullest implementation, with the notable exception of fragmented MPEG-4, which QuickTime doesn't yet support.

One very useful format feature is reference movie. In a reference movie, all or parts of the media tracks can just be references to actual media tracks in other files. This way you can have multiple versions of the same content, all referencing the same source files, without having to duplicate that content. For example, you can trim the head and tail off a file, and instead of having to export a whole new file, you can save a reference movie that simply points to the original, unmodified file in a few KB. This is what Final Cut Pro exports when "Self-Contained" is not checked.

QuickTime Tracks

The basis of many of QuickTime's unique features is the track. A track is a piece of media with a start time and end time. If it's a visual track, it appears on the screen. Those visual tracks can be smaller than the movie, and can overlap each other including with transparency. Most track attributes can be changed by tween (for "in-between") tracks, which interpolate values like size and shape over time.

Video

A video track is anything implemented as a "native" QuickTime codec. This includes everything you can select in an "Export to QuickTime Movie."

Audio

Audio tracks are the same as video (Figure 28.1).

Hint

A hint track contains instructions for the server on how to break up the video and audio bitstreams into packets (Figure 28.2). Their use is described later in this chapter. Note that they're pretty big, and ignored when not streaming, so you shouldn't have them for any file that will be delivered via progressive download.

You can hint any track type. However, optimal results require a native packetizer for the particular codec, which allows the codec to guess at missing content; native packetizers are supported for all modern delivery codecs in QuickTime.

MPEG-1

When QuickTime imports simple file formats, like AVI, the content shows up as standard QuickTime audio and video tracks. MPEG-1 is a different case, as QuickTime didn't natively support B-frames until QT7. So, instead of teaching QuickTime how to think like MPEG-1, MPEG-1 support was implemented as an file handler (sometimes called an import component). While QT7 added B-frame support, older codecs and components have largely not been updated to support it.

Name	Start Time	Duration	Data Size	Data Rate	Format
levi's in-out CIF 500-128.mov	00:00:00.00	00:01:02.52	4.22 MB	567.30 kbits/sec	-NA-
☑ Sound Track	00:00:00.00	00:01:02.52	969.31 KB	126.98 kbits/sec	AAC
☑ Video Track	00:00:00.00	00:01:02.52	3.28 MB	440.32 kbits/sec	H.264

A

Name	Start Time	Duration	Data Size	Data Rate	Format
levi's in-out CIF 500-128.mp4	00:00:00.00	00:01:02.52	4.22 MB	567.30 kbits/sec	-NA-
☑ Sound Track	00:00:00.00	00:01:02.52	969.31 KB	126.98 kbits/sec	AAC
☑ Video Track	00:00:00.00	00:01:02.52	3.28 MB	440.32 kbits/sec	H.264

B

Figure 28.1 (A) and (B) The video and audio track structures are identical for .mov (28.1A) and .mp4 (28.1B).

Name	Start Time	Duration	Data Size	Data Rate	Format
levi's in-out CIF 500-128_streaming.mp4	00:00:00.00	00:01:02.52	4.56 MB	611.93 kbits/sec	-NA-
☑ Sound Track	00:00:00.00	00:01:02.52	969.31 KB	126.98 kbits/sec	AAC
☑ Video Track	00:00:00.00	00:01:02.52	3.28 MB	440.32 kbits/sec	H.264
☐ Hinted Video Track	00:00:00.00	00:01:02.52	273.20 KB	35.78 kbits/sec	Hint
☐ Hinted Sound Track	00:00:00.00	00:01:02.52	67.57 KB	8.85 kbits/sec	Hint

Figure 28.2 Track structure of a hinted movie.

Name	Start Time	Duration	Data Size	Data Rate	Format
levi's in-out_MPEG1.mpg	00:00:00.00	00:02:04.91	4.35 MB	291.89 kbits/sec	-NA-
☑ MPEG1 Muxed Track	00:00:00.00	00:02:04.91	4.35 MB	291.89 kbits/sec	MPEG1 Muxed

Figure 28.3 MPEG − 1 is its own track type, with video and audio muxed together.

It wasn't until QT7.6 that you could even export the audio from an MPEG-1 movie. QuickTime doesn't have built-in MPEG-2 support (even though Macs have long included DVD playback). You can purchase a reasonably capable MPEG-2 playback component from Apple for $19.95. Note that it's Main Profile only, and so can't decode 4:2:2. Support for 4:2:2 MPEG-2 production formats are installed with Final Cut Pro. See Figure 28.3.

Text

A text track is simply a series of lines of text with a font, color, and location. Each line of text has a start and a stop time. Because many codecs don't compress text legibly, text tracks provide a very high-quality, very low-bandwidth way to provide perfect text on the screen. Providing subtitles as text instead of including them as part of the video can provide a much better experience. And, of course, you can have different text tracks in different languages, without having to re-encode the video for each audience.

To author a text track, you can use a tool that supports making them, or write text files in a special format that QuickTime knows how to import. QuickTime Player doesn't include authoring or editing text tracks, but can play them back with fonts, styling, and specified positioning. Here's an example:

```
{QTtext}{font:Geneva}{plain}{size:12}
{textColor: 65535, 65535, 65535}{backColor: 0, 0, 0}
{justify:center}{timeScale:600}{width:160}{height:48}
{timeStamps:absolute}{language:0}{textEncoding:0}
[00:00:00.000]
Here is my first line of text
[00:00:03.000]
Make sure to include the timestamp before the text you want to use
[00:00:04.000]
Typing this into any old text editor isn't that bad, but I'd rather use Magpie
Figure: Magpie
```

Chapters in QuickTime are implemented as a special kind of text track. Most podcasting tools can set these correctly. See Figure 28.4.

Figure 28.4 You can click on the QuickTime chapter indicator and select a chapter to jump to.

Music

Music tracks use QuickTime as a MIDI synthesizer, telling it when and how to play a series of notes. QuickTime is pretty decent as a software synth, and typically requires less than 1 Kbps for a full score of music. You can import existing MIDI files into QuickTime to make a music track. QuickTime has pretty full support for modern MIDI, including being able to import sound fonts.

Music tracks haven't seen significant use in recent years, but they were all too often embedded as a hidden autoplay movie on the early web, without any good way to turn them off.

QuickTime VR

QuickTime took part in the virtual reality craze of the mid-1990s. QuickTime VR (QTVR) debuted in QuickTime v2.5 as a component of the QuickTime Media Layer (along with the less-fortunate QuickDraw 3D). A QuickTime VR track enabled the viewer to navigate within and between panoramas.

QuickTime 5 introduced the long-awaited cubic panorama. Where the original cylindrical panoramas couldn't allow the user to look too far up or down (no top to the cylinder), cubic panoramas model the scene by mapping the stitched images onto the six sides of a cube.

This cubic approach allows panoramas in which the viewer can actually look straight up and straight down.

There are a variety of tools from different vendors that allow you to "stitch" together a series of photographs into a panorama—a topic beyond the scope of this book.

Sprites

Sprite tracks had an inauspicious beginning in QuickTime v2.5. They were simply a means to make icon graphics move around the screen. With QuickTime Media Layer came Wired Sprites, which were programmable. They formed the basis for interactivity in QuickTime.

Unfortunately, Apple pulled back from interactivity in QuickTime around 2000, and no one has released any updates to tools for it for some years (Totally Hip's LiveStage Pro led this market). It seems that each new release of QuickTime introduces features that limit the utility of what used to work, particularly with scripting and interactivity as opposed to simple animation.

Flash

Believe it or not, was possible to include a Flash .swf as a QuickTime track. QuickTime 4 added support for Flash 3, with Flash 4 in QuickTime 5 and Flash 5 in QuickTime 6. However, by the time QuickTime 7 rolled around, Flash was a media playback platform in its own right, competing with QuickTime. The Flash component hasn't been updated, and was turned off by default for security reasons as of QuickTime 7.3.1.

Skins

A Skin track replaces the standard UI of QuickTime Player with a new, custom one. This is useful for creating branded experiences in the player. They haven't seen significant use in ages Apple hasn't even updated their web samples in more than seven years.

Delivering Files in QuickTime

QuickTime not only predates the web, it actually predates the CD-ROM as a standard feature in personal computers! However, the excellent work done on QuickTime's original design has enabled it to evolve into the world of the web with relative ease.

QuickTime for CD-ROM

QuickTime was the dominant file format of the golden age of the CD-ROM (with AVI a close second). Macs dominated multimedia production back then, and Macromedia Director, the leading CD-ROM authoring tool, always had better native support for QuickTime than any other format. Back in the day, whenever a client allowed we'd always include the QuickTime for Windows installer on our cross-platform CD-ROM titles, since QT was more reliable than VfW. If we couldn't do that, we'd use AVI files that QuickTime on Mac could play, like Cinepak + IMA.

CD-ROM doesn't get much use anymore. MPEG-1 is probably the best cross-platform media format available on every Mac and Windows PC. Director 11.5 added MPEG-4 H.264 decodes as a built-in feature, so no OS decoders are needed anymore.

QuickTime for Progressive Download

QuickTime pioneered the progressive download model for web video, and it was used for most QuickTime web video even after v4 finally introduced RTSP. The lack of a bitrate switching MBR kept RTSP from being appropriate for long-form content; progressive almost always offered a better experience.

Every QuickTime movie has a movie header, an index to the locations of media in the file which the player requires to start playback. Because that structure can't be known until the file is completely encoded, the movie header used to be appended to the end of the file. As CD-ROMs offer random access, this wasn't a problem before the web. However, when doing progressive download via FTP or HTTP, the file is read front to back (this is pre-byterange requests). Because none of the file can be shown until the movie header is available, progressive download wouldn't work. So, in QuickTime v2.1, Apple introduced the Fast Start movie, which is a fancy name for a movie with the header at the start of the file.

Fast Start is the default selection on "Export to QuickTime." However, if you do any editing in QuickTime Player, even simple things like changing annotations, on Save the header gets moved to the back of the file again, and Fast Start no longer works. When this happens, you need to re-flatten the movie, either by doing a Save As, or by doing a batch flatten with a Apple-supplied AppleScript or other tool.

QuickTime 3 introduced the compressed movie header. Because there was a lot of redundancy in the header, Apple used traditional lossless compression to make it perhaps 80 percent smaller. While this didn't reduce the size of the total file more than a few percent, it significantly shortened startup delay.

MPEG-4 in QuickTime

QuickTime v6 added integration of MPEG-4 within QuickTime. The MPEG-4 file format is based on QuickTime, which enabled Apple to do more than just add a MPEG-4 file handler. Instead, QuickTime treats an MPEG-4 file as a QuickTime movie. Thus, you can open an MPEG-4 file in QuickTime, and it will show up as a movie with MPEG-4 codecs in the media tracks. MPEG-4 codecs also show up as options inside the standard QuickTime export dialogs.

However, there are some differences between the QuickTime and MPEG-4 file formats, so a QuickTime file with the right codecs still isn't a quite a legal MPEG-4 file. This requires a remux, QuickTime Player Pro can do this by selecting MPEG-4 in the Export dialog and setting video and audio to pass-through.

QuickTime before v7 didn't natively support B-frames, which are part of Advanced Simple, thus QuickTime only supports Part 2 Simple Profile, not the more common ASP (regularly used in DivX/Xvid content).

The sprite and other interactive movie features of QuickTime are radically different than the BIFS system of MPEG-4. Native MPEG-4 interactivity never caught on anyway, so this hasn't been a signicant issue of interoperability. The main places people still use .mov instead of .mp4 is to access those features.

QuickTime and Darwin Streaming Server support native MPEG-4 streaming as does QuickTime as a client.

Lastly, Apple's QuickTime Broadcaster is a live broadcasting application that supports MPEG-4 as well as QuickTime streams.

QuickTime for RTSP

QuickTime 4 introduced RTSP and the hint track, which gives the server the information it needs to stream the file. QuickTime can hint anything that lives in a movie. However, codecs designed to be used in RTSP have native packetizers, which gives the server much better information for how to stream the file, and makes them more robust on lossy networks (see Table 28.1).

Table 28.1

Track type	Native packetizers
Audio	Sorenson Video v2, Sorenson Video v3, H.261, H.263, MPEG-4, H.264
Video	QDesign Music v2, Qualcomm PureVoice, AAC, CELP, AMR
Other	MPEG-1 files

You'll almost always be using H.264 with AAC for desktop playback. The other options are mainly useful when targeting phones.

Apple provides the branded QuickTime Streaming Server for Mac OS X Server only. However, you can also download binary and source-code versions of Darwin Streaming Server for many operating systems. Darwin is the same server without the Mac GUI.

QuickTime for Live Broadcasting

QuickTime has facilities for live broadcasting similar to those of other formats.

Apple provides QuickTime Broadcaster for free—a live compression tool for Mac OS X. It can broadcast via both MPEG-4 and QuickTime. It is simple and functional, although without much depth.

Since QuickTime can consume standard MPEG-4 RTSP, most live encoding targeting QuickTime just uses live MPEG-4 products, which can offer much better quality and workflow than QT Broadcaster.

HTTP Live Streaming

QuickTime X and the iPhone recently added Apple's adaptive streaming. It's already implemented in the iPhone 3.0 software and QuickTime X, but Apple's said nothing about Windows or AppleTV support.

Apple has made an informational submission to the Internet Engineering Task Force (IETF) documenting it in more detail. Contrary to some reports, that submission is not a proposal for standardization, but an informative document describing the implantation for interoperability—which isn't to say that Apple might not go through a standards process down the road.

Apple's documentation calls it "HTTP Live Streaming," so I'll call it AHLS for short. The name is a bit misleading, as it clearly supports on-demand playback as well.

The basic architecture of AHLS is like other adaptive streaming technologies, with video delivered in a series of small files. It uses MPEG-2 as the file format, which like fMP4 can be authored with byte ranges easily sliced into independent fragments. M2TS is also widely supported in existing live broadcast encoders, which could make for easy interoperability.

Apple's documentation assumes that a separate stream segmenter will be responsible for converting the live transport or UDP stream in to individual .ts chunk files. The index (what's called a manifest in Smooth Streaming) is ".m3u8" a modification of the standard .m3u MP3 playlist format. For live video, the index needs to get updated every chunk, which seems like a lot of overhead. Here's a VOD sample they give:

```
#EXTM3U
#EXT-X-TARGETDURATION:10
#EXTINF:10,
http://media.example.com/segment1.ts
#EXTINF:10,
http://media.example.com/segment2.ts
#EXTINF:10,
http://media.example.com/segment3.ts
#EXT-X-ENDLIST
```

A live stream won't have the ENDLIST, and the last two entries are placeholders that can't be sought into until the index is updated and the second-to-last chunk becomes third-to-last and thus seekable.

Only the range in the current index is seekable, pause and random access is limited by the number of items in the list. As chunks get shorter, the index gets bigger as it contains more entries to cover the same time period. Apple recommends 10 seconds as a good chunk duration, quite a bit longer than other adaptive streaming solutions. Their examples all give fixed cadence, but there no apparent reason why chunk duration couldn't vary.

10-second chunks with suggested settings yields about a 30-second broadcast delay, quite a bit higher than other streaming technologies, adaptive and otherwise.

Multibitrate encoding is supported, of course. Apple says the current implementation has been tested for 100–1600 Kbps, appropriate to the iPhone. But many desktop users would be able to get much higher than 1600 Kbps. Apple hasn't cracked the MBR audio problem, either, specifying that the same audio bitrate be used for all bitrate alternates.

Here's a sample multibitrate playlist:

```
#EXTM3U
  #EXT-X-STREAM-INF:PROGRAM-ID = 1,BANDWIDTH = 1280000
  http://example.com/low.m3u8
  #EXT-X-STREAM-INF:PROGRAM-ID = 1,BANDWIDTH = 2560000
  http://example.com/mid.m3u8
  #EXT-X-STREAM-INF:PROGRAM-ID = 1,BANDWIDTH = 7680000
  http://example.com/hi.m3u8
```

The current supported codecs are as follows:

- Video: H.264 Baseline Level 3.0
- Audio:
  - HE-AAC or AAC-LC up to 48 kHz, stereo audio
  - MP3 (MPEG-1 Audio Layer 3) up to 48 kHz, stereo audio

Suggested settings for a three-stream encode are as follows:

- H.264 Baseline 3.0 video
- HE-AAC (version 1) stereo audio at 44.1 kHz
- Low—96 Kbps video, 64 Kbps audio
- Medium—256 Kbps video, 64 Kbps audio
- High—800 Kbps video, 64 Kbps audio

Those make sense for the iPhone, basically mapping to 3 G, and Wi-Fi.

And a few other things I've noted:

- Audio can be delivered as a sequence of segmented .mp3 files.
- The live encode can easily be archived and converted to VOD.
- Encryption is supported, with references to the encryption key files in the index file. However, it's not clear how they could prevent a network sniffer from grabbing the key on the local machine.
- They have proxy caching in mind, explicitly warning against using https for the chunks as it isn't cachable.
- Inlet (Spinnaker 7000) and Envivio (4Caster C4) have both announced support for AHLS.

The Standard QuickTime Compression Dialog

The most basic way to encode video with QuickTime is through the standard Export dialog, supported in QuickTime Player Pro and many other applications. While this isn't a professional encoder, there is a lot of stuff it can be used for, including exporting a file for later processing in another application. Other applications like Compressor, Squeeze, and Carbon access the same parameters through the QuickTime API.

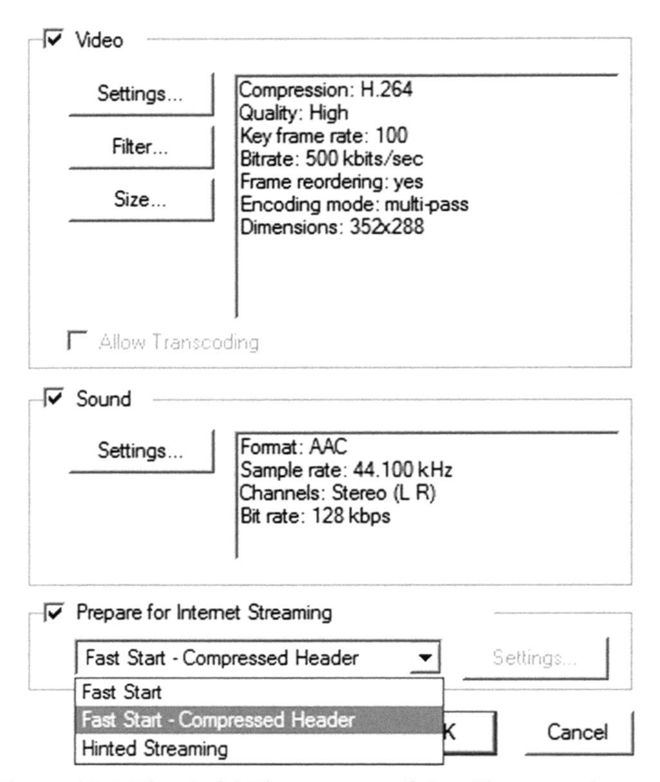

Figure 28.5 The QuickTime export dialog file type selector.

The QuickTime dialog lets you specify your audio and video settings, size, and Internet streaming options. See Figure 28.5. For progressive and other non-streaming video, you'll want Fast Start – Compressed Header. This makes the file slightly smaller, and a lot faster in startup time. For RTSP streaming, you'll choose hint tracks. This automatically generates the hint tracks along with the audio and video files.

In the video option, you pick your codec and other settings. The Quality slider maps to QP, and doesn't do anything in most codecs if data rate has been set. If you don't specify a data rate, or are using a codec like the JPEG that doesn't have a data rate control, the quality slider controls the quality of the video, and hence the file size.

There's also a secret Temporal Quality slider in some interframe codecs, including H.264. You reach that by holding down Control and Alt (Mac) or Option (Win) in the dialog. Temporal Quality nominally controls the quality of non-keyframes. However, I don't have any idea how well-tuned it is anymore; I may be the last person who remembers it. Setting it doesn't appear to do anything for H.264 in current versions.

Figure 28.6 QuickTime's H.264 settings.

Frames Per Second should generally be Current, so the frame rate of your source is used.

Keyframe Every X refers to the minimum number of keyframes. A natural keyframe, which is normally inserted whenever there is a video cut, resets this counter, so the Keyframe Every value might not matter for video with rapid cuts.

The options button opens up the special features of the codec, if any. Recent codecs like H.264 offer everything in a single dialog (Figure 28.6).

The Sound settings are a lot simpler. They let you choose your codec, sample rate, and channels. The Quality control determines the quality (and speed) of the resampling filter used when changing sample rates, channels and bit depths. See Figure 28.7.

Detailed descriptions of the H.264 and AAC dialogs are found in Chapters 13 and 14.

QuickTime Alternate Movies

QuickTime has a Multiple Bit Rate system, called Alternates. Unlike MBR in Windows Media and RealMedia, QuickTime doesn't bundle multiple versions of the data in a single file. Instead, authors create a number of different files, only one of which is played for a given user.

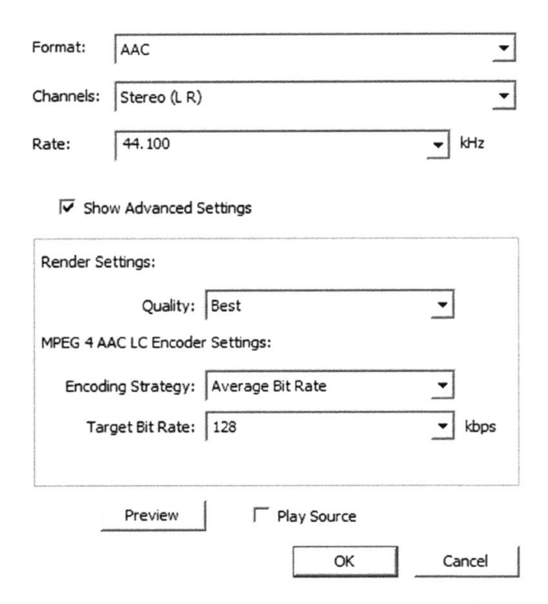

Figure 28.7 QuickTime's AAC–LC settings.

The traditional implementation didn't support any kind of bitrate switching, only an initial selection as the file starts. So it's not well suited to longer content where available bandwidth may vary during playback.

Master Movie

The key to alternates is the Master Movie. A master movie stores a list of the available files, the paths to them, and properties on which to base the decision of which to use.

Alternates Parameters

Each movie linked to the reference movie has a number of properties used to determine which file is played. Apple's free MakeRefMovie utility (Figure 28.8) is the most common way to set these. There's an older version for Windows, but only the Mac one is up to date. Back in QuickTime's heyday, Cleaner was the dominant QuickTime compression tool, and had by far the most complete implementation of alternates.

QuickTime selects the alternate with the highest "quality" value and bitrate that's compatible with the player and at or below the current bitrate:

- The clip is at the target data rate,
- none are at the correct bandwidth, at the next lowest,

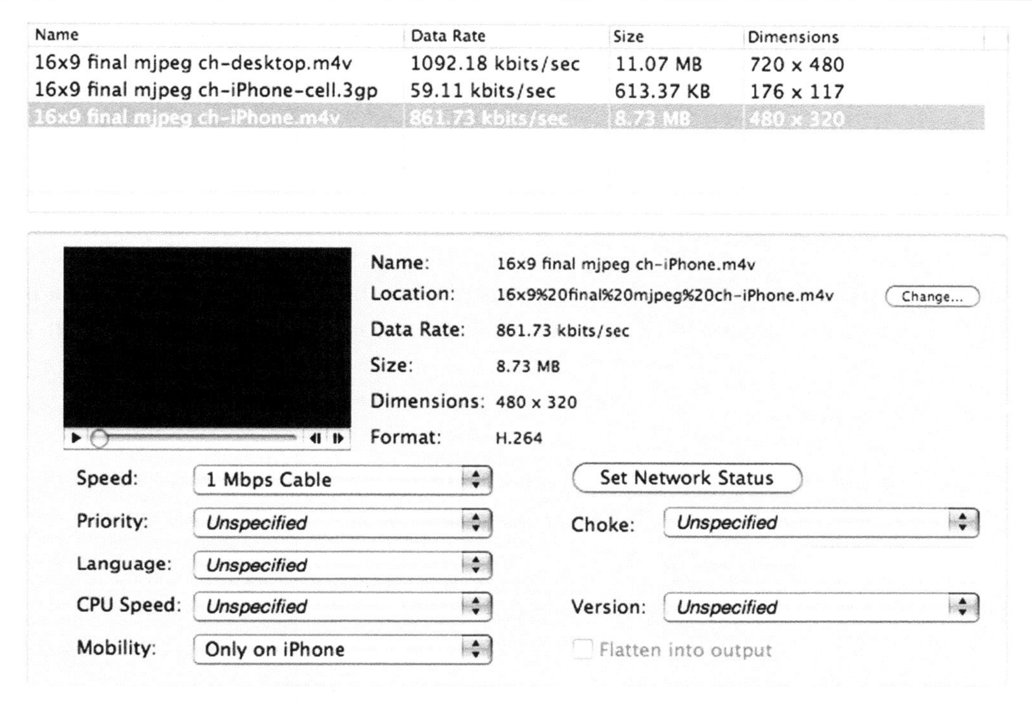

Figure 28.8 MakeRefMovie, with the default output from "Export for web." The GUI was actually better 10 years ago.

- none are lower, those at the lowest available bandwidth, or
- play the file with the highest quality setting matching the specified parameters.

Connection

Connection speed required manual configuration by the user until QuickTime 7, limiting its utility dramatically. Most users never even knew it was there to be set. QuickTime 7 added a quick bandwidth measurement test, so the optimal connection speed can be picked based on current download speeds. It doesn't have any ability to switch bitrates once playback starts, however, reducing its utility, particularly for real-time streaming, as content gets longer.

Language

This allows you to specify the language of the file. It should be left blank unless you're doing multilingual video, but is very handy if you are.

Priority

This specifies the rank of priority for movie playback. If multiple files meet the requirements for playback, the highest priority clip will be played back.

CPU speed

This is somewhat poorly defined. The values seem to roughly correlate to the MHz clock speed of 604e processors, with a "5" around 500 MHz (and thus slower than an iPhone, let alone any Mac of the last decade). Just leave unspecified.

Mobility

This specifies a file as being iPhone-only, otherwise unspecified. Useful to make sure a file of too high a bitrate for an iPhone to decode doesn't get sent to it regardless of bandwidth.

Authoring Alternates

Alternates are a pain to create compared to other formats. Each alternate is a unique encode, so targeting four bitrates means the file needs to be compressed four separate times.

QuickTime Player Pro does have one Export preset that makes a set of alternates. It's very simple, with only three checkbox options and no configurations:

- iPhone: H.264 1 Mbps
- iPhone (Cellular): H.264 80 Kbps
- Computer: H.264 up to 5 Mbps

See Figure 28.9.

I keep expecting Apple to do something with Alternates, but other than the surprise addition of "Export for web" and bandwidth measurement in QT7, this hasn't been changed since the 1990s.

16x9 final mjpeg ch

Name	Date Modified	Size	Kind
16x9 final mjpeg ch-desktop.m4v	Yesterday, 6:58 PM	11.1 MB	MPEG-4 Video File
16x9 final mjpeg ch-iPhone-cell.3gp	Yesterday, 6:48 PM	624 KB	3GPP Movie
16x9 final mjpeg ch-iPhone.m4v	Yesterday, 6:55 PM	8.7 MB	MPEG-4 Video File
16x9 final mjpeg ch-poster.jpg	Yesterday, 6:45 PM	192 KB	JPEG image
16x9 final mjpeg ch.mov	Yesterday, 6:45 PM	4 KB	QuickTime Movie
ReadMe.html	Yesterday, 6:45 PM	8 KB	HTML document

Figure 28.9 The files that go with the MakeRefMovie demo.

QuickTime Delivery Codecs

QuickTime has always had a broad selection of codecs for all kinds of tasks. I won't try to give a complete list, instead focusing on the ones you're most likely to encounter or want to use.

H.264

QuickTime supports decode of progressive scan H.264 in the Baseline, Main, and High 8-bit 4:2:0 profiles. H.264 is the best delivery video codec in QuickTime by a wide margin. Before QuickTime X, playback is software-only, so many systems may struggle to play a H.264 file in Mac OS X that the same machine booted into Windows wouldn't break a sweat doing via DXVA hardware acceleration. QuickTime X added accelerated decode to some recent Mac models, although those were probably fast enough to decode in software anyway.

QuickTime's H.264 encoding is a lot more limited than its decoder, and tuned to optimize decode performance more than compression efficiency. It is covered in full detail in the H.264 chapter.

The upside of the relatively simple encoder is that even relatively modest machines can play back QuickTime H.264 pretty well; bear in mind that Apple was still selling PowerPC G4 laptops when QuickTime 7 shipped. But achieving the same quality can take quite a bit higher bitrate than a well-tuned High Profile encoder. Of course, as QuickTime is High Profile–compatible, content targeting QuickTime can use other, more efficient encoders. Just make sure to test their output on the minimum target platform to make sure performance is adequate. CABAC seems particularly slow in QuickTime; I've seen it double-decode CPU load. CAVLC should be used for higher bitrate content targeting QuickTime.

If QuickTime-specific features are needed, QuickTime can open an H.264 .mp4 file and losslessly Save As to .mov.

Legacy Video Delivery Codecs

MPEG-4

Apple's "MPEG-4" codec is MPEG-4 part 2 Simple Profile. It was introduced with QuickTime 6, before Apple fixed the architectural limitation preventing B-frames in codecs, and thus Apple only supported encoding and decoding Simple Profile. Even though QuickTime 7 fixed this issue (required for H.264), Apple hasn't added fuller part 2 support. Thus QuickTime isn't natively compatible with most DivX/Xvid content. Moreover, even though QuickTime can demux AVI, and decode MP3 and at least MPEG-4 SP, it won't play an AVI file containing MPEG-4 SP and MP3, excluding most Xvid/DivX content it theoretically would be compatible with.

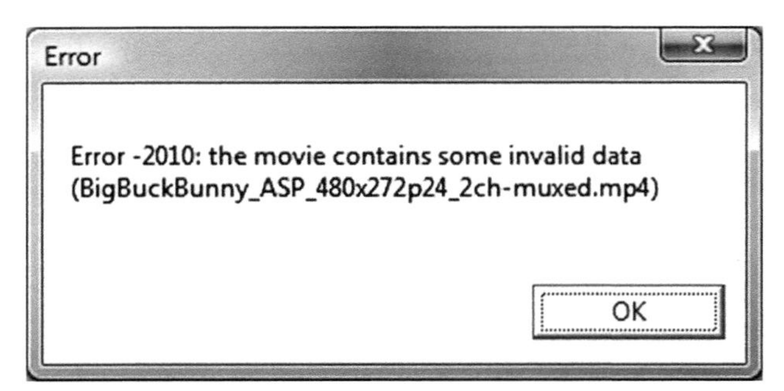

Figure 28.10 This error you get trying to open up .mp4 with MPEG – 4 part 2 Advanced Simple Profile lacks Apple's legendary user–friendliness.

There are third-party plug-ins (Perian is a popular choice) that do decode part 2 ASP. However, users have to manually install those in advance. Otherwise they're greeted with the singularly unhelpful "Error -2010: the movie contains some invalid data."

With QuickTime 6, it sometimes made sense to use MPEG-4 SP for QuickTime-targeted cross-platform content, as older codecs (including the otherwise superior Sorenson Video 3) needed to be encoded with different gamma to look the same on Macs and Windows. But even then, it was preferable to use a higher-quality SP implementation like Squeeze or Compression Master's (now Episode). As an encoder, Apple's is a very limited 1-pass only with a lot lower efficiency than other implementations.

H.263

H.263 is a videoconferencing codec, and the basis of MPEG-4 part 2 (where generic H.263 is the "short header" profile). It's only available by default in the "Movie to 3 G" export mode.

Sorenson Video 3

Long-hyped and long-delayed, Sorenson Video 3 (SV3) was the most-used QuickTime codec in the QuickTime 5 and 6 eras, only dethroned by QuickTime 7 and H.264. It was the last widely used proprietary codec in QuickTime.

QuickTime built in a basic implementation of the encoder, but professional content used the higher quality, faster, and vastly more flexible Pro version. Among other things, it could do 2-pass VBR via Squeeze or Cleaner.

Many pages of the first edition of this book lovingly covered the many fine features of SV3 Pro and its alpha channels, clever hack to enable B-frames, streaming robustness features, and deblocking filter.

But I can't imagine why anyone would make new content with it. If you have a penchant to do so, it's still included with Sorenson Squeeze.

Sorenson Video 1/2

The original Sorenson codec, used the same bitstream in both Sorenson Video 1 (SV1) and Sorenson Video 2 (SV2), was the leading codec for QuickTime 3 and 4.

While decent for a Clinton administration–era codec, it seems laughably primitive today. It's notable as the last codec using the YUV-9 color space, where there is only one color sample per each 4 × 4 pixel block. This meant edges in colorful content looked hideously blocky.

Video

The Apple Video codec was in the original QuickTime 0.9 back in 1991. I include it here as a warning; picking "Video" to encode Video was an all-too-common mistake.

Video's code name was Road Pizza (4CC is "rpza"). I imagine this was because it left the video a flattened, crushed mess vaguely reminiscent of the original.

It was designed for 25 MHz computers, and thus is a weird duck compared to modern codecs. It was quality-limited, not data rate–limited, and encoded in 5-bit-per-channel RGB (15-bit color). It had an extremely fast encoder and decoder, so it was used for doing comps and that sort of thing. But even DV and JPEG were fast enough a decade ago for real-time comps on a modern computer.

QuickTime Authoring Codecs

Unlike the other major web formats, QuickTime is as much a content creation platform as a delivery platform, and includes a lot of codecs designed for capturing and editing.

An authoring codec is one whose features make it useful for the acquisition, editing, and storage of full-quality content. An authoring codec's output can be recompressed with minimal loss. Some codecs, like MPEG-2 and Apple Animation, can be used for both authoring and delivery when different settings are applied. A number of authoring codecs are installed with Final Cut Pro, and so indicated below. Sometimes Mac editors forget which codecs came with FCP, and send files using those to users without FCP and hence without the ability to decode them.

ProRes

ProRes is Apple's big new authoring codec, introduced with Final Cut Studio 2 and upgraded with FCS3. It's fast (at least on Mac), flexible, and high quality. In order of compression ratio, these are the ProRes modes:

- ProRes 4444: Up to 4:4:4 12-bit RGB or Y′CbCr, with alpha. Useful for capturing and storing animation, film, and dual-link SDI content, but unnecessary overkill for 4:2:2 sources. Introduced with FCS 3.

- ProRes 422 (HQ): The top quality of 4:2:2 mode. Fine to use for capture, multigenerational editing, archiving, and intermediates.
- ProRes 422: Still visually lossless, but less mathematically accurate than HQ. Fine for an intermediate, but I'd prefer HQ as source for 10-bit image processing.
- ProRes 422 (LT): A lighter-weight version meant for transcoding from 8-bit interframe sources like AVCHD. Shouldn't be used in 10-bit workflows.
- ProRes 422 (Proxy): Not mastering-quality, but allows CPU and storage efficient editing in full resolution for a later conform with the higher-quality sources.

Apple has a downloadable ProRes decoder for Windows, but no encoder, and so far lacks support for the newer 4444 mode.

DV/DVCPRO

QuickTime supports DV25 as a QuickTime codec, and can also open a raw DV files and AVI files containing DV as well.

By default, DV is shown in QuickTime in a preview-quality mode, only displaying a single field. If you want to see the file as it actually is, you'll need to set the High Quality flag for the video track, as described in Chapter 6. Most compression tools will do that automatically for QuickTime sources.

DV25 is a fine acquisition format, but as we discussed in Chapter 5, its use of 4:1:1 makes it a poor choice as a production format, so you should never compress to it.

DVCPRO50 (via Final Cut)

When Final Cut Studio is installed, DV50 is available as a codec. It uses twice the bitrate of DV25 and 4:2:2, so it's a fine intermediate codec

DVCPROHD (via Final Cut)

The DVCPROHD codecs (also called DV100) are HD variants of DV, and popular with Mac users due to their long heritage of excellent Final Cut support.

HDV (via Final Cut)

Also installed with Final Cut, QuickTime can decode to the standard HDV formats.

MPEG IMX (Final Cut)

IMX is an older Sony MPEG-2 I-frame-based format targeting tape.

XDCAM EX (Final Cut)

XDCAM EX is Sony's current VBR long GOP capture format meant for flash memory.

Motion-JPEG

Motion-JPEG is the granddaddy authoring codec in QuickTime. Essentially, it takes JPEG, makes it 4:2:2 instead of 4:2:0, and offers both interlaced and progressive modes.

The M-JPEG A and B codecs were invented by Apple to provide interoperability between the many varieties of digital video capture hardware of the mid-1990s. Two main chipsets were used in all these systems, and Apple got the vendors to provide the information necessary to make universal file formats. The A and B flavors of M-JPEG are based around the chipsets used by different cards, but can be losslessly converted between.

Because JPEG is a long-time free standard, many non-Apple decoders can handle M-JPEG in .mov. The video mode of still cameras is often progressive M-JPEG.

Animation

Using Run-Length Encoding (RLE), Animation's compression depends on long horizontal lines of identical pixels, and as many as possible identical lines between frames. Animation is lossless at the default 100 Quality; reducing the Quality slider for higher compression largely works by flattening out these lines, which can be pretty devastating to quality.

Animation does provide interframe compression, which some QuickTime transcoding products don't work well with.

I find PNG to be a superior codec to store RGB content in, even though it's I-frame only.

PNG

The PNG (pronounced "ping") codec is just the Portable Network Graphics lossless RGB format as a QuickTime codec, which each frame a PNG file. PNG is always lossless, and offers much better compression efficiency than Animation for stuff not amenable to RLE.

PNG is ideal for intermediate files of RGB content like screen shots (almost all the graphics in this book were PNG files at one point in their life), and for interoperation with RGB-based applications like After Effects. It has alpha channel ("Millions+") and grayscale (8-bit luma only) modes and a variety of less useful indexed color options.

In the Options dialog, you have different choices for compression efficiency versus speed. Although they're not so labeled, they're roughly in order of quality versus speed. I normally just park it on Best, which selects the most effect option for each pixel. It encodes quite a bit slower, but saves quite a bit of disc space.

None

The None codec is just uncompressed RGB, and thus very fast but very wasteful in bits.

QuickTime Audio Codecs

QuickTime, and the Mac OS before it, has had audio codecs since the late 1980s, well before other PCs even had sound by default.

AAC

QuickTime 6 introduced AAC-LC support as a standard audio codec. AAC-LC is easily the strongest audio codec in QuickTime 6 and 7. Its myriad rate control options are documented in Chapter 13.

QuickTime X adds encode and decode of HE AAC v1 and Low Delay modes.

AMR Narrowband

AMR is Apple's implementation of the Adaptive Multi-Bitrate codec widely used in 3GPP. It's the best low-bitrate speech codec in QuickTime, targeting 4.75 to 12.2 Kbps in 8 KHz mono. It includes a VBR nice silence detection mode where it turns the bitrate down further when things are quiet.

Still, 8 KHz is low even for speech-only content; it's not good for much outside of audioconferencing unless you need to get a *lot* of audio in a very small file.

Apple Lossless

Apple Lossless is, yep, a lossless audio codec. It's fine for storing content, but won't offer any predictable data rate. AAC-LC at 320 Kbps is going to sound just as good for any practical purpose, and most people can't hear a difference between lossless and 192 Kbps AAC-LC.

iLBC

iLBC is the most recent QuickTime codec, an implementation of the royalty-free "Internet Low Bitrate Codec." It's used in other products like Google Talk and Yahoo! Messenger. It offers 8 KHz mono at 15.2 and 13.3 Kbps.

Legacy Audio Codecs

Like video, QuickTime's long heritage includes a huge range of audio codecs, most of which shouldn't be used anymore.

QDesign Music 2

QDesign Music 2 was an enhancement to the original QDesign Music codec that added RTSP support and improved quality. Originally introduced in 1999, it was the best streaming-compatible codec before QT6 introduced AAC. Quality was unprecedented for music at low bitrates, but it never did voice well (including singing), and had a pretty low quality ceiling irrespective of bitrate; it was much better than MP3 at 32 Kbps, and much worse at 128 Kbps.

There is a basic version of the QDesign encoder bundled with QuickTime, but it only goes up to 48 Kbps with awful quality. QDesign itself went out of business years ago, although they released a Mac OS X–compatible version of Pro on their way out.

QDesign Music

QDesign Music was the original implementation of QDesign, and offered lower quality yet and a much more complex Pro encoder version.

Qualcomm PureVoice

Qualcomm PureVoice is a QuickTime implementation of an early cell phone codec. As would be expected, it produces audio that sounds like a cell phone—speech is intelligible, anything else sounds lousy.

PureVoice is always 8 KHz mono and offers 13 and 7 Kbps rates.

MP3

QuickTime supports MP3 as an audio codec for playback, although hasn't ever included an MP3 compressor directly.

MP3 soundtracks were popular for progressive download before QuickTime 6 and AAC, but AAC-LC outperforms it handily.

IMA

The IMA audio codec (created by the long-defunct Interactive Multimedia Association) was the dominant CD-ROM codec of the Cinepak era. It offered a straight 4:1 compression over uncompressed, so where a 44.1 kHz mono track would be 705.6 Kbps, an IMA version of that would be 176.4 Kbps. Audio quality was quite good at 44.1 kHz, but quickly degraded at lower data rates.

None

As with the other formats, QuickTime's None mode is uncompressed PCM. As such, it's mainly used for authoring. Ancient QuickTime files that predate IMA (before QuickTime v2.1) will use the None codec, either in 16-bit or (heaven forbid) 8-bit.

MACE

MACE (Macintosh Audio Compression and Expansion) 3:1 and 6:1 are 1980s-era speech codecs from Mac OS that predate even QuickTime. They're natively 8-bit, and provide strikingly bad quality. Of course, they were optimized for real-time compression and playback on 8 MHz computers. Never use MACE for new content.

μ-Law

μ-Law, (that's the Greek letter, pronounced "mew" or "moo" depending on whom you ask) is an old-school telephony codec, offering 2:1 compression, and hence 16-bit quality over 8-bit connections. It sounds okay for speech, but the poor compression efficiency makes it an unusual choice for any modern applications. It is also called "U-law" (for those who can't easily type the "μ" character) and G.711.

A-Law

A-Law is quite similar to μ-Law, being a 2:1, 16-bit telephony codec. It is less commonly used, and offers no significant advantages.

QuickTime Import/Export Components

QuickTime has broad format extensibility via component structures that allow other formats to be read, played, and exported. All QuickTime apps then have access to installed components for reading and writing (as long as they expose it). There's a wide variety of components out there, but there's a few I wanted to highlight.

Flip4Mac

As discussed in more detail in the Windows Media chapter, Flip4Mac is a component that supports (depending on version) WMV playback, importing, and encoding from QuickTime for Mac.

The free version is playback only. But this fills an ancient need for PowerPoint users: a good video format that can be embedded in PowerPoint and play on Mac and Windows. Until now, MPEG-1 was the best (and weak) choice. But as long as Mac users install Flip4Mac and Mac Office 2008, all is good.

Perian

Perian is a Mac-only component porting much of ffmpeg into QuickTime. It includes support for the wide variety of ffmpeg-supported codecs, formats, and features like the following:

- MKV
- FLV
- MPEG-4 ASP (with B-frames!)
- Huffyuv
- FRAPS
- Ogg Vorbis
- DTS
- Dolby Digital
- Subtitles

XiphQT

XiphQT is a Mac and Windows component supporting playback and export of the various Ogg codecs like Vorbis, FLAC, and Theora, described in Chapter 19. Perian can often play those as well, but can't export. See Figure 28.11.

Flash Encoding

For Flash 6-9, Adobe provided a QuickTime export component for FLV encoding, including basic VP6 support. Since it was installed for free along with Flash (including the trial version…), it was widely used by those who didn't want to buy a full commercial tool or license.

On2 also sells their high-end exporter as a QuickTime component. Both are covered in Chapter 15.

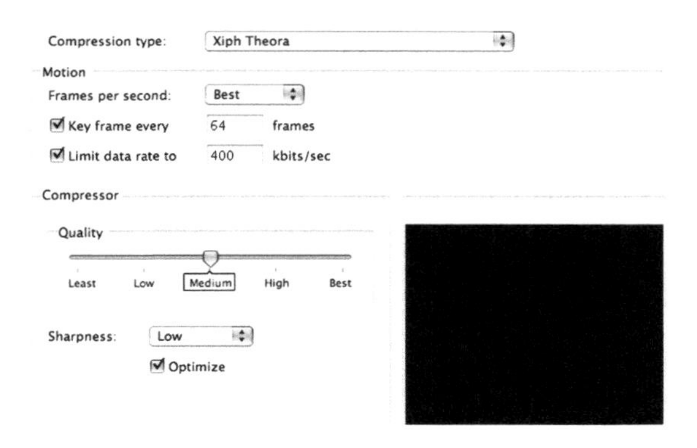

Figure 28.11 XiphQT's Theora output settings.

QuickTime Authoring Tools

QuickTime has had a powerful, open API for a decade now, and unsurprisingly a very healthy third-party industry built up around using QuickTime as both an authoring and delivery technology.

However, Apple's broader embrace of MPEG-4 has left many just using .mp4 for QuickTime and Mac playback instead of supporting .mov and any of its unique features.

These are all covered in much broader detail in Compression Tools.* (visit the companion website to view this bonus content).

QuickTime Player Pro

QuickTime Player Pro is the essential Swiss Army knife of QuickTime authoring, and cheap at $28.99. If you use QuickTime, the time you save with Pro will more than cover its cost.

From a QuickTime perspective, it's at least as valuable for lossless tasks, like trimming and merging files, generating reference movies, importing image sequences, swapping out audio tracks, and remixing to other formats and with/without hint tracks. Note that the QuickTime Player in QuickTime X doesn't provide the same features; QuickTime Pro 7 can be installed side-by-side, and is still needed for most authoring tasks.

Compressor

Apple's Compressor is the most full-featured QuickTime encoder on the market. It sits on top of the same codec as any other QuickTime product, but offers more workflow management and much more advanced preprocessing.

Episode

Telestream's Episode couples its own codecs with both .mp4 and .mov formats. For H.264 and part 2, this offers better quality with the same compatibility. Note that it doesn't enforce any constraints on setting to make them QuickTime-compatible, so it'll allow part 2 to be set to ASP and H.264 to turn on interlacing even when .mov is the output format.

Episode Pro also includes encoding to .mov with IMX and XDCAM payloads for FCP interoperability.

Sorenson Squeeze

Squeeze started out as a QuickTime-only encoder to deliver on the high-end functions of the Sorenson Video 3 Pro codec.

*Visit the companion website at www.focalpress.com/9780240812137 to view this bonus chapter.

It still implements full support for the QuickTime API and its many codecs, and bundled Sorenson Video 3, but they're no longer the focus of the tool.

While it doesn't expose Sorenson's superior MPEG-4 encoder while making .mov, it does allow MP3 as the audio track of .mp4 files.

ProCoder/Carbon

Carbon doesn't have any special .mov support, but has a full implementation of the QuickTime API, including export components. It uses the FLV Export components for its FLV and 3 GP support.

QuickTime Player X

QuickTime X (Time Ten") is scheduled to be released as part of Mac OS X 10.6 enabled by 10.6 to media playback update plans, if any. A core goal of QT X is to bring the media pipleline improvements created by. Plans for Windows, older versions of Mac OS, and PowerPC Macs (10.6 is Intel-only) are unknown.

The new QuickTime Player has some, but not all, Pro features. There are indications that QuickTime Player 7 Pro may still be required for more advanced editing and exporting features.

Public details are scant at this point, but it is known that QT X includes:

- New clean UI for QuickTime Player
- GPU accelerated video decode, finally, on NVidia 8400M GPUs
- A new "HTTP live streaming" technology, presumably Apple's adaptive streaming technology
- Screen recording
- Easy transcoding for devices and the web, including YouTube
- Limited editing functionality like trimming (previously in QuickTime Pro only)
- Accurate color using ColorSync (hopefully fixing the color space transocoding bugs)

Tutorial: Reference Movie from Image Sequence

The MPEG-4 related tutorials all make QuickTime-playable content, so I want to hit QuickTime as a workflow technology here.

Scenario

We're at a high-end 3D animation studio, working on a short film we're hoping will be up for an Oscar in a couple of years.

This being high-end animation, each frame can take the better part of an hour to render, and they are rendered out as individual DPX frames at 2048 × 1024 using 32-bit floating point. Animation is hard, so we want our master to be inarguably perfect.

However, these files are massive and hard to browse. We have a script that tracks the current version of each frame, makes a 1920 × 1080 PNG out of it, and stores them in a directory with the frame number in the file name.

Our marketing and invester relations folks are constantly bugging us about making one off demos file of where we are now to show to VIPs. It's burning a bunch of animator time, so we want to make it easy for marketing's video guy (just some kid intern who got the job because his uncle knows somebody, and with minimal chops) to do quick encodes to whatever he wants, be it DVD, MPEG-4, Blu-ray, Smooth Streaming, whatever. He's got different tools on Mac and Windows, even. We don't care—we just want him to go away so we can finish this movie!

Three Questions

What Is My Content?

An image sequence of PNG files in a directory. 14,316 of them, today.

Plus a .wav file of the current temp soundtrack.

Who Is My Audience?

This kid in marketing who probably thinks DPX is some kind of mountain bike.

What Are My Communication Goals?

We want to make it easy for him to get sources so he goes away. I'm sure he'll take all the credit, though. Figures.

Tech Specs

We're going to teach the kid how a QuickTime reference movie that points to our PNG files and audio.

Settings in QuickTime Player Pro

This is going to take QuickTime Player Pro. We're going to make a reference movie combining the PNG images and .wav file so that it works like a single movie.

(*Continued*)

First, we select "Import Image Sequence" and select the first file in our PNG sequence. And hit okay. This is 24p exactly; we're *film*makers—none of the crazy 23.976 video stuff for us (Figure 28.12).

14K images will take a while this is a fine time to make the kid buy me coffee and listen to some Ed Catmull stories.

And behold! QuickTime shows us our PNG files as a movie, the correct duration and everything.

We now open the .wav file too, so both the PNG movie and WAV are open at once. The next part is needs to be done precisely:

- With the WAV track selected:
    - Select All.
    - Copy All.
- Switch to the movie window.
- Hit the leftmost button in the transport controls to make *sure* we're on the first frame.
- Select "Add to Movie." Not Paste! See Figure 28.13.

And that's it. "Add to Movie" pastes what's in the clipboard as a new track at the playhead. Since the audio started on the first frame, we should now have perfect sync. See Figure 28.14.

We can now save our movie in a few ways:

- Self Contained will copy all the data into the file into the file we write, but not recompress anything. It's fast and big.
- Reference Movie won't even do that. It'll just have links to the video and audio. Since we're using network storage, it'll need to resolve a relative path to the referenced files. The easiest way to ensure that is to stick the reference movie in the same directory of the other files, and it will be openable by any Mac or Windows tool that uses the QuickTime API. See Figure 28.15.

Figure 28.12 If you want your audio to be in sync, get this right!

(*Continued*)

Figure 28.13 Add to Movie. Accept no substitutes—particularly not Paste.

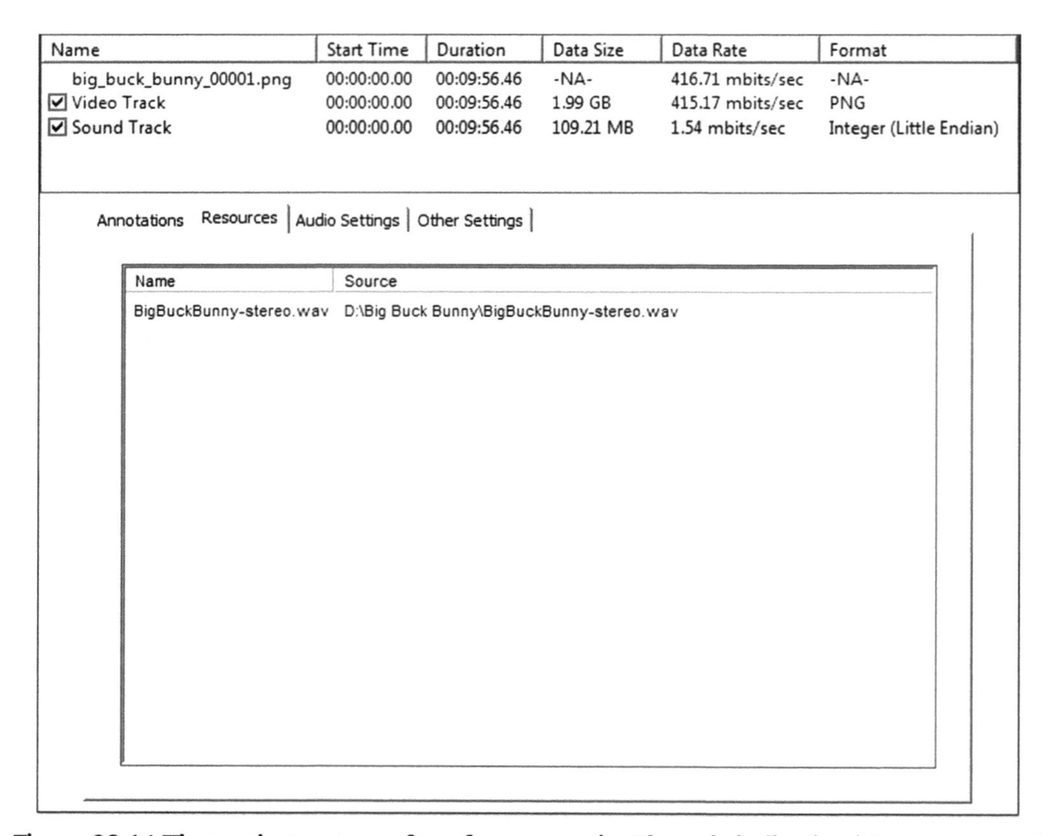

Name	Start Time	Duration	Data Size	Data Rate	Format
big_buck_bunny_00001.png	00:00:00.00	00:09:56.46	-NA-	416.71 mbits/sec	-NA-
☑ Video Track	00:00:00.00	00:09:56.46	1.99 GB	415.17 mbits/sec	PNG
☑ Sound Track	00:00:00.00	00:09:56.46	109.21 MB	1.54 mbits/sec	Integer (Little Endian)

Annotations | Resources | Audio Settings | Other Settings

Name	Source
BigBuckBunny-stereo.wav	D:\Big Buck Bunny\BigBuckBunny-stereo.wav

Figure 28.14 The track structure of a reference movie. The only indication it's not a normal movie is the Resource data indicating where the data really lives.

(Continued)

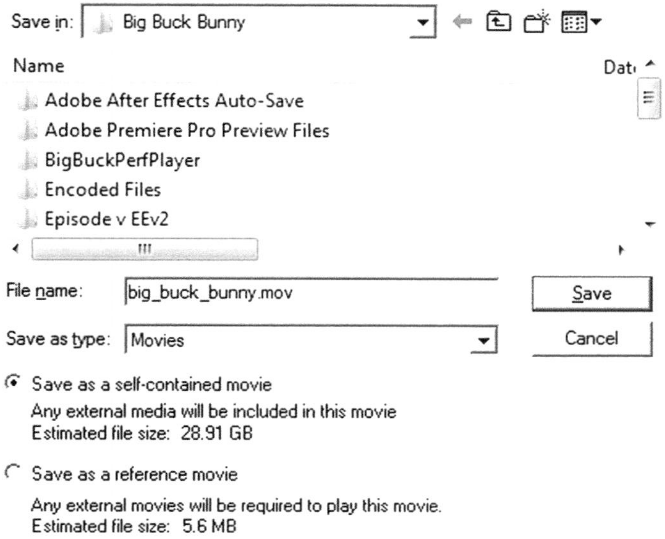

Save in: Big Buck Bunny
Name
Adobe After Effects Auto-Save
Adobe Premiere Pro Preview Files
BigBuckPerfPlayer
Encoded Files
Episode v EEv2
File name: big_buck_bunny.mov
Save as type: Movies
Save
Cancel
Save as a self-contained movie
Any external media will be included in this movie
Estimated file size: 28.91 GB
Save as a reference movie
Any external movies will be required to play this movie.
Estimated file size: 5.6 MB

Figure 28.15 There's a big difference between 5.6 MB and 28.91 GB; reference movies can save a lot of time and space.

- We could also Export as QuickTime Movie to something easiser to play back, like ProRes 422.

And that's it!
And I guess the kid isn't so bad after all.

Index

Q

T

U

V

W

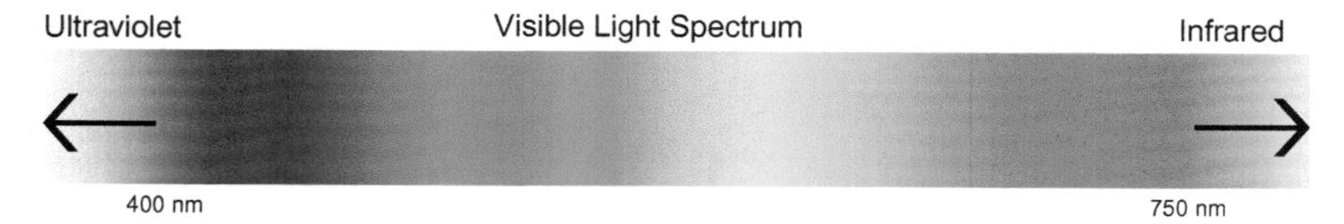

Figure C.1 The visual spectrum, from red at the lowest wavelength to violet at the highest.

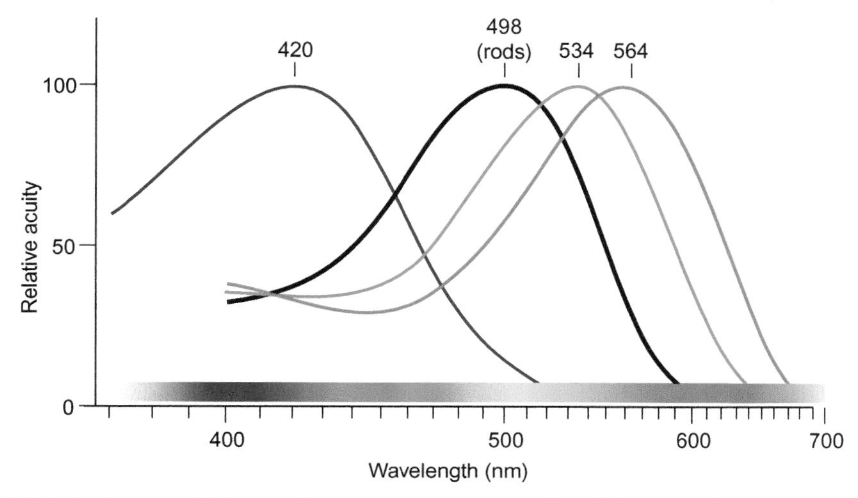

Figure C.2 Relative sensitivity of the eye's receptors to the different primary. The gray line is the rods; the other lines the three cones.

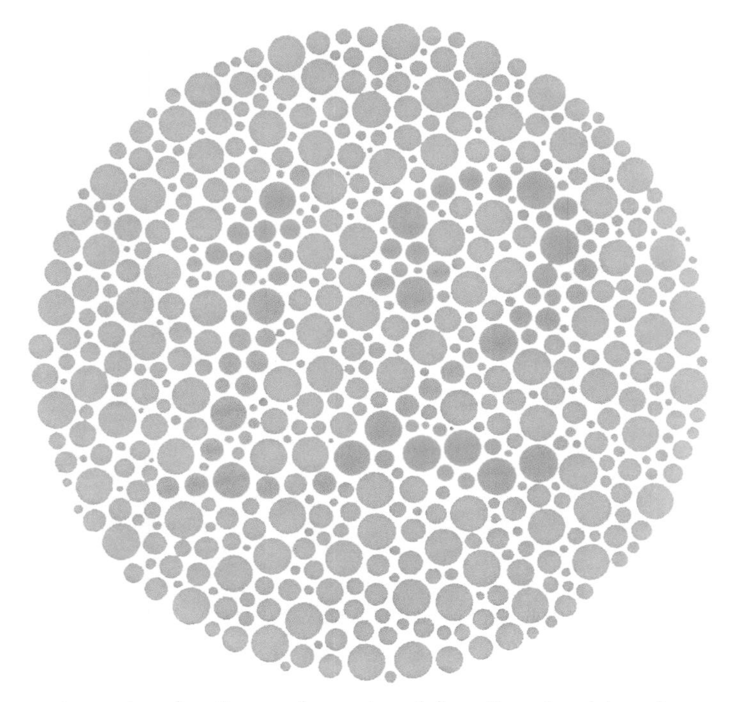

Figure C.3 Ishihara chart showing for testing color vision. People with red-green color blindness aren't able to see the number in this image.

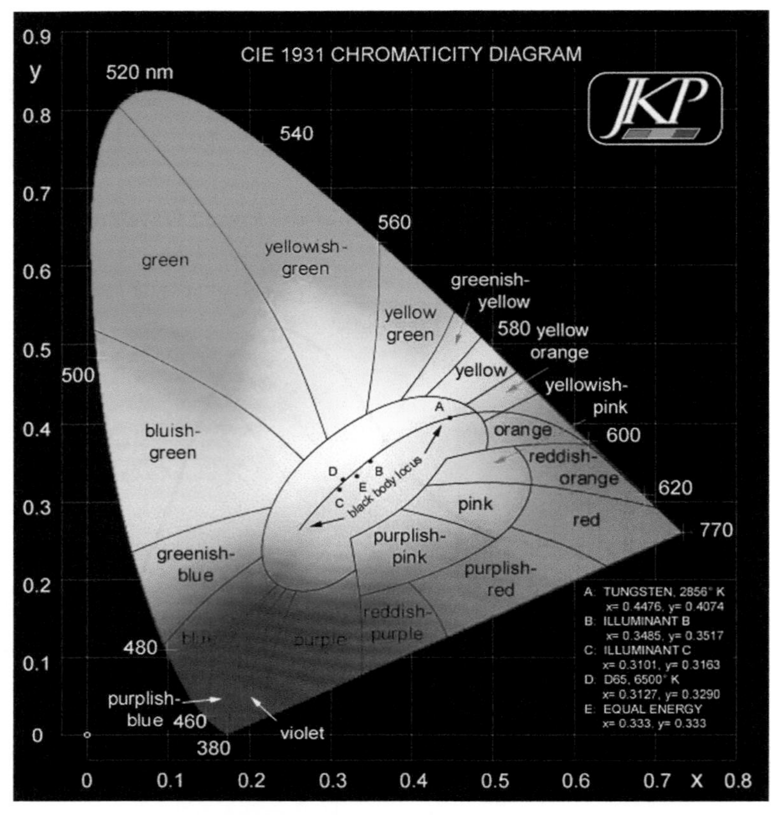

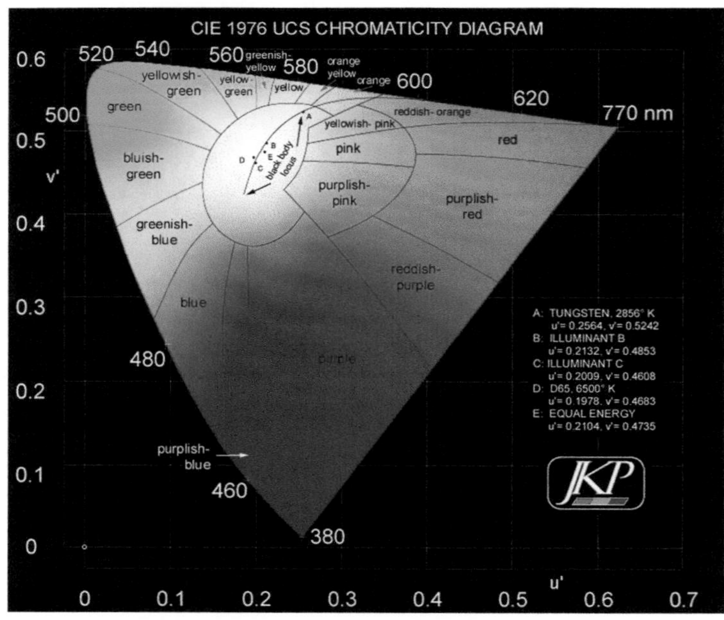

Figure C.4 and C.5 The 1931 and 1976 CIE color chromaticity diagram. The visible spectrum is the outside edge, with mixtures of the primaries as they go towards the white center. (Courtesy of Joe Kane Productions.)

Figure C.6 The same image in full range, just luma, and just chroma.

Figure C.7 In this image, the inside of the different color squares vary by the surrounding color. But if you cover the surrounding shape, it's clearly just white.

Figure C.8 *Paris Street: A Rainy Day* by Gustave Caillebotte. This painting uses both shading and converging lines to convey perspective.

Figure C.9 The same image with progressively coarser sampling. Even in the finest sampling the text on the cable card disappears, but by the final image it's not clear it's a cable car at all. C.9A Original image (1535 × 1464). C.9B Sampled at 256 × 192. C.9C Sampled at 128 × 96. C.9D Sampled at 64 × 48.

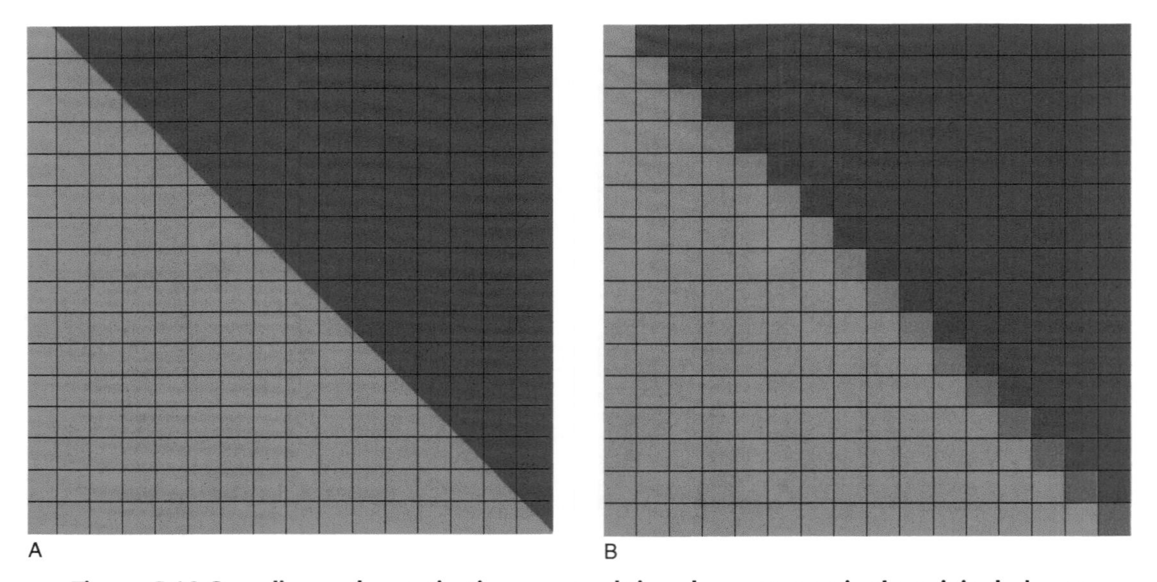

Figure C.10 Sampling and quantization can result in colors not seen in the original, due to averaging. On the left, we see how purple emerges from samples that were red or blue. This is actually the most accurate and visually pleasing result.

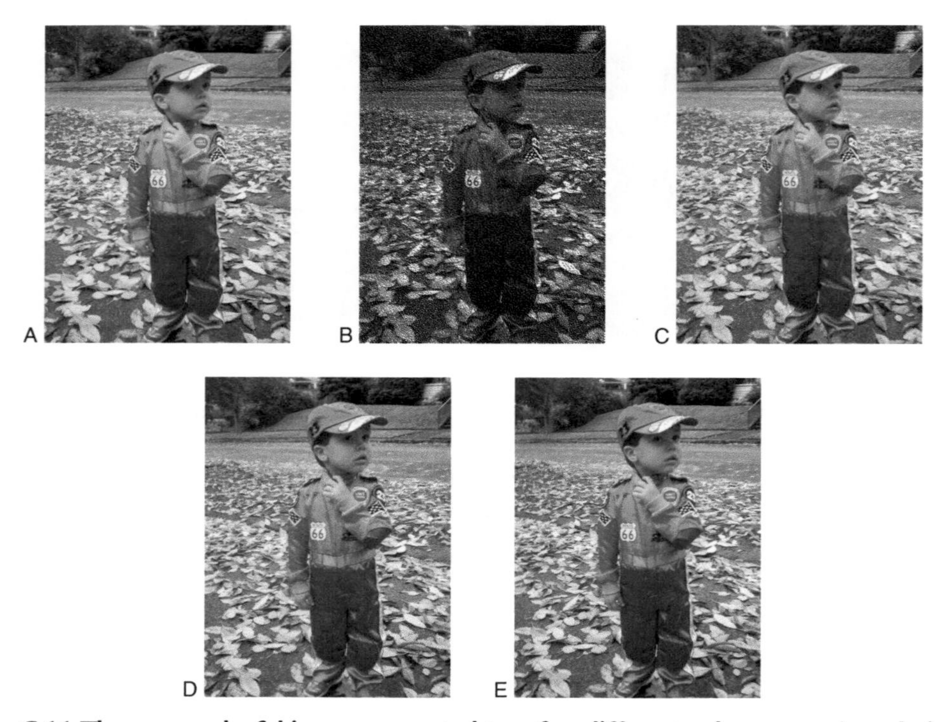

Figure C.11 The same colorful image converted to a few different color spaces (at relatively low resolution to show the details). C.11A The source C.11B Dithered to a 1-bit image. C.11C Dithered to the classic 8-bit "web safe" palette. C.11D Dithered to a custom 8-bit palette for the image. C.11E 15-bit (5 bit per channel).

Figure C.12 Two images compared in source, high quality, low quality, and at the quality that yields the same file size. The more complex image takes more bits at the same quality, and has lower quality with the same bits.

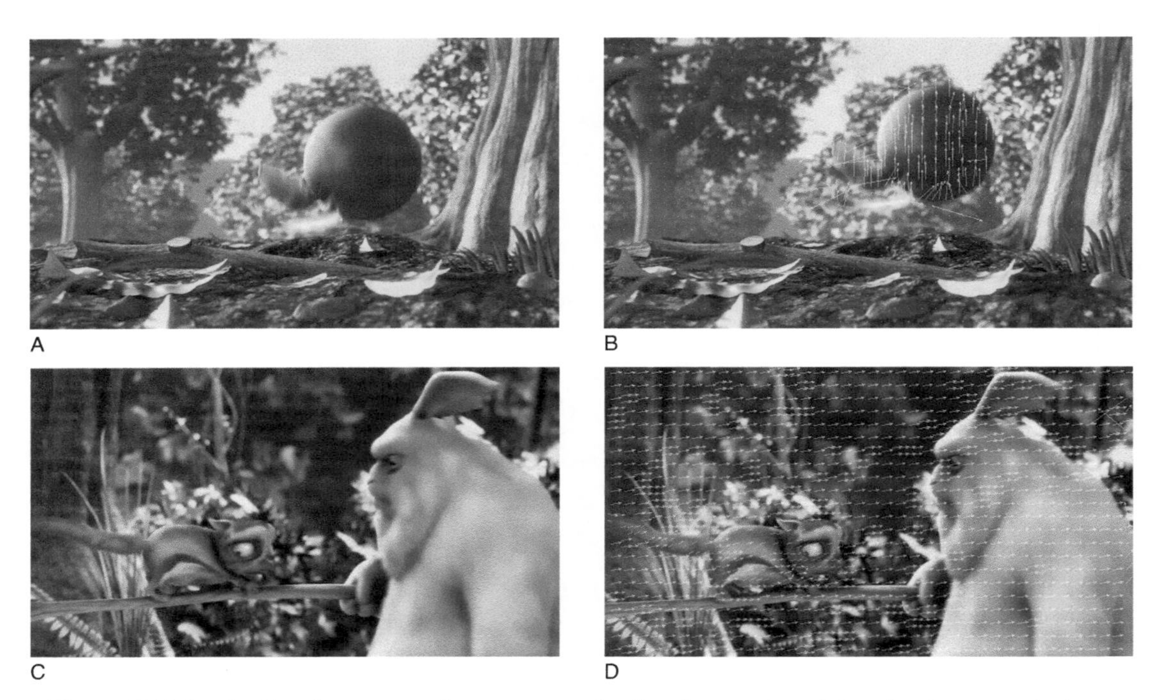

Figure C.13 In the first image, motion vectors exist for the moving object. In the second, there are motion vectors for the whole frame as the whole frame is in motion.

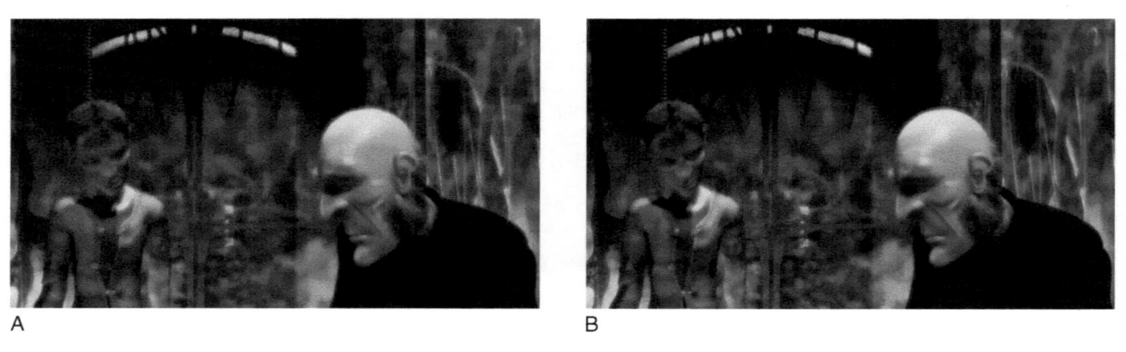

Figure C.14 Differential quantization lowers compression in more visually important blocks, while increasing compression in parts of the image where texture detail can better hide the artifacts.

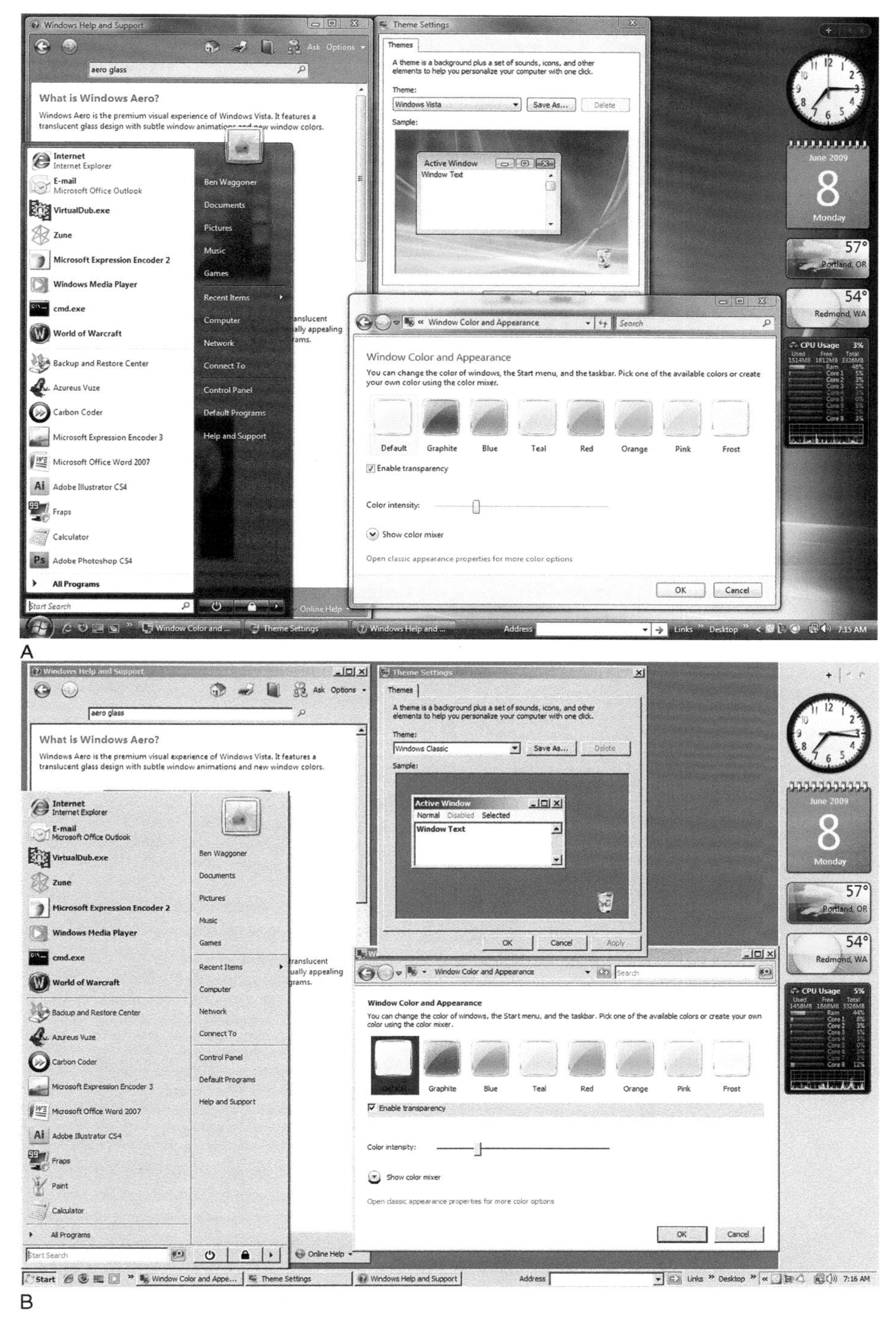

Figure C.15 The same screen rendered in the Aero Glass (15A) and Classic (15B) themes. Aero Glass has smoother edges better for DCT style compression, while Classic has lots of flat areas for easy RLE-like compression.

Figure C.16 A high-motion HDV frame showing bad blocking (from an interlaced source, deinterlaced for clarity).

Figure C.17 Getting 601 and 709 right matters, particularly with skin tones and while details. C.17A is correctly converted to 709. C.17B shows when 601 to 709 correction is applied instead of 709 to 601. C.17C shows the cumulative effect of a double conversion.

A

The same source frame

B

Preprocessed by cropping letterboxing

C

Correcting aspect ratio

D

Applying inverse telecine

Figure C.18 The quality advantage gets even bigger after compression even though the preprocessed video is only 800 Kbps compared to the unprocessed's 1000 Kbps.

Figure C.19 A good dither (C.19A) can do a dramatic job of eliminating banding in this kind of subtle gradients (Example courtesy of Stacey Spears)

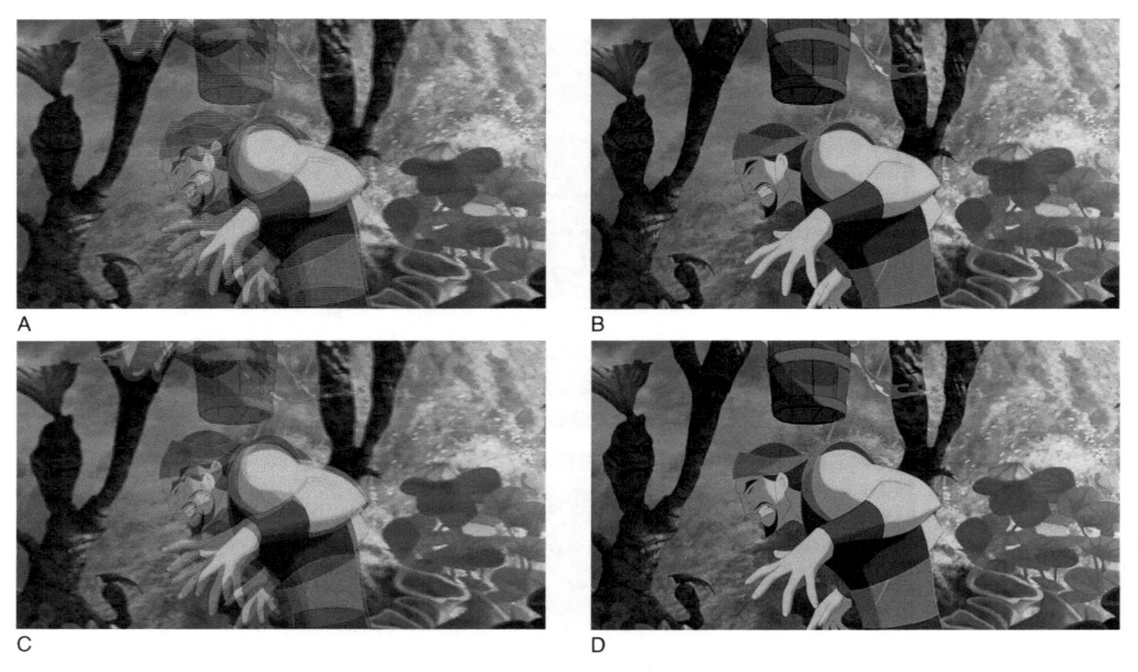

Figure C.20 The same frame as interlaced (C.20A), with a field elimination deinterlace (C.20B), a blend (boo!) deinterlace (C.20C), and finally with full reconstruction via inverse telecine (C20D).

24 frame a second content is shot in the film camera, alternating as Red and Blue. During the telecine process, it is slowed down 0.1% to 23.976, and converted to 59.94 fields a second in a 3:2 pattern of fields per frame. For "3" frames, the first and third fields are identical (the third, repeated field drawn as darker). Any frame with red and blue lines is interlaced, any frame of the same color is progressive.

On inverse telecine, the 3:2 pattern and repeated fields are detected, the 12 repeated fields out of 60 removed, and remainng 48 unique fields out of 60 are reassembed into their original 24 progressive frames, now running at 23.976 fps."

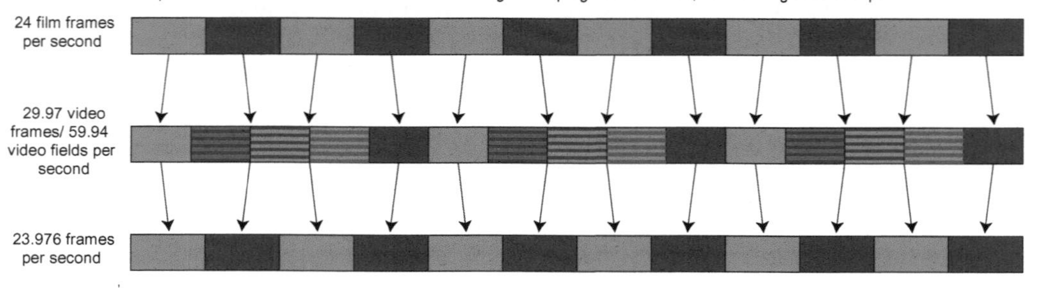

Figure C.21 An illustration of how 24p goes into 30i and back again. Also, an illustration of why I'm a compressionist and not an illustrator.

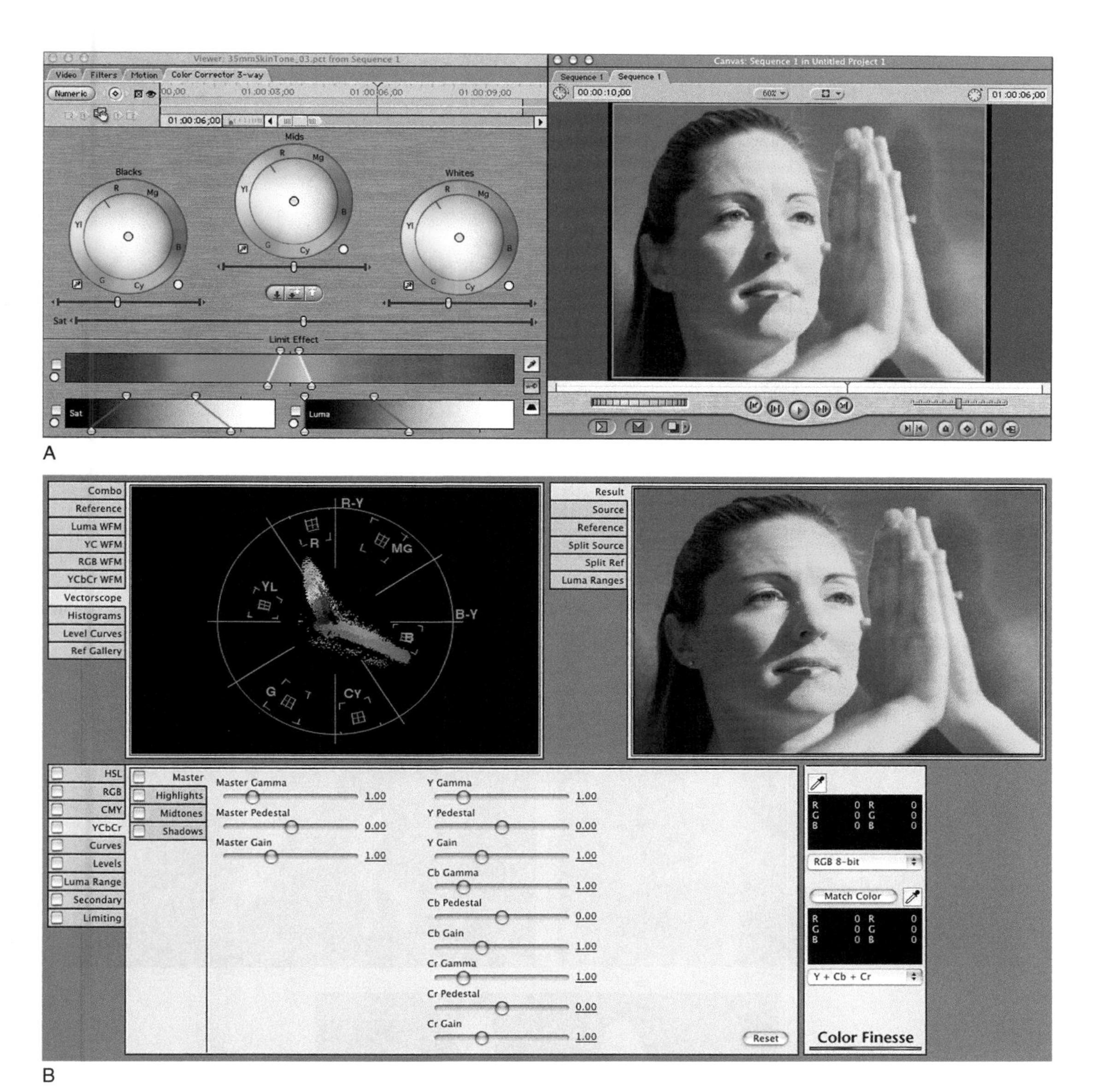

Figure C.22 Typical interfaces for color correction. Note the line around 10:30 on the color wheel, indicating the normal hue of skin tones.

C

Figure C.22 (Continued)

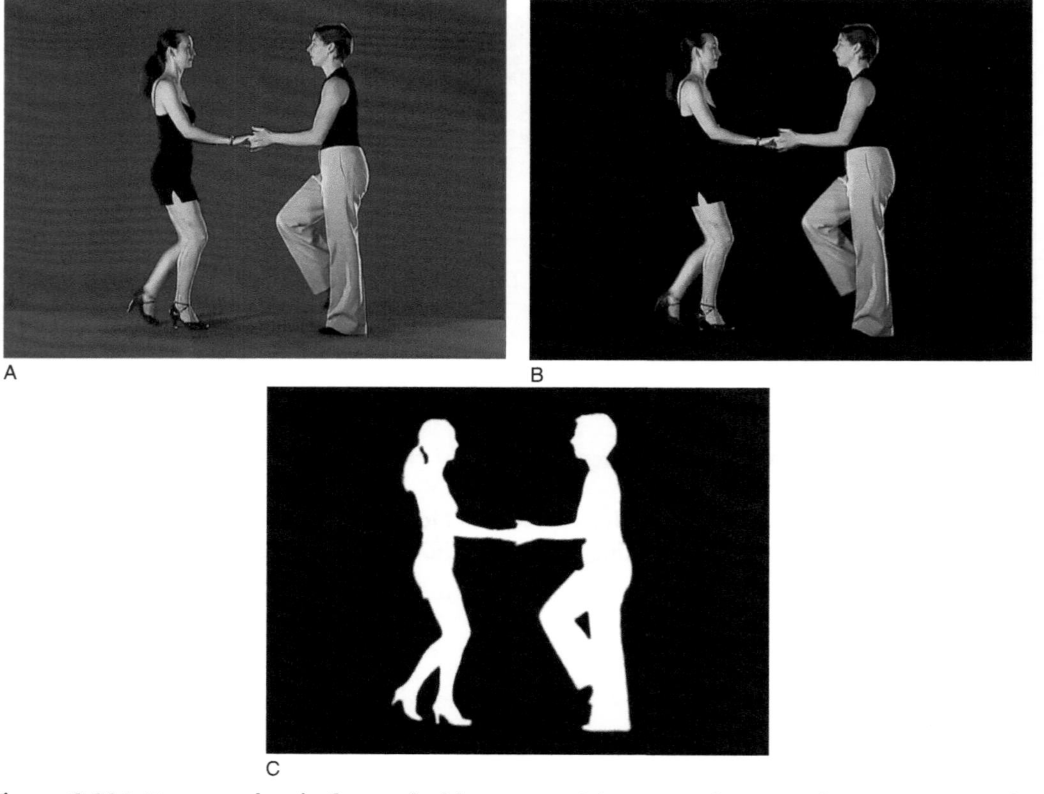

A B C

Figure C.23A Footage shot in front of a blue screen. It's a great first step, but no compression tool is going to do a good job with it. C.23B The visual part of a well-keyed frame. No noise, just the foreground visible. C.23C And the critical part, the alpha channel itself, showing which pixels are image and which aren't. That's what you need to have in the file to key.